T0142308

Smart Innovation, Systems and Technologies

Volume 51

Series editors

About this Series

The Smart Innovation, Systems and Technologies book series encompasses the topics of knowledge, intelligence, innovation and sustainability. The aim of the series is to make available a platform for the publication of books on all aspects of single and multi-disciplinary research on these themes in order to make the latest results available in a readily-accessible form. Volumes on interdisciplinary research combining two or more of these areas is particularly sought.

The series covers systems and paradigms that employ knowledge and intelligence in a broad sense. Its scope is systems having embedded knowledge and intelligence, which may be applied to the solution of world problems in industry, the environment and the community. It also focusses on the knowledge-transfer methodologies and innovation strategies employed to make this happen effectively. The combination of intelligent systems tools and a broad range of applications introduces a need for a synergy of disciplines from science, technology, business and the humanities. The series will include conference proceedings, edited collections, monographs, handbooks, reference books, and other relevant types of book in areas of science and technology where smart systems and technologies can offer innovative solutions.

High quality content is an essential feature for all book proposals accepted for the series. It is expected that editors of all accepted volumes will ensure that contributions are subjected to an appropriate level of reviewing process and adhere to KES quality principles.

More information about this series at http://www.springer.com/series/8767

Suresh Chandra Satapathy · Swagatam Das
Editors

Proceedings of First International Conference on Information and Communication Technology for Intelligent Systems: Volume 2

Editors
Suresh Chandra Satapathy
Department of Computer Science
and Engineering
Anil Neerukonda Institute of Technology
and Sciences
Visakhapatnam
India

Swagatam Das
Indian Statistical Institute
Jadavpur University
Kolkata
India

ISSN 2190-3018 ISSN 2190-3026 (electronic)
Smart Innovation, Systems and Technologies
ISBN 978-3-319-80919-9 ISBN 978-3-319-30927-9 (eBook)
DOI 10.1007/978-3-319-30927-9

Printed on acid-free paper

This Springer imprint is published by Springer Nature
The registered company is Springer International Publishing AG Switzerland

Preface

This SIST volume contains the papers presented at the ICTIS 2015: International Conference on Information and Communication Technology for Intelligent Systems. The conference was held during November 28–29, 2015, Ahmedabad, India and organized communally by Venus International College of Technology, Association of Computer Machinery, Ahmedabad Chapter and supported by Computer Society of India Division IV—Communication and Division V—Education and Research. It targeted state-of-the-art as well as emerging topics pertaining to ICT and effective strategies for its implementation for engineering and intelligent applications. The objective of this international conference is to provide opportunities for the researchers, academicians, industry persons, and students to interact and exchange ideas, experience and expertise in the current trend and strategies for information and communication technologies. Besides this, participants were enlightened about vast avenues, current and emerging technological developments in the field of ICT in this era and its applications were thoroughly explored and discussed. The conference attracted a large number of high-quality submissions and stimulated the cutting-edge research discussions among many academic pioneering researchers, scientists, industrial engineers, and students from all around the world and provided a forum to researcher. Research submissions in various advanced technology areas were received and after a rigorous peer-review process with the help of program committee members and external reviewer, 119 (Vol-I: 59, Vol-II: 60) papers were accepted with an acceptance ratio of 0.25. The conference featured many distinguished personalities like Dr. Akshai Aggarawal, Hon'ble Vice Chancellor, Gujarat Technological University, Dr. M.N Patel, Hon'ble Vice Chancellor, Gujarat University, Dr. Durgesh Kumar Mishra, Chairman Division IV CSI, and Dr. S.C Satapathy, Chairman, Division V, Computer Society of India, Dr. Bhushan Triverdi, Director, GLS University, and many more. Separate invited talks were organized in industrial and academia tracks during both days. The conference also hosted few tutorials and workshops for the benefit of participants. We are indebted to ACM Ahmedabad Professional Chapter, CSI Division IV, V for their immense support to make this conference possible in such a grand scale. A total of 12 sessions were organized as a part of *ICTIS 2015* including nine technical, one plenary, one inaugural, and one valedictory session. A total of 93 papers were presented in the nine technical sessions with high discussion insights. The total number of accepted submissions was 119 with a focal point on ICT and intelligent systems. Our sincere thanks to all sponsors, press, print, and electronic media for their excellent coverage of this conference.

November 2015

Suresh Chandra Satapathy
Swagatam Das

Conference Committee

Patrons

Mr. Rishabh Jain, Chairman, VICT, Ahmedabad, India
Dr. Srinivas Padmanabhuni, President, ACM India

General Chairs

Dr. A.K. Chaturvedi, Director, VICT, Ahmedabad, India
Mr. Chandrashekhar Sahasrabudhe, ACM India
Dr. Nilesh Modi, Chairman, ACM Ahmedabad Chapter

Advisory Committee

Dr. Chandana Unnithan, Victoria University, Australia
Dr. Bhushan Trivedi, Professional Member, ACM
Mustafizur Rahman, Endeavour, Agency for Science Technology and Research, Australia
Mr. Bharat Patel, Professional Member, ACM
Dr. Durgesh Kumar Mishra, Chairman, Division IV, CSI
Mr. Hardik Upadhyay, GPERI, India
Dr. Dharm Singh, PON, Namibia
Mr. Ashish Gajjar, Professional Member, ACM
Hoang Pham, Professor and Chairman, Department of ISE, Rutgers University, Piscataway, NJ, USA
Mr. Maitrik Shah, Professional Member, ACM
Ernest Chulantha Kulasekere, Ph.D., University of Moratuwa, Sri Lanka
Ms. Heena Timani, Professional Member, ACM
Prof. S.K. Sharma Director PIE, PAHER, Udaipur, India

Shashidhar Ram Joshi, Ph.D., Institute of Engineering, Pulchowk Campus, Pulchowk, Nepal
Mr. Bhaumik Shah, Professional Member, ACM
Subhadip Basu, Ph.D., Visiting Scientist, The University of Iowa, Iowa City, USA
Prof. Lata Gohil, Professional Member, ACM
Abrar A. Qureshi, Ph.D., University of Virginia's College at Wise, One College Avenue, USA
Dr. Aditya Patel, Professional Member, ACM
Dr. Muneesh Trivedi, Professor, ABES, Ghaziabad, India
Dr. Ashish Rastogi (CSI), IT Department, Higher College of Technology, Muscat-Oman
Dr. Kuntal Patel, Professional Member, ACM
Dr. Aynur Unal, Standford University, USA
Prof. Rutvi Shah, Professional Member, ACM
Dr. Malaya Kumar Nayak. Director IT Buzz Ltd., UK
Dr. Nisarg Pathak, Professional Member, ACM

Organising Committee

Chair
Prof. Jayshree Upadhyay, HOD, CE, VICT, Ahmedbad, India

Secretary
Mr. Mihir Chauhan, VICT, Ahmedabad, India

Members
Prof. Mamta Jain, VICT, Ahmedabad, India
Mr. Zishan Shaikh, VICT, Ahmedabad, India
Mr. Jalpesh Ghumaliya, VICT, Ahmedabad, India
Prof. Nehal Rajput, VICT, Ahmedabad, India
Prof. Murti Patel, VICT, Ahmedabad, India
Prof. Jalpa Shah, VICT, Ahmedabad, India
Ms. Lajja Vyash, VICT, Ahmedabad, India
Ms. Ruchita Joshi, VICT, Ahmedabad, India
Mr. Aman Barot, VICT, Ahmedabad, India
Mr. Maulik Dave, VICT, Ahmedabad, India
Ms. Priyadarshini Barot, VICT, Ahmedabad, India

Technical Program Committee

Chair
Dr. Suresh Chandra Satapathy, Chairman, Division V, CSI

Co-Chair
Prof. Vikrant Bhateja, SRMGPC, Lucknow

Members
Dr. Mukesh Sharma, SFSU, Jaipur
Prof. D.A. Parikh, Head, CE, LDCE, Ahmedabad
Dr. Savita Gandhi, Head, CE, Rolwala, GU, Ahmedabad
Dr. Jyoti Parikh, Associate Professor, CE, GU, Ahmedabad
Ms. Bijal Talati, Head, CE, SVIT, Vasad
Dr. Harshal Arolkar, Member ACM
Dr. Pushpendra Singh, JK Lakshimpath University
Dr. Sanjay M. Shah, GEC, Gandhinagar
Dr. Chirag S. Thaker, GEC, Bhavnagar, Gujarat
Mr. Jeril Kuriakose, Manipal University, Jaipur
Mr. Ajay Chaudhary, IIT Roorkee, India
Mr. R.K. Banyal, RTU, Kota, India
Mr. Amit Joshi, Professional Member, ACM
Dr. Vishal Gour, Bikaner, India
Mr. Vinod Thummar, SITG, Ahmedabad, India
Mr. Nisarg Shah, 3D Blue Print, Ahmedabad, India
Mr. Maulik Patel, Emanant TechMedia, Ahmedabad, India

CSI Apex Committee

Prof. Bipin V. Mehta, President, CSI
Dr. Anirban Basu, Vice-President, CSI
Mr. Sanjay Mohapatra, Hon-Secretary, CSI
Mr. R.K. Vyas, Hon-Treasurer, CSI
Mr. H R Mohan, Immediate Past President, CSI
Dr. Vipin Tyagi, RVP, Region III, CSI

Conference Track Managers

Track#1: Image Processing, Machine Learning and Soft Computing—Dr. Steven Lawrence Fernandes
Track#2: Software and Big Data Engineering—Dr. Kavita Choudhary
Track#3: Network Security, Wireless and Mobile Computing—Dr. Musheer Ahmad

Contents

Part I
ICT Based Security and Privacy Applications

POS Word Class Based Categorization of Gurmukhi Language Stemmed Stop Words

Kaur Jasleen and R. Saini Jatinderkumar

Abstract Literature in Indian language must be classified for its easy retrieval. In Punjabi literature classifier, five different categories: nature, romantic, religious, patriotic and philosophical, are manually populated with 250 poems. These poems are pre-processed through data cleaning, tokenization, bag of word, stop word identification and stemming phases. Due to unavailability of Punjabi stop words in public domain, manual collection of 256 stop words are done from poetry and articles. After stemming, 184 unique stemmed words are identified. Based on part of speech tagging, 184 stop words are categorized into 98 adverbs, 7 conjunctions, 43 verbs, 24 pronouns and 12 miscellaneous words. These unique 184 stemmed words are being released for other language processing algorithm in Punjabi. This paper concentrates on providing better and deeper understanding of Punjabi stop words in lieu of Punjabi grammar and part of speech based word class categorization.

Keywords Adverb · Conjunction · Verb · Pronoun · Part of speech · Punjabi · Stop word

1 Introduction

With the advent of World Wide Web and Unicode encoding, Indian language content is increasing on the web day by day. In today's internet era, people prefer to use their regional language or mother language to communicate their thoughts. So this data must be classified for its easy retrieval and usage. Text Classification is an act of assigning natural language text into predefined categories [1]. India, being a

K. Jasleen (✉)
Uka Tarsadia University, Bardoli, Gujarat, India
e-mail: sidhurukku@yahoo.com

R. Saini Jatinderkumar
Narmada College of Computer Application, Bhaurch, Gujarat, India
e-mail: saini_expert@yahoo.com

S.C. Satapathy and S. Das (eds.), *Proceedings of First International Conference on Information and Communication Technology for Intelligent Systems: Volume 2*, Smart Innovation, Systems and Technologies 51, DOI 10.1007/978-3-319-30927-9_1

multilingual country, consists of wide number of languages and rich literature. Out of these languages, 22 languages are recognized as regional languages [2]. Punjabi, one of them, is widely spoken language in Punjab (India) as well as in Pakistan [3]. Punjabi Language belongs to Indo-Aryan Language Family. Punjab is known for its rich culture and literature. Poem is one form of literary art. Poetry is always imaginative in nature with a message to its reader [4]. Poetry always has a strong association with feelings, thoughts and ideas. An automatic poetry classification is a text classification problem. Input to classifier is poem in Punjabi language and classifier will assign a category on the basis of its content. This paper is focused on the analysis of stop words from grammatical point of view.

2 Indian Language Based Text Classifier

India is a multilingual country. Many languages are being used in India. Indo-Aryan (consists of Hindi, Gujarati, Bengali, Punjabi, Marathi, Urdu, and Sanskrit) and Dravidian (Telugu, Tamil, Kannada) are major language families spoken in India [5]. Brief survey about the text classification works done in Indian languages is given below.

2.1 Text Classification in Indo-Aryan Language Family

Statistical techniques using Naïve Bayes and Support Vector Machine are used to classify subjective sentences from objective sentences for Urdu language. As Urdu language is morphological rich language, this makes the classification task more difficult. The result of this implementation shows that accuracy, performance of Support Vector Machines is much better than Naïve Bayes classification techniques [6]. For Bangla text classification, n-gram based algorithm is used and to analyze the performance of the classifier. Prothom-Alo news corpus is used. The result show that with increase in value of n from 1 to 3, performance of the text classification also increases, but from value 3 to 4 performance decreases [7]. Sanskrit text documents have been classified using Sanskrit Word net. Semantic based classifier is method is built on lexical chain of linking significant words that are about a particular topic with the help of hypernym relation in Word Net [8]. Very few works in literature are found in field of text classification in Punjabi Language. Domain based text classification is done by Nidhi and Vishal [9]. This classification is done on sports category only. Two new algorithms, Ontology based classification and Hybrid approach are proposed for Punjabi text classification. The experimental results conclude that Ontology based classification (85 %) and Hybrid approach (85 %) provides better results. Sarmah et al. [10] presented an approach for classification of Assamese documents using Assamese WordNet. This approach has accuracy of 90.27 % on Assamese documents.

2.2 Text Classification in Dravidian Language Family

Naïve Bayes classifier has been applied to Telugu news articles to classify 800 documents into four major classes. In this, normalized term frequency-inverse document frequency is used to extract the features from the document. Without any stop word removal and morphological analysis, at the threshold of 0.03, the classifier gives 93 % precision [11]. For morphologically rich Dravidian classical language Tamil, text classification is done using vector space model and artificial neural network. The experimental results show that Artificial Neural network model achieves 93.33 % which is better than the performance of Vector Space Model which yields 90.33 % on Tamil document classification [12]. A new technique called Sentence level classification is done for Kannada language; in this sentences are analyzed to classify the Kannada documents. This Technique extended further to sentiment classification, question answering, text summarization and also for customer reviews in Kannada blogs [13].

3 Pre-processing Steps Involved in Punjabi Poetry Classifier

Before classification, data must be pre processed to remove unwanted words and noise [14]. Pre-processing phase of poetry classifier consists of Data Cleaning, Feature Extraction and Feature Selection. As Punjabi is resource scarce language, there is no publicly available corpus, so manual collection of poetry is done. Initially, Data is collected into 5 different categories: ਕੁਦਰਤ [*kudarata*] 'Nature', ਪ੍ਰੀਤ [*prīta*] 'Romantic', ਧਾਰਮਿਕ [*dhāramika*] 'Religious', ਦੇਸ਼ਭਗਤੀ [*dēśabhagatī*] 'Patriotic' and ਦਾਰਸ਼ਨਿਕ [*dāraśanika*] 'Philosophical'. Initially, these categories are populated with 50 poems in each category. Implementation of various subphases (as discussed below) is done in Visual Basic.Net using Microsoft Visual Studio 2010 as front end and Microsoft Access 2007 at back end using Unicode characters [15].

3.1 Cleaning

Preprocessing step is involved to remove the noise from data so that this noisy data don't penetrate into the next higher levels. It includes special symbol deletion. Symbols like: comma (,), dandi (।), double dandi (॥), sign of interrogation (?) and sign of exclamation (!) are present in poems. In case of Punjabi language, dandi (।) is used in place of full stop. Double dandi (॥) is generally used in ancient Punjabi writings like religious poetry.

3.2 Feature Extraction and Feature Selection

Feature Extraction phase consists of tokenization, unique words and its term frequency calculation, stop word removal and stemming. In tokenization, each poem is tokenized and 'bag of word' model is created. Data structures used for implementation are hash tables, files and arrays. Unique words are identified from poems and its frequency is calculated from tokenized words. After this, stop words are eliminated from unique words. Stop words are most common words occurring in the text which are not significant for classifier. 256 stop words are identified from poetry, news articles and other Punjabi stories. Stemming is way of converting a written text into its root form [16]. Gupta [17] developed different rules for handling stemming for verbs, adverbs and pronouns. These stemming rules are manually applied to 256 identified stop words. After stemming, 184 unique stemmed stop words are identified and presented in Table 1. This table consists of Columns: word in Punjabi (C1), its transliteration in English (C2) and its meaning in English (C3) [18, 19].

Table 1 List of stemmed stop words

S. no.	C1	C2	C3	S. no.	C1	C2	C3
1	ਇਸ	[isa]	This	2	ਜਸਿ	[jisa]	Who, what, which
3	ਵਚਿ	[vica]	In the	4	ਨ	[na]	No
5	ਤਕ	[taka]	Up	6	ਹੁਣ	[huṇa]	Now
7	ਵੀ	[vī]	Too	8	ਜਨਿਾਂ	[jināṁ]	Whom
9	ਉਤੋਂ	[othon]	Upon	10	ਨਾਲ	[nāla]	With
11	ਨਹੀ	[nahīṁ]	No	12	ਚਾਹੇ	[cāhē]	Either
13	ਭੀ	[bhī]	Too	14	ਕਸਿ	[kisa]	What
15	ਵਲੋਂ	[valōṁ]	By	16	ਪਛਿੋਂ	[pichōṁ]	After
17	ਇਹ	[iha]	This	18	ਏਧਰ	[ēdhara]	Around
19	ਏ	[iha]	This	20	ਨੂੰ	[nū]	To
21	ਜਦੋ	[jadōṁ]	When, while	22	ਅਜਹਿੇ	[ajihē]	Such
23	ਕਈ	[ka'ī]	Many	24	ਹੀ	[hī]	Only
25	ਤੱਦ	[tada]	Then	26	ਕੇ	[kē]	By
27	ਅੰਦਰ	[andar]	Within	28	ਹਾਂ	[hain]	Yes
29	ਉੱਤੇ	[utē]	Upon	30	ਬਹੁਤ	[bahuta]	Much
31	ਸਾਬੁਤ	[sābuta]	Complete	32	ਕਾਫ਼ੀ	[kāfī]	Enough
33	ਕਦੀ	[kadī]	Sometime	34	ਹੁਣੇ	[huṇē]	Now
35	ਨੇ	[nēṁ]	The	36	ਲਈ	[la'ī]	For
37	ਜੀ	[jī]	Respect	38	ਕਿ	[ki]	That
39	ਕਸਿੇ	[kisē]	Someone	40	ਮਗਰ	[magara]	Behind
41	ਪੂਰਾ	[pūrā]	Complete	42	ਦਾ	[dā]	Of
43	ਨੇ	[nē]	The	44	ਤਰ੍ਹਾਂ	[tar'hāṁ]	Like
45	ਹੋਵੇ	[hovē]	If	46	ਫੇਰ	[phēra]	Later

(continued)

Table 1 (continued)

S. no.	C1	C2	C3	S. no.	C1	C2	C3
47	ਜੇਕਰ	[jēkar]	Just in case	48	ਵੇਲੇ	[vēlē]	Times
49	ਦੇ	[dē]	Of	50	ਉੱਥੇ	[othē]	There
51	ਜਹਿੜਾ	[jēhara]	Which	52	ਕਤਿ	[kitē]	Somewhere
53	ਬਾਅਦ	[bā'ada]	After	54	ਇੱਥੇ	[ithē]	Here
55	ਸਾਰਾ	[sārā]	all,whole	56	ਜਨ੍ਹਿੂੰ	[jinhanu]	Whom
57	ਚੋ	[cho]	Out	58	ਜਦ	[jad]	When
59	ਕਦੀ	[kadē]	Never	60	ਵਾਂਗ	[vāṅga]	Like
61	ਸਭ	[sab]	All	62	ਦੌਰਾਨ	[doraan]	During
63	ਤਾਂ	[tan]	When	64	ਵਰਗਾ	[varagā]	Like
65	ਕਿ	[ki]	That	66	ਜੋ	[jō]	That
67	ਲਾ	[la]	To attach	68	ਕਰਕੇ	[karkē]	Because
69	ਪੂਰਾ	[pura]	Complete	70	ਬਲਿਕੁਲ	[bilkul]	Absolutely
71	ਨਾਲੇ	[naale]	Also	72	ਐਹੋ	[eho]	Such
73	ਤੋ	[ton]	From	74	ਕੌਣ	[kaun]	Who
75	ਹੋਣਾ	[hona]	Be	76	ਫਰਿ	[pher]	Then
77	ਪਾਸੋ	[paso]	From	78	ਤਦ	[tad]	Then
79	ਜਹਿਾ	[jeha]	Little	80	ਕੋਲੋਂ	[kolon]	From
81	ਏਸ	[ēs]	This	82	ਕਨਿਾ	[kina]	How much
83	ਜਨ੍ਹਿਾਂ	[jina]	Who	84	ਜਵਿੇ	[jivē]	Such as
85	ਕੁਝ	[kujh]	Some	86	ਹੇਠਾਂ	[hethan]	Below
87	ਦੁਆਰਾ	[dobara]	By	88	ਸਾਰੇ	[sarē]	All
89	ਸਦਾ	[sada]	Forever	90	ਜੱਥਿੇ	[jithē]	Where
91	ਏਥੇ	[ethē]	Here	92	ਕੋਈ	[koi]	Someone
93	ਬਾਰੇ	[barē]	About	94	ਕੀ	[ki]	What
95	ਕਦ	[kad]	When to	96	ਜੀ	[je]	Please
97	ਕਦੇ	[kadē]	Never	98	ਦੀਆਂ	[dī'āṁ]	Of
99	ਹੋਏ	[hoye]	Happen	100	ਚਲਾ	[chala]	Goes
101	ਰਹੇ	[rahē]	Are	102	ਲੈ	[lai]	Take
103	ਬਣੋ	[bano]	Become	104	ਆਖ	[aakh]	Say
105	ਦੇਣੀ	[dēṇī]	Give	106	ਬਣ	[baṇa]	Made
107	ਪਆਿ	[pi'ā]	Lying	108	ਕਰ	[kara]	Do
109	ਹੋਇਆ	[hō'i'ā]	Happened	110	ਪੈਣ	[pain]	Falling
111	ਗਈ	[ga'ī]	Gone	112	ਕਹਿ	[kēh]	Say
113	ਲਗ	[laga]	Seem	114	ਚੁਕੇ	[chukē]	–
115	ਹੁੰਦਾ	[hudā]	Happen	116	ਕਹਿਾ	[keha]	Said
117	ਜਾਂਦਾ	[jāndā]	Going	118	ਕਰਵਾਈ	[karvayei]	Conducted
119	ਵੇਖ	[vēkha]	See	120	ਬਣਾਏ	[banaye]	Created
121	ਸੁਣ	[suṇa]	Hear	122	ਕੀਤਾ	[kitta]	Carried out
123	ਆਈ	[ā'ī]	Occurred	124	ਜਾਵਣ	[javan]	Going
125	ਸਕਦੇ	[sakdē]	Can	126	ਦੇਖ	[dēkh]	See

(continued)

Table 1 (continued)

S. no.	C1	C2	C3	S. no.	C1	C2	C3
127	ਜਾਵੇ	[javē	Go	128	ਆਦਿ	[ādi]	So on
129	ਜਾਂਦਾ	[janda]	Going	130	ਲਿਆ	[li'ā]	Taken
131	ਕਰਣ	[karana]	Doing	132	ਆ	[ā]	Come
133	ਲਗਾਉਦਾਂ	[lagoda]	Not involving	134	ਰਹਿਾ	[reha]	Going
135	ਆਵੇ	[aavē]	Arrives	136	ਗਿਆ	[geya]	Been
137	ਕਰੀ	[kari]	Do	138	ਉਠ	[otha]	Arise
139	ਲਾਇਆ	[laeya]	Attach	140	ਰਹੀ	[rahi]	Been
141	ਰਹਿ	[reh]	Living	142	ਉਸਨੇ	[usnē]	He
143	ਉਹ	[uha]	He, she	144	ਤੁਸੀ	[tusi]	You
145	ਸਾਂ	[sāṁ]	Was	146	ਮੇਰਾ	[mera]	My
147	ਸਭ	[sabha]	All	148	ਉਸਦੀ	[usdi]	His
149	ਹਨ	[hana]	Are	150	ਤੇਰਾ	[tera]	Your
151	ਤੂੰ	[tu]	You	152	ਉਸ	[us]	His
153	ਸੀ	[si]	Was	154	ਉਏ	[oyē]	Person
155	ਹੋ	[ho]	Are	156	ਆਪ	[aap]	you
157	ਤੈਨੂੰ	[tēnu]	You	158	ਸਨ	[san]	Was
159	ਤੁਸਾਂ	[tusa]	You	160	ਮੈ	[mein]	I
161	ਹੈ	[hain]	Are	162	ਤੁਸੀ	[tusi]	You
163	ਹੈ	[hai]	Is	164	ਅਸੀ	[assi]	We
165	ਆਪਣਾ	[apna]	My	166	ਪਰ	[par]	but
167	ਜੇ	[jē]	If	168	ਤੇ	[tē]	And
169	ਅਤੇ	[aatē]	And	170	ਤਾਂ	[tāṁ]	So
171	ਜਾਂ	[jāṁ]	Or	172	ਭਾਵੇ	[bhāvēm]	Although
173	ਕੁਲ	[kal]	Total	174	ਅਗਲੀ	[aagali]	Next
175	ਵਗ਼ੈਰਾ	[vaġairā]	Etc.	176	ਵਰਗ	[varg]	Category
177	ਰੱਖ	[rakh]	Put	178	ਆਮ	[āma]	Common
179	ਲੱਗ	[laag]	Take	180	ਲਾ	[lā]	Apply
181	ਗੱਲ	[gal]	Thing	182	ਹਾਲ	[hāla]	Condition
183	ਪੀ	[pī]	Drink	184	ਇੱਕ	[ek]	One

On lieu of Punjabi Grammar and Part of Speech (POS) based word class categorization, stop words are categorized into 4 different word classes: Adverbs [20], Conjunctions [21], Verbs [20], Pronouns [20] and other miscellaneous words. Any word which is not suitable for first four categories is assigned to miscellaneous one. 98 different adverb forms, 43 different verbs, 24 pronouns, 7 conjunctions are identified from 184 stemmed stop words. And remaining 12 stop words are assigned to miscellaneous category.

Adverb forms in Punjabi language are classified into 2 categories: by function and by form [20]. By function, adverb clauses are categorized into following subclasses: Adverb clause of time: ਜਦ [*jad*] ‘when’, ਅੱਜ [*ajj*] ‘today’. Adverb clause of place: ਉੱਪਰ [*uppar*] ‘upon’, ਉੱਤੇ [*uttē*] ‘over’. Adverb clause of purpose: ਨੂੰ [*nu*] ‘to’, ਲਈ [*laii*] ‘for’. Adverb clause of manner: ਜਿਵੇ [*jive*] ‘as’, ਉਵੇ [*ove*] ‘custom’. Condition clause: ਅਗਰ [*agar*] ‘if’, ਜੇਕਰ *[jekar]* ‘in case’. Result clauses: ਨਾਲੋ [*naalo*] ‘concurrent’, ਤੋ [*to*] ‘from’. Adverb clause of degree: ਬਹੁਤ [*bahut*] ‘much’, ਕਾਫੀ [*kafi*] ‘*enough*’, ਸਾਬੁਤ [*sabut*] ‘complete’. By form, adverb clause is divided into subgroups like derived adverbs, pure adverbs, phrasal adverbs, clausal adverbs, reduplicated adverbs and particles. Few examples are like ਇਥੇ [*ethe*] ‘here’, ਉਥੇ [*othe*] ‘there’, ਕਿਥੇ [*kithe*] ‘where’, ਜਿਥੋ [*jitho*] ‘where’, ਕਿਥੋ [*kitho*] ‘where’. List of Adverbs are shown from serial number 1–98 in Table 1.

Verbs found among stop words are shown in Table 1 from serial number 99–141. For example: ਆਉਣਾ [*aaouna*] ‘to come’, ਜਾਣਾ [*jaana*] ‘to go’.

Pronouns are shown from serial number 142–165. For example ਉਸਦਾ [*usda*] ‘his’, ਅਸੀ [*assi*] ‘we’.

Conjunctions are used to join words, phrases, and clauses [21]. For example, ਅਤੇ [*atē*] ‘and’, ਜਾਂ [*jāṃ*] ‘or’. Serial number 166–172 presents conjunction list.

Miscellaneous words are present from serial number 173–184. For example: ਵਰਗ [*varg*] ‘category’, ਵਗੈਰਾ [*vagera*] ‘etc’.

4 Conclusion

An automatic poetry classifier is used to classify poems according to its content. Before classification starts, these poetries must have to pass through various pre-processing phases. 256 stop words are identified from poetry and news articles written in Punjabi. These 256 stop words are stemmed to its root form using Punjabi stemming rules. Analysis of 184 stemmed stop words from grammatical point of view is discussed in this paper. These stop words are categorized into adverbs, pronouns and conjunctions. In this paper, 184 stemmed stop words are presented for future use in other NLP task in Gurmukhi script. This paper provides enhanced understanding of stop words in light of part of speech tags in Punjabi language.

References

1. Sebastiani, F.: Machine learning in automated text categorization. ACM Comput. Surv. **34**, 1–47 (2002)
2. Languages of India.: http://en.wikipedia.org/wiki/Languages_of_India#Prominent_languages_of_India
3. Punjabi Language.: http://en.wikipedia.org/wiki/Punjabi_language
4. Poem.: http://oxforddictionaries.com/definition/english/poem
5. Kaur, J., Saini, J.R.: A study and analysis of opinion mining research in Indo-Aryan, Dravidian and Tibeto-Burman Language families. Int. J. Data Mining Emerg. **4**(2), 53–60 (2014)

6. Ali, R.A., Maliha, I.: Urdu text classification. In: 7th International Conference on Frontiers of Information Technology, ACM New York, USA, (2009). ISBN 978-1-60558-642-7, doi:10.1145/1838002.1838025
7. Mansur, M., UzZaman, N., Khan, M.: Analysis of N-Gram Based Text Categorization for Bangla in a Newspaper Corpus. Center for Research on Bangla Language Processing. BRAC University, Dhaka, Bangladesh (2006)
8. Mohanty, S., Santi, P.K., Mishra, R., Mohapatra, R.N., Swain, S.: Semantic based text classification using wordnets: Indian language perspective. In: 3rd International Wordnet Conference (GWC 06). pp. 321–324 (2006). doi:10.1.1.134.866
9. Nidhi., Gupta, V.: Domain based classification Punjabi text documents. In: International Conference on Computational Linguistics, pp. 297–304 (2012)
10. Sarmah, J., Saharia, N., Sarma, S.K.: A novel approach for document classification using assamese wordnet. In: 6th International Global Wordnet Conference, pp. 324–329 (2012)
11. Murthy, K.N.: Automatic Categorization of Telugu News Articles. Department of Computer and Information Sciences, University of Hyderabad, Hyderabad (2003). doi:202.41.85.68
12. Rajan, K., Ramalingam, V., Ganesan, M., Palanive, S., Palaniappan, B.: Automatic classification of Tamil documents using vector space model and artificial neural network. Expert Syst. Appl. **36**(8), 10914–10918 (2009)
13. Jayashree, R.: An analysis of sentence level text classification for the Kannada language. In: International Conference of Soft Computing and Pattern Recognition, pp. 147–151 (2011)
14. Gupta, V., Lehal, G.S.: Preprocessing phase of Punjabi language text summarization. In: International Conference on Information System for Indian languages, vol. 139, pp. 250–253 (2011)
15. Unicode Table. http://www.tamasoft.co.jp/en/general-info/unicode-decimal.html
16. Stemming. http://en.wikipedia.org/wiki/Stemming
17. Gupta, V.: Automatic stemming of words for Punjabi language. In: Advances in Signal Processing and Intelligent Recognition systems, Advances in Intelligent Systems and Computing, vol. 264, pp. 73–84 (2014)
18. Google Translation. https://translate.google.co.in/#auto/en/%E0%A8%AA%E0%A8%8F
19. Transliteration and Translation. http://www.shabdkosh.com/pa/
20. Bhatia, T.K.: Punjabi: a cognitive-descriptive grammar. Rout ledge Descriptive Grammar Series (1993)
21. Overview of Punjabi Grammar. http://punjabi.aglsoft.com/punjabi/learngrammar/?show=conjunction

Techniques and Challenges in Building Intelligent Systems: Anomaly Detection in Camera Surveillance

Dinesh Kumar Saini, Dikshika Ahir and Amit Ganatra

Abstract Security is tedious, complex and tough job in today's digitized world. An attempt is made to study and propose an intelligent system for surveillance. Surveillance camera systems are used for monitoring and controlling the security. Anomaly detection techniques are proposed for designing the intelligent control system. In the paper challenges in detection and processing of anomaly in surveillance systems are discussed and analyzed. Major components related to an anomaly detection technique of camera control system are proposed in the paper. Surveillance data is generated through camera, and then this data is transmitted over the network to the storage. Processing is to be done on real time basis and if there is any anomaly detected, the system must produce an alert. This paper is an attempt to study soft computing approaches for anomaly detection.

Keywords Surveillance · Systems · Camera · Control · Anomaly · Detection

1 Introduction

Surveillance cameras are used extensively for security purposes. Access and control of these cameras through a remote computer over the web improves the security aspects of the system. The camera control system consists of a set of cameras located at different locations and the cameras are controlled remotely. Multimedia is

D.K. Saini (✉)
Faculty of Computer and Information Technology, Sohar University,
Sohar, Oman
e-mail: dinesh@soharuni.edu.om

D. Ahir · A. Ganatra
Charotar University of Science and Technology, Changa,
Gujarat, India
e-mail: 14pgce001@charusat.edu.in

A. Ganatra
e-mail: amitganatra.ce@charusat.ac.in

S.C. Satapathy and S. Das (eds.), *Proceedings of First International Conference on Information and Communication Technology for Intelligent Systems: Volume 2*,
Smart Innovation, Systems and Technologies 51, DOI 10.1007/978-3-319-30927-9_2

the combination of different media like text, still images, audio, video, animation and graphics [1]. Generally multimedia content is bulky, so the media storage and the cost of transmission are noteworthy. To solve this, media are compressed in file for both streaming and storage. Streaming is a means of sending multimedia information over the internet so that the recipient plays it as it is being transmitted. Multimedia involves buffering mechanism (temporary storage). In that the limited segments of the streamed information is temporary stored for continuous play. Streaming avoids the copy and save/store an entire file. Without store entire file user can play data [2]. In the camera control systems anomaly detection can be used as one of the mechanism for building the intelligent system.

2 Anomaly Detection

Anomaly detection in data is the identification of patterns that do not match with normal behavior. And these anomalous patterns are also referred to as exceptions, outliers, novelties, noise, deviations, discordant observation in different application domain [3]. Anomaly detection in data is important because many times the detected anomalies translate the important, actionable information in many application domains. Here are some examples, in computer network is abnormal traffic pattern is found then it mean that sensitive information is transfer by hack computer to an unauthorized receiver [4].

An anomaly means something unusual, irregular or unexpected behavior that does not match to the normal behavior. Because of many kind of reasons anomalies could be occur in data. For example network intrusion, cyber intrusion, or terrorist activity and break down of a system.

Figure 1 shows a simple example of anomaly in two-dimension data set. In which R1 and R2 are two normal regions. The points which are outside normal region are considered as anomalies. Here point's a1, a2 and points in region a3 are anomalies.

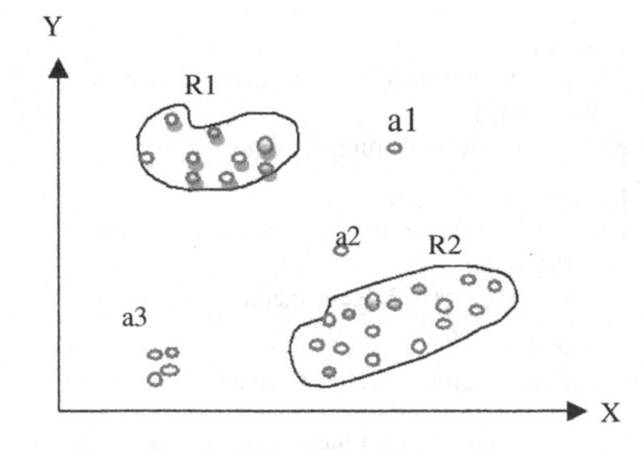

Fig. 1 Anomalies in a two-dimensional data set [5]

2.1 Challenges in Anomaly Detection

A straight forward approach to anomaly detection is defining a region that represents a normal behavior and remaining part that not belong to normal region declared as anomalies. But some factors make this approach very difficult [5].

- It's hard to define a region that comprehends all probable normal behavior. Because the boundary of anomalous and normal behavior is not much accurate. Thus an anomaly observation that situated close to the boundary may be normal or abnormal.
- While anomaly is aeries due to the malicious action, these malicious rivals often accommodate themselves to appear abnormal behavior as normal so the task of labeling normal behavior region becomes more difficult.
- By the time in various application domains, normal behavior is keep growing so the current conviction of normal behavior is may not be enough in future.
- Another challenge is, for different application domain the exact view of anomaly is dissimilar. For example, a small variation in normal reading might be diseases in medical domain. Whereas, in stock market domain small variation might be consider as normal. This makes it complex to apply a particular domain technique to another domain.
- When using recorded and real world datasets, the major issue is labeled data availability.

Sometimes noise is similar to anomaly and it is difficult to distinguish between noise and an anomaly because the data contains noise is analogous to actual anomalies hence it is hard to recognize and remove. Anomalies and noise removal are related to but both are two different things. Since the difference between noise and anomalies is lies in the interest of analyst. Noise can be refer as something unwanted and obstacle to data analyst. While anomalies are consider as something meaningful to data analyst.

3 Different Aspect of Anomaly Detection Issues

3.1 Input Data Nature

Input data nature is core aspect of Anomaly detection. In general the input consists of set of data instances which can call as sample, objects, record, point, vector, entity, pattern, event or case. And all data instance can be defined by a set of attributes. This attributes are also referred as variable, characteristics, feature, field or transmission. There are different types of attributes such are binary, categorical or continuous [5].

In choice of anomaly detection technique the characteristics of attributes have a major impact. For example, in statistical techniques the underlying model is depends on whether the attributes data types is continuous or categorical. Likewise, in nearest-neighbor-based technique, the distance measure used in technique is determined by the characteristics of attributes.

There is another way of categorizing the input data. It is based on the relationship between the data instance. Mostly the anomaly detection technique deals with single point data. In point data there is no relationship among the data instance.

3.2 *Data Labels*

Data labels describe whether the data instance is normal or anomalous. Data labeling is expensive and usually it is done by human expert.

The systems are becoming extremely complex and dynamic and to understand the systems behavior it requires exploratory data analysis and descriptive modeling.

3.3 *Types of Anomaly*

3.3.1 Point Anomaly

Most common type of anomaly is point anomaly and has been focus on most of research. Point anomaly can be defined as an individual entity which is considered as abnormal with regards to other data. In Fig. 1 point a1, point a2 and points in region a3 are point anomalies as they are exterior to the normal region.

3.3.2 Contextual Anomaly

Contextual anomaly is also known as conditional anomaly. It can be defined as in some specific context if a data instance is anomalous then it is contextual anomaly. These types of anomalies are usually found out in spatial data and time series data.

3.3.3 Collective Anomaly

Collective anomaly can be defined as a set of related data instances which is anomalous with regard to remaining data in data set. Individual entities may not be anomalies by themselves in collective anomalies but their occurrence to gather is considered anomalous.

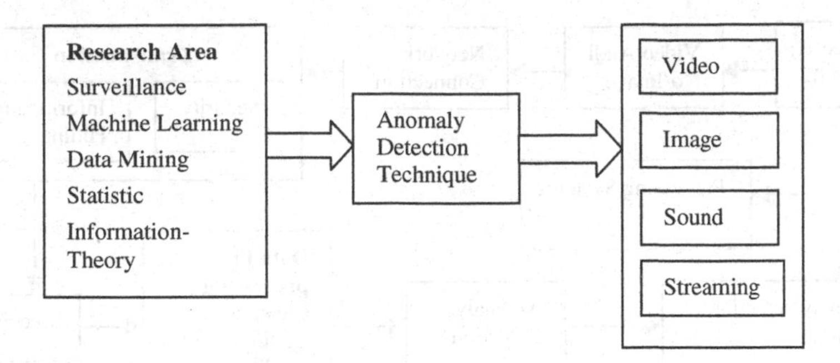

Fig. 2 Key component related to an anomaly detection technique of surveillance camera

4 Application Domain of Anomaly Detection

There are varieties of application domain for anomaly detection; such are network intrusion detection, industrial damage detection, machine learning, statistic, fraud detection, image processing, video surveillance, intrusion detection, medical science and public health. Figure 2 illustrates the component of camera control system associate with anomaly detection techniques [6]. In camera control system Input data for anomaly techniques are video data, image data, sound, or streaming. How this system work is described in next section.

5 Camera Control System

Figure 3 shows the basic block diagram for camera control system. Source in surveillance system is camera and the target is the entity or entities. The target consists of entities in which anomaly detection technique is used to find anomalies. Example of target can be crowds, road traffic, individuals, or network traffic. Data is taken from source and stored at server. From server streaming (streaming is transferring of data) of data is next part. Data can be type of text, image, voice, animation, or video. Streaming can be done in real-time/online or offline.

At server processing of data is done, this data can be recording or imaging. But this system is only focus on image and video part. Image processing is one part of this system. Processing of image is done by mathematical operations. At receiver side data preprocessing is done [7]. Data preprocessing includes editing, cleaning, modify, and feature extraction. Feature extraction is an essential preprocessing step. Feature extraction is used to reduce the required amount of resource to

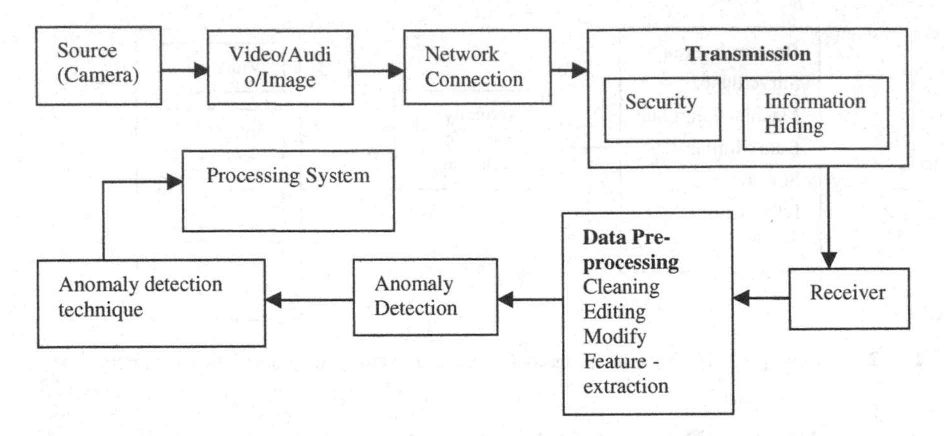

Fig. 3 Block diagram for camera control system an intelligent system

describe the large amount of data. It starts from initial set of calculated data and then it constructs resultant values. Feature extraction include edge detection, corner detection etc. From feature extraction the data is derived [8, 9].

Modeling and simulation predict the behavior of data that under what condition and in which way a part could fail and how it will behave. Soft computing technique is used in simulation and modeling. Soft computing techniques are statistical, mathematical, neural network, fuzzy logic, Bayesian, genetics. Soft computing techniques are statistical, mathematical, neural network, fuzzy logic, Bayesian, genetics.

6 Soft Computing Approaches and Learning

Various Soft computing techniques can be implemented in anomaly analysis. Those are Neural network, Artificial neural network, Fuzzy logic, Clustering, Decision tree, Genetic algorithm.

Fuzzy logic in anomaly detection can be used to manage ambiguity in normal verses abnormal determination. It can also be used in allocating a membership to a class which determines abnormal behavior. Artificial neural network is the mathematical model for machine learning. In which individual node are identified as neurons which are linked together in network [10]. Three layers are input, output and hidden. SVM (support vector machine) method is used in ANN as classification method to find linear separating in two dimensional planes between two classes of data.

6.1 Soft Computing Learning Methods

Learning methods in soft computing for anomaly detection are

1. Supervised learning,
2. Unsupervised learning
3. A priori knowledge application [11].

6.1.1 Supervised Learning

In supervised learning method an algorithm learns a model from training data (representative set of data). All these element of set is labeled with its class. Four methods are includes in supervised learning method for anomaly detection [11]:

1. Learn only normal events consisting from data.
2. Learn only an anomalous events consisting from data.
3. Learn both events.
4. Learn multiple classes of events.

Among this method 1 and method 2 lies in one-class classification problem. Method 3 lies in two-class classification problem. And method 4 is for multiclass problem.

1. **Learn normal events**

This is the most common approach for anomaly detection usually used in automated surveillance system. It trains an algorithm for normal event and then all the events which are outside the class are classify as anomalous. Benefit of this method is that it doesn't require data set from anomalous events. However, problem with this method is it may suffer from high rates of false positives because an event may be detected as anomalous which is not adequately represented in training data set.

2. **Learn anomalous events**

This is the least common approach in supervised learning method. Reason for avoidance is that there is a high risk in missed detection of abnormal events which are not fit into the learned pattern. Zhang et al. [1] used this method of learning for a system to retrieving anomalous events recorded in video surveillance.

3. **Learn normal and abnormal events**

This is a two-class approach. In this method normal and abnormal events are learned. The training data consist of the labeled examples of normal and anomalous events. This method can work well when the anomalous events are correctly defined and well characterize in the data set. Success to this approach is depending on the previous assumption of definition of anomalousness.

4. **Learn multiple classes of events**

This is multiple class approach in which behavior classification operation is performed. In this approach detection anomaly is found out by rules defined in initial classification. An issue to this approach is that only learned event can recognize reliably.

6.1.2 Unsupervised Learning

An unsupervised learning method involves normal and anomalous training data without labels. A simple assumption is made in this approach. The events which occur more frequently are declared as normal, and the events which occur rarely are declared as anomalous. Benefit of this method is that it doesn't require labeling of training data set. Unsupervised learning methods typically take clustering approach for anomaly detection [7]. In clustering approach, anomalies are detecting by distance of unobserved data points from closest cluster.

6.1.3 A Priori Modeling

There is no requirement of training data (labeled or unlabeled) in this approach. This method creates models or rules by applying external knowledge to domain for normal and anomalous classes. A key point to success of this technique is the use of external knowledge to given target. Precision and usefulness of knowledge to given domain is important part for designing proactive defense [12].

7 Classification of Anomaly Detection Techniques

Figure 4 shows the classification of anomaly detection techniques by classes. Main classes of anomaly detection are nearest neighbor based, classification based, clustering based, spectral, statistical, information theoretic [13]. Every technique has its own pros and cons. There are several anomaly detection techniques in variety of application domain but this paper only focus on those techniques which are used in video surveillance domain. Within surveillance domain there are different application areas like public area, land transport, maritime, under water, land transport, air. These techniques are used to identify abnormality in data.

Automated anomaly detection processor is proposed by Kraiman et al. [3] that use multi surveillance data and sensor data to identify and describe events and objects of military area. AADP use clustering algorithm SOM to training data set to classify new observation as normal or anomalous. AADP use Gaussian mixture model [5] and Bayesian analysis [5] for automate the anomaly detection.

Ye et al. [2] proposed a wireless video surveillance system for maximizing the quality of received video. In this system it uses the Gaussian mixture model to detect

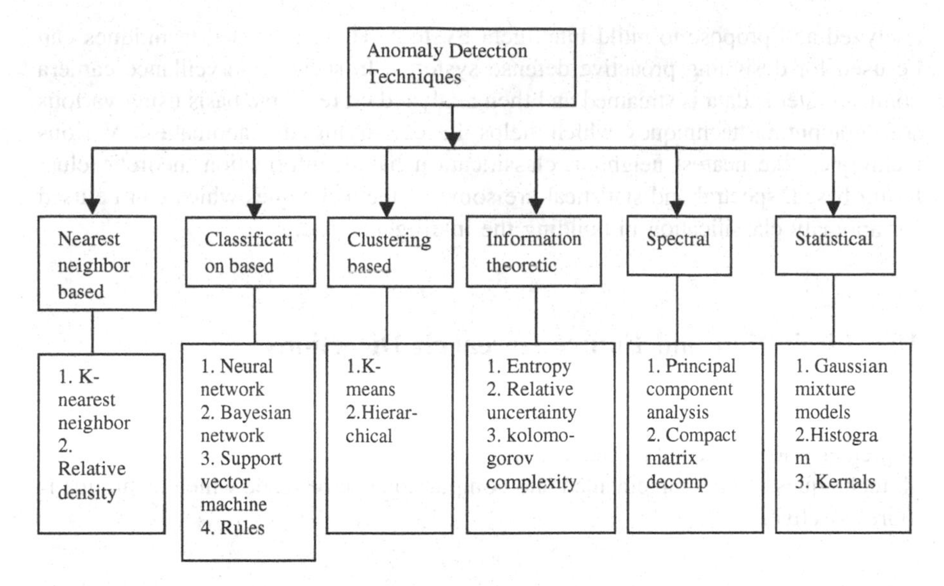

Fig. 4 Classification of anomaly detection techniques

abnormality. It is temporal learning process that models different condition of pixels at different position. Value of each Gaussian model is updated and the pixels which are not matching with any background model are detected and consider as anomaly.

8 Anomaly Detection Post Processing

Clustering and Aggregation is used for designing system level handling of false alarm. Aggregation can be used for analyzing flows of detected anomalies. After aggregation ranking can be done for the anomaly based on its severity. The post-processing mechanism allows constructing easy-to-interpret and easy-to-visualize ranked results, providing insights and explanations about the detected anomalies to the network operator. This allows us to prioritize its time and reduce its overall workload on the proposed system [14]. Post processing is an important issue that reduce lot of overhead in generating alert system.

9 Conclusion

Various Techniques are used in building Surveillance security systems. In this paper anomaly detection technique is used for detection of an abnormal behavior of the system to predict and generate alerts. Various soft computing approaches are analyzed and propose to build Intelligent System. These proposed techniques can be used for designing proactive defense systems. In security surveillance camera control systems data is streamed and then analyzed on real time basis using various soft computing techniques which helps in classifying the anomalies. Various techniques like nearest neighbor, classification based, information theoretic clustering based, spectral and statistical are some of the techniques which can be used for anomaly classification in building the intelligent systems.

10 Limitation and Future Research Directions

In this paper we are proposing various techniques of anomaly detection but implementation on actual testbed is not carried out. Real testbed can set up and then all techniques can be implemented and comparison can be made which technique is more effective.

References

1. Zhang, C., Chen, W., Chen, X., Yang, L., Johnstone, J.: A multiple instance learning and relevance feedback framework for retrieving abnormal incidents in surveillance videos. J. Multimedia **5**(4), 310–321 (2010)
2. Ye, Y., Ci, S., Katsaggelos, A.K., Liu, Y., Qian, Y.: Wireless video surveillance: a survey. In: Access, IEEE, vol. 1, pp. 646–660 (2013). doi:10.1109/ACCESS.2013.2282613
3. Kraiman, J.B., Arouh, S.L., Michael, L.W.: Automated anomaly detection processor. In: Proceedings SPIE 4716, Enabling technology for Simulation Science VI, p. 128, 15 Jul 2002
4. Saini, D.K.: Security concerns of object oriented software architectures. Int. J. Comput. Appl. **40**(11), 41–48 (2012)
5. Chandola, V., Banerjee, A., Kumar, V.: Anomaly detection: a survey. ACM Comput. Survey **41**, 15 (2009)
6. Wang, Y.-K., Fan, C.-T., Cheng, K.-Y., Deng, P.S.: Real-time camera anomaly detection for real-world video surveillance. In: Proceedings of Machine Learning and Cybernetics (ICMLC) Conference IEEE, vol. 4, pp. 1520–1525, Jul 2011
7. Li, H., Achim, A., Bull, D.: Unsupervised video anomaly detection using feature clustering. IET Signal Proc. **6**, 521–533 (2012)
8. Maybury, M.: Information fusion and anomaly detection with uncalibrated cameras in video surveillance. In: Multimedia Information Extraction: Advances in Video, Audio, and Imagery Analysis for Search, Data Mining, Surveillance and Authoring, vol. 1, pp. 201–216. Wiley-IEEE Press (2011). doi:10.1002/9781118219546.ch13
9. Li, W., Mahadevan, V., Vasconcelos, N.: Anomaly detection and localization in crowded scenes. IEEE Trans. Pattern Anal. Mach. Intell. **36**(1), 18–32 (2014). doi:10.1109/TPAMI.2013.111
10. Xiao, T., Zhang, C., Zha, H.: Learning to detect anomalies in surveillance video. Signal Process. Lett. IEEE **22**(9), 1477–1481 (2015). doi:10.1109/LSP.2015.2410031
11. Nguyen, V., Dinh P., Duc-Son, P., Svetha, V.: Bayesian nonparametric approaches to abnormality detection in video surveillance. Ann. Data Sci. **2**(1), 21–41 (2015). doi:10.1007/s40745-015-0030-3

12. Saini, D.K., Saini, H.: Proactive cyber defense and reconfigurable framework for cyber security. Int. Rev. Comput. Softw. (IRCOS) **2**(2), 89–98 (2007) (Italy)
13. Steinwart, I., Hush, D.R., Scovel, H.: A classification framework for anomaly detection. J. Mach. Learn. Res. (2005)
14. Mazel, J., Casas, P., Fontugne, R., Fukuda, K., Owezarski, P.: Hunting attacks in the dark: clustering and correlation analysis for unsupervised anomaly detection. Int. J. Netw. Manage. **25**(5), 283–305 (2015)

Ensemble Recommendation for Employability Using Cattell's Sixteen Personality Factor Model

Ashima Sood and Rekha Bhatia

Abstract In corporate world, recruiting the best candidate for an organization is a big challenge these days. Whilst it is agreed that basic skills, technical abilities and expertise in a particular domain of interest are central to employability, there are arguably a number of additional factors which guide employability. To fill the void between the employee and the recruiter, is the prime idea behind the current research work. The paper highlights the personality traits of potential employees identified through Cattell's 16 Personality Factor Model and elaborates the areas which the present corporate world should focus upon. The paper will show direction to the corporate employers especially seeking long term relationship with the employees regarding selection of such workforce which can sustain in the challenging and vibrant global work environment. The new approach of using psychology with ensemble learning will also provide new insights into the research works related to emotional intelligence.

Keywords Cattell's 16PF · Data mining · Employment · Machine learning · Ensemble learning

1 Introduction

In today's competitive world the need of hour is to produce output at a tremendous rate. This fast growing requirement needs self-sufficient employees. To strengthen this work, personality traits of the person has played a major role which is the foundation of the current research work. One of the efficient personality model

A. Sood (✉) · R. Bhatia
Punjabi University Regional Centre, Mohali, India
e-mail: soodashima91@gmail.com

R. Bhatia
e-mail: r.bhatia71@gmail.com

S.C. Satapathy and S. Das (eds.), *Proceedings of First International Conference on Information and Communication Technology for Intelligent Systems: Volume 2*, Smart Innovation, Systems and Technologies 51, DOI 10.1007/978-3-319-30927-9_3

available is the Cattell's 16 Personality Factor model [1] has diverse sixteen characters that analyze the deep personality of the individual. The data in the form of personality characters form data sets on which data mining is to be done. Excel was used for data cleansing and extracting the viable data [2]. After the data cleansing process, machine learning and data mining models come into play. Also, secondary data set has been used for the training and testing of the machine. Hence, supervised machine learning is being done in R programming environment.

Individually, machine learning and data mining models can perform with good accuracy. But, single machine learning model can get restricted when introduced to different data sets and accuracy may fluctuate. So, for the consistency of results the concept of ensemble modelling has been used. Ensemble modelling is the technique of combining two or more machine learning models based on their consistent performance checked by K-Fold validation.

The organization of the paper has been systematically set. Following sections are self-explanatory and set the platform for clear work flow. Following Sect. 2 handles the details of Cattell's 16 Personality Factor Model. Section 3 covers the related work, covering all the corresponding research in this field. Section 4 handles the implementation part, which is further divided into 6 parts namely, data set description, data cleansing, hypothesis generation, machine learning, k-fold validation and ensemble modelling. Finally, the research has been concluded with the recommendation of the best ensemble model for the purpose of employability prediction.

2 Cattell's 16 Personality Factor Model

The sixteen personality factor model is a panoramic and broad gauged measure of personality traits of a person which is being widely used in many sectors. This personality test which includes experimental and factual research over the decades was developed by Raymond Cattell [1]. He believed that these fundamental characters can prove to be the building blocks to predict the personality of one's individual. The test is conducted on the basis of these adjectives, people are asked questions related to these factors and then the rating is to be done on the scale of five in the form of answers of the people.

Table 1 gives the detail description of each of the sixteen personality factors.

Now, to make the picture more clear the current personality model can be compared with the other generalized one named Big Five. Big five personality questionnaire comprise of five character traits namely Openness, Conscientiousness, Extrovertness, Agreeableness and Neuroticism making the acronym as O.C.E.A.N [3]. These characters are the generalized form of all the explanatory traits of sixteen personality factor model and the comparison of the same has been elaborated in the table 2.

Table 1 Cattell's 16 PF description

Characters	Low score	High score
Warmth	Reserved, careful	Close to people
Reasoning	Lower IQ	Higher IQ
Emotional stability	Emotionally weak	Strong personality
Dominance	Cooperative, avoid conflicts	More commanding
Liveliness	Quiet, serious, mature	High spirit, energetic, social person
Rule-consciousness	Lack standards	More dutiful, follow rules
Social boldness	Shy personality	Social contact with an ease
Sensitivity	Tough at situations, realistic	Emotional and illogical
Vigilance	More trusting	Can misunderstood people
Abstractedness	More practical	Own fantasies, creative
Privateness	Transparent personalities	Conservative
Apprehension	Self confident	Self-doubting
Openness to change	Less flexible	Free-minded, experimental
Self-reliance	Dependent	Individualistic personality
Perfectionism	Casual, disorder behavior	Meticulous personality
Tension	More relaxed, patient personality	Frustrated, irritable

Table 2 16 PF versus big five personality model

S. no.	Big five characters	Corresponding 16 personality factors
1.	Openness	Sensitivity, abstractedness, openness to change, warmth, reasoning
2.	Conscientiousness	Rule-consciousness, perfectionism, liveliness, abstractedness, reasoning
3.	Extrovertness	Warmth, liveliness, social boldness, privateness, self-reliance
4.	Agreeableness	Dominance, self-reliance, vigilance
5.	Neuroticism	Tension, apprehension, emotional stability, dominance, privateness

3 Related Work

These days personality traits has been widely used for predicting the user's behavioral characteristics which can be used in various fields like medicines, mental disorders, relationship statuses, various industries, occupational careers and many more [1]. In some European countries the 16 PF traits has been used in amalgamation with handwriting analysis to assess the personality of the people [4]. Moreover, software engineering which is a broad field has mapped the personality traits of the software engineer along with the soft skills in the various stages of software engineering to result better employability. For example, in software design phase person has to be more creative likewise in programming phase person should be more logical and analytical [5]. The previous studies have shown that job satisfaction is very important for the personal growth of the employee which can be achieved if the right person is employed in the right kind of job which is suitable according to the behavioral needs [6].

Further, data mining has been used in cleansing the raw data in excel [2]. There have been many algorithms used in data mining like SVM, PageRank, Apriori, EM, CART [7]. Different machine learning algorithms have been used for multiple purposes and analysis [7–12]. Ensemble models also play a major role when finding the efficiencies of two or more machine learning and data mining models [13, 14]. Neural systems have high acknowledgment capacity for loud information, high exactness and are best in data mining. Concentrates on the information mining procedure taking into account neural system [12]. Earlier personality has been predicted using the standard mobile phone logs using support vector machine classifier [8].

4 Implementation

The implementation part starts from the collection of the raw data. Raw data can be of primary type or the secondary type. Considering the fact that raw data consists of noisy stuff, i.e. the data which is irrelevant from the machine learning point of view. So, this data has to be filtered out by cleaning the data in excel. Once the data has been cleaned, data processing including hypothesis is done to make the file suitable for machine learning. After this, bagging is done to make the K-Fold validation possible for checking the consistency of the models. Once the top five models are found, combinations are made by different permutations for checking the accuracies via ensemble modelling. Finally the recommendation is done on the basis of combinations tested.

4.1 Raw Data Description

The secondary data set of personality traits has been taken from reliable source [15] to perform the analysis and research related to personality prediction used for employability. The data set has been generated by taking the test by different people and the same is compiled when users give their test which is in the form of questionnaire. There are 16 characters and each character has 10–13 questions (total 163) based on that specific character trait. People rate their answers from 1 to 5 labelled 1-emphatically dissent, 2-deviate, 3-neither concur nor deviate, 4-concur, 5-firmly concur. When people give their online test and give answers to their questions, their record is saved for the data set purposes. This record has been taken from 49,159 people from different 253 countries and one proxy. Raw data consisted of all the rating done by different people, country code, gender, age, time duration of the test, and how the participant came to know about the test. This test also gives right to the participant to give his information or not i.e. answers given by them can be used for the research purposes or they may decline this right.

4.2 Data Cleansing

Data Cleansing is the initial and crucial part of data mining which is done to remove the noisy data. Noisy data is the data which is not required in the data processing and further in machine learning tool. So, for this data like country name, gender category, age was removed and only rating of people was saved for further analysis.

4.3 Hypothesis Generation

Table 3 shows the hypothesis of the each character in which range has been set and against every character it has been calculated that for a particular trait some people are employed which come under the hypothesis range. The person whose rating lies in that acceptance region is set to 1 else 0. After hypothesis it has been observed that out of total 49,159 people, 1258 people are employable.

4.4 Machine Learning

In the current paper R programming environment has been used for the machine learning process. Data is uploaded in the machine and partitioning of data has been done in 60–40 ratio which represents the training and testing partition. Along with this it takes the random seed value which process data in all mixed environment.

Table 3 Hypothesis generation

S. no.	16 PF characters	Acceptance region (scale 1 to 5)	People qualifying out of 49,159
1.	Warmth	$2 < A \leq 5$	48,405
2.	Reasoning	$2 < B \leq 5$	48,523
3.	Emotional stability	$2 < C \leq 5$	48,346
4.	Dominance	$0 < D \leq 3$	19,456
5.	Liveliness	$2 < E \leq 5$	47,836
6.	Rule-consciousness	$2 < F \leq 5$	48,416
7.	Social-boldness	$2 < G \leq 5$	48,501
8.	Sensitivity	$0 < H \leq 3$	26,704
9.	Vigilance	$0 < I \leq 3$	23,312
10.	Abstractedness	$0 < J \leq 3$	16,751
11.	Privateness	$1 < K \leq 4$	48,244
12.	Apprehension	$0 < L \leq 3$	18,950
13.	Openness to change	$2 < M \leq 5$	48,338
14.	Self-reliance	$1 < N \leq 4$	44,243
15.	Perfectionism	$2 < O \leq 5$	48,464
16.	Tension	$0 < P \leq 3$	25,109

Classification model divides the executable file which is being uploaded in 16 rows which are all the sixteen characters and 17th row is the target value which is set as employability numeric. The seed value taken is 673,475 which is again a random value. After this evaluation is done on some parameters essential for analysis. It is clearly shown in Table 4 in which first column represents various machine learning models, showing results in their own magnitude values. Second column gives the methods and packages being deployed. Next column gives the accuracy of different models which shows how much accurate the model is while predicting the results. Fourth column gives the region of curve (ROC) which compares false positive rate to the true positive rate. Fifth column gives the degree of precision or can be mentioned as risk factor. Last, sixth column represents the confusion matrix also termed as error matrix gives the measure of error of each model. Less the measure of error, better for the machine learning model.

Table shows high accuracies of all the models with Random Forest being the highest one. The reason for high accuracies with the testing of different machine learning models can be explained by the hypothesis. Since the ranging set in hypothesis are linear in their form so it was easy for the models to crack the equations through the training data.

4.5 K-Fold Validation

K-Fold validation is the bagging of data with different seed values. This is done to test the data set in all possible ways to check the consistency of all the results. Considering there are multiple models, and the selection of the models which are performing consistently with least variations are to be chosen for the ensemble model purpose.

In this testing is repeated ten times (K = 10) with various seed values as 42, 673,475, 111, 271,903, 333, 444, 555, 619,013, 175,313, 697,349. Figure 1 shows the accuracies of the different models.

Table 4 Evaluation results of seed value 673,475

Model name	Method/package	Accuracy	ROC value	Precision/recall	Confusion matrix
Support vector machine	svm/ksvm	98.2659	0.9876	0.988	0.0939256
Ada boost	ada/ada	99.9695	1.0000	1.000	0.0058365
Decision tree	rpart/rpart	99.9746	0.9999	1.000	0.0048732
Neural network	neuralnet/neuralnet	99.3186	0.9853	0.985	0.0676936
Random forest	rf/randomForest	99.9847	1.0000	1.000	0.0029354
Linear model	lm/glm	97.3047	0.8881	0.890	0.3936422

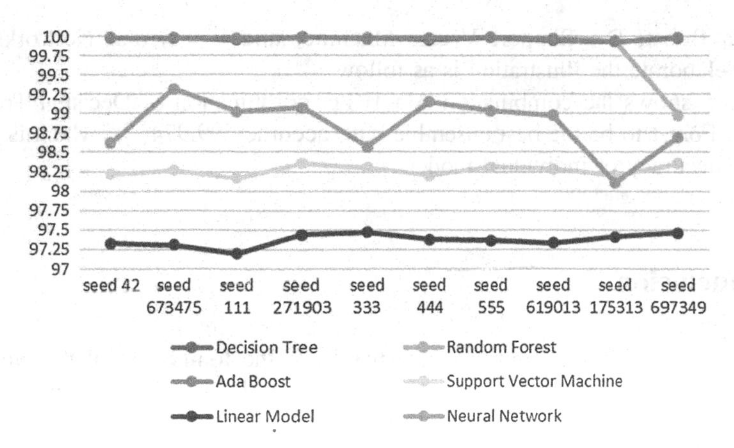

Fig. 1 K-fold validation

Table 5 Combinations and comparison of ensemble models

S. no.	Combinations	Accuracy (%)	S. no.	Combinations	Accuracy (%)
1.	A + B + C + D + E	99.4864	14.	A + B + D	99.7610
2.	B + C + D + E	99.4864	15.	A + B + C	99.9695
3.	A + C + D + E	99.4864	16.	D + E	99.1914
4.	A + B + C + E	99.6440	17.	C + E	99.6440
5.	A + B + C + D	99.7559	18.	C + D	99.7559
6.	C + D + E	99.4864	19.	B + E	99.6542
7.	B + D + E	99.4965	20.	B + D	99.7712
8.	B + C + E	99.6440	21.	B + C	99.9695
9.	B + C + D	99.7559	22.	A + E	99.6440
10.	A + D + E	99.4864	23.	A + D	99.7610
11.	A + C + E	99.6440	24.	A + C	99.9695
12.	A + C + D	99.7559	25.	A + B	99.9746
13.	A + B + E	99.6440			

4.6 *Ensemble Modelling*

It has been observed that after the ten validations all models has shown consistency in their performance except the Linear Model. So, top five models have been further used in ensemble modelling. Ensemble modelling is done to increase the accuracy of the models when they are used in combinations. So to enhance this, models are combined using permutations of different models and further their accuracies are calculated.

According to the dataset, total 25 combinations of the top five models were designed with the seed value of 673,475 and partitioning of 60–40 model as training and testing part. Considering A = Decision Tree, B = Random Forest,

C = Ada Boost, D = Support Vector Machine, and E = Neural Networks. The table to elaborate the illustration is as follow.

Table 5 shows the combination A + B i.e. combination of Decision Tree and Random Forest to be the best ensemble with accuracy 99.9746 % which is higher than the accuracy of individual models.

5 Conclusion

The prediction of employability which has been the foundation of the work has been achieved as per the assumptions by using the psychology and machine learning in combination. The personality traits act as a catalyst while employing the best suitable candidate. In the previous work of predicting user's personality from the emails has use the single-label classifier which is support vector machine which has given the accuracy of 69 % that is far better than the approach of machine learning use in the current work [7]. The results concluded the Decision Tree and Random Forest to be the most effective combination of machine learning models. In future this work can be used by the companies who demand meticulous employees with stable personality along with other required soft skills. Moreover, the current work can be divided into diverse divisions of various employment sectors where recruiters can employ the individuals as per their personality traits. This dataset can also be used for research in emotional intelligence.

References

1. Dorfman, W.I., Hersen, M.: Understanding Psychological Assesment. Sringer Science +Business Media, LLC, pp. 187–215 (2001)
2. Tang, H.: A simple approach of data mining in excel. In: 4th International conference on Wireless Communications Network and Mobile Computing, pp. 1–4. IEEE, 12–14 Oct 2008
3. Goldberg, L.R.: The development of markers for the big-five factor structure. Psychol. Assess. **4**(1), 26–42 (1992) (American Psychological Association Inc)
4. Bushnell, I.W.R.: A comparison of the validity of handwriting analysis with that of the Cattell 16PF. Int. J. Selection Assess. **4**(1), 12–17 (1996)
5. Capretz, L.F., Ahmed, F.: Making sense of software development and personality types. IEEE Comput. Soc. IT Prof. **12**(1), 6–13 (2010)
6. Watson, D., Slack, A.K.: General factors of affective temperament and their relation to job satisfaction over time. Organ. Behav. Human Decis. Process. **54**(2), 181–202 (1993)
7. Shen, J., Brdiczka, O., Liu, J.: Understanding Email Writers: Personality Prediction from Email Messages, vol. 7899, pp. 318–330. Springer, 10–14 June 2013
8. de Montjoye, Y.-A., Quoidbach, J., Robic, F., Pentland, A.: Predicting Personality Using Novel Mobile Phone-Based Metrics. vol. 7812, pp. 48–55. Springer (2013)
9. Liaw, A., Wiener, M.: Classification and regression by randomforest. R News. **2**(3), 18–22 (2002)

10. Burges, J.C.: A tutorial on support vector machines for pattern recognition. Bell Laboratories. Lucent Technologies. Data Mining and Knowledge Discovery, vol. 2, pp. 121–167. Springer (1998)
11. An, T.K., Kim, M.H.: A New Diverse Adaboost Classifier. In: International Conference on Artificial Intelligence and Computational Intelligence, pp. 359–363, IEEE (2010)
12. Wang, S.: Research on a new effective data mining method based on neural network. In: IEEE International Symposium on Electronic Commerce and Security, pp. 195–198, 3–5 Aug 2008
13. Xu, L., Adam, K., Ching, Y.S.: Methods of combining multiple classifiers and their applications to handwriting recognition. IEEE Trans. Syst. **22**(3) (May–June 1992)
14. Polikar, R.: Ensemble based systems in decision making. IEEE Circuits and systems Magazine. Third Quarter (2006)
15. https://goo.gl/SlArZP

Search Logs Mining: Survey

Vivek Bhojawala and Pinal Patel

Abstract Search engine process millions of query and collect data of user interaction every day. These huge amount of data contains valuable information through which web search engine can be optimized. Search engine mostly relies on explicit judgement received from domain experts. To survive the competition search engine must understand user's information needs very well. Search logs provide implicit data about user's interaction with search engine. Search logs are noisy, they contain data of both successful search and unsuccessful search. The challenge is to accurately interpret user's feedback to search engine and learning the user access patterns, such that search engine will better be able to cater the user's information needs. User feedback can be used to re-rank the search result, query suggestion and URL recommendation.

Keywords Search logs · Click-through · Implicit feedback · Search engine · Web search ranking · Search destination · Entropy · Information goal · User behavior · Search history · Learning from user behavior

1 Introduction

Web search engine mostly consists of explicit judgement received by domain expert. Web is dynamic, new websites are created every day and also new queries with different information need arrives. To prepare such relevance judgement and maintain it up to date is an expensive and time consuming task. Search logs contains implicit user feedback which can be used as relevance judgement. To use implicit feedback as relevance judgement we need to understand how user interacts with search engine. From search log we can: (1) Re-rank search result: search engine uses

V. Bhojawala (✉) · P. Patel
Government Engineering College Gandhinagar, Gandhinagar, India
e-mail: vivek141@gecg28.ac.in

P. Patel
e-mail: pinalpatel@gecg28.ac.in

S.C. Satapathy and S. Das (eds.), *Proceedings of First International Conference on Information and Communication Technology for Intelligent Systems: Volume 2*, Smart Innovation, Systems and Technologies 51, DOI 10.1007/978-3-319-30927-9_4

implicit feedback of user to rank the result. (2) Find Ambiguous query: User queries are often ambiguous. Single query contains multiple meanings. (3) Positional bias of results: How User interacts with top 10 results returned by a Search engine. (4) Interpreting User behavior: Differentiating user's behavior for successful and unsuccessful search. (5) modeling user's implicit feedback: it can be modeled at query level, session level or task level. (6) Dynamic search result: based on past search of user search result contains both old and new results based on relevancy. (7) Measure Efficiency of search engine: Search engine is producing relevant result to query. (8) Query suggestion: based on implicit feedback similar queries are suggested to user for (9) URL suggestion: based on URL visited by other user for same query. (10) Prediction of user action: based on most recent interaction search engine can make prediction of upcoming information need of user.

2 Web Search and Search Logs

This section contains: (1) comparison of classic information retrieval system and classical information retrieval augmented for web search. (2) Classification of query. (3) web search behavior of user. (4) Example of anonymized search log. (5) Measuring entropy of search logs.

2.1 Web Search Fundamental [1]

In information retrieval system user is having specific information needs [1]. That information need is converted into query and submitted. Submitted query is matched against collection of documents (corpus) with certain rules and most relevant documents are returned back to the user.

Web search contains a little bit different structure from information retrieval system. In web search main difference is users perform tasks rather than specific information search, each task requires some information and that information need is converted into verbal form and then it is submitted as a query to search engine. Now search engine will return results based on some rules and relevance of the documents to the query submitted by the user. User examines the results and now user will compare relevance of URL returned by search engine to the information need. If result matches with information need then user is satisfied and stops searching for that topic otherwise user will reformulate the query and repeats the process until user gets required information need. After a fair amount of query reformulation user may abandon the search task.

Web queries can be: (1) Navigational: User is searching for particular web site URL. Like Login page of Gmail, official website of android Marshmallow, home page of apple iPhone etc. Navigational query is generally satisfied with single URL click. (2) Information: User search for information which is based on facts that can be present on multiple websites. Like height of the Mount Everest, capital of India, planned cites of India etc. Information queries are generally satisfied by multiple URL clicks. (3) Transaction: Transactional query involves searching for URL such that user can perform more action on that website. Like 'buy nexus 5' will return multiple e-commerce websites which sells nexus 5, user selects particular website and performs other interaction with website like applying coupon, giving address and contact information for shipping of nexus 5 and at last user selects one of the payment option available on the website. Payment option may redirect user to a particular bank website for online payment.

3 Understanding the Search Logs

This section contains: (1) Interpreting click through data. (2) Measuring retrieval quality of search engine. (3) Relationship between searcher's query and information goal (4) Features representing user behavior.

3.1 Interpreting Click Through Data [2]

Using search logs as implicit feedback is difficult to interpret correctly and it includes noisy data. To evaluate reliability of click through data, study was conducted to analyze how users interact with Search engine result page and compare their implicit feedback with explicit feedback. To analyze following experiments were performed (1) user views result page from top to bottom? How many abstract do they read before clicking? (2) How implicit feedback matches with explicit feedback constructed by domain experts. For experiment users were asked questions which contained both informational and navigational query. Eye tracker was used to capture eye fixations which is defined as concentration of eye at particular part of web page for 200–300 ms approximately. Fixation measures interestingness of URL.

Results of experiment shows: (1) user views first two links of result page equally but number of clicks for 1st links are very high compared to second link. Same behavior observed for 6th and 7th link. (2) User scans results from top to bottom. (3) User does not observe abstracts of all links but more likely to observe abstract of clicked link and link above and below the clicked link. To better understand user behavior on first two links second experiment was carried out. In second experiment, each user was assigned one of the following three conditions: (1) Normal: User was given results directly received from Google search engine. (2) Reversed: Results returned by user were reversed. (3) Swapped: First two results returned by search engine were swapped. Result of experiment shows: (1) in reversed list of result user viewed more abstracts compared to normal shows that order in which relevant results presented does not matter, user views abstracts and clicks according to query. (2) When top two results were swapped even if 2nd result was more relevant to query, most user clicked 1st link showing trust bias. Trust bias is user's trust on particular search engine.

Important deductions from experiments. (1) If user clicks on particular URL it means that URL is examined by user and it is relevant to query issued by the user. (2) If URL is clicked and URL above that is skipped means that user examined the URL and it is not relevant to query. (3) The rank at which user clicks on the link is also important and it shows relevance to particular user.

3.2 Measuring Retrieval Quality of Search Engine Using Search Logs [3]

To measure retrieval quality of search engine search logs can be directly provided as input to feedback system. User's satisfaction is ultimate goal of search engine, so in order to better serve users search engine needs to measure its own retrieval quality form click through data. Following are the absolute metrics to measure retrieval quality. (1) Abandon rate: It is measured in number of times user issued a query and didn't clicked on any results. (2) Reformulation rate: it is measured in part of query used by successive query in the same session. (3) Clicks per query: Mean no of results clicked for each query. (4) Time to first click: Mean time between queries submitted by user and first click of results. (5) Time to last click: Mean time between queries submitted by user and last click of results.

3.3 Relationship Between Search's Query and Information Goal [4]

For rare and complex information goal user behavior changes significantly, click through rate decreases and query reformulation increases. User query can be specific or general to information needs, success of search depends on search engine's ability to interpret the information needs. Efficiency of search engine can be measured in session length. When search engine is unable to produce relevant result to user query session length increases. Search session contains sequence of queries in chronological time. Session ends when 30 min of inactivity. A detailed observation in search logs leads to observation that user issues more than one queries which are interrelated for single information goal.

Example shows that at time t0 user issued query to buy Samsung mobile online and then user clicks on Samsung mobile home page and from there user picked one particular model, click at time t2 is URL Click indicating that visited URL is not from SERP. After finding particular model of mobile user searched for review of that mobile. After reading review, at t4 user decided to buy a phone at t5 and after seeing result of query user reformulate the query to buy phone from particular e-commerce website at time t6 and finally bought the phone. Click at time t8 is checkout click to fill out payment and shipping details from particular e-commerce website. So we can conclude that user executed all this queries. Search engine returns results according to user's query and query reflects information goal. It is difficult to know precisely user's satisfaction. User's satisfaction can be measured by examining URL visited by user at the end of the session. In this session example, we can say that user is satisfied by examining last URL click which is of checkout which proves that user has bought the mobile which was user's ultimate goal. Other concern about user behavior is parallel loading URLs in browser tabs without reading the content of URL. To identify sessions of parallel tabs, dwell time is observed. If dwell time is below threshold for more than one clicks then user may have this scenario.

To study post query behavior of user for rare information goal two weeks of search engine data was collected. Tail query is defined as queries and URLs observed in second week that were not observed in first week for the information goal initiated in first week, all other queries are non-tail. Results in comparison of tail and non-tail query shows that: (1) Query reformulation rate for tail query is higher than non-tail query because tail query represents rare and specific information need. Length of reformulation represents search engine's ability to understand rare information need of user. (2) When user reformulate the query to be more specific then query length increased as compared to initial submitted query.

3.4 Features Representing User Behavior [5]

Features representing user behavior can be (1) Query Text Features, (2) Browsing Features, (3) Click Through features. Query feature includes: (1) query length: Numbers of words in query. (2) Next query overlap: number of words common with next query. (3) Domain overlap: words common with query and domain. (4) URL overlap: words common with query and URL. (5) Abstract overlap: words common with query and abstract. Browsing feature includes: (1) Dwell time: Time spent on URL. (2) Average dwell time: Average time on the page for single query. (3) Dwell time deviation: Deviation from overall average dwell time on page. Click Through feature includes: (1) Position: Rank at which URL clicked. (2) Click frequency: Number of clicks for this query, URL pair. (3) Click relative frequency: Relative frequency of click for this query and URL. (4) Click Deviation: deviation from observed number of clicks.

4 How People Recall, Recognize, and Reuse Search Results [6]

When a user issues the query, user has certain expectations about how search results will be returned. These expectations can be based on information goal, and also based on knowledge about working of search engine. Such as where relevant results are expected to be ranked based on previous searches of individual user for a specific topic. When a result list of URL is changed according to particular modeling schema, users have trouble re-using the previously viewed content in the list. However study has shown that new relevant result for same query can be presented where old results have been forgotten, making both old and new content easy to find.

4.1 *Recall*

Two main factors affecting how likely result was to remember (1) position at which result was ranked. (2) Whether or not the result was clicked. Results that were clicked were significantly more likely to be recalled. 40 % of clicked result were remembered, compared with 8 % of results that were not clicked. Among the clicked results, last Clicked result in the list appears more memorable than previous result. How user memorized rank of search results was studied to understand how user re-find the same result. The recalled rank differed from actual rank 33 % of time. Users correctly identified initial results 90 % of time and accuracy dropped as rank increased.

4.2 *Recognizing Change in Result List*

Most of the time user recognizes results that are different from initial result, but study showed that very different list can be recognized as the same if they maintain consistency in recalled aspect. To study how user recognizes as same or different when list is different form initial list it is constructed in following ways: (1) *Random Merge*: Four results viewed previously were randomly with top six results of new list. (2) *Clicked Merge:* Results clicked during session 1 were ranked first, followed by new results. The exact no of results preserved varied as a function of how many results were clicked. (3) *Intelligent Merge:* Old and new results were merged with an attempt to preserve the memorable aspects of the list during previous session. (4) *Original Merge:* The result list was exactly same as the originally viewed list. (5) *New:* The list was comprised of entirely new results. For intelligent merge user voted highest 81 % of time results returned by search engine were same.

4.3 Reusing the Search Results

Reusing the search result focuses on finding same result again which were found during previous session. To observe this user's history of clicking the result were captured by proxy in session 1 and during session 2 user had to re-finding and New-finding the results based on results visited during first session. For new finding tasks intelligent merge gains lowest mean and median task time, and for re-finding task original merge gains lowest mean and median task time. For re-finding task intelligent merge performs closest to original merge compared with other methods.

5 Click Through Bipartite Graph: [7]

Web search engine does not only retrieves the document relevant to query but it also ranks the documents such that most relevant document appears at the higher position of search result. When user is unable to formulate query to satisfy required information goal search engine provides query suggestion based on user's current query. With the help of click through bipartite graph we can efficiently measure query-document, document-document and query-query similarity. Document ranking can be assigned using click through bipartite graph using no of clicks on query URL pair. Highly clicked URL will be positioned on top of the SERP. Query suggestion is given based on overleaping URLs between queries in click through bipartite graph. Challenge with click through bipartite graph is that query-document relevancy is not calculated based on only no of click to the URL because we saw in previous section that due to positional bias higher ranked URL may get more click even if both URL contains same similarity to query. In order to avoid that multiple feature of web search log is taken into consideration (Fig. 1).

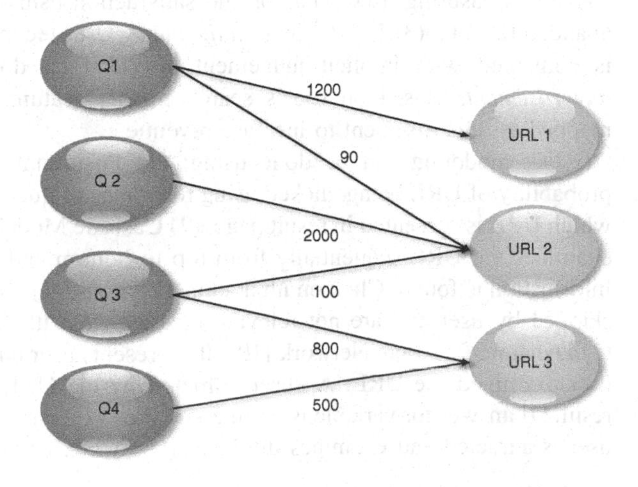

Fig. 1 Click-through bipartite shows query and url as nodes and link shows click rate [7]

Query document bipartite graph consist of triplet (q, d, t) where q denotes query d denotes document and t denotes no of times clicked. For considering multiple feature query and document are represented as vectors in query space Q_i and document space D_i where Q_i and D_i are sub spaces of Euclidian distance. Q_i and D_i may or may not be the same space.

Click through graph consists of query, URL and number of times URL clicked for a query. Relationship between query and URL can be identified by: (1) Euclidean distance [7]: it maps Query q and its feature in query space Q_i, and URL u and its features in URL space U_i. Based on Query q in Q_i and URL u in U_i Euclidean distance Di is calculated. (2) Co-visited method [8]: If two URLs are visited by same query it is possible that both URL presents similar information are called as co-visited. Number of times URL clicked for same query is used for calculating similarity between URLs. (3) Iterative Algorithm [8]: Iterative algorithm compared with co-visited method also checks for query similarity based on URLs visited by two or more queries. Based on derived relation from query and URL similarity, Iterative algorithm iteratively derives new relation which were not discovered in previous iterations. (4) Learning the query intent [9]: It is a semi-supervised method to learn query intent based on small amount of manually labeled queries and related URLs. When new query arrives which is not present in labeled queries it classifies URL to +/− based on labeled data.

6 Click Modeling [10]

User clicks are known to be biased based on position presented on result page. Main challenge is to model user click unbiasedly. Click model provides user search behavior which can be compared with judgements of web document and can be used in following ways: (1) *Automated ranking alteration:* Highly ranked results are altered with user preference to achieve user satisfaction. (2) *Search quality metrics:* Measuring user behavior and satisfaction using query-reformulation rate, abandon rate etc. (3) *Validation of judgement:* User feedback as explicit judgement is compared with implicit judgement provided by domain expert. (4) *Online advertisement:* Based on user's search history feature clicks are predicted and mapped as advertisement to increase revenue.

Click modeling can be done using: (1) Positional model [10]: it calculates probability of URL being clicked using relevance of query and URL and position at which URL is presented in result page. (2) Cascade Model [10]: It assumes that user examines the URL sequentially from top to bottom and stops as soon as relevant information is found. Click on ith document means: (1) URLs above ith position are skipped by user and are not relevant to query. (2) Ith URL is relevant to query. (3) Dynamic Bayesian Network [10]: It represents user click into three variables E_i: user examined the URL, A_i: user attracted by the URL, S_i: user satisfied by the result? If answer for variable is yes then it takes value as 1 otherwise 0. It says: (1) if user is attracted and examines the URL then Click of URL occurs. (2) Attraction dependent on URL's relevancy to query. (3) User examines abstracts of results from top to bottom and clicks on URL with certain probability of being satisfied. (4) As soon as user is satisfied by ith URL, probability of examining URL ith below is 0. (4) Click Chain Modeling (CCM) [11]: in CCM each URL on result page contains its own probability of examination, probability of click and relevance to query. Main difference here is probability of current URL is connected with all previous URLs and their probability.

6.1 Context Aware Ranking in Web Search [12]

Context of a search query gives meaningful information about search intent. For example user raises query "apple" after searching "Samsung mobile phone", it is very likely that user is searching for apple mobile phone rather than apple fruit. There are two main challenges in context aware ranking: (1) Using context to rank result. (2) Using different types of contexts such as user query, URL clicked. Context aware ranking principles: (1) Reformulation: user reformulates the query to be more specific about information needs. (2) Specialization: User issues specialized query to see results that are more specific about user's intent. (3) Generalization: User may ask a query more general than previous one to gain more general knowledge. (4) General association: When query is generally associated with its context, context may help to narrow down the user's search intent.

7 Short Term and Long Term User Interests

This section contains: (1) Predicting short term user interests. (2) Using Long term search history to improve search accuracy.

7.1 Predicting Short Term User Interests: [13]

Short term search is limited to only single search session containing consecutive queries. Query context is pre-query activities that includes previous query and page visited last in past. Developing and evaluating user interest model for current query, its context and their combination is called as intent. Based on past query and URL visited within session context is created. When user issues new query Q3, intent is constructed from context for current query Q3 and based on that intent optimal result is returned to user. Short term user interests can be modeled in: (1) Query model: Open Directory Project (ODP, dmoz.org) provides human-edited directory of web. It contains categorized list of URLs. In this model categories for top 10 result are retrieved from dmoz.org. Based on URL clicks of user their categories are mapped and user preference for particular category is saved. When same query is again raised by user search engine assigns higher ranks to URLs that belongs to category which user previously visited. (2) Context model: Context model also categorizes URLs visited by user in categories provided by ODP. Weight to particular category is assigned based on dwell time on the page. If user visits URL for more than 30 s than that URL contains interesting contents. (3) Intent model: Intent model is combination of query model and context model. Since query model includes information from current query and context model includes information about user's activity in current session, combination of both information gives more accurate results. (4) Relevance model or Ground truth: Relevance model predicts future actions. It assigns higher weights for most recent actions of user. This model captures most recent user action as more valuable for constructing context. This model generates best result based on observation that each user action leads closer to information goals.

7.2 Long Term Search History to Improve Search Accuracy: [14]

Most existing retrieval system including search engine offers generalized web search interface which is optimal for all web users. Retrieval of document is made based on only the query submitted by user and ignoring user's preferences or the search context. When user submits query "python" is ambiguous and search result may contain mixed content which is non-optimal for the user. Instead of using query only as retrieval option, user search context can be used to match with user's intended information needs. There is wide variety of search contexts like bookmarks, user interests in particular categories, user's long term search history etc. In long term search history logs of user's search history is maintained based on URL clicked by user. For example user has searched for "debugging" and "Java code" and currently searching for "python" suggests that user is searching for python related to programming context. Second optimization can be done based on user's past searches for example if user searched for "Perl programming" in past and visited some web pages and if same search is repeated then based on user's past visited URL current SERP can be re ranked based on user preference.

8 Search Trail and Popular URLs

This section contains: (1) Search trails. (2) Evaluating effectiveness of search trails. (3) Using Popular URLs to enhance web search.

Table 1 Shows session example

Time	Action	Value
t0	Query	Buy Samsung mobile online
t1	SERP click	http://www.samsung.com
t2	URL click	http://www.samsung.com
t3	Query	Samsung galaxy s6 edge review
t4	SERP click	http://www.in.techradar.com
t5	Query	Buy Samsung galaxy s6 edge
t6	Re-query	Buy Samsung galaxy s6 edge flipkart
t7	SERP click	http://www.flipkart.com/
t8	URL click	https://www.flipkart.com/checkout/

8.1 Search Trail [15]

User with certain information goal submits query to search engine and visits URL presented on result page. User also visits URLs that are presented on web page whose URL address is returned by search engine. Search trail consists of both URL presented on result page and URL that were not presented on result page. As shown in session example in Table 1 search trail for "buying a Samsung mobile phone" at time t1 clicks on URL presented on result page and at time t2 user clicks on link from web page visited at time t1 which was not presented on result page. At time t3 based on information gained from URL visited at time t1 and t2 user formulates new query at time t3 and finally search trail ends when user performs payment of buying a mobile at time t8. Search trail may span to multiple session.

8.2 Evaluating Effectiveness of Search Task Trails [16]

Experiment conducted to measure effectiveness of search task trails on large scale dataset from commercial engine shows results that (1) User tasks trails are more accurate as compared to session and query modeling. (2) Task trails provides unambiguous user information needs. (3) Task trail based query suggestion performs well as compared with other models. Query task clustering approach: Queries which belong to same task can be combined into a single cluster. Based on observation consecutive query pairs are more likely belong to same task rather than non-consecutive ones. User search interests are measured by using mapping URL belongs to same task into categories provided by ODP on dmoz.org. Dwell time, Hidden Markov Model to measuring success rate of search and number of clicks on URL are taken as user implicit feedback. Query suggestion models: (1) Random walk, (2) Log likelihood, (3) co-occurrences is used to measure the performance of task trails modeling.

8.3 *Using Popular URLs to Enhance Web Search: [6]*

Query suggestion offers similar query to current query of user. Query suggestion allows user to express query more specifically leading to improved retrieval performance. Search engines gives query suggestion based on query reformulation of users. Examining the most common URLs visited by majority of user for a given query is referred as popular URLs. Popular URLs may not be ranked in result list of query or may not contains words similar to user query. This approach gives user a shortcut to reach information goal. To examine the usefulness of destinations four system were used in study: (1) baseline web search system with no explicit support for query recommendation, (2) A search system with a query suggestion method that recommends additional query (3) query destination which suggest popular URL destination for given query. (4) Session destination which suggests endpoint of session trails. Among four system query destination achieves highest positive user feedback and mean average time to complete task was minimum.

9 Summary

This paper provides survey on search log mining for web search, with focus on accurately interpreting user feedback and various methods to model user's implicit feedback. By modeling user's implicit feedback search engine results can be re-ranked to improve retrieval quality of search engine, query suggestion and URL recommendation.

References

1. Broder, A.: A taxonomy of web search. In: ACM Sigir forum, vol. 36, no. 2, pp. 3–10. ACM (2002)
2. Joachims, T., Granka, L., Pan, B., Hembrooke, H., Gay, G.: Accurately interpreting clickthrough data as implicit feedback. In: Proceedings of the 28th Annual International ACM SIGIR Conference on Research and Development in Information Retrieval, pp. 154–161. ACM (2005)
3. Radlinski, F., Kurup, M., Joachims, T.: How does clickthrough data reflect retrieval quality? In: Proceedings of the 17th ACM Conference on Information and Knowledge Management, pp. 43–52. ACM (2008)
4. Downey, D., Dumais, S., Liebling, D., Horvitz, E.: Understanding the relationship between searchers' queries and information goals. In: Proceedings of the 17th ACM Conference on Information and Knowledge Management, pp. 449–458. ACM (2008)
5. Agichtein, E., Brill, E., Dumais, S., Ragno, R.: Learning user interaction models for predicting web search result preferences. In: Proceedings of the 29th Annual International ACM SIGIR Conference on Research and Development in Information Retrieval, pp. 3–10. ACM (2006)
6. Teevan, J.: How people recall, recognize, and reuse search results. ACM Trans. Inf. Syst. (TOIS) **26**(4), 19 (2008)

7. Wu, W., Li, H., Xu, J.: Learning query and document similarities from click-through bipartite graph with metadata. In: Proceedings of the Sixth ACM International Conference on Web Search and Data Mining, pp. 687–696. ACM (2013)
8. Xue, G.-R., Zeng, H.-J., Chen, Z., Yu, Y., Ma, W.-Y., Xi, W.S., Fan, W.G.: Optimizing web search using web click-through data. In: Proceedings of the Thirteenth ACM International Conference on Information and Knowledge Management, pp. 118–126. ACM (2004)
9. Li, X., Wang, Y.-Y., Acero, A.: Learning query intent from regularized click graphs. In: Proceedings of the 31st Annual International ACM SIGIR Conference on Research and Development in Information Retrieval, pp. 339–346. ACM (2008)
10. Chapelle, O., Zhang, Y.: A dynamic bayesian network click model for web search ranking. In: Proceedings of the 18th International Conference on World Wide Web, pp. 1–10. ACM (2009)
11. Guo, F., Liu, C., Kannan, A., Minka, T., Taylor, M., Wang, Y.-M., Faloutsos, C.: Click chain model in web search. In: Proceedings of the 18th International Conference on World Wide Web, pp. 11–20. ACM (2009)
12. Xiang, B., Jiang, D., Pei, J., Sun, X., Chen, E., Li, H.: Context-aware ranking in web search. In: Proceedings of the 33rd International ACM SIGIR Conference on Research and Development in Information Retrieval, pp. 451–458. ACM (2010)
13. White, R.W., Bennett, P.N., Dumais, S.T.: Predicting short-term interests using activity-based search context. In: Proceedings of the 19th ACM International Conference on Information and Knowledge Management, pp. 1009–1018. ACM (2010)
14. Tan, B., Shen, X., Zhai, C.X.: Mining long-term search history to improve search accuracy. In: Proceedings of the 12th ACM SIGKDD International Conference on Knowledge Discovery and Data Mining, pp. 718–723. ACM (2006)
15. Bilenko, M., White, R.W.: Mining the search trails of surfing crowds: identifying relevant websites from user activity. In: Proceedings of the 17th International Conference on World Wide Web, pp. 51–60. ACM (2008)
16. Liao, Z., Song, Y., He, L.-W., Huang, Y.: Evaluating the effectiveness of search task trails. In: Proceedings of the 21st International Conference on World Wide Web, pp. 489–498. ACM (2012)

An Effective Model for Face Detection Using R, G, B Color Segmentation with Genetic Algorithm

Devesh Kumar Srivastava and Tarun Budhraja

Abstract Face detection is a grave concern in digital image processing and automatic face recognition system. This research work proposed a complete mechanism for face detection using R, G, B color segmentation and search optimization scheme with Genetic Algorithm, also refer a discrete technique that is appropriate for combinatorial problems. In this paper we tried to build an R, G, B color range that will shelter a skin part from an image and handover a best fitted solution as fitness function for GA to perform further operation to detect images in complex background. In this paper our tryout are to enrich detection accuracy in lesser computational time. The evaluation shows that this algorithm is capable to detect the face from complex background conditions and for side faces too. This algorithm is tested on a wide number of test images. All the simulation has been done on MATLAB.

Keywords Face recognition and detection · Fitness function · Genetic algorithm (Search optimization scheme)

1 Introduction

Face detection is the foremost problem in digital image processing. In some Automatic Face Recognition (ASR) systems, it can be debated that the system user will cooperate to provide a face image with a side faces pose. However it is not practical to assume that user will always have control over the multifaceted background and surrounding conditions (e.g. lighting) during the face verification system. So detection of a face before trying to recognize saves lot of human work. In static image based automatic recognition system (ARS) try to find a portion of

D.K. Srivastava (✉) · T. Budhraja
Department of CSE and IT, Manipal University, Jaipur, India
e-mail: devesh988@yahoo.com

T. Budhraja
e-mail: budhraja.tarun123@gmail.com

S.C. Satapathy and S. Das (eds.), *Proceedings of First International Conference on Information and Communication Technology for Intelligent Systems: Volume 2*, Smart Innovation, Systems and Technologies 51, DOI 10.1007/978-3-319-30927-9_5

the entire face by removing the background area and hairs that are not essential for recognition system most of the background and other areas of an individual's head such as hair that are not necessary for the face recognition.

In human-computer interaction (HCI) face detection have found considerable attention it also provides an efficient way to communicate b/w humans and computer. Detection of face from an image sequence is also a challenging problem in HCI surveillance systems; video conferencing, secure access control, forensic applications [1]. Knowledge based face detection approach for face detection using skin color segmentation with color models (RGB, YCbCr and HSV) with threshold helps to remove non skin part from a static image [2]. Face detection using GA and back propagation Neural Network (NN) where NN used for feature extraction and classification is carried out using GA [3]. Human face detection use adaptive and non-adaptive RGB color space to detect feature like eye and lips [4]. Video based face detection with skin segmentation where face is identified from a video [5]. Real time face detection of profile faces using genetic algorithm to augment the coordinates of rotation angel of profile face using flipping scheme [6]. Proposed Eigen face technique was used to determine the fitness of the face regions and limited the searching space to the eyes and determines existence of the different facial features [7]. In this paper we key objectives is to build a face detection model by using R, G and color space with help of Genetic Algorithm The rest of paper described as follows: (1) Proposed methodology 2.1.2. Genetic Algorithm 2.2.1. Proposed fitness algorithm, (2) Experimental results, (3) Discussion, (4) Conclusion and future work.

2 Methodology

2.1 *R, G and B Color Space for Skin Detection (Background)*

The apparent human color fluctuates as a function of the relative path to the illumination. Using normalize color histogram it is easy to collect skin pixel and intensity of color be changed on dividing by luminance [8]. The RGB model is characterized by a 3-D dice with (red, green and blue) colors. Main purpose of RGB color model is to represent an image in electronic form. RGB color model don't usage a palette. RGB color model are widely used in video camera and color monitors (Figs. 1 and 2).

2.1.1 Genetic Algorithm: (Background)

Genetic algorithm is a well-known serial search optimization strategy based on principal of natural selection. In GA decision variable of search operation encoded

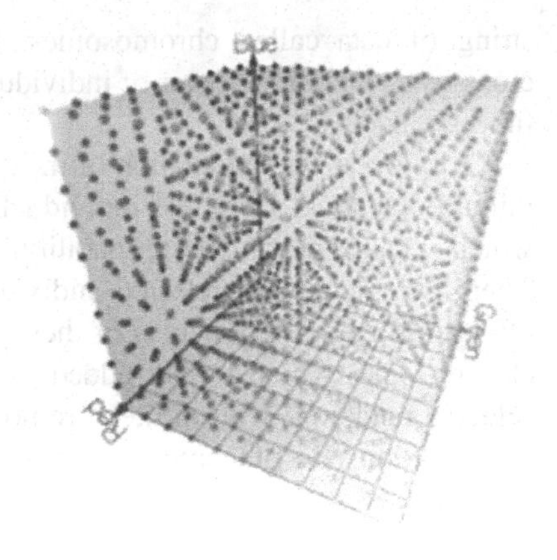
Fig. 1 RGB colorspace

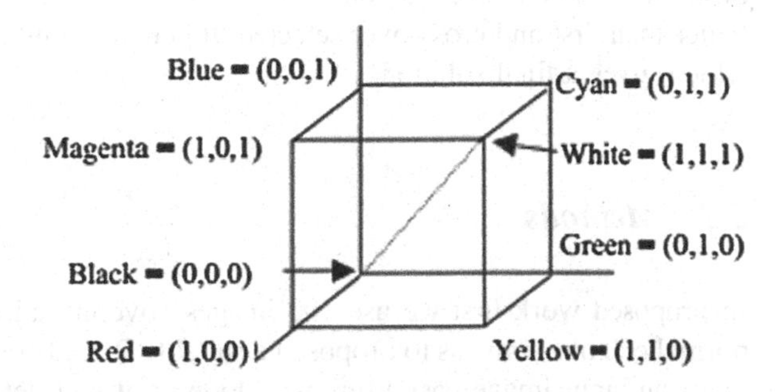

Fig. 2 RGB color cube

into a finite length of string and candidate solution to the search problem are referred as a chromosomes for example in popular travelling salesman problem chromosomes represent route and city is represented by genes. GA is also helpful to solve both constrained and unconstrained problem which is rely on natural selection as well as simulation of biological model. GA is best suited for addressing the problems of mixed integer programming, where some components are constrained to be integer-valued. We can also differ Genetic Algorithm from a classical derivative-based optimization algorithm in two main ways (Table 1).

GA first selects a random population where the population size, which is a user-defined parameter, is an important factor which result performance of Genetic Algorithm.

Selection: Genetic algorithm takes a list of parameters which can be used drive an evaluation procedure. This list of parameters typically represented as a simple

Table 1 Difference between GA and classical derived based optimization

Genetic algorithm	Classical algorithm
Generate a population of points at each repetition, and best point approaches an optimal solution	Generate a single point at each repetition
Generate a single point at each repetition	Next point is selected by deterministic computation

string of data called chromosomes. Using any generic operator e.g. selection, cross-over, mutation, a pair of individual are selected for to endure based on higher fitness value.

Cross-over: The population of most fitted individual images is selected from the initial population. Here two individual are selected randomly with a probability p_c which is called cross-over probability. A uniform number is generated and if r < $p_{c.}$ Then two randomly selected individual endure recombination. Otherwise two off-springs are simply copies of their parents. Cross-over outcomes two new child chromosomes which further added to the second generation pool. The ranges are selected randomly with in the chromosomes for exchanging the contents.

Mutation (Re-placement): in this step new individual chromosomes accidentally mutated with probability of mutation p_m which generates a random vale range b/w [0, 1] for each chromosomes. If this random value is less than PM then chose a bit at random position. The aim of mutation is to generate next generation which is better than first and cross-over selected. It generates until an individual is produced which gives a final solution.

2.2 Methods

In proposed work first we use 100 images covering all types of fair person, dark, normal etc. our aim was to proposed a suitable R, G, B range for skin part in image. Once an input image pass from pixel looping it will detect only the defined range part and eliminate rest of part from an image, because R, G, B values for skin part will be diverse from tiring part in an image as well as from background also. By spending lots of time with math simulation we came with some range for a skin part in image which is shown in (Table 2).

Table 1 show that for a skin part, intensity of green color is lesser as compare to red and blue as well as red intensity is higher than from both green and blue. So there are several possibilities to think and calculate a suitable skin part in an image. Our tryout were to find detect appropriate range for skin part and we came with five assumption.

In image (1) values of R, G, B for skin part is taken from 1 to 95 chosen experimentally and observed values are

Table 2 Range of R, G and B for skin part in image

Type person	R	G	B
Extra fair person	210–250	155–210	90–180
Fair person	180–230	130–170	90–140
Normal	150–200	110–170	70–80
Dark	140–180	90–130	60–80
Extra dark	70–100	40–70	30–50

- Value of R > 95 && Value of G > 40 && Value of B > 20………….. (I)
- Maximum of (R, G, B) – Minimum of (R, G, B) > 15………………..(II)
- (R-G) >20 where R > G && R >B……………………………………..(III)
- (R-B)>28 where R>G && R>B………………………………………..(IV)
- (B+G-R)>12…………………………………………………………..(V)

We calculated these ranges by using each image and enchanting R, G, B values that lies on face part and with some mathematics equations build a relationship of R, G, and B value for a skin part in an image. We tabled value of R, G and B color for a skin part in image.

2.2.1 Proposed Face Detection Algorithm Based on R, G and B Color Segmentation

1. Read an image file (I)
2. Calculate R, G, and B from the input file.
3. Calculate Rows and Columns.
4. K=size(I,1-2)
5. Perform pixel wise looping for input image I.
6. For i=1 to size(I,1)
7. For j=1 to size(I,2) //scan each pixel for an input image//
8. If (R>95 && G>40 && B>20)
   If Maximum of (R, G, B) – Minimum of (R, G, B) >15
   If (R-G)>20 where R>G && R>B
   If (R-B)>28 where R>G && R>B
   If (B+G-R)>12
   Set k=1

In proposed face detection algorithm our tryout was to detect a face through intensity of R, G and B color which falls on face regions. This algorithm is pretty straight forward to understand because in this algorithm we just pass an input images from proposed range and the detected the skin appearance where these ranges exists.

Experiment results shows that defined R, G and B ranges efficiently detecting skin part from image we test this algorithm on 110 images and found correct 78 correct detected faces. This algorithm came with one drawback of detecting extra dark person images. When we input an extra dark person it fails because for an extra dark person value of these R, G and B ranges are not satisfying.

2.2.2 Proposed Algorithm for R, G, and B Color Using Genetic Algorithm

GA takes an initial population which is a user defined function. In this case we are using GA to find accurate value b/w certain range that we are assigning with fitness function, once we gave fitness function that is also refer as a problem statement where we try to find our best result. By running each input range of R, G and B distinctly we found that our last calculated range is showing better results as compare to rest four. For GA MATLAB gives an easy toolbox to operate (Fig. 3).

ALGORITHM FOR GA WITH BEST SUITED FITNESS FUNCTION

Step 1: Read an image file (I)
Step 2: Calculate R, G, and .B from the input file.
Step 3: Calculate Rows and Columns.
Step 4: K=size (I, 1-2)
Step 5: Perform pixel wise looping for input image I.
for i=1 to size (I, 1); for j=1 to size (I, 2)
Step 6: If (abs (B+G) - R) > input)) //set a fitness function for GA; Evaluation starts //
Step 7: Option structure for GA
g= gaoptimset ('Generations', 20, 'Display', 'iter')
// create a list of parameters with valid value //
Step 8: m= ga ('best suited range as input ', '12', '20', g)

Moreover, Instead of covering all R, G and B ranges we took best suited range and now keeping best suited range in hand we starts evaluation of GA. Objection function with input refer as a fitness function or a problem statement for Genetic Algorithm. Without passing the actual value of selected range we pass an input. Genetic algorithm starts evaluation by selecting random generations and perform several step mutation, cross-over and provide a best value for this specified range e.g. here we put the input values between a range instead of a fix value so in this range GA detect best suited value for this range and enrich the detection accuracy. We can also pass multiple ranges for GA but it will directly imitate computational time.

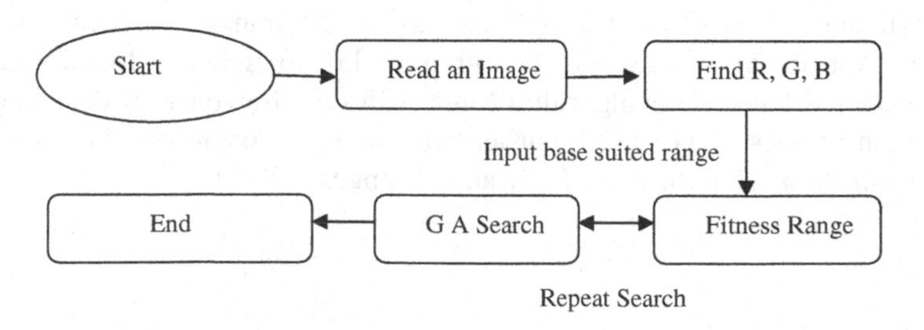

Fig. 3 Proposed methodology for GA

3 Experimental Set up and Results

Here, we intended to show the outcomes of our novel face detection approach by using GA and R, G and B color segmentation. We test our algorithm on MATLAB environment using a local database of images which consists 110 types of different images. First we will see the outcome from R, G and B color range algorithm which we proposed for face detection.

3.1 Input Images

(See Fig. 4).

3.2 Output Given by R, G and B Range Algorithm

By giving input images for first algorithm it seems to be good but not so much efficient because in image 4, 5 and 6 we found some bad result covering less part of skin and more part than face. We achieve good 80 % detection rate by using these projected range algorithm. Using this methodology of range algorithm we achieve good rate in respect of fair person and normal person too & it fails on extra dark person e.g. fourth image. Furthermore we will assign our best suited range for GA as a fitness function and try to treasure some GA operation that can detect faces more accurately. Here rectangle labeling is done on detected part only (Fig. 5).

Fig. 4 Input images

Fig. 5 Output images by skin detection using RGB range

3.3 Results Using Ga with R, G and B Color Range

GA creates a sequence of new population and uses individuals in the current generations to create next population by scoring each member of the current population using their fitness function. Here table shows fitness count for each generation with best and mean fitness count. Table also shows that fitness value is more than generations.

In our case we gave a best suited range to GA as a fitness function and trying to find more accuracy for compare to previous work. In GA we also defined some parameters like number of iterations that we want and a certain range by which GA rings his search with aim of finding best suited number.

Generations	f-count	Best f(x)	Mean f(x)	Stall generations
1	40	9	14.52	0
2	60	9	11.43	1
3	80	9	10.75	2
4	100	9	9.953	3
5	120	9	9.953	4
6	140	9	9	5
7	160	9	9	6
8	180	9	9	7
9	200	9	9	8
10	220	9	9	9
11	240	9	9	10
12	260	9	9	11
13	280	9	9	12
14	300	9	9	13
15	320	9	9	14

3.4 Outcome Given by GA

Here results demonstrate that unlike R, G and B color algorithm here we found good detection rate in less computational time. GA takes best suited range with some user definite values by which it generate its best suited result for a particular image. Results also clearly show the faces of dark person exactly and also for next two images in which our defined range was unable to cover exactly (Fig. 6).

Fig. 6 Output images by proposed range with Genetic algorithm

4 Conclusion

In this manuscript our key objective is to build a complete mechanism for face detection that exists in complex background using RGB color model and Genetic algorithm. Results show that proposed algorithm gives precise outcomes in lesser computational time. Result of this paper also showed that GA can be used in many computer visions where our goal is to produce good solution from many outcomes and for human faces. GA is best which gives good approximation in lesser computational time. Our simulation was implemented on MATLAB. Using proposed work we got 92 % accurate results. In future optimization work can also be presented by using PSO (Particle swarm optimization) which comes from agent oriented prototype and allows greater assortment and assessment over a single population.

References

1. Bhaiswar, R., Kshirasagar, P., Salodkar, A.A.: A noval approach for face detection in complex. In: Background using Genetic Algorithm, (IJEIT), vol. 1, (3), March 2012
2. Prashanth Kumar, G., Shashidhara, M.: Skin color segmentation for detecting human face region in image. Communication and Signal Processing (ICCSP) (2014)
3. Sarawat A., Md. Islam, S., Kashem, M.A., Islam, M.N., Islam, M.R., Islam, M.S.: Face recognition using back propagation neural network. In: Proceedings of the International MultiConference of Engineers and Computer Scientists 2009 Vol I IMECS 2009, 18–20 Mar 2009
4. Shubban, R., Mishra, M.: Rule-based face detection in color images using normalized RGB color space—A comparative study. Computational Intelligence & Computing Research (ICCIC) (2012)
5. Gajame, T., Chandrakar, C.L.: Face detection with skin color segmentation and recognition using genetic algorithm. **3**(9), 1197–1208. ISSN 2231-1297 (2013)
6. You, M., Akashi, T.: Profile face detection using flipping scheme with genetic algorithm. In: 2015 10th Asian Control Conference (ASCC)
7. Lam, K.-M., Siu, W-C.: An efficient algorithm for face detection and facial feature extraction under different conditions. Pattern Recogn. **34,** 1993–2004 (2001)
8. Crowley, J.L., Coutaz, J.: Vision for man machine interaction. Robot. Auton. Syst. **19**, 347–358 (1997)

Analysing Performance Disruption of MANET Under Explicit Flooding Attack Frequency

Priyanka Wadhwani and Sourabh Singh Verma

Abstract The number of routing protocol is available in Mobile Ad hoc Networks (MANETs), but none of them is perfect as it is hard to achieve the security in it. The MANETs is in vulnerable of different attacks because the network is scalable and has very dynamic mobile nodes. The performance of protocols is severely affected in the presence of malicious nodes as these causes routing information to be erroneous and introduces excessive traffic load and inefficient routing. In this paper, we analyse the network performance extensively using Ad hoc On Demand Distance Vector (AODV) routing protocol in the presence of a flooding attack with specific frequency rate. The NS2 network simulator is used to analyse this flooding attack on AODV and its impact are shown using various performance metrics like Packet Delivery Ratio (PDR), throughput with variable flooding rates and malicious nodes etc.

Keywords Flooding · Attack · Flood frequency · AODV · RREQ flooding · MANET

1 Introduction

Mobile ad hoc networks (MANETs) are significantly different and more complex from the wired networks as it is composed of autonomous wireless nodes. These nodes are mobile thus topology of the network gets changed over the period of time and due to this node are susceptible to malicious attacks [1]. There are a large number of known attacks against MANET like flooding, black hole, wormhole, sinkhole, etc. These attacks cause hazards on the network by manipulating the

P. Wadhwani (✉) · S.S. Verma
Mody University of Science and Technology, CET, Lakshmangarh, Rajasthan, India
e-mail: wadhwanipriyanka@ymail.com

S.S. Verma
e-mail: ssverma.cet@modyuniversity.ac.in

S.C. Satapathy and S. Das (eds.), *Proceedings of First International Conference on Information and Communication Technology for Intelligent Systems: Volume 2*, Smart Innovation, Systems and Technologies 51, DOI 10.1007/978-3-319-30927-9_6

parameters of routing messages and traversing the packet in the wrong direction. Among these attacks, we have evaluated RREQ flooding attack. During a flooding attack [2], attacker floods the entire network by sending a number of fake messages to unknown destination nodes. Such an attack can be categorized as RREQ flooding, data packet flooding and Hello message flooding, explained as follows.

RREQ flooding: During the route discovery process of the routing protocol, malicious node floods fake RREQ message and broadcast them through intermediate nodes in the network till the destination is reached which is non-existent. Unnecessarily forwarding these fake RREQ packets results in network congestion and an overflow of route table. Due to which intermediate nodes in the network are busy to transmit such control packets and data packets remains unsent or may be dropped. This degrades throughput and increases consumption of energy [3].

Data Flooding: Route discovery process towards the destination node of routing protocol is maintained by the attacker (malicious nodes) and then frequently sends a large number of useless data packets along the path. On receiving excessive packets from the attacker, will result in wastage of bandwidth and thus nodes were unable to communicate efficiently.

Hello Flood: Nodes broadcast hello packets at specific interval to know all its neighbouring. On receiving a Hello message from a neighbour node, route tables are updated so that it may also not contain any stale entry. Flooding of hello message with high frequency makes nearby nodes unable to process the data. This result increases routing overhead.

The rest of the paper is organized as follows: Sect. 2 gives a literature survey, Sect. 3 contains the simulation parameters used followed by the simulation results and Sect. 4 concluding remarks.

2 Literature Survey

Number of researches [4, 5] has been made in finding out malicious node attacks. In a MANET, different types of devices exist and work together in a cooperative manner while it is quite unfair to restrict all these devices with some threshold value.

Reference [3] discussed implementation and analysis of different attacks in AODV and how these attacks effect packet efficiency and throughput of the network. Study of routing attacks in MANETs by making AODV work maliciously, call it malicious AODV. Routing in ad hoc networks [6] has been a challenge since wireless network came into existence and hence dynamic routing protocols are needed to function properly (e.g. AODV, DSR). As AODV [7], on demand routing protocol discovers a route to a destination only when required. A malicious node abuses route discovery mechanism of AODV to result in Denial of Service attack and these nodes prevent other nodes from establishing a path by sending fake routing information.

Due to network load imposed by fake RREQ and useless data packets [8], a non-malicious node cannot serve other nodes and leads to wastage of bandwidth, overflow of routing table entries and wastage of nodes' processing time. A malicious node may be responsible for these attacks; due to flooding [9] the network with large number of route request to invalid destination creates dramatic impact over the performance of the protocol. Such as AODV performs worse when packet loss increases with increase in the number of fake RREQ packets and as the mobility decreases i.e., the pause time increases the packet efficiency improves, but not substantially.

References [10, 11] discussed how flooding affects the performance, particularly with variable duration of flooding nodes considering a different number of malicious nodes. It is observed if flood node is active for more time than it shows drastic effect on Quality of service (QOS) parameters of the network. Further, it was shown that how throughput and bandwidth consumed by flood RREQ are inversely proportion to each other.

In [12] discussed how route disruption, resource consumption effect AODV protocol over performance metrics as the number of data packets sent and received. In [13] influence of flooding attack is analysed under the circumstances of different parameter on the entire network, including number of attack nodes, network bandwidth and number of normal nodes. Reference [14] provides a common set of security needs for routing protocols that are subject towards attack. These attacks can harm the network operations or the individual user as a whole. This paper discussed about the attacks against well-considered, well-defined implementation of routing protocols. Reference [15] provides a comparison of all routing protocols' performance and determines which protocol performs best under different number of network scenarios. Traditional TCP/IP structure is being employed by MANETs and each layer in TCP/IP model requires modification to function efficiently in it.

In our paper, we took an approach which shows flooding effect with variable mobile node speed, constant bit rate (CBR) in packets per second and connections between nodes considering over different flood frequency and malicious nodes.

3 Simulation Results

In order to simulate the impact of flooding attack in MANET performance, the AODV routing protocol was modified to add malicious nodes. In our evaluation, we are tracing performance metrics with varying speed of nodes, connections between nodes and CBR rates over a different number of flooding nodes and variable flood rates.

3.1 Performance Metrics

PDR (Packet Delivery Ratio)—the number of delivered data packet to the total packets to be delivered by the node. The larger number of pdr means better performance of the nodes.

Throughput—total amount of data in terms of number of bytes received by the destination per second measured in kbps. For better performance of nodes in the network, throughput should be larger with less mobility of nodes.

3.2 *Simulation Setup*

All evaluation is done using NS2 [16]. Our simulation uses following setup: (Table 1).

We have run the simulations as shown in Table 1 various times by using all the parameters mentioned and log the traffic of our created network in number of conditions and results are processed for further evaluation.

3.3 *Results and Discussion*

PDR over number of flooding nodes (2, 4) with varying nodes' speed: Our result in Fig. 1 shows on increasing speed of nodes, greater % of PDR results for network with less malicious nodes and lesser % of PDR for network with more malicious nodes at constant flooding rate. Thus, due the impact of request flooding attack number of malicious node at greater speed increases packet loss and decreases efficiency of the routing protocol.

Throughput over number of flooding nodes (2, 4) with varying speed of nodes: Our result in Fig. 2 shows on increasing mobility speed throughput gets decreased for more number of malicious nodes as wastage of bandwidth gets more due impact of flooding attack in the network. As it is shown that at 10 m/s speed of nodes, throughput is 69 kbps for 2 malicious nodes and 13 kbps for 4 malicious nodes.

Table 1 Simulation parameters and their values

Simulation parameters	Value
Simulation time	50 s
No. of nodes	50
Area	500 × 500 m
Traffic	CBR (constant bit rate)
CBR rate	5, 10, 15, 20, 25, 30
Motion	Random
Routing protocol	AODV
No. of flooding nodes	2, 4
Flooding rates	0.05, 0.1
Transport layer	UDP
Node motion	Random
Node max speed	10, 20, 30, 40, 50 m/s

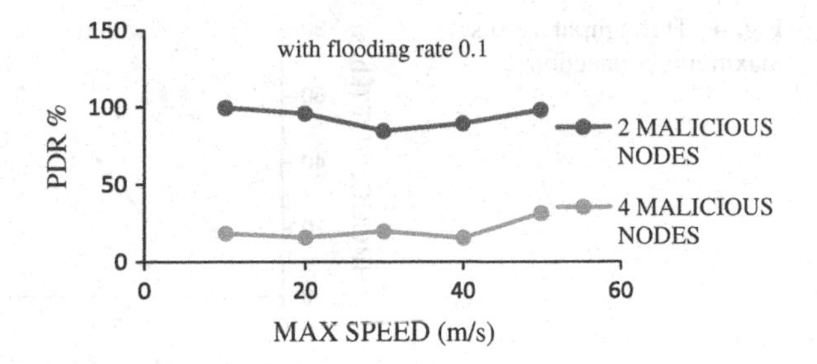

Fig. 1 PDR versus max speed

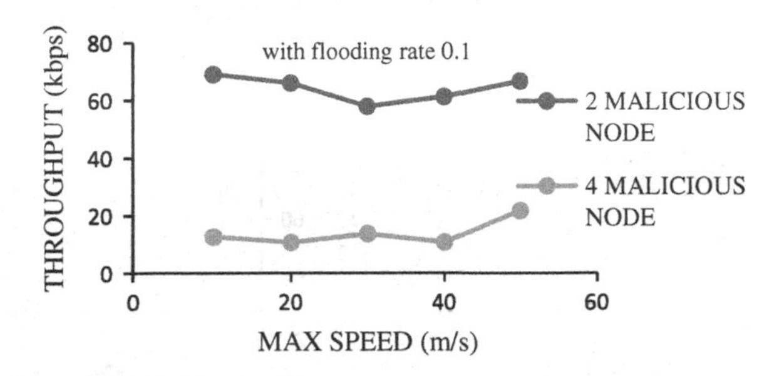

Fig. 2 Throughput versus max speed

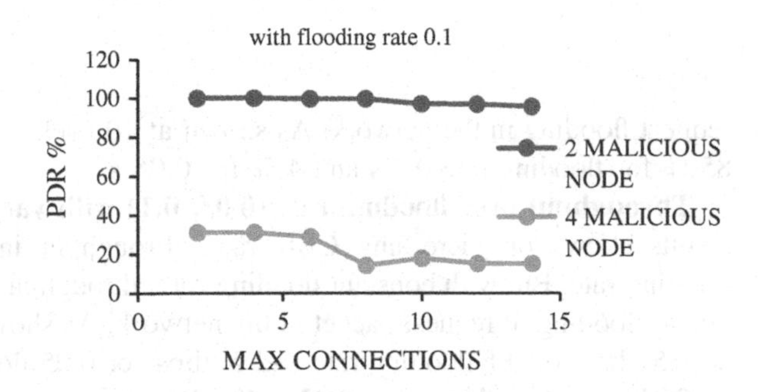

Fig. 3 PDR versus maximum connections

PDR over number of malicious node (2, 4) with varying connections between nodes: Our result in Fig. 3 shows on increasing connections between nodes PDR decreases with increase in malicious nodes at same flooding rate. As it is shown that PDR with 5 connections is 100 % for 2 malicious nodes and 30 % for 4 malicious nodes.

Throughput over number of malicious nodes (2, 4) with varying connections between nodes: In Fig. 4 result shows on increasing connections between nodes throughput decreases with increase in malicious node in the network. As shown that throughput with 5 connection is 35 kbps for 2 malicious node and 11 kbps for 4 malicious nodes at 0.1 flooding rate.

PDR over flooding rates (0.05, 0.1) with varying CBR rate: In Fig. 5 results shows that on increasing CBR rate, PDR increases with increase in flooding rate. But at constant flooding rate PDR decreases with increase in CBR due to impact of

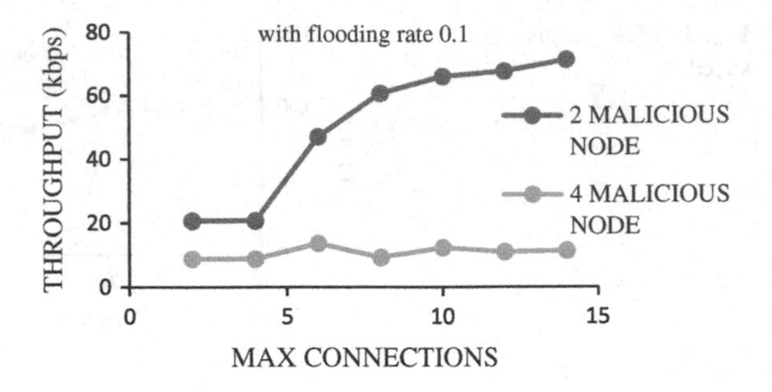

Fig. 4 Throughput versus maximum connection

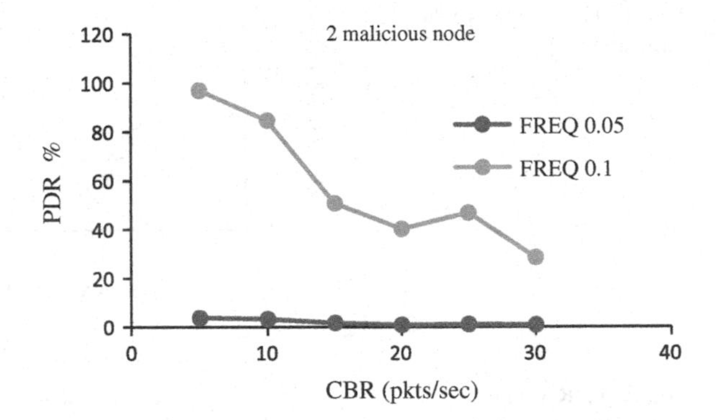

Fig. 5 PDR versus CBR rate

request flooding in the network. As shown at 10 packets per second CBR, PDR is 85 % for flooding rate 0.1 s and 4 % for 0.05 s.

Throughput over flooding rate (0.05, 0.1) with varying CBR rate: In Fig. 6 results shows on increasing CBR rate, throughput increases with increase in flooding rate. But with constant flooding rate, throughput increases with CBR rates due to flooding of request packet in the network. As shown at 10 CBR, throughput is 115 kbps for 0.1 flooding rate and 5 kbps for 0.05 flooding rate.

PDR over flooding rates (0.05, 0.1) with varying connections between nodes: In Fig. 7 results shows on increasing number of connection between nodes PDR increases with increase in flooding rate. As PDR is nearby 100 % for flooding rate 0.1 and 5 % for flooding rate 0.05 with increase in connection between nodes.

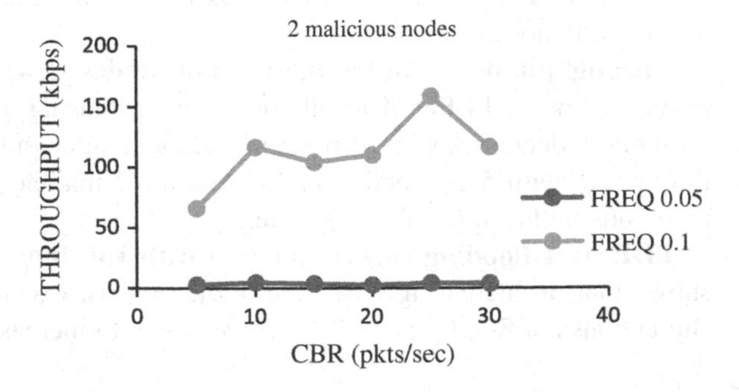

Fig. 6 Throughput versus CBR

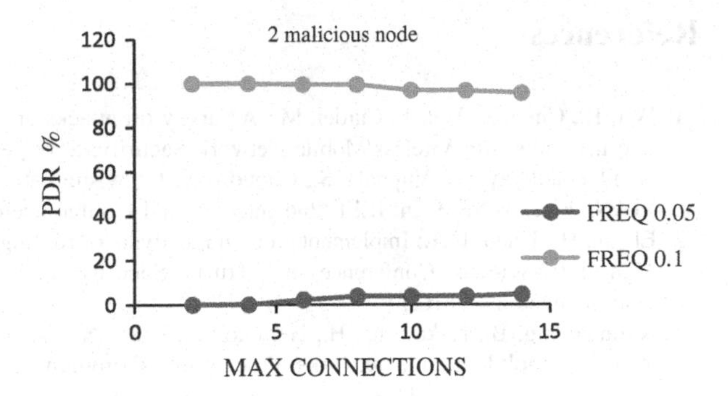

Fig. 7 PDR versus maximum connections

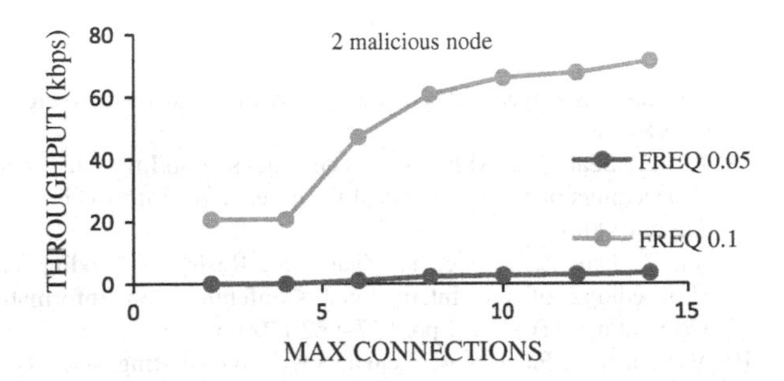

Fig. 8 Throughput versus maximum connections

Throughput over flooding rates (0.05, 0.1) with varying connections between nodes: Fig. 8 result shows on increasing connections between nodes throughput drastically increases with increase in flooding rate in presence of malicious nodes. As shown with 10 connections throughput is 3 kbps at 0.05 flooding rate and 66 kbps at 0.1 flooding rate in presence of 2 malicious nodes.

4 Conclusion

By identifying the impact of RREQ flooding attack on AODV routing protocol in MANET using NS2- network simulator, it was noticed that the presence of malicious flooding nodes can affect the performance of the overall wireless network as it introduces a fake route request and can act as one of the major security threat. From the simulation it can be concluded that due to the extensive flooding in the network average percentage of packet loss and bandwidth increases, which decreases packet delivery ratio and throughput with variable increase in parameters as speed, constant bit rate and connections between nodes. If flood nodes are more in number then there is drastic impact on the performance metrics of the routing protocol in the network. In future work, we will study and assess the effect of various types of attacks on MANET and further some novel security scheme will be proposed to detect and avoid any malicious nodes.

References

1. Wu, B., Chen, J., Wu, J., Cardei, M.: A survey on attacks and countermeasures in mobile ad hoc networks. In: Wireless/Mobile Network Security. Springer, Berlin (2008)
2. Bandyopadhyay, A., Vuppala, S., Choudhury, P.: A simulation analysis of flooding attack in MANET using NS-3. In: IEEE 2nd International Conference on Wireless VITAE (2011)
3. Ehsan, H., Khan, F.A.: Implementation and analysis of routing attacks in MANETs. In: IEEE 11th International Conference on Trust, Security and Privacy in Computing and Communications (2012)
4. Kannhavong, B., Nakayama, H., Nemoto, Y., Kato, N., Jamalipour, A.: A survey of routing attacks in mobile ad hoc networks. Proc. Wirel. Commun. IEEE **14**(5), 85–91 (2007)
5. Patel, M., Sharma, S.: Detection of malicious attack in MANET a behavioural approach. In: Advance Computing Conference (IACC), IEEE 3rd International (2013)
6. Corson, S., Macker, J.: Mobile ad hoc networking (MANET): routing protocol performance issues and evaluation considerations. Internet request for comment RFC 2501 (1999)
7. Perkins, C., Royer, E.M.: Ad hoc on demand distance vector (AODV) routing. Internet draft (1998)
8. Eu, Z., Seah, W.: Mitigating route request flooding attacks in mobile ad hoc networks. In: Proceedings of the International Conference on Information Networking (ICOIN'06), Sendai, Japan (2006)
9. Yi, P., Dai, Z., Zhong, Y., Zhang, S.: Resisting flooding attacks in ad hoc networks. In: Proceedings of the International Conference on Information Technology: Coding and Computing (ITCC'05), pp. 657–662 (2005)
10. Verma, S.S., Patel, R.B., Lenka, S.K.: Investigating variable time flood request impact over QOS in MANET. In: 3rd International Conference on Recent Trends in Computing 2015 (ICRTC-2015)
11. Verma, S.S., Patel, R.B., Lenka, S.K.: Analyzing varying rate flood attack on real flow in MANET and solution proposal: real flow dynamic queue (RFDQ). Int. J. Inf. Commun. Technol. (in press) Inderscience, **7** (2015)
12. Shandilya, S.K., Sahu, S.: A trust based security scheme for RREQ flooding attack in MANET. Int. J. Comput. Appl., **5**(12), 4–8 (2010)
13. Ning, P., Sun, K.: How to misuse AODV: a case study of insider attacks against mobile ad-hoc routing protocols. Ad Hoc Netw. **3**(6), 795–819, Elsevier (2005)
14. Murphy, S., Yang, Y.: Generic threats to routing protocols. In: IETF RFC4593. Status Informational (2006)
15. Abolhasan, M., Wysocki, T., Dutkiewicz, E.: A review of routing protocols for mobile ad hoc networks. Technical report, Telecommunication and Information Research Institute, Australia (2003)
16. Fall, K., Varadhan, K.: NS manual. The VINT Project

An Efficient Agro-Meteorological Model for Evaluating and Forecasting Weather Conditions Using Support Vector Machine

Baghavathi Priya Sankaralingam, Usha Sarangapani and Ravichandran Thangavelu

Abstract Weather prediction is an essential area of analysis in everyday life. Climate forecasting is one of the highly relevant attributes affecting agricultural sectors and industries. Predicting climate conditions is necessary for diverse areas. Metrological department facing the greater challenge to predict the state of the environmental temperature to forecast the weather conditions based on the present, future time for expecting the rainfall. This paper majorly focuses on handling Weather data using big data statistical analysis and for effective forecasting. Support Vector Machine (SVM) predictive based modeling is used for classifying the weather dataset by using regression analysis and thereby forecasting for predicting weather conditions which is suitable for agriculture. Experiment the input dataset parameters of weather like mean temperature, mean dew point, max_sustained wind speed, mean sea level pressure, mean station pressure max_ temperature, min_ temperature, precipitation amount, max_wind gust, snow depth. The results are compared with single decision tree.

Keywords Weather forecasting · Support vector machine · Statistical analysis · Single decision tree

B.P. Sankaralingam (✉) · U. Sarangapani
Rajalakshmi Engineering College, Thandalam, Chennai,
TamilNadu 602105, India
e-mail: baghavathipriya.s@rajalakshmi.edu.in

U. Sarangapani
e-mail: usha.s@rajalakshmi.edu.in

R. Thangavelu
SNS College of Technology, Vizhiyampalayam, Coimbatore,
TamilNadu 641035, India
e-mail: dr.t.ravichandran@gmail.com

S.C. Satapathy and S. Das (eds.), *Proceedings of First International Conference on Information and Communication Technology for Intelligent Systems: Volume 2*, Smart Innovation, Systems and Technologies 51, DOI 10.1007/978-3-319-30927-9_7

1 Introduction

Weather forecasting is required not only for future development in agriculture sector and industries but also in several other areas similar to military, canyoneering, transportation and aerospace steering etc. [1]. It also plays a very crucial part in deciding the activities of day to day life. Weather prediction is very much important for making decisions in various applications such as monitoring climate conditions, helps during drought situations thereby it assist the farmers to improve greater productivity in the agricultural sector and also in other industries [2]. Precise weather information assists the farmer the finest instance to choose the best crops to plant. Abnormal weather conditions may cause (i) spoil or damage to crops and (ii) leads to soil erosion. Prediction of rainfall is one of the greatest challenges for planning the agriculture all over the world.

Agricultural meteorology is concerned with various interdisciplinary departments such as meteorological, hydrological, pedagogical, geophysics and biological factors that affect agricultural production and with the interaction between agriculture and the atmosphere. Meteorology is study of atmospheric climate of a particular location, which also includes various branches like atmospheric physics and chemistry. Power generation, transportation, agriculture and construction are the various applications of meteorology [3]. Establishing proper multidisciplinary structure is to bring together climate and agricultural details, particularly predicting systems, for agriculture management in order to plan the successful agriculture [4]. The Susceptible agriculture zones can efficiently utilize the changing climatic conditions and weather information provided to effectively manage for long term strategic decisions.

The International Center for Agricultural Research in the Dry Areas (ICARDA) uses weather forecasting. IT field assist to forecast weather by providing various tools, for meteorological stations and global information systems (GIS), so that researchers can collect the complete data to deal with the issues that the countryside communities in waterless areas frontispiece from the climatic stress. Weather stations gather every day climatic data that are analyzed by analyst to find out appropriate planting, transplanting increasing the harvest, atmospheric uncertainty evaluation. GIS technology aids discover budding fields where latest technical advancements and scientific tools developed by technicians will be used to increase farming yield raise agriculture earnings and improve efficient natural resource management [5].

The purpose of scientific tool is for assessing the deterioration of land and collection of water for agriculture that consider dissimilar attributes, as well as financial and social factors. This information will help for successful improvement of farming and a number of committed and innovative researchers and techniques are building their finest hard work to obtain the revenue of information technology transformation to rural underprivileged [5]. Naturally, in order to handle various activities like planning the productivity and managing the water supply, the details of weather information are needed. The yielding of crop will be highly reduced and affected at the crucial period during the stages of crop growth due to the prolonged drought circumstances or heavy rainfall. The financial system of the country is mostly based on agriculture and based on agriculture productivity. Thus the agriculture relies on the prediction of rainfall. In the national and regional level, various rainfall forecast techniques are handled to predict the weather [6].

2 Related Work

Nowadays, Big Data Analytics plays a vital role for Sustainable Agriculture. In particular it focuses on diverse ecological conditions' data like rainfall, wind, temperature. Big data analytical tool is used to obtain some significant message obtained from the model that may be effectively applied by farmers for deliberate and flourishing crop growing. Weather firm focuses on distinctive yield details and insurance for farmers. The farmers will be provided details of agricultural processing such as suitable time to watering crops, spreading fertilizers, supplements and right time to harvesting [7].

Back Propagation Algorithm-An Approach is a step by step depiction of how the classification and prediction of weather forecasting is analyzed, and thereby providing a means for describing the designing phase of Classification and Prediction technique [8]. It is basically developed for forecasting weather and handing out information related to cultivating various seasonal crops to yield maximum productivity. In this weather forecasting kit, the data is sent by using a proper cellular medium. This system provides the information of future weather subsequent to some period of time by varying some of the input parameters or possibility of what will be the cause on other supporting parameter, if there is a transform in one parameter later than some period of time [9].

In order to obtain the simulated results from the developed model, the correct weather data can be provided. The various statistical methods are analyzed to interpret the data. To select and apply appropriate statistical methods, the trained empirical skills are required. The analysis of the results can be done effectively by deeper understanding of statistical science [10]. The level of profit in agriculture production is accomplished based on proper analysis of weather conditions and selecting the best requirements for crop growth. In addition to this, it also relies on supervision view point is to preclude the crops from serious climatic circumstances. Generally, the agricultural progress will be dependent on weather and climate. The tremendous yielding of various crops is achieved by means of meteorological conditions [11].

Support Vector Machine algorithm depends on statistical learning theory. In order to make best possible hyper plane in the latest space by applying in different (non-linear) mapping function to plot the actual data 'A' into a feature space 'S' with high range. Classification and regression can be effectively handled through SVM techniques. Formation of hyperplane in categorization is done by separating into two classes of data whereas in retrodradation a hyperplane is to be built up that presents close to many points as possible [12]. Now the climate forecasting is constantly rising based on the traditional users, such as farming, air traffic services and other sectors like power, surroundings, that need trustworthy information on the current and upcoming weather conditions. In adding together to this, the forecasters have to deal with raising the amount of data, especially from statistical Weather forecast models; meteorological satellites, radars and other surveillance systems such as AWS, wind speed and radiometers etc. Forecaster's concentrates on climate forecasting ability instead of trailing time accessing the information, improve and upgrading their forecasting skill with many types of data arrangement. Forecaster's knowledge can be shared with their colleagues and convey their know-how to junior forecasters, look through, imagine and animate all the obtainable data instinctively with a partial number of events, take benefit of multi-screens through a GUI (Graphical User Interface) based on multi-windows, Use of collaborative tools for graphical creation of knowledge data. Particular forecast of weather parameters is a very difficult job due to dynamic environment. Different methods like linear and auto regression, Multi-Layer Perception (MLP) [13] are applied to analyze approximately atmospheric parameters like, rainfall, meteorological pollution, temperature, wind speed etc.

A Simple Weather Forecasting Model was developed using mathematical regression. This model was simplified because of the usage of easiest numerical equation. That equation was written using Multiple Linear Regression (MLR). Through this model the farmers have obtained average knowledge. The data which was obtained from the particular station was recorded as time series data. The various weather attributes are max_temperature, min_temperature and relative humidity have been used [1]. The time series of max_temperature, min_temperature and relative humidity can be used to predict the rainfall. The regression equations co-efficient are used to estimate the future climate conditions. The future weather conditions can be estimated using this model. In order to obtain optimum size of the period for rainfall prediction is by means of evaluating the relative humidity either by 15 week once or 45 weeks. Piyush kapoor et al. predicted future weather conditions based on sliding window algorithm. The fortnight of previous year weather conditions were matched with current year climate conditions. The best matched window was used for predicting weather conditions [14].

3 Frame Work of the Proposed System

The proposed system is developed by using supervised learning algorithm known as Support Vector Machine (SVM), as recognized as kernel machine [15, 16]. The SVM is based on statistical learning theory. This supervised learning algorithm is used to evaluate data and identify patterns for classification and regression techniques [3]. The methodology of SVM is that it tries to plot the primal data 'A' into a feature space termed as 'S' with a high range all the way through a non-linear mapping function and thus builds the best possible hyper plane in an original space.

Fig. 1 Hyper planes

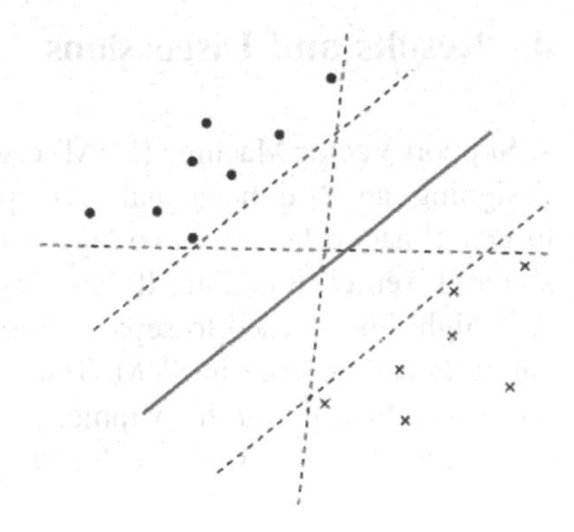

The weather prediction was done using support vector machine [2]. Time series data of max_temperature at different stations on a daily basis is calculated so as to predict the utmost high temperature of the succeeding day at that location depending on the daily max_temperatures for duration of foregoing 'n' number of days based on the categorize of the input. The results of the method are evaluated for a arrangement of make spans of a week by using of the paramount potential values of the kernel [13]. The computational complication of SVMs does not depend on the range of the input space. SVMs are less prone to over fitting [17].

There can be several hyper planes (shown as dashed lines) shown in Fig. 1 that separate a dataset, but intuitively the one with the largest margin (the red line) has the best generalization [18]. In this system, weather dataset can be given as input. The input is preprocessed, classified and compared with single decision tree algorithm. The output of the comparative study is given to agro meteorological advisors to make effective decisions for the various development of agriculture sector.

3.1 SVM Program Procedure [19]

Step 1: Training sets and prediction models
Step 2: Empirical risk and true risk
Step 3: Choosing the set of functions
Step 4: VC dimension— maximum number of points
Step 5: Capacity of set of functions— Classification
Step 6: Capacity of set of functions— Regression
Step 7: Capacity of hyperplanes

4 Results and Discussions

A Support Vector Machine (SVM) categorizes the input data into two categories by designing an *N*-dimensional hyperplane. SVM illustrations are more similar to neural networks. Like two-layer perceptron neural network, SVM model uses a sigmoid kernel function. It has target variables with two predictor variables. A straight line is used to separate the collection of points. Two or more predictor variables can be used in SVM. The accuracy of an SVM model is fully dependent on the combination of the parameters. DTREG—Predictive Modeling Software has two searches such as grid search and pattern search. These two schemes are used for finding optimal parameter values. The other names of pattern search are "compass search" or a "line search". The pattern search tries to find and constructs trial steps for each parameter in each route. The linear search is repeated until the search core reaches to the latest point if no improvement in the investigation of the pattern. The repetition of the exploration is carried out in order to decrease the step size. If the decreased step size reaches a particular threshold, then the investigation of the pattern is stopped [20].

Weather conditions dataset is taken from www.ncdc.noaa.gov/pub/data/gsod [21] which consists of 11 attributes shown in Table 1 (10-input attributes, 1-target attribute) and are shown in Fig. 3. This dataset is loaded into the DTREG predictive modeling software (Table 2).

The SVM training data is compared with single decision tree training data shown in Table 3. The performance is analyzed based on the accuracy to predict the rainfall. The analysis results are given below.

4.1 Input Data for Single Decision Tree and SVM

Input data file: C:\Users\Administrator\Desktop\towork\towork.csv
Number of variables (data columns): 22
Data subsetting: Use all data rows
Number of data rows: 40
Total weight for all rows: 40
Rows with missing target or weight values: 0
Rows with missing predictor values: 0

The Fig. 2 represents single decision tree in which each node shows "test" on an attribute. The classification rule is applied to represent the path from root node to leaf node [3]. The decision tree is generated based on the attributes such as max_temperature, min_temperature, precipitation and maximum wind speed from the weather dataset.

Table 1 Attributes of weather dataset

Number	Variable	Class	Type	Missing rows	Categories
1	STN	Predictor	Categorical	1	Constant value
2	WBAN	Predictor	Categorical	1	Constant value
3	YEARMODA	Predictor	Categorical	40	
4	TEMP	Predictor	Categorical	38	
5	FT	Predictor	Categorical	1	Constant value
6	DEWP	Predictor	Categorical	37	
7	FD	Predictor	Categorical	1	Constant value
8	SLP	Predictor	Categorical	38	
9	FS	Predictor	Categorical	7	
10	STP	Predictor	Categorical	1	Constant value
11	FST	Predictor	Continuous	1	Constant value
12	VISIB	Predictor	Categorical	20	
13	FV	Predictor	Continuous	1	Constant value
14	WDSP	Predictor	Categorical	37	
15	FW	Predictor	Continuous	1	Constant value
16	MXSPD	Predictor	Categorical	21	
17	GUST	Predictor	Continuous	15	
18	MAX	Predictor	Categorical	30	
19	MIN	Predictor	Categorical	24	
20	PRCP	Target	Categorical	19	
21	SNDP	Unused	Categorical		
22	FRSHTT	Unused	Categorical		

Table 2 Attributes and its description

Number	Attributes	Description
1	STN	Station number
2	WBAN	Weather Bureau Air force Navy
3	YEARMODA	The year, month and day
4	TEMP	Mean temperature
5	DEWP	Mean dew point
6	SLP	Mean sea level pressure in millibars
7	STP	Mean station pressure
8	VISIB	Visibility
9	WDSP	Mean wind speed in knots
10	MXSPD	Maximum sustained wind speed in knots
11	GUST	Maximum wind gust
12	MAX	Maximum temperature
13	MIN	Minimum temperature
14	PRCP	Precipitation
15	SNDP	Snow depth in inches
16	FRSHTT	Fog rain snow hail thunder Tornado

Table 3 Project parameters of single decision tree and support vector machine

Project Parameters		
Parameters	Single decision tree	Support vector machine
Target variable	PRCP	PRCP
Number of predictor variable	19	19
Type of model	Single tree	Support vector machine
Type of SVM model	–	C-SVC
SVM kernel function	–	Radial basis function (RBF)
Maximum splitting levels	10	–
Type of analysis	Classification	Classification
Splitting algorithm	Gini	–
Category weights (priors)	Data file distribution	Data file distribution
Misclassification costs	Equal (unitary)	
Variable weights	Equal	
Minimum size node to split	10	
Minimum rows allowed in a node	5	
Max. Categories for continuous predictors	1000	
Misclassification costs	–	Equal (unitary)
Tree pruning and validation method	Cross validation	Cross validation
Number of cross-validation folds	10	10
Tree pruning criterion	Minimum cost complexity (0.00 S.E.) [20]	

4.2 Support Vector Machine

(See Table 4)

4.3 SVM Parameters

Type of SVM model: C-SVC
SVM kernel function: Radial Basis Function (RBF)
SVM grid and pattern searches found optimal values for parameters:
Search criterion: Minimize total error
Number of points evaluated during search = 148
Minimum error found by search = 0.725000

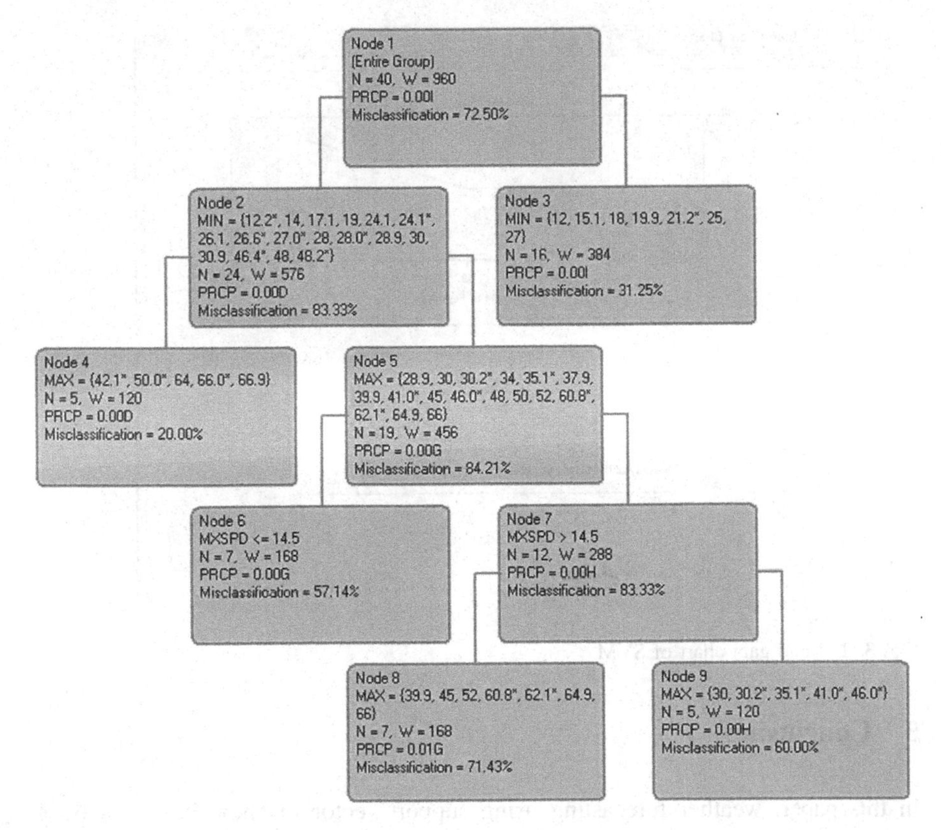

Fig. 2 Single decision tree—tree diagram

Table 4 Continuous variables

Variables	# Rows	Minimum	Maximum	Mean	Std. Dev
FST	40	0.00000	0.00000	0.00000	0.00000
FV	40	24.00000	24.00000	24.00000	0.00000
FW	40	24.00000	24.00000	24.00000	0.00000
GUST	40	15.00000	999.90000	416.24250	476.57949

Parameter values:
Epsilon = 0.001
C = 0.1
Gamma = 0.001
Number of support vectors used by the model = 39

The performance of SVM is represented by using gain and lift chart shown in Fig. 3.

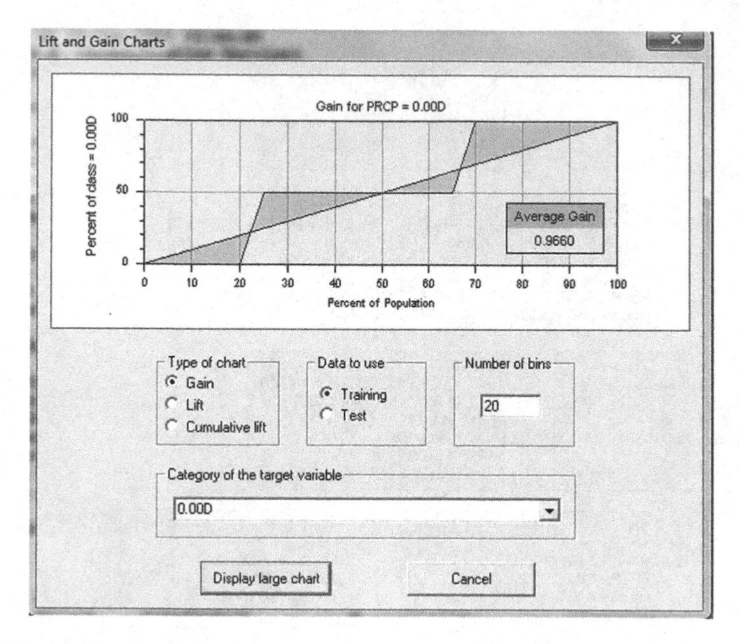

Fig. 3 Lift and gain chart of SVM

5 Conclusion

In this paper, weather forecasting using support vector machine has been done effectively. The results are compared with Single Decision tree. The results also show that SVM produces improved performance than single decision tree. The parameters are carefully chosen in SVM's case that has produced efficient prediction.

References

1. Paras, A., Mathur, S.: Simple weather forecasting model using mathematical regression. Indian Res. J. Extension Educ. Special Issue **1**, 161–168 (2012)
2. Radhika, Y., Shashi, M.: Atmospheric temperature prediction using support vector machines. Int. J. Comput. Theor. Eng. **1**, 55–58 (2009)
3. Wikipedia, http://WWW.wikipedia.org
4. Stone, R.C., Meinke, H.: Weather, climate, and farmers: an overview. Meteorol. Appl. **13**, 7–20 (2006). (Royal Meteorological Society)
5. Zahedi, S.R., Zahedi, S.M.: Role of information and communication technologies in modern agriculture. Int. J. Agri. Crop Sci. **4**(23), 1725–1728 (2012)
6. World Academy of Science, Engineering and Technology, http://WWW.waset.org
7. Waga, D., Rabah, K.: Environmental conditions big data management and cloud computing analytics for sustainable agriculture. IEEE Prime Res. Educ. (PRE) **3**(8), 605–614 (2013)

8. Lee, M.C., To, C.: Comparison of support vector machine and back propagation neural network in evaluating the enterprise financial distress. Int. J. Artif. Intell. Appl. (IJAIA) **1**, 31–43 (2010)
9. Sawaitul, D., Wagh, K.P., Chatur, P.N.: Classification and prediction of future weather by using back propagation algorithm: an approach. IJETAE **2**, 110–113 (2012)
10. Stephenson, D.B.: Data analysis methods in weather and climate research (2005)
11. Maini, P., Rathore, L.S.: Economic impact assessment of the Agrometeorological Advisory Service of India. Curr. Sci. **101**, 1296–1310 (2011)
12. Rani, R.U., Rao, T.K.R.K.: An enhanced support vector regression model for weather forecasting. IOSR J. Comput. Eng. (IOSR-JCE) **12**, 21–24 (2013)
13. Rao, T., Rajasekhar, N., Rajinikanth, T.V.: An efficient approach for weather forecasting using support vector machines. In: International Conference on Computer Technology and Science (ICCTS) IPCSIT, vol. 47, pp. 208–212 (2012)
14. Kapoor, P., Bedi, S.S.: Weather forecasting using sliding window algorithm. ISRN Signal Process. 2013:1–5 (2013)
15. Haykin, S.: Neural networks: a comprehensive foundation. Prentice Hall, New Jersey (1999)
16. Cortes, C., Vapnik, V.: Support vector networks. Mach. Learn. **20**, 273–297 (1995)
17. Saffarzadeh, S., Shadizadeh, S.R.: Reservoir rock permeability prediction using support vector regression in Iranian oil field. J. Geophys. Eng. **9** (2012)
18. Computer Science and Engineering, https://WWW.cseweb.ucsd.edu/~akmenon/ResearchExam.pdf
19. http://WWW.arxiv.org
20. Predictive Modeling Software, http://WWW.dtreg.com
21. National Climatic Data Center, http://WWW7.ncdc.nowa.gov—GSOD Sample Data

Simulation of a Model for Refactoring Approach for Parallelism Using Parallel Computing Tool Box

Shanthi Makka and B.B. Sagar

Abstract Refactoring is the process of retaining the behavior of a program by making changes to the structure of a program. Initially refactoring is used only for sequential programs, but due to highly configurated architectural availability, it also aids parallel programmers in implementing their parallel applications. Refactoring provides many advantages to parallel programmers, in identifying independent modules, in refining process of programs, it also helps in separating concerns between application and system programmers, and it reduces the time for deployment. All mentioned advantages benefit the programmer in writing parallel programs. The approach for refactoring using multi core system is already developed. Hence all these advantages made us to thought of a system to develop refactoring approach for parallelism which uses heterogeneous parallel architectures which uses combination of both Graphic Processing Unit (GPU) and Central Processing Unit (CPU). A Tool in MATLAB, Parallel Computing Toolbox can be used to execute programs on multiple processing elements simultaneously with local workers available in the toolbox, which takes benefit of GPUs. This tool box uses complete processing speed of multi core system to execute applications on local workers without changing the code. Our suggested model can be simulated by using Parallel Computing Toolbox.

Keywords Refactoring · MATLAB · Parallel computing toolbox · GPU · CPU · Heterogeneous parallel architecture

---

S. Makka (✉) · B.B. Sagar
BITs, Mesra (Noida Campus), Sector-15, Noida, India
e-mail: shanthi_makka@yahoo.com; shanthi.makka@jre.edu.in

B.B. Sagar
e-mail: drbbsagar@gmail.com

S. Makka
JRE Group of Institutes, Plot no-5,6,7,8, Greater Noida 201308, Uttar Pradesh, India

S.C. Satapathy and S. Das (eds.), *Proceedings of First International Conference on Information and Communication Technology for Intelligent Systems: Volume 2*, Smart Innovation, Systems and Technologies 51, DOI 10.1007/978-3-319-30927-9_8

1 Introduction

According to Moore's law [1], for every two years the number of transistors on Integrated Circuit is getting double, i.e., the speed of processor increases double. For decades, programmers relied on Moore's Law [2] to improve the performance of their applications. With the advent of multi cores, programmers are forced to exploit parallelism if they want to improve the performance of their applications, and if they want to enable new applications and services that were not possible earlier i.e., to have enhanced user experience and better quality of service. Haskell [3], intel's multi core architecture has eight cores by default. In future computer hardware is likely to have even more cores, with many cores. This makes programmers to think parallel, i.e., we should move away from conventional (sequential) programming approach to parallel approach.

There are two different approaches for parallelizing a programs, one is to rewrite it from scratch, it is very expensive and time consuming and second is parallelize a program incrementally, one piece at a time. Each small segment can be seen as a behavior preserving transformation, i.e., a refactoring. Mostly Programmers prefers second approach because it is safer and economical because it provides workable and deployable version of the program. MATLAB is proprietary programming language [4] for implementing mathematical computations and it is also used for development of algorithms, simulation of models, data reduction, testing and evaluation process and it also provides an excellent platform to create an accessible parallel computing environment. A Parallel Computing Toolbox [5], a tool in MATLAB can be used to run applications on a multi core system where the local workers available in the toolbox, can run applications concurrently by taking the benefit of GPUs.

2 Refactoring

The term refactoring was introduced by William Opdyke in his PhD Thesis [6]. Refactoring is process of changing lines of program without changing its originality or in other words changing of internal structure without altering its external behavior to improve understandability and maintainability of code. Refactoring improves [7] its internal structure.

The different activities are to be performed during the refactoring process are:

a. Identify the segments in a program where the refactoring can be applied.
b. Determine what type of refactoring i.e., extract method, renaming etc. Should be applied to the identified segments.
c. Make sure that the applied refactoring preserves application behavior.
d. Apply the refactoring.
e. Assess the effect of refactoring on quality characteristics of the software.
f. Maintain the consistency between the refactored program and other software artifacts.

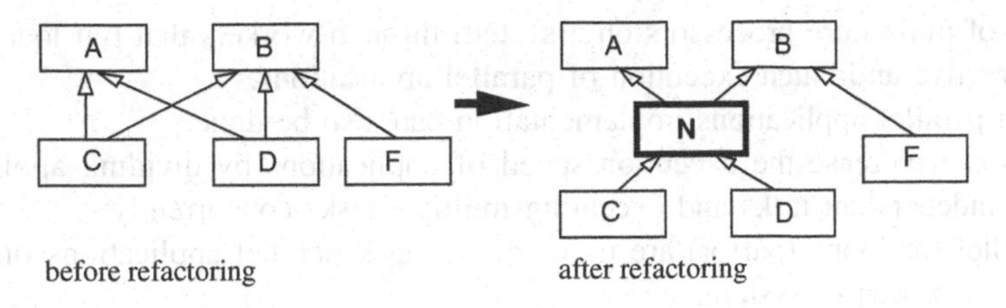

Fig. 1 Example of a refactoring. before refactoring. after refactoring

A refactoring approach for parallelism is already developed for multi core system. Then we thought of to extend our vision to develop a new refactoring approach for parallelism using heterogeneous parallel architectures. Key issues include dealing with advanced *heterogeneous* parallel architectures, involving combinations of GPUs and CPUs; providing good hygienic abstractions that cleanly separate components written in a variety of programming languages; identifying new high level *patterns* of parallelism; developing new rule based mechanisms for rewriting (refactoring) source-level programs based on those patterns etc.

Consider a below example for refactoring "the class A is derived from C and D and class B is derived from C, D, and F. That means the class A can use features of C and D and class B can use features of C, D, and F. Instead of making copy of features of C and D twice once for A and second time for B, We can make new class N which has feature of C and D and that can be used when class A and B requires those features" (Fig. 1).

The refactoring tools [8, 9] or automatic refactoring compilers can improve programmer productivity, performance of applications, and program portability and currently the toolset supports various refactoring techniques for following:

a. To increase throughput, a sequential program can be divided into threads,
b. To make programs thread safe, and
c. To improve scalability of applications.

3 Parallel Computing Toolbox

Parallel Computing Toolbox [5] can solve computationally intensive as well as and data intensive problems on multi core processor environment, GPUs, and computer clusters. This toolbox enables us to parallelize applications without Compute Unified Device Architecture (CUDA) and MPI programming environment. Simulink is a simulator which simulates suggested model which runs multiple parts of the model parallel.

Key Features

a. Parallel Computing Toolbox provides GPU Array, which is associated with certain functions through which we can perform computations on CUDA.

b. Use of multi core processors on a system through workers that run locally.
c. Interactive and batch execution of parallel applications.
d. Data parallel applications implementation can also be done.
e. You can increase the execution speed of applications by dividing application into independent tasks and executing multiple tasks concurrently.
f. Parallel for loops (parfor) are used for run task parallel applications on multi core or processor system.

4 Refactoring for Parallelism Using GPUs and CPU

(See Fig. 2).

Algorithm

1. Program Dependence Graph can be used to represent an application.
2. Independent modules can be identified using refactoring technique.
3. Those modules can be directed (using appropriate mapping technique) to heterogeneous pool during running time of applications.

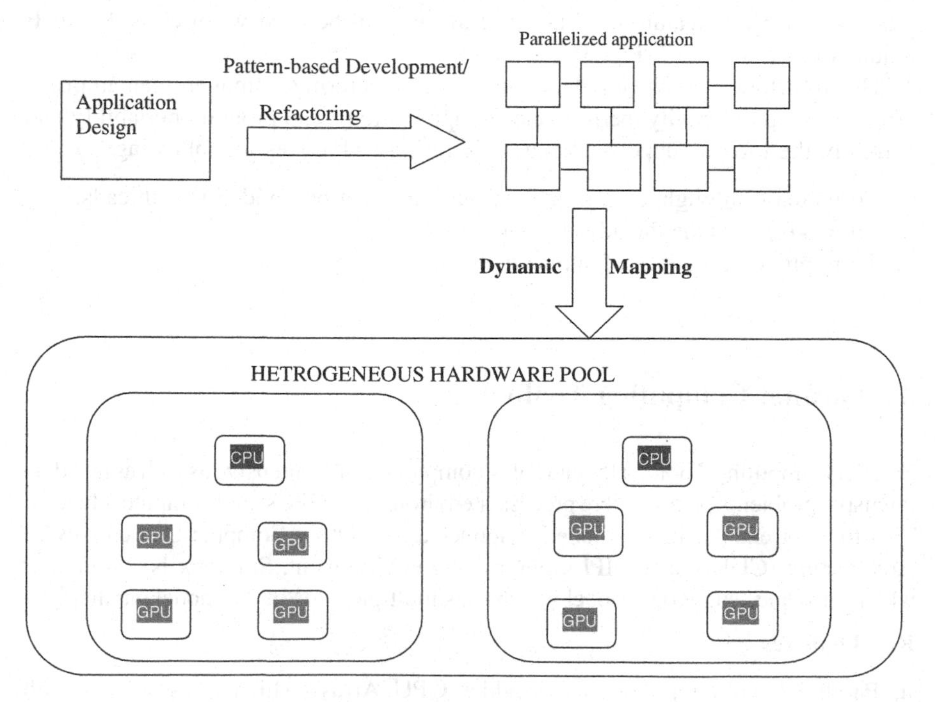

Fig. 2 refactoring approach for parallelism using combination of GPU and CPU

4.1 Why Refactoring

Refactoring [10, 11] is the process of preserving program behavior by changing its structure which makes program more readable, understandable and maintainable. After a refactoring, program should not give any syntactic errors. Refactoring never change the behavior of a program; i.e., if any program called twice before and after refactoring the results must be same for same set of inputs.

While refactoring do not change the external behavior of a program, it can support for designing of software in very effective way and also in evolution process by restructuring a program in the way that allows other modifications to be made more easily and rapidly. Complicated changes to a program can require both refactoring and additions. Refactoring is behavior preserving so that, whenever the preconditions are met it do not break the program.

A refactoring approach has many advantages in parallel programming: it helps the programmer in the process of identification of independent modules, it also guides the programmer in the refining process of a parallel program and it also reduces time for deployment.

Applications of refactoring:

1. Refactoring [12] aims to improve software design.
2. It encourages good program design.
3. It makes software easier to understand.
4. It improves readability of the code.
5. A good program design is not only guarantees rapid rate, but also accurate software development.

4.2 Why the Combination of Both GPUs and CPU

1. GPU are designed for highly parallel architectures [13] which allows large blocks of data to be processed.
2. GPUs does similar computations are being made on data at the same time (rather than in a sequence one after other).
3. The GPU has emerged as a computational accelerator [14] that dramatically reduces the time to discovery in High End Computing (HEC).
4. GPU can easily reduce the execution time of a parallel code.

CPU is required to coordinate all GPUs which are executing segments concurrently or to manage communication overhead between GPUs, and also to execute lighter segments of program. GPUs are very expensive, so it should be used when there essential requirement.

5 Related Work

The refactoring is different for both software development and optimization. The frame work chosen for refactoring is depending on overall program design and knowledge representation. The Object oriented refactoring was first introduced by Opdyke [6] in his Ph.D. thesis. The transformation of program into modules is being described in [15]. There are certain automatic compilers exist which can make transformations automatically by acting either on source level programs or their intermediate language representations or during parsing of program, this approach has been discussed in [16]. Burstall and Darlington [17] mentioned algorithms for transformations of source program for both sequential and parallel machines. In critical situations, non executable modules may transformed into an executable program. Later work is exemplified by the relational approach of Bird and de Moor [18]. A catalogue for refactoring is given by Fowler in [11] an website and this is also kept up to date at www.refactoring.com, which also has links to tools and other resources. The most widely known tool for refactoring is the Refactoring Browser Smalltalk [19].

6 Simulation of Suggested Model Using Parallel Computing Toolbox

The suggested model can be simulated by using parallel computing toolbox as follows: After identification of independent modules which can be executed simultaneously or parallel through refactoring approach can be assigned to local workers, which makes use of processing speed of GPUs. Scheduling of these modules can be done by Job Scheduler (CPU). All these local workers produce solution to their individual modules and finally merging of these solutions gives the solution for original problem (Fig. 3).

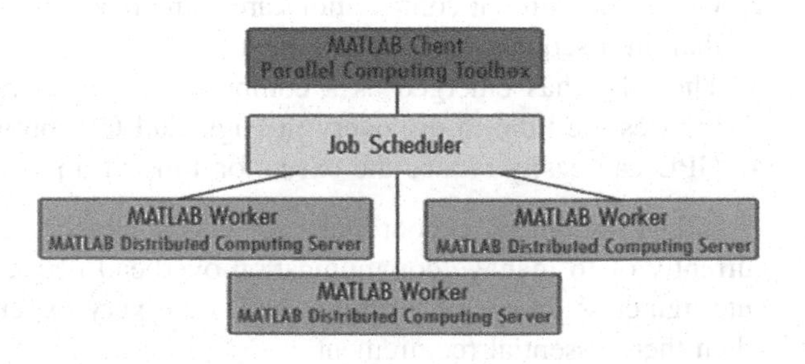

Fig. 3 Parallel computing tool box in MATLAB

7 Conclusion

a. To increase the execution speed of applications, if we use single processor system to run applications at higher clock speeds which consumes lots of power and which also generates large amount of heat. To avoid such circumstances, programmer should think parallel.
b. Refactoring is an approach through which we can identify modules which can be executed simultaneously.
c. To execute applications parallel, we can use multi core system, which has been already developed.
d. Finally we thought of to develop a system which uses refactoring approach to identify modules for simultaneous execution on heterogeneous parallel architectures which uses the combination of both GPUs and CPUs.
e. A parallel computing toolbox can be used to simulate above said Model.

References

1. Moore, G.E.: Readings in computer architecture: Cramming more components onto integrated circuits. Morgan Kaufmann Publishers Inc., San Francisco, CA, USA, 56–59 (2000)
2. Dig, D.: A refactoring approach to parallelism. Software, IEEE **28**(1), 17–22 (2011)
3. Brown, C., Loidl, H.W., Hammond, K.: Paraforming: forming parallel haskell programs using novel refactoring techniques. Trends in Functional Programming. Springer, Berlin Heidelberg, 82–97 (2012)
4. Kim, H., Mullen, J., Kepner, J.: Introduction to parallel programming and pMatlab v2. 0. Massachusetts Inst Of Tech Lexington Lincoln Lab (2011)
5. http://in.mathworks.com/products/parallel-computing
6. Opdyke, W.F.: Refactoring: a program restructuring aid in designing object-oriented application frameworks. University of Illinois at Urbana Champaign, (1992)
7. Tom, M., Tourwé, T.: A survey of software refactoring. IEEE Trans. on Softw. Eng. 30(2), 126–139 February 2004
8. Liao, SW., Diwan, A., Bosch Jr., R.P., Ghuloum, A., Lam, M.S.: Suif explorer: an interactive and interprocedural parallelizer. In PPoPP'99 7th symposium on Principles and practice of parallel programming ACM SIGPLAN, 37–48 (1999)
9. Dig, D., Marrero, J. Ernst, M.D.: Refactoring sequential Java code for concurrency via concurrent libraries. In 31st International Conference on Software Engineering. IEEE Computer Society, 397–407 (2009)
10. Murphy-Hill, E., Black, A.P.: Breaking the barriers to successful refactoring. 30th International Conference on ACM/IEEE Software Engineering ICSE'08. IEEE (2008)
11. Fowler, M., Refactoring: Improving the design of existing code. Addison-Wesley Longman Publishing Co., Inc (1999)
12. Berthold, H., Pére, J., Mens, T.: A case study for program refactoring. Proc. of the GraBaTS Tool Context (2008)
13. Rofouei, M., Stathopoulos, T., Ryffel, S., Kaiser, W., Sarrafzadeh, M.: Energy-aware high performance computing with graphic processing units. In Workshop on Power Aware Computing and System, December 2008

14. Huang, S., Xiao, S., Feng, W.C.: On the energy efficiency of graphics processing units for scientific computing. International Symposium on. Parallel & Distributed Processing IPDPS. IEEE (2009)
15. Partsch, H., Steinbrüggen, R.: Program transformation systems. ACM Computing Surveys, 15 (3). September 1983
16. Jones, S.L.P.: Compiling Haskell by program transformation: a report from the trenches. In European Symposium on Programming (ESOP'96), April 1996
17. Burstall, R.M., Darlington, J.: A transformation system for developing recursive programs. J. ACM **24**(1), 44–67 (1977)
18. Bird, R., De Moor, O.: Algebra of Programming. Prentice-Hall, (1997)
19. Roberts, D., Brant, J., Johnson, R. A.: Refactoring Tool for Smalltalk. Theory and Prac. of Obj. Sys. (TAPOS). Special Issue on Software Reengineering, 3(4),253–263 (1997) see also http://st-www.cs.uiuc.edu/users/brant/Refactory/

Dynamic Congestion Analysis for Better Traffic Management Using Social Media

Sujoy Chatterjee, Sankar Kumar Mridha, Sourav Bhattacharyya, Swapan Shakhari and Malay Bhattacharyya

Abstract Social media has emerged as an imperative tool for addressing many real-life problems in an innovative way in recent years. Traffic management is a demanding problem for any populous city in the world. In the current paper, we explore how the dynamic data from social media can be employed for continuous traffic monitoring of cities in a better way. To accomplish this, congestion analysis and clustering of congested areas are performed. With the term congestion, we denote co-gatherings in an area for two different occasions within a defined time interval. While doing so, we introduce a novel measure for quantifying the congestion of different areas in a city. Subsequently, association rule mining is applied to find out the association between congested roads. To our surprise, we observe a major impact of various gatherings on the disorder of traffic control in many cities. With additional analyses, we gain some new insights about the overall status of traffic quality in India from the temporal analysis of data.

S. Chatterjee (✉)
Department of Computer Science and Engineering, University of Kalyani, Nadia 741235, West Bengal, India
e-mail: sujoy@klyuniv.ac.in

S.K. Mridha · S. Bhattacharyya · S. Shakhari · M. Bhattacharyya
Department of Information Technology, Indian Institute of Engineering Science and Technology, Shibpur, Shibpur 711103, West Bengal, India
e-mail: msankar@it.iiests.ac.in

S. Bhattacharyya
e-mail: bhattacharyya.sourav4@gmail.com

S. Shakhari
e-mail: swapanshakhari@gmail.com

M. Bhattacharyya
e-mail: malaybhattacharyya@it.iiests.ac.in

S.C. Satapathy and S. Das (eds.), *Proceedings of First International Conference on Information and Communication Technology for Intelligent Systems: Volume 2*, Smart Innovation, Systems and Technologies 51, DOI 10.1007/978-3-319-30927-9_9

1 Introduction

Efficient control of traffic movement is one of the major challenges in any populous country. The unavoidable assembly of people or anomalies in roads due to various reasons causes disorders in the traffic flow [1]. Under such conditions, it is often necessary to find out the optimal path by which we can reach to our destination in a limited time. However unfortunately, as we do not have any prior knowledge about the sudden traffic irregularities, therefore, filtering out the congested roads to find out the optimal way is not quite easy. Various social networking sites (like Facebook, Twitter, etc.) have become increasingly popular because billions of people have accepted them as an easier and user-friendly platform to connect with each other. Recent studies highlight that they can be also employed as a powerful tool for solving and analyzing different complex real-life problems, except from conventionally being used as a medium of connecting people. As for example, better traffic management through the use of social media has drawn immense attention of researchers in recent years [1, 2]. Congestion of traffic is a serious problem in many of the populous cities both financially and in terms of the wastage of time [3]. In this paper, we explore the utility of social media in analyzing the congestion and other important factors for a better management of traffic flow.

The issues like traffic maintenance and governance have been successfully addressed by social media analytics in recent times. Social media has played important roles in maintaining eGovernance by the traffic police of different populous cities [4]. Detection and tracking of events have also been shown to be feasible by studying the contents posted in different social media [5–7]. In the same line, studying the social networking pages (likewise in Facebook) of popular city traffic polices might be helpful in understanding the traffic scenarios of a country. In this study, we have analyzed some Facebook traffic pages of some populous cities in India to study the dynamic behavior of different city traffic conditions. Normally, the traffic officials do some alternative arrangements when a particular road is going to be heavily loaded due to anomalies. They regulate all the vehicles to move through the possible alternative paths and post such information online. We are therefore able to find out the alternating paths/roads that might be free in a given time interval. This gives us the clue about using congestion analysis for a better traffic control. This is the principle aim of this paper.

2 Related Works

Significant contributions have been made earlier to discover traffic patterns from different types of real-life data. This comprises large-scale and high resolution spatio-temporal traffic sensor data [8], human mobility data [1], data from hand-held devices, etc. However, the inclusion of social media in this genre is the latest accomplishment.

How the online social platforms (basically a microblogging service) like Twitter can be used for enhancing the emergency situation awareness has been recently studied through data mining and natural language processing [9]. In this, a burst-detection module is developed that continuously monitors a Twitter feed to identify unexpected incidents. A classification model is also employed for the impact assessment of incidents. Additionally, an online incremental clustering algorithm is used to automatically group similar tweets (based on cosine similarity and Jaccard similarity) into event-specific topic clusters.

A study by Endarnoto et al. has shown that traffic conditions can be understood by analyzing the tweets posted in Twitter [2]. This work determines the traffic conditions in Jakarta, Indonesia by processing tweets through tokenization, then applying the rule based approach to extract the parts-of-speech, and finally plotting the locations on the Google map. In a different study, traffic anomalies have been recognized by mining representative terms from the tweets posted by people [1]. Such anomalies are modeled as a subgraph with abnormal routing patterns in a road network. The anomalies may be accidents, disasters, control, protests, celebrations, sport events, etc. This study uses the social networking site WeiBo (like Twitter) for analyzing the tweets posted while an anomaly takes place. This helps to eliminate irrelevant posts based on the location and time information obtained from the anomalous graph.

There are a lot more scopes to use social media as a tool for better traffic management. Congestion analysis is one of such promising areas of exploration. Unfortunately, there are limited studies in the literature dealing with the explicit problem of controlling congestion by using the social media. In this paper, we propose a novel measure of congestion and use this to analyze the pattern of gatherings in a city by mining the corresponding Facebook page of traffic police.

3 Methodologies

Let us introduce some terminologies that will be used in the paper. We basically work on road networks that are represented as dynamic directed graphs. A road network can be formally defined as a graph $G_\tau = (V, E, W_\tau)$ for a given time interval τ, where $V = \{v_1, v_2, \ldots v_n\}$ denotes the set of terminal points, $E \subseteq V \times V$ denotes the roads (areas) connecting the terminal points, and the weight parameter $W : E \rightarrow \mathbb{R}$ signifies the amount of congestion in a road.

A road network can be used to reflect the traffic scenarios of a region (e.g., a city) over the change of time. To practically formulate a road network, we introduce a novel measure, hereafter termed as the congestion factor ($\mathcal{CF}$), to quantify the congestion in a road for a given time interval. This is defined in terms of the normalized difference of co-gatherings, in the same road within a time interval τ^*, as follows.

$$\mathcal{CF}_{\tau^*}(E^*) = \begin{cases} 1 - \dfrac{|\mathcal{G}_{t1}(E^*) - \mathcal{G}_{t2}(E^*)|}{\mathcal{G}_{t1}(E^*) + \mathcal{G}_{t2}(E^*)}, & \text{if } |t_1 - t_2| \leq \tau^* \\ 0, & \text{otherwise} \end{cases}$$

Here, $\mathcal{G}_t(E^*)$ denotes the gathering in the road E^* (or at the terminal points of a road) at a given time point t. Note that, we consider t as the time of commencement of the occasion due to which the gatherings might have happened.

Based on this, we can identify the roads (having high congestion factor) that are safe to avoid for a given time period. We consider the time interval in terms of a single day. A higher value of $\mathcal{CF}$ signifies a higher amount of congestion (co-gathering). Interestingly, $\mathcal{CF}$ ranges between [0, 1].

Note that, the normalized measure $\mathcal{CF}$ has been devised in a way such that it quantifies the amount of co-gatherings, i.e., the possible collision among two different gatherings on the same road within a fixed time interval. Hence, if a large gathering occurs in a road with no other gatherings, it is not a congestion (as per our consideration), and therefore $\mathcal{CF}$ becomes 0. For the brevity of the analysis, which also guarantees a better traffic management, we consider that no more than two gatherings can happen in a road for a given time interval. From this perspective, the following lemma becomes important.

Lemma 1 *The adjacent road of a congested road in a road network is not necessarily congested.*

Proof Consider a congested road (v_i, v_j) $(v_i, v_j \in V)$ in a road network $G_\tau = (V, E, W_\tau)$. So, the gatherings are either at the terminal points v_i or v_j or both. Now, even though v_i and v_j have a gathering, the other terminal point might have no gathering within the time interval τ. Hence the lemma.

For the cluster analysis, we consider the number of gatherings and average strength per gathering as the features of a road. By using these features, we cluster a road network to obtain the roads having similar kinds of gatherings. This provides us the regions of common interest. We use the k-means clustering approach for the said purpose. The iterations are continued until the process converges.

To uncover the relations between the congestion pattern among various roads, association rule mining is applied. This helps to identify the flow of traffic at the time of possible co-gatherings at a particular time interval. An association rule can be defined as $A \longrightarrow B$, where A is the antecedent and B is the consequent. Generally the possible meetings or procession follow a particular route originating from one end and terminating at another. But this congestion of a particular route can implicitly affect another neighbouring route as people need to choose alternative paths to reach the destination in timely manner. This motivates us to find out the different association of various roads by applying wellknown Apriori algorithm on the dataset obtained from different posts of the corresponding Facebook pages.

4 Configuration Details

We create a GUI-based online interface to access the dynamic real-time data from the Facebook pages of city traffic polices (using the Facebook API) in India. The Facebook API helps to extract different attributes of online pages dynamically to explore the status and quality of traffic control in real time. The environment is platform-independent (implemented in PHP) and it dynamically works on the data. Additional statistical details are put together for providing a comparative traffic status of all the major cities in India. The gatherings, congestion factors, and other details are represented in tabular form. The Google map is also connected to the analysis pages for each of the cities, namely, Delhi, Bangalore, Kolkata, Chennai and Mumbai.

5 Empirical Studies

To understand the traffic scenarios from the contents of Facebook pages of city traffic polices in a better way, we first recognize the areas of principle attention. To accomplish a comprehensive analysis of congestion, we propose a novel approach to recognize the congested areas in a city. Then we perform cluster analysis to group them together and finally try to find our their associations in a formal way. These are successively described hereunder.

5.1 Congestion Analysis

We initially determine the hub areas where traffic problems (mainly gatherings followed by congestion) happen very frequently. Conversely, this should enable us to identify the topmost safe roads in a city. With this goal, we sort out the roads based on the number of gatherings happened in a road network for a given time period. The areas with frequent gatherings in Kolkata and Delhi are shown in Fig. 1a, b, respectively. Interestingly, as it can be noted from these two figures, the number of gatherings is a dynamic behavior that might change over time. However, they reflect the difference of loads between the cities at a larger scale. So, the dynamic change of this difference can lead to the recognition of dynamic behavior of traffic conditions of various cities.

Frequency of gatherings gives us an idea about the areas where congestion might happen. If these roads can be avoided then the optimal (and safe) path for reaching the destination can be determined easily. Again, we observe that people share their individual experiences about the jam and load of a particular road when they opt for possible alternative paths. The behaviors and patterns of different meetings and

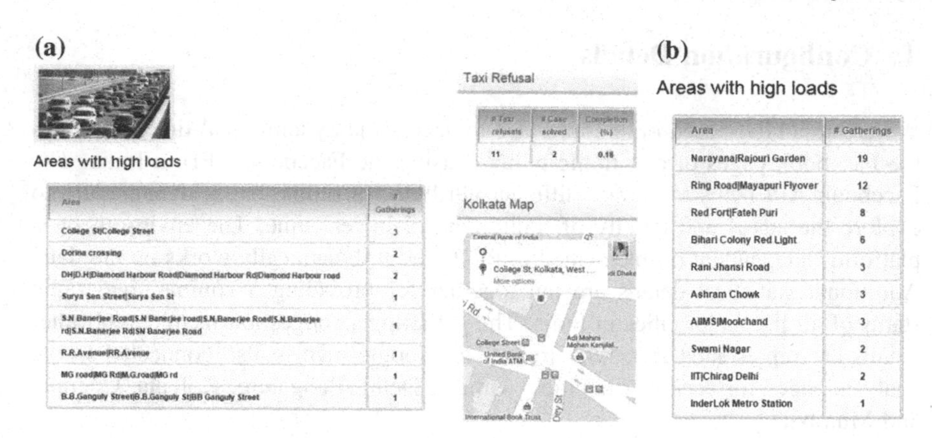

Fig. 1 High-load areas in decreasing order of frequent gatherings during January-February of 2015 in **a** Kolkata and **b** Delhi

processions that take place in most of the important places might help in the overall traffic management.

Traffic regulation is one of the important tasks that are regularly maintained by traffic polices to avoid the heavy traffic congestion in highly populated cities like Mumbai, Kolkata, Delhi, Bangalore, etc. Generally, the prior information about the possible meetings and procession become available through the social media along with the estimation of gatherings. When the people try to reach from one place to their destination in a timely manner they need to find out the most optimal way that are basically low congested. Therefore, it is useful to find out the amount of congestion between the terminal points that are connected to each other by a road.

As we have already selected the roads with frequent gatherings, we therefore calculate the congestion factors (using the measure given in Sect. 3) for different roads. The congestion factors obtained for the roads in Kolkata for the duration of a month are shown in Fig. 2a. It becomes evident from this figure that the amount of congestion can be very high (as the $\mathcal{CF}$ values reach to a high level) in reality. Certainly, the congestion of an entire chain of roads (i.e., a path) can also be estimated by this approach.

5.2 *Cluster Analysis*

By applying *k*-means algorithm on the collected road information, we obtain the clusters of high and low congestion. The number of gatherings and average strength per gathering are both considered as features of different roads. Based on this, the road clusters are obtained from the dynamically extracted data. The number of clusters is set to be three (i.e., $k = 3$). As for example, the clusters obtained for Kolkata for a single month are shown in Fig. 2b.

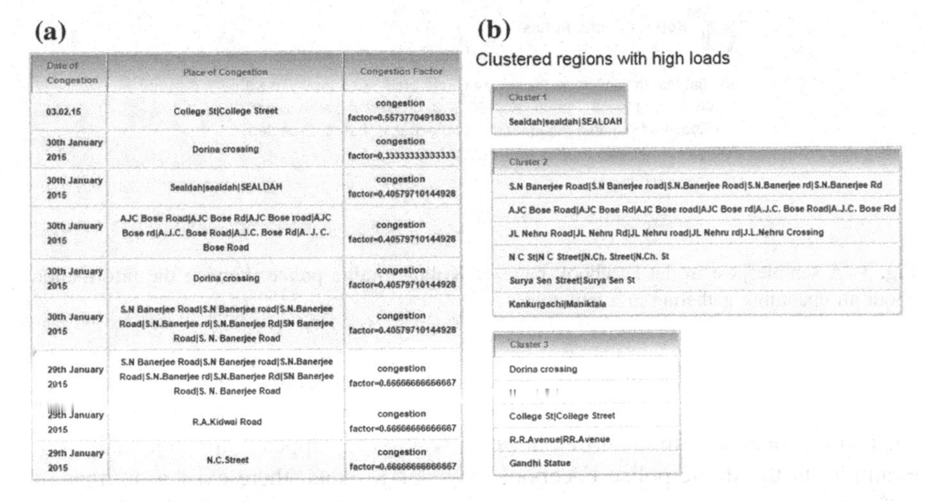

(a)

Date of Congestion	Place of Congestion	Congestion Factor
03.02.15	College St|College Street	congestion factor=0.55737704918033
30th January 2015	Dorina crossing	congestion factor=0.3333333333333
30th January 2015	Sealdah|sealdah|SEALDAH	congestion factor=0.40579710144928
30th January 2015	AJC Bose Road|AJC Bose Rd|AJC Bose road|AJC Bose rd|A.J.C. Bose Road|A.J.C. Bose Rd|A. J. C. Bose Road	congestion factor=0.40579710144928
30th January 2015	Dorina crossing	congestion factor=0.40579710144928
30th January 2015	S.N Banerjee Road|S.N Banerjee road|S.N.Banerjee Road|S.N.Banerjee rd|S.N.Banerjee Rd|SN Banerjee Road|S. N. Banerjee Road	congestion factor=0.40579710144928
29th January 2015	S.N Banerjee Road|S.N Banerjee road|S.N.Banerjee Road|S.N.Banerjee rd|S.N.Banerjee Rd|SN Banerjee Road|S. N. Banerjee Road	congestion factor=0.66666666666667
29th January 2015	R.A.Kidwai Road	congestion factor=0.66666666666667
29th January 2015	N.C.Street	congestion factor=0.66666666666667

(b)

Clustered regions with high loads

Cluster 1
Sealdah|sealdah|SEALDAH

Cluster 2
S.N Banerjee Road|S.N Banerjee road|S.N.Banerjee Road|S.N.Banerjee rd|S.N.Banerjee Rd
AJC Bose Road|AJC Bose Rd|AJC Bose road|AJC Bose rd|A.J.C. Bose Road|A.J.C. Bose Rd
JL Nehru Road|JL Nehru Rd|JL Nehru road|JL Nehru rd|J.L.Nehru Crossing
N C St|N C Street|N.Ch. Street|N.Ch. St
Surya Sen Street|Surya Sen St
Kankurgachi|Maniktala

Cluster 3
Dorina crossing
College St|College Street
R.R.Avenue|RR.Avenue
Gandhi Statue

Fig. 2 Congestion analysis of different high-load areas having frequent gatherings with a time interval of a single day during January-February of 2015 in Kolkata (**a**) The clustered areas of Kolkata, on the basis of number of gatherings and congestion, obtained after applying the k-means algorithm (**b**)

Formally, we cluster the edges of the road network to obtain clusters of roads. These clusters group together the roads having similar gathering patterns in a given road network. When a road becomes heavily congested what other roads simultaneously get loaded can be identified from such clusters [8]. Moreover, as the algorithm executes dynamically, the traffic polices can get an idea about what alternative streets or locations (areas) to be used for a smoother traffic management during the period of a meeting or occasion reported earlier. Again, based on the pattern of the clusters (whether the cluster is highly congested or not), the traffic officers can decide how many polices are to be posted for the corresponding road to better control the traffic. As can be seen from Fig. 2b, Sealdah is a singleton cluster among the clusters obtained for the congested areas of Kolkata. This indicates that finding an alternative exit from this place in a crisis condition is challenging. Therefore, maximum possible support is to be provided by the traffic police while managing an event happening in Sealdah. On the other hand, the other two clusters consisting of a large number of roads helps to identify the alternative paths for a better traffic management. In this way, the cluster analysis of congestion might help in efficient balancing of loads during a crisis condition.

5.3 *Association Rule Mining*

To derive how much the various roads are associated to each other, we apply the well-known Apriori association rule mining algorithm [10]. We prepare a dataset

Fig. 3 A sample post in the Facebook page of Kolkata traffic police showing the information about an upcoming gathering in a few areas

containing the information about various roads, especially the chain of roads through which a high amount congestion is found. The information about different roads are collected from the different posts where they appear simultaneously. For example, in the traffic police Facebook page of Kolkata, there exist a number of different posts containing a sequence of roads where a meeting is scheduled with an approximate number of gatherings. As can be found in a sample post (see Fig. 3), there is a sequence of road names, namely Sri Santashi Maa Mandir—D.H. Road—Chandi Maa Mandir—Sakharbazar, where an activity will happen that might involve roughly 450–500 people.

We are collecting such information as how many times the various combinations of roads are occurring concurrently in a post. Here, this information about the different combinations of a road from a particular post has been mapped as corresponding transaction. The different roads have been mapped into a corresponding integer value by appropriate encoding. We have listed twenty different roads obtained from the posts in the Facebook page of Kolkata Traffic Police. In this way, we have a list of sequence of roads for further processing. After preparing this, we apply the well-known Apriori algorithm with varying minimum support and minimum confidence value [10] to obtain the association between congested roads.

We start by considering the minimum support value as 5 and minimum confidence value as 0.3. Thereafter, we increase this up to a minimum support of 12 and a minimum confidence of 0.9. We have seen that no additional information is obtained by increasing this value further. It is seen that in the most extreme case (when minimum support = 12 and minimum confidence = 0.9) there are only two

Table 1 Rules generated after applying the Apriori association rule mining algorithm to find out association between several combinations of roads (a) with minimum support = 12 and minimum confidence = 0.7 (b) with minimum support = 9 and minimum confidence = 0.7

(a)		(b)	
Antecedent	Consequent	Antecedent	Consequent
Sealdah	Howrah	S.N. Banerjee road	Dorina crossing
Howrah	Sealdah	S.N. Banerjee road	Sealdah
		Sealdah	Howrah
		Howrah	Sealdah

tuples that reflect the strong association between Sealdah and Howrah (as shown in Table 1a). This reveals there is a high probability that if a procession origins at Sealdah then it might be ended at Howrah. So, the traffic police can take necessary actions towards the possible path from Sealdah to Howrah. Again, if we reduce the minimum support to 9 and minimum confidence to 0.7 (to allow more rules) then we get four tuples that reflect some additional information that S.N. Banerjee road is highly associated to Sealdah and Dorina crossing. The different rules for various roads have been represented in Table 1b. This information might help them to be aware of the hazardous roads to avoid.

5.4 Popularity Analysis

There are a few quantifiable features in Facebook that can be used to find out the activeness and popularity of different city traffic pages in Facebook. How better is the city traffic police working can be highlighted through the number of total 'likes' obtained in their social pages. Similarly, the posts against the call for action might signify the responsiveness of the authority. A comparative analysis of the actions (or prosecutions) taken against the refusals or other cases in different major cities and the count of 'likes' is shown in Table 2. This is interesting because the number of 'likes' are not correlated with the duration of activity (i.e., the time elapsed since the traffic pages have been created). The popularity can also be precisely highlighted by the average number of likes against the posts (e.g., 4.8 likes/post in Kolkata). However, some of the important features are unique to some cities. E.g., number of breakdowns is reported only for Delhi (which is 4 for the duration of the study). However, we obtain an overall view of the traffic scenarios in different cities in India.

Table 2 Comparative analysis of actions taken against the taxi refusals or other cases and the popularity factors of social pages of different city traffic polices

Factors	Delhi	Bangalore	Kolkata	Chennai	Mumbai
#Breakdowns	4	NA	NA	NA	NA
#Refusals	NA	NA	11	NA	NA
#Actions	34	2	2	NA	NA
#likes	203244	254185	63620	1261	285
#Likes/post	5.21	NA	4.8	NA	NA
#Comments/post	2.33	NA	2.33	NA	NA

6 Conclusion

In this article, we proposed a novel method for studying the congestion pattern from social networking sites. The current study highlights a major impact of social media analytics on the better side of traffic management and eGovernance. Again, the association rule mining method has been applied to derive how a pair of roads are associated with each other based on their pattern of congestion. Hence, the traffic police as well as general people might be benefitted from those analysis. A limitation of our study lies in the consideration of only a pair of gatherings in a road for a time interval. However, in real cases it might be more although this can be restricted for a better traffic management. Still if it is more, the computation of congestion factor can be suitably revised to handle this problem. Though the current paper deals with only main four city traffic pages of India but the interface can be applicable for any city-traffic pages all over the world. While computing of congestion factor, the time interval of commencement of two meetings has been calculated with respect to a single day but it can also be generalized for a couple of days. The granularity of a day for monitoring the traffic does not fully reflect the overall congestion scenario of a road but from this approach the different traffic regulations can be easily performed. Every page has their own distinct patterns that prevent from developing a general framework of analysis. As an example, we observe that the Facebook page of Delhi Traffic Police comprises a significantly higher number of videos in comparison with other cities, whereas in Kolkata the principle awareness is on taxi refusals. So, it is more challenging to devise a common evaluation approach to compare between the traffic conditions in different cities.

References

1. Pan, B., Zheng, Y., Wilkie, D., Shahabi, C.: Traffic anomalies based on human mobility and social media. In: Proceedings of the 21st ACM SIGSPATIAL International Conference on Advances in Geographic Information Systems, pp. 344–353, Orlando, Florida, USA (2013)
2. Endarnoto, S.K., Pradipta, S., Nugroho, A.S., Purnama, J.: Traffic condition information extraction & visualization from social media twitter for android mobile application. In: Proceedings of the International Conference on Electrical Engineering and Informatics, pp. 1–4, Bandung, Indonesia (2011)
3. Arnott, R., Small, K.: The economics of traffic congestion. Am. Sci. **82**:446–455 (1994)
4. Mainka, A., Hartmann, S., Stock, W.G., Peters, I.: Government and social media: a case study of 31 informational world cities. In: Proceedings of the 47th Hawaii International Conference on System Science, pp. 1715–1724, Hawaii Island (2014)
5. Sakaki, T., Okazaki, M., Matsuo, Y.: Earthquake shakes twitter users: real- time event detection by social sensors. In: Proceedings of the 19th international conference on World wide web, pp. 851–860, Raleigh, North Carolina (2010)
6. Sayyadi, H., Hurst, M., Maykov, A.: Event detection and tracking in social streams. In Proceedings of the International AAAI Conference on Weblogs and Social Media, pp. 311–314, San Jose, California (2009)

7. Zhou, X., Chen, L.: Event detection over twitter social media streams. VLDB J. **23**(3), 381–400 (2014)
8. Banaei-Kashani, F., Shahabi, C., Pan, B.: Discovering traffic patterns in traffic sensor data. In: Proceedings of the 2nd ACM SIGSPATIAL International Workshop on GeoStreaming, pp. 10–16, Chicago, Illinois, USA (2011)
9. Yin, J., Lampert, A., Cameron, M., Robinson, B., Power, R.: Using social media to enhance emergency situation awareness. IEEE Intell. Syst. **27**(6), 52–59 (2012)
10. Agrawal, R., Srikant, R.: Fast algorithms for mining association rules in large databases. In: Proceedings of the 20th International Conference on Very Large Data Bases, pp. 487–499, Santiago, Chile (1994)

An Intelligent Optimization Approach to Quarter Car Suspension System Through RSM Modeled Equation

M.B.S. Sreekar Reddy, S.S. Rao, P. Vigneshwar, K. Akhil and D. RajaSekhar

Abstract This paper present on minimum value of rider comfortness vibration values to obtain maximum rider comfortness during riding. The simulation model was being achieved with the help of MATLAB/Simulink for further process to Genetic Algorithm through Response surface methodology modeled equation. As the response surface methodology is a long established technique in optimization for experimental process. Recently a new intelligent approach to the quarter car suspension system has been tried with response surface methodology and genetic algorithm which is new in the computational field. For the Response surface methodology, an experimental design was chosen in order to order to obtain the proper modelling equation. Later this modeled equation was served as evaluation function or objective function for further process into genetic algorithm. In Genetic algorithm case, the optimality search was carried without the knowledge of modelling equations between inputs and outputs. This situation is to choose the best values of three control variables. The techniques are performed and results indicated that technique is capable of locating good conditions to evaluate optimal setting, to reduce comfortness vibrations for maximum comfortness.

Keywords Matlab simulink · Optimization · Response surface methodology · Genetic algorithm · Comfortness

M.B.S. Sreekar Reddy (✉) · S.S. Rao · P. Vigneshwar · K. Akhil · D. RajaSekhar
Mechanical Department, KL University, Vaddeswaram, Guntur, India
e-mail: mbssreddy@kluniversity.in

S.S. Rao
e-mail: ssrao@kluniversity.in

P. Vigneshwar
e-mail: mallanjulap1@gmail.com

K. Akhil
e-mail: Akhilaug31@gmail.com

D. RajaSekhar
e-mail: drajasekhar283@gmail.com

S.C. Satapathy and S. Das (eds.), *Proceedings of First International Conference on Information and Communication Technology for Intelligent Systems: Volume 2*, Smart Innovation, Systems and Technologies 51, DOI 10.1007/978-3-319-30927-9_10

1 Introduction

A good automotive suspension system should provide maximum rider comfortness having satisfactory damping capability to handle the uneven surfaces on roads includes (i.e. cracks, uneven pavement and pot holes), the vehicle body should have large damping co-efficient or resistance to avoid large oscillations. Axis and the angular accelerations of the front and rear car body are the two parameters defines the rider comfortness, therefore the angular acceleration and numerical axis must be minimise in order to attain higher ride comfort. Optimization of rider comfort is based on spring stiffness (Ks), damping coefficient (b) mass of spring (m_s). Empirical approaches are generally used to develop these kinds of systems. The optimization requirements in intermission systems and the literature review of this field in the last decennium, are very much effective, There is a large literature in this area and would like to mention those that are, in our opinion, most relevant for reasons of print-space restrictions. It should be noted, however, that the available literature is vast and only a small portion of it can be presented here. This paper includes the well-known optimization technique for basic vehicle suspension systems.

Tak and Chung [1] approach which is considered as systematic was proposed to meet the optimum geometric conditions to obtain the best design for maximum comfortness by taking geometric parameters with varied design variables for the best optimum rider comfortness. Koulocheris et al. [2] combined the deterministic optimization combined with the stochastic optimization vehicle suspension parameters and it has been noticed that this type of optimization was faster and flexible in rider comfortness. Weirs and Dhingra [3] designed the rear suspension system of racing motorcycle drive, and their designed increased the rear tire traction. Mitchell et al. [4] obtained the genetic algorithm optimization for best optimum parameters by considering the description of a suspension model through model known as scoring. Raghavan [5] reduced the non-linear function complexity, by the process of scaling, which is done up or down from shape optimization process.

Gągorowski [6] analyzed the complete system which influences the human body vibrations, by considering various sub-systems. Sun [7] used automotive seat and suspension systems by making alone the vehicle operator in order to avoid the different circumstances, which may affect the rider. For further improving the rider comfortness, an optimization technique through genetic algorithms is used to determine both the active mechanical control parameters combined with passive mechanical parameters. Response surface methodology uses the various mathematical, graphical, and statistical techniques to develop, to improve and also to optimize the experimental process. It is often adequate for the process improvement in an industrial setting by empirical approach. By careful design of experiment and also several factors of independent variables (input variables) the objective of optimization of output variables will be approached [8].

Optimization problems are generally solved by intelligent exploitation of random search by using Genetic algorithms. Discovering the parameters to optimize the model which is being done by the mathematical modelling of real world problem for the optimal functioning of a system is common in engineering [9]. Search space or state space for selection the solution which is best among the others is found in accessing for the feasible solution while solving the problem. Genetic algorithm selects for the best resultant value from the group of results which is indicated represented by unique point in the search field. Obtaining some optimum value is equal to searching a solution in the search field whether it may be the minimum objective function or the maximum objective function. In genetic algorithm in evaluating the other results produces the other set of traces as evaluation continues points [10–12].

2 Formulation of Matlab/Simulink Model

Designing a passive suspension entity for a car is a fascinating design problem. Suspension system is designed by using a quarter car model which simplifies the problem to a one dimensional spring damper entity.

From Fig. 1 the traditional differential formulae were derived corresponding to both the masses i.e. sprung and unsprung respectively

$$m_u\ddot{x} = k_s(y - x) + b(\dot{y} - \dot{x}) - k_w(x - r) \tag{1}$$

$$m_s\ddot{x} = -k_s(y - x) - b(\dot{y} - \dot{x}). \tag{2}$$

Equations (1) and (2) are worked out by the Simulink drag and drop box which is provided in the Matlab ver, 2014. From this model RMS acceleration or the rider comfort RC is being solved for the various levels of data. From Table 1 spring stiffness (K_s) damping coefficient (b) and sprung $M_{u,}$ as shown below.

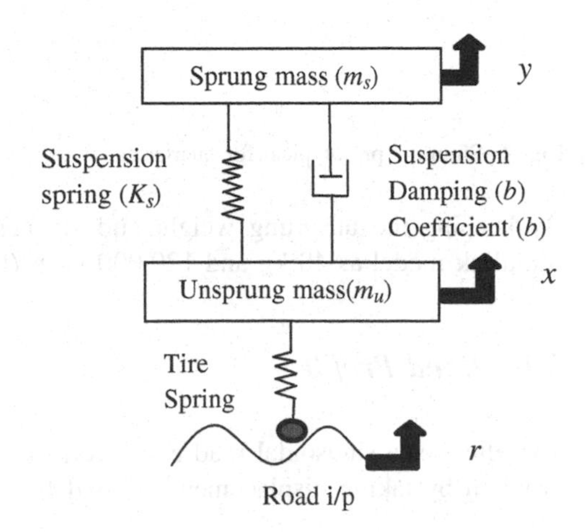

Fig. 1 Vehicle suspension system

Table 1 Three level factors for Box-Behnken design

S. no	Parameters	Level-1	Level-2	Level-3
1	Spring stiffness	8000	14,000	20,000
2	Suspension damping	500	1000	1500
3	Sprung mass	80	190	300

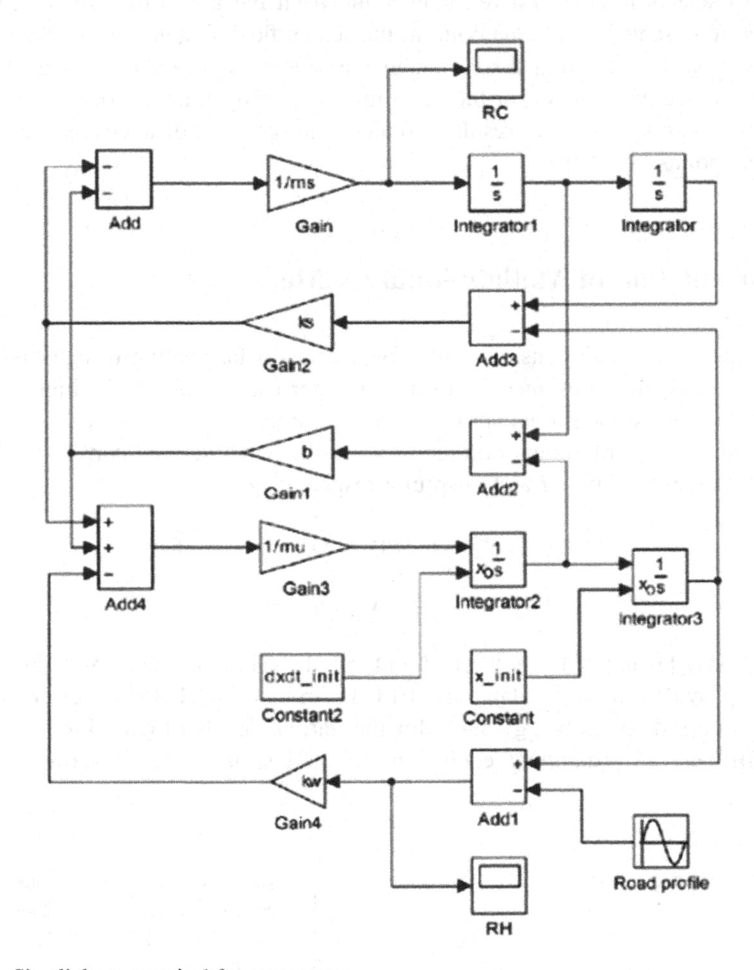

Fig. 2 Simulink prototypical for quarter car

By keeping the unsprung weight and stiffness of tire as constant values in the Simulink model as 40 kg and 120,000 N/m (Fig. 2).

2.1 *Road Profile*

The effect of a sinusoidal road excitation of the ride comfort is considered in this problem by taking displacement of road holding into account, for various bump

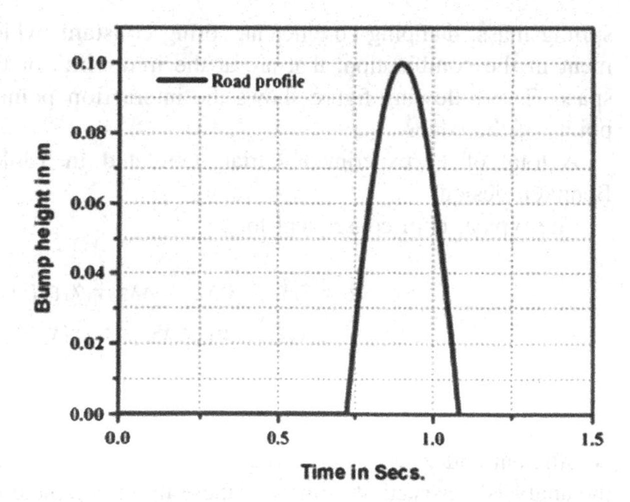

Fig. 3 Road profile

width L with different velocities V of approaching to bump profile which was formulated with time frame t. The time frame t represents the running time of simulation, when the vehicle is about to approach the generated road profile. In this problem the mathematical model for the bump profile is considered from the previous works [8]. The minimum and maximum bump height considered in this problem is clearly seen from this Fig. 3. With 0.0–0.10 m. The function following below a sine wave road profile was being developed for optimum conditions.

$$
\begin{aligned}
r &= 0; \quad when\, t < \frac{d}{v} \\
&= h \sin\left[\frac{\pi v}{L}(t - \frac{d}{v})\right] \quad for\, \frac{d}{v} \le t \le \frac{d+L}{v} \\
&= 0 \quad when\, t > \frac{d+L}{v}
\end{aligned} \tag{3}
$$

2.2 Design and Methodology of Box Behn-Ken Design

One of the mathematical forms like design of experiment systematically plans and control scientific studies that changes the experimental variables to benefit best amount information on cause and effect relationship contacts with a least possible sample size. In this work Box Behn-Ken design which is an Response Surface Methodology was done. Since Box Behn-Ken does not require the embedded factorial (or) fractional factorial design, as it is a self-reliant of rectilinear design. It is well known that it requires only 3-factors, these 3-factors used in this work were sprung mass, damping co-efficient, spring constant, which are useful in the treatment in the combination, that are at the midpoints of the edges of experimental space. The following figure shows the interaction points on the cube at the mid points of the edges.

A total of 15 experimental trials tabulated in Table 1 are required for box Behnken design.

The typical is of consequent for

$$\begin{aligned} y = & \alpha_0 + \alpha_1 x_1 + \alpha_2 x_2 + \alpha_3 x_3 + \alpha_{11} x_1^2 + \alpha_{22} x_2^2 \\ & + \alpha_{33} x_3^2 + \alpha_{12} x_1 x_2 + \alpha_{13} x_1 x_3 + \alpha_{23} x_2 x_3 \end{aligned} \quad (4)$$

where y is the predicted response, α_0 model constant; x_1, x_2 and x_3 are the independent variables α_1, α_2 and α_3 are linear co-efficient; α_{12}, α_{13}, α_{23} are cross product co-efficient and α_{11}, α_{22}, α_{33} are quadratic co-efficient. With the experimental data the analysis consisted of estimating these three parameters for a first order model, if the first order model demonstrates any statistical lack of it, a second order model was developed. Therefore a second order empirical regression model is required when the true response function is non-linear and unknown.

From various literary works [1–7], Stiffness (K_s), damping (b), sprung mass (m_s) are the three factors as sighted in Table 1, are extracted for the study. The SIMULINK model was generated to carry out the simulation as shown in Fig. 2 as per the three factorial Box-Behnken design generated in DOE++. The run time simulation results obtained are sighted in the Table 2.

Table 2 Box-Behnken design observation table

Run order	Stiffness	Damping	Mass	Response
1	8000	1000	80	2.45321
2	14,000	1500	300	0.61342
3	20,000	500	190	1.32451
4	14,000	1000	190	1.23453
5	8000	1500	190	1.50326
6	14,000	500	80	2.20012
7	14,000	500	300	0.61342
8	14,000	1500	80	3.46332
9	20,000	1000	300	0.77845
10	20,000	1500	80	1.71345
11	8000	500	300	0.74561
12	14,000	1000	190	1.23453
13	14,000	1000	190	1.23453
14	20,000	1000	80	3.16745
15	8000	1000	300	0.73214

2.3 Coded Co-efficient

First a target rate and the maximum of the response rate are to be contributed, in order to minimize the response. If the rate of response is about the target value then the ability of deservedness is considered as zero, and if value is less than the target value then it is unity. As the response approaches to the target value, the desirability reaches one. From the model summary of the statistics it is clear seen that the regression model of the quadratic is most fitted one when compared to linear (Table 3).

2.4 Data Analysis of RC

In this present paper, Minitab-17 an excellent statistical package is used for analysis. From the statistical summary model it is clearly understood that the response is good to have its nature in polynomial equation which is being formed by the three independent variables, the final suggested regression equation in un-coded values is provided in the Eq. (5).

$$\begin{aligned} RC = &+1.85055 + 7.79328E - 005k_s + 1.6125E - 004b - 0.017164m_s \\ &- 3.07258e - 008k_s b - 2.53004E - 007k_s m_s - 4.60132E - 009bm_s \\ &+ 1.18385E - 009k_s k_s + 1.78235bb + 4.1794E - 005m_s m_s \end{aligned} \quad (5)$$

The fitness for the good regression equation is examined with the following conditions:

- Quantitative checking for fitness between anticipated model and recognized model.
- Visual verification of the validated assumptions of random errors.
- Checking out for the R, Adjusted R Square, Predicted R square values in linear, quadratic.

Table 3 Coded coefficients of three parameters

Terms	Effect	Coef	SE coef	t-value	p-value	VIF
Constant		1.2428	0.0247	50.42	0.000	
Stiffness	0.3874	0.1937	0.0181	10.68	0.000	1.00
Damping	0.6024	0.3012	0.0181	16.60	0.000	1.00
Mass	−2.1367	−1.0683	0.0181	−58.89	0.000	1.00
Stiffness*stiffness	0.1453	0.0727	0.0266	2.73	0.034	1.01
Mass*mass	0.9471	0.4735	0.0266	17.78	0.000	1.01
Stiffness*damping	−0.1844	−0.0922	0.0257	−3.59	0.011	1.00
Stiffness*mass	−0.3340	−0.1670	0.0257	−6.61	0.001	1.00
Damping*mass	−0.6316	−0.3158	0.0257	−12.31	0.000	1.00

Table 4 Classical summary data

Source	Std. dev.	R-squared	Adjusted R-squared	Predicted R-squared	PRESS	
Linear	0.34	0.8640	0.8327	0.7485	2.85	
2FI	0.34	0.8995	0.8392	0.6214	4.29	
Quadratic	0.050	0.9985	0.9965	0.9753	0.28	Suggested

The assessable verification of the fitment of anticipated model is shown in Table (4). It is clear from the table that the determined or assured coefficient R2 is 99.85 % indicates that 99.85 % fortunate, the output matches with the observed values. It is recognized that the adapted R2 value (99.65 %) is very near to the R2 value (99.85 %) intimating that the model is does not contain any excess variable. The predicted or anticipated R2 value (97.53 %) assuring that 97.53 % times, the process of optimization for further is being validated by the anticipated response for the given input variables through regression model.

3 Optimization Using GA

Genetic Algorithm is a search technique which is generally used in computing to optimize or to find the true or approximate solutions and search problems. Genetic Algorithm is the technique inspired by evolutionary biology, which are a distinct class of evolutionary algorithms, such as inheritance, mutation, selection, and crossover. These are categorized as global search heuristics. It is a computer simulation, implemented in which the chromosomes of possibility solution evolve to better solution, through optimization problem. Generally GA starts from the randomly generated individuals, which will be the parents to the next generations. In each new generation, every individual fitness is evaluated, by the fitness function, based on their fitness value the individuals may modified to form new generations. The so formed generations are used to the next iterations. The simulation terminates, when it reaches maximum generations, or satisfactory fitness level has been selected from the populations. To implement GA simulation, one need to provide the evaluation function, which gives the individuals a score based on how well they perform the given task.

In this present paper, the objective is to determine the best controlling variables from the given variables, and also their best values in evaluation for the maximum rider comfortness. From the introduction to genetic algorithm it clearly says that when one need to perform the GA simulation, need to provide evaluation or objective function. Inorder to perform the simulation, Eq. (5) was used as the objective function, which is generated from the RSM modelled equation, which is given as follows: Minimize

$$\begin{aligned} RC = & + 1.85055 + 7.79328E - 005k_s + 1.6125E - 004b - 0.017164m_s \\ & - 3.07258e - 008k_sb - 2.53004E - 007k_sm_s - 4.60132E - 009bm_s \\ & + 1.18385E - 009k_sk_s + 1.78235bb + 4.1794E - 005m_sm_s. \end{aligned}$$

In this present work GA with same parameters, It starts from the initialization where early various individuals, initial population are produced by the random generation of solutions. Generally the population size varies from the problem to problem it may range from several hundreds to thousands the more the population the more the time complexity, in this problem population size is constrained to 50, 100, 150. The other important criteria need to be considered is selection process, where during each successive generation, a proportion of the existing population is selected to breed a new generation, followed by the crossover probability, mutation functions, where crossover probability is 0.8 with feasible mutation function is maintained during simulation. Trial error method was used to parameter settings and also for alterations to obtain the best results several combinations of the rider conditions have been tried using Matlab Optimization tool box. It leads to minimum value of rider comfortness vibration values.

The following is the code implemented

PSUEDO code for GA:

1: Initialize the population for *Ks*, m_s, *b*
2: Generations = 50, 100, 150.
3: While Generations <max_generations do
4: Create offspring by crossover and mutation
5: fitness function is used to evaluate the offspring
6: Two individuals in the pópulation are chosen, so may be altered by the offspring
7: choose if this/these individuals will be retrieved.
8: Generations = generations + 1
9: end while
10: Return population (optimized)

4 Results and Discussions

Combination of input variable settings are identified by optimization by genetic algorithm which optimizes a single response or a set of response. Thus optimal solution is obtained for the combination of control input variables, with an optimization plot of best fitness and best mean. The plot related to optimization is

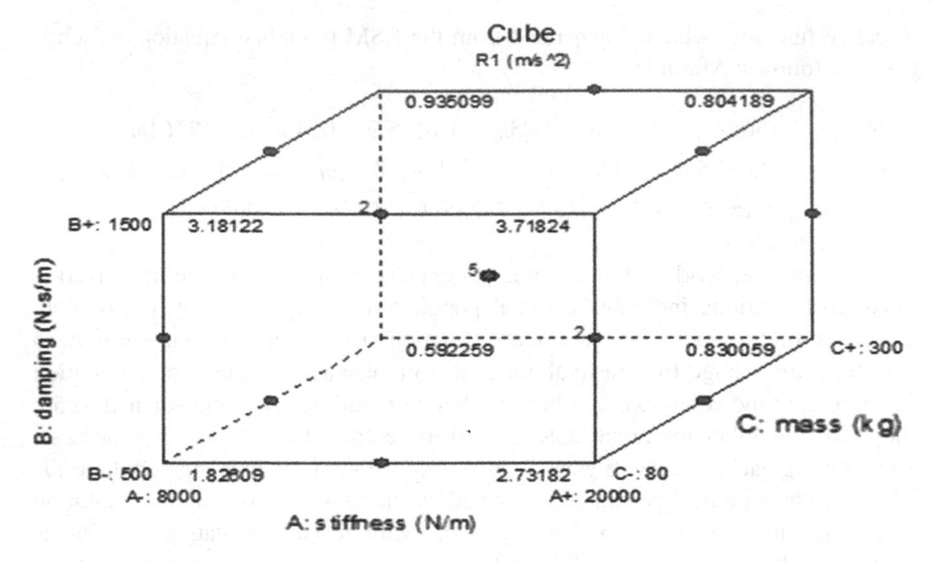

Fig. 4 Three factorial Box-Behnken design for quarter car parameters

shown in the Fig. 4 is interactive, for various control unstable frame work on the plot for local and global solution by searching for more desirable one's. The combined factor settings for the local solution can be found from the initial point, from where it begins. The desired response is achieved by the best combination of factor settings and global solution. The chromosomes sizes presented in this work includes, population size 50, 100 and 150, tournament selection, crossover probability is 0.8 and mutation function is adaptive feasible (Table 5).

The results obtained above commend that GA can be preferred in complex and realistic designs often met in the engineering from the modelled equation generated through response surface methodology (Fig. 5, Table 6).

Table 5 Optimization input values for best response

GA parameters	Value
Population size	50, 100, 150
No of generations	500
Fitness scaling	Rank
Crossover technique	Heuristic
Probability of crossover	0.9
Mutation technique	Uniform
Generation gap	0.9
Lower boundary	[8000 500 80]
Upper boundary	[20,000 1500 300]
Objective function accuracy	$1e^{-12}$

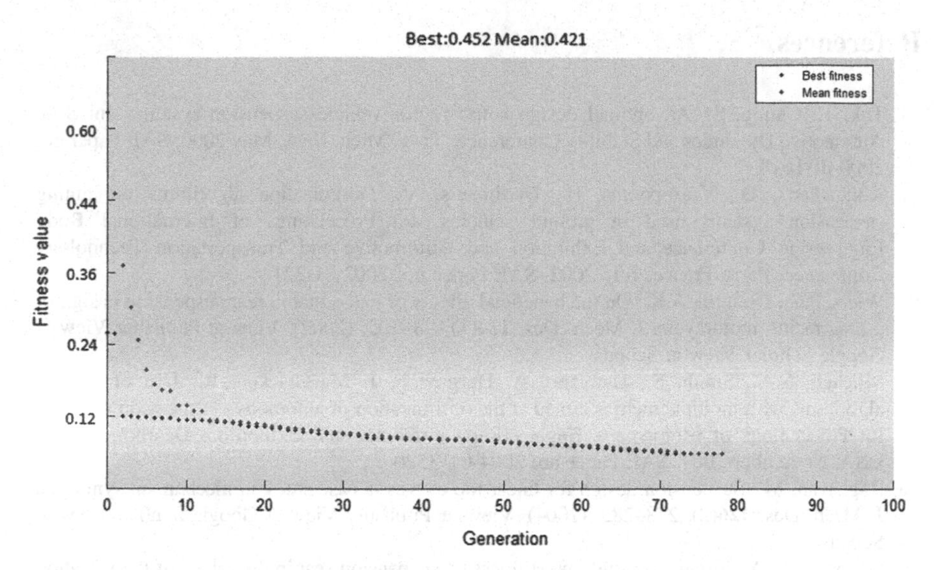

Fig. 5 Best fitness and mean fitness plots

Table 6 Optimal parameters

Parameters	Stiffness (K_s) N/m	Damping (b) N s/m	Mass (m_s) kg
Initial parameters	8000	500	80
Global solution	10,000	500	293.5

From the above graph it is clearly understood that the minimization of Response of the system was given as the best value as RC = 0.452, and the mean value was given as 0.421, which is very appreciable minimization in the case of genetic algorithm.

GA parameters, such as population size and mutation probability, may be adapted more effectively to further improve efficiency and consistency in the results. The GA results obtained shows that there is potential to incorporate global optimization methods for suspension system design. The model is optimized by GA through RSM modelled equation for getting factors to obtain optimized ride comfort. The factors setting are stiffness, Ks = 10,000 N/m, damping, Cs = 500 N-s/m and mass, ms = 293.5 kg for the value of RC equal to 0.452.

Acknowledgments This research was supported/partially supported by Prof. C.h. Ratnam, Mechanical Department, College of Engineering, Andhra University. We thank our Head of Department of Mechanical Engineering, Dr. Y.V. Hanumanth Rao. Dean of Research and Development, Dr. K.L. Narayana. And authors P. Vigneshwar and Rajasekhar would like to special thanks to Dr. Sumathi, Department of Humanities KL University, P. Manjula Reddy, P. Malla Reddy, and D. Kesava Reddy.

References

1. Tak, T., Chung, S.: An optimal design software for vehicle suspension systems. In: SAE Automotive Dynamics and Stability Conference, Troy, Mich, USA, May 2000, SAE Paper no: 2000-01-1618
2. Koulocheris, D., Vrazopoulos, H., Dertimanis, V.: Optimization algorithms for tuning suspension systems used in ground vehicles. In: Proceedings of International Body Engineering Conference and Exhibition and Automotive and Transportation Technology Conference, Paris, France, July 2002, SAE Paper no: 2002-01-2214
3. Wiers, P.C., Dhingra, A.K.: On the beneficial effects of anti-squat in rear suspension design of a drag racing motorcycle. J. Mech. Des. **124**(1), 98–105, (2002). View at Publisher View at Google Scholar View at Scopus
4. Mitchell, S.A., Smith, S., Damiano, A. Durgavich, J. MacCracken, R.: Use of genetic algorithms with multiple metrics aimed at the optimization of automotive suspension systems. In: Proceedings of Motorsports Engineering Conference and Exhibition, Dearborn, Mich, USA, November 2004, SAE Paper no: 2004-01-3520
5. Raghavan, M.: Suspension design for linear toe curves: a case study in mechanism synthesis. J. Mech. Des. **126**(2), 278–282, (2004). View at Publisher View at Google Scholar View at Scopus
6. Gągorowski, A.: Simulation study on stiffness of suspension seat in the aspect of the vibration assessment affecting a vehicle driver. Logistics Transp. **11**(2), 55–62 (2010)
7. Sun, L.: Optimum design of 'road-friendly' vehicle suspension systems subjected to rough pavement surfaces. Appl. Math. Model. **26**(5), 635–652 (2002). doi:10.1016/S0307-904X(01)00079-8
8. Box, G.E.P., Draper, N.R.: Empirical Model-Building and Response Surfaces, p. 74. Wiley, New York (1987)
9. Gundogd, O.: Optimal seat and suspension design for a quarter car with driver model using genetic algorithms. Int. J. Ergonomics **37**(4), 327–332 (2007). doi:10.1016/j.ergon.2006.11.005
10. Abbas, W., Abouelatta, O.B., El-Azab, M.S., Megahed, A.A.: Application of genetic algorithms to the optimal design of vehicle's driver-seat suspension model. In: Proceedings of the World Congress on Engineering (London, UK, June 30–July 2, 2010), 1630–1635 (2010)
11. Hada, M.K., Menon, A., Bhave, S.Y.: Optimisation of an active suspension force controller using genetic algorithm for random input. Defense Sci. J. **57**(5), 691–706 (2007)
12. Abbas, W., Abouelatta, O.B., El-Azab, M.S., El-Saidy, M., Megahed, A.A.: Optimization of biodynamic seated human models using genetic algorithms. SCIRP J. Eng. **2**, 710–719

MEAL Based Routing in WSN

Tejas Patalia, Naren Tada and Chirag Patel

Abstract Wireless sensor network is growing area. Characteristics of WSN (Wireless Sensor Network) is such that nodes are running on limited battery power. As each device are now trying to connect to WAN, this field has caught many eyes of researchers. We are modifying LEACH in such a way that it include advantages of both the energy leach (E-LEACH) and multi-hop leach (MH-LEACH) in proposed protocol. We have deployed our proposed MEAL (Multi-hop Energy Aware Leach) in our college campus, and we have monitored our campus environment by CO2 level and temperature of each department. Our protocol giving best output with respect to energy usage compare to LEACH. EL-LEACH and ML-LEACH. We have used TINY-OS and TINY-VIZ environment for configuring our motes.

Keywords LEACH · WSN · Energy leach · Multi-hop LEACH · MEAL

1 Introduction

These days we all are observing a drastic change in lifestyle of people. We are more dependent on technology, and we have started believing in machines (automation). And wireless sensor will play a huge role in the future automation industry. Wireless sensor nodes are tiny computers that are designed to perform a specific task with very high efficiency. The efficiency is in terms of power usage. Lower the usage—longer the life. WSNs can greatly simplify system design and operation [1].

T. Patalia (✉) · N. Tada · C. Patel
VVP Engineering College, Rajkot, India
e-mail: pataliatejas@rediffmail.com

N. Tada
e-mail: narentada@gmail.com

C. Patel
e-mail: cecrp@vvpedulink.ac.in

S.C. Satapathy and S. Das (eds.), *Proceedings of First International Conference on Information and Communication Technology for Intelligent Systems: Volume 2*, Smart Innovation, Systems and Technologies 51, DOI 10.1007/978-3-319-30927-9_11

This field is rapidly developing. We can attach any sensor in these nodes. Thus, this technology can be applied in a variety of applications like security, maintenance observation, agricultural, traffic monitoring and many more. WSN is a collection numerious tiny devices with low cost and low power equipement.

1.1 Motivation and Objective

In this paper, we want to introduce implementation and results of MEAL based routing in WSN.

1.2 Leach Variants

1.2.1 Multi-Hop Leach

Due to single base station LEACH becomes inefficient when network diameter increased beyond some limit. Network nodes require high amount of power to send the data when base is too far. This situation will lead to decrease the lifetime of a sensor node that is not feasible for wireless sensor network. It also decrease the life span of entire network. To overcome this problem, author develop Multi-Hop LEACH. In this variant of LEACH, the author has introduced a method that supports long distance between base station and sensor node without affecting the life of the network. In this alternative, some nodes from the network will play as Cluster Head and form their cluster (a group of nearby sensor nodes). Now all nodes from the cluster will communicate with CH only. And then CH handles forwarding that data towards Base Station. This is called setup phase. After setup phase, the network goes to steady state phase. Multi-Hop LEACH supports two types of communication operations [3]. (1). Inter-cluster communication and (2). Intra-cluster communication. In intra-cluster communication, WSN is divided into multiple clusters, each has their cluster head. This CH handles communication between all nodes in the cluster; In a wireless sensor network CH gathers all the data from their territory area and redirects it to the base station after performing data aggregation. Most of the times CH. Found nearer to the base station but what if not? Mutlihop approach is useful in which data relay between more than one CH will reach to the base station.

1.2.2 Energy Leach (E-Leach)

Multihop leach has great advantage over leach protocol. But it suffer from one major problem to select CH node in second round onwards. It has been seen that CH node to suffer about more battery power compare to other member. To solve

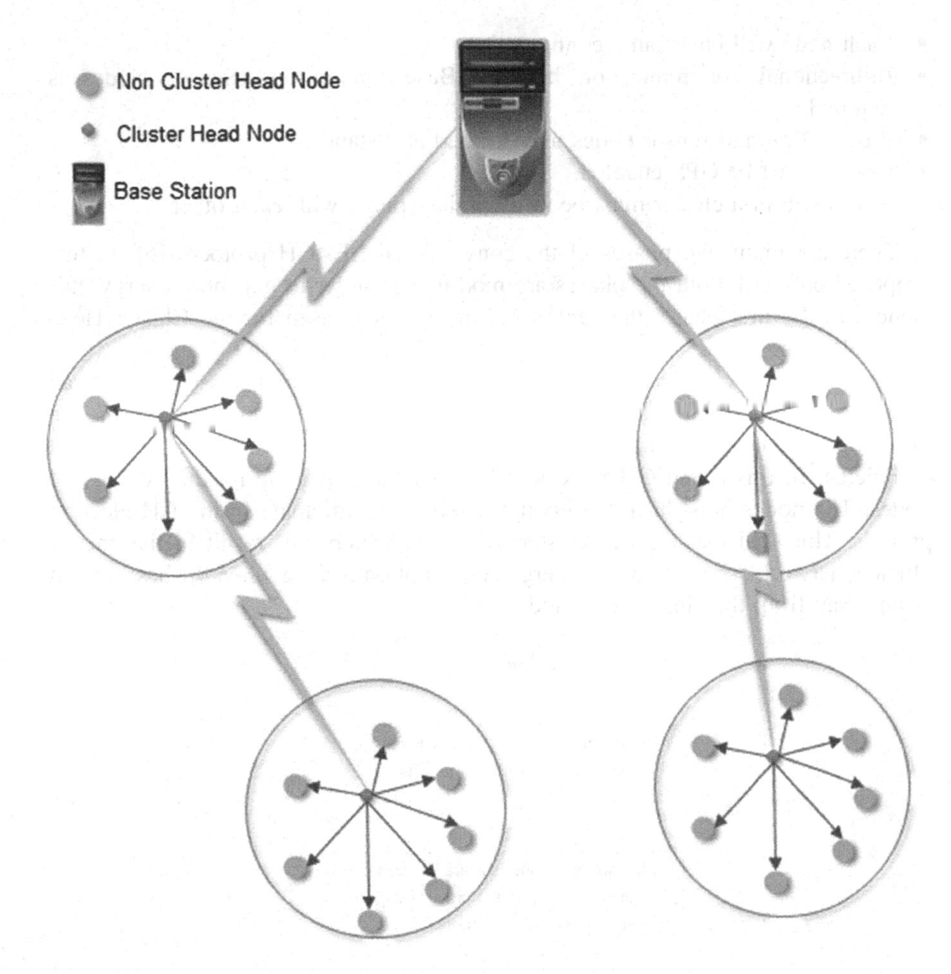

Fig. 1 Multi Hop Leach

this problem Energy LEACH has been proposed [4]. It uses residual energy for cluster head selection process. In second round onwards CH is selected based on its residual energy after the first round. As a result, nodes that have more residual energy will become CHs probably more than nodes with less residual energy (Fig. 1).

2 Proposed Approach

Evaluating different descendants of LEACH protocol [5], it is shown that each protocol somehow makes the improvement in the original one. We have implemented hybrid technique. Our system having following assumption:

- Each node will have same configuration.
- Bidirectional communication between Base Station and Sensor Nodes is required.
- Base station and sensor nodes may located at distance.
- Device must be GPS enabled.
- Nodes within a cluster must be in a wireless range with each other.

There are main two phases of the conventional LEACH protocol [6]. In this proposed protocol, both the phases are modified in order to get more energy efficiency. In the first phase, the centralized algorithm is used for the Cluster Head node selection.

Second phase would be data communication phase. In first phase cluster head will be selected as per the traditional leach protocol. Data communication done via multihop base method. There is a E-leach method applied After the first round completed. In this round CH node will be selected as per the residual energy of nodes. The nodes must have the enough energy to participate in the CH election process. This will make better cluster formation which will result in the energy efficient network. The other main energy consumption is done when the base station is far away from the cluster head node.

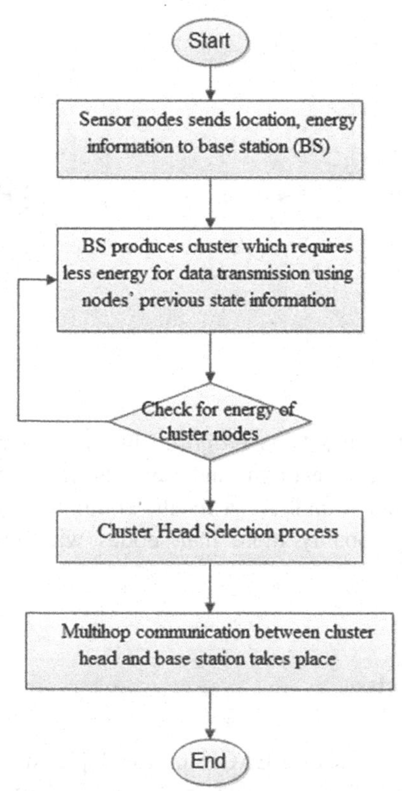

3 Simulation Environment

To test our protocol in the real environment, we have implemented it in our college. Different nodes are installed at each department and the entire area is around 50 acres. Nodes are installed on the ground floor, first floor and second floor. The distance between two nodes may vary based on obstructing objects in-between. We have considered following parameters to implement test environment and we have used TelosB motes as wireless sensor nodes. We are expecting improvement in the cluster head selection process and in steady state phase (Fig. 2).

Here initial energy is considered 100 JL for each node. The comparison between LEACH, ELEACH, MHLEACH and MEAL is depicted in the chart. And MEAL outperformance all other variants of LEACH in terms of rounds. MEAL last for around 450 rounds, which is far better than E leach and MH Leach. So based on simulation results we can conclude that our proposed MEAL protocols perform better compare to LEACH, MH-leach and E-leach. As we increase the simulation time, MEAL shows higher network lifetimes and execute more rounds.

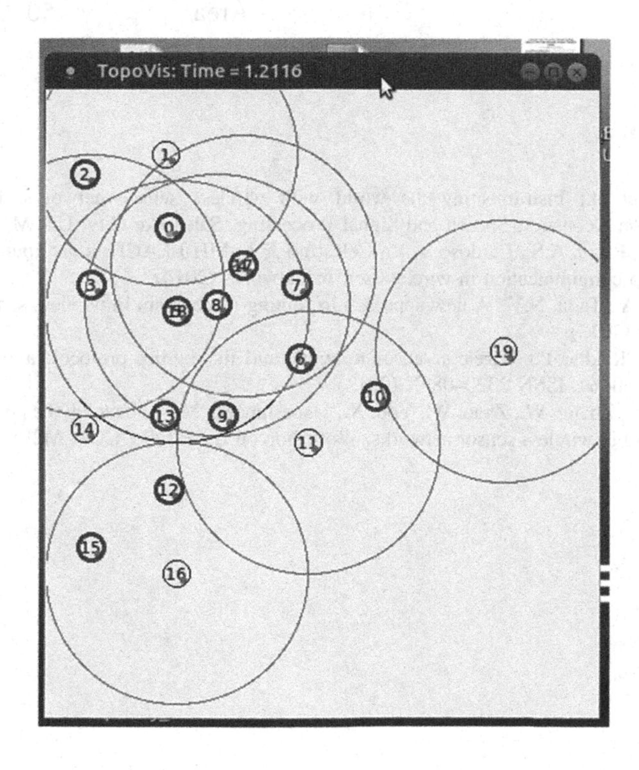

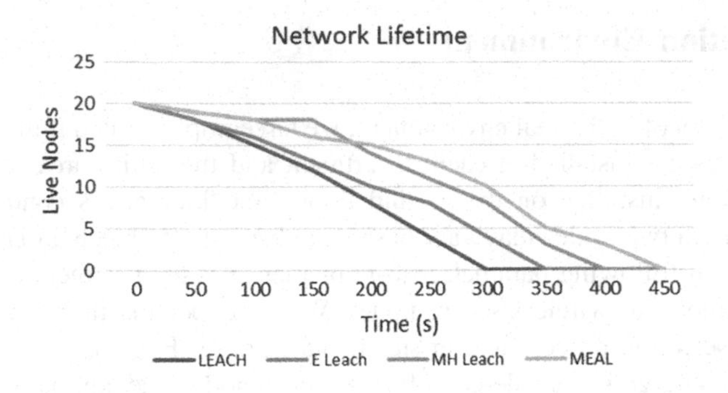

Fig. 2 Simulation of MEAL in nasC

Sr No	Parameter	Value
1	No. of Motes	20
2	Protocol	MEAL
3	Initial Energy	100 JL
4	Base Station	1
5	Link Layer	802.11
6	Area	50 Acres

References

1. Estrin, D., et al.: Instrumenting the world with wireless sensor networks. International Conference on Acoustics, Speech and Signal Processing, Salt Lake City, UT, May 2001
2. Neto, J.H.B., Rego, A.S., Cardoso, A.R., Celestino Jr J.: MH-LEACH: a distributed algorithm for multi-hop communication in wireless sensor networks (2014)
3. Vaishnav, P.A., Tada, N.V.: A new approach to routing mechanism in wireless sensor network environment (2013)
4. Reena, S.P., Rekha, P.: Wireless sensor network and its routing protocol: a survey paper. IJCSMC, 2, 60–64, ISSN 2320–088X (2013)
5. Li, Y., Yu, N., Zhang, W., Zhao, W., You, X., Daneshmand, M.: Enhancing the performance of leach protocol in wireless sensor networks. Workshop on IEEE INFOCOM M2MCN, 223–228 (2011)

Intelligently Modified WFQ Algorithm for IEEE 802.16 Resource Allocations

Akashdeep

Abstract IEEE 802.16 standard more commonly known as WiMAX is an upcoming standard popularized by WiMAX Forum. Quality of service in WiMAX is provided with help of five scheduling services with different properties. WiMAX standard does not define provisioning for providing bandwidth allocation to different scheduling services. Most algorithms focus on traffic classes having rigid time constraints and classes without real time requirements are neglected This paper proposes modified Weighted Fair Queuing Algorithm using concepts of fuzzy logic for grant of bandwidth to all types of traffic. It enables fair distribution of network resources to low priority traffic classes and helps to improve their performance. Simulations have been done for performance justification and results are encouraging.

Keywords Scheduling · Bandwidth allocation · Simulation · Wimax · IEEE 802.16

1 Introduction to IEEE 802.16 and WiMAX

Growth of rich media applications as a result of mounting popularity of smart phones had awakened wireless community to popularize IEEE 802.16 standard, also known as WiMAX [1, 2]. The inherent architecture of WiMAX is such that it can support real and non real time applications in both wired and wireless scenarios. It must be able to accommodate these application requirements. QoS is IEEE 802.16 is provided with number of mechanisms like Time division duplexing, frequency division duplexing, frame error control, and orthogonal frequency division multiplexing. Each packet in IEEE 802.16 is associated with service flow that

Akashdeep (✉)
Panjab University, Chandigarh, India
e-mail: akashdeep@pu.ac.in

S.C. Satapathy and S. Das (eds.), *Proceedings of First International Conference on Information and Communication Technology for Intelligent Systems: Volume 2*,
Smart Innovation, Systems and Technologies 51, DOI 10.1007/978-3-319-30927-9_12

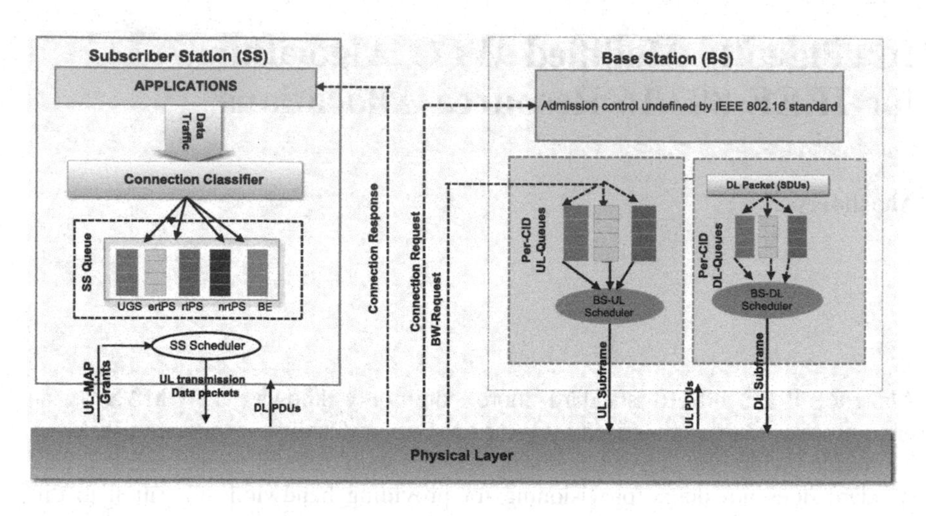

Fig. 1 Scheduling and QoS Framework in IEEE 802.16

helps to implement connection oriented MAC. Base stations and subscribers communicate with help of unidirectional connections. Qualtiy of service framework in WMAx is shown in Fig. 1. Each user application is checked for service to be offered and a specific scheduling service is attributed to handle this flow. The allocation strategy in WiMAX is request grant which is supported with help of five different scheduling service types: unsolicited grant service (UGS), real-time polling service (rtPS), extended real-time polling service (ertPS), non real-time polling service (nrtPS), and best effort (BE). These traffic classes have been designed considering various applications supported by WiMAX and every traffic class has its own priority and quality of service parameters. Priorities for real time scheduling types are more as compared to non real types as specified by standard.

The signaling mechanism in WiMAX allows for implementation of request and grant based algorithm for bandwidth allocation. The subscribers estimated its requirements and communicated it to the base station for required allocations to be made. Base station gathers requirements of all subscribers and keeping in view available bandwidth makes allocations to different subscribers employing a suitable algorithm. Figure 1 defines quality of service framework employed in IEEE 802.16 systems. The standard has only defined framework for quality of service implementation but implementation of specific algorithm has been left to device manufactures. This paper has implemented a scheme to intelligently modify traditional weighted fair queuing algorithm using fuzzy logic for allocating resources to various subscribers in IEEE 802.16 networks. The coming sections explore proposed method.

2 Related Work

This section summarizes some of recent studies for IEEE 802.16 resource allocation problem. The author itself have contributed survey of recent approaches found here [3]. Ball et al. [4] has proposed a round robin variant of scheduler while Jalali et al. [5] has proposed proportional fairness (PF) scheme. Other studies based on fairness allocation of resources have been proposed by Demres [6] and Andrews [7]. Juliana Freitag et al. [8] categorized traffic into three queues according to priority and served these queues using strict priority algorithm. However migration of requests from low level queue to high level queue was allowed for requests with approaching deadlines.

Mukul et al. [9] implemented a dynamic schedulers that used prediction of traffic patterns for rtPS packets. Pheng et al. [10] considered queue length factor and Lagrange's Interpolation function for estimating amount of rtPS packets. Hwang [11] proposed an adaptive allocation strategy based on study of various queue parameters. Fathi et al. [12] proposed another channel aware scheduler in which bandwidth was allocated using law of moving averages. In spite of the availability of large number of algorithms our work is different from previous work as it implements a dynamic system. Fuzzy logic has been exploited to inculcate fairness in bandwidth allocation to all types of traffic classes. Studies discussed in this section indicate that schedulers tend to overlook non-real time traffic resulting in performance dop of non-real time applications and that of overall network degrades significantly. Scheduler shall consider requirements of non real time traffic for decision making. The research presented in this paper addresses the following questions from preceding sections:

- Implementing an allocation scheme taking care of request grant allocation scheme used by WIMAX scheduler
- Modifies WFQ algorithm using uncertainty principles of fuzzy logic
- Improves overall performance of non real time applications

3 Proposed Scheme

Allocation of resources in WiMAX is multifaceted problem as scheduler has the task of catering to requirements of various applications with diverse requirements. WiMAX employs request grant allocation mechanism in a scenario where traits of traffic can suddenly vary in different applications. Traditional WFQ algorithm fails to serve all service classes fairly because real time applications tend to absorb more amounts of resources as their traffic increases. In order to provide fairness to all classes, scheduling strategy shall be able to adapt as there are changes in incoming

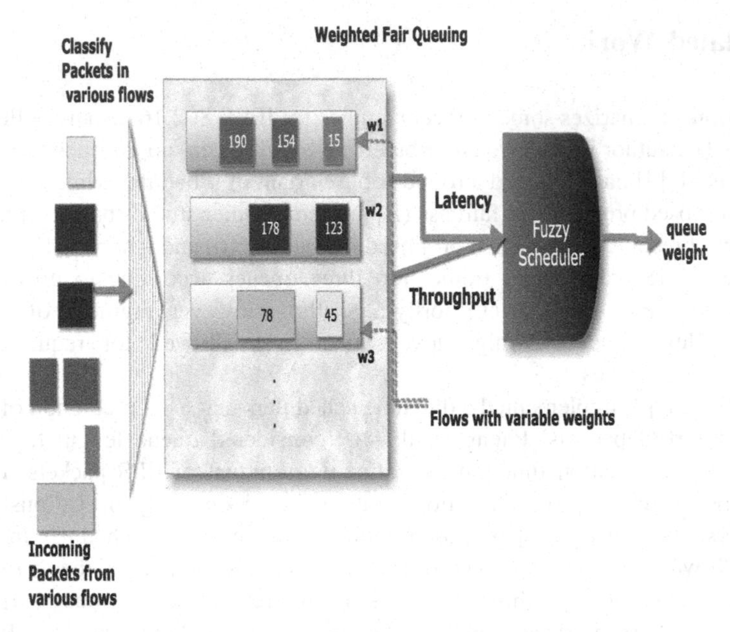

Fig. 2 Proposed Methodology

traffic patterns. Proposed scheme has used vague principles of fuzzy logic to implement a dynamic WFQ algorithm. The algorithm is aimed at providing fairness to all traffic classes in WiMAX. WiMAX scheduling services can be categorized into two major classes: real time having rigid time (latency) requirements like UGS, ertPS, rtPS and non real time class having minimum throughput requirements. Proposed approach extracts information of these two requirements from incoming traffic and modifies weights of queues serving these classes in WFQ algorithm. Both latency and throughput requirements act as input variables of fuzzy based system with weight as output variable. The output value given by fuzzy system is used to alter weight of serving queues in WFQ algorithm. Figure 2 shows the implemented methodology while Figs. 3, 4, 5 defines implemented membership functions used in study. There are total of five linguistic levels used in the study NB, NS, Z, PS, PB. Different variables supports different number of linguistic levels. Table 1 shows the proposed rule base for our fuzzy inference system.

$$\sum\nolimits_{i=0}^{n} w_i = 1 \quad 0.001 \leq w_i \leq 1 \tag{1}$$

All flows having weight w_i in a packet switched network shall satisfy constraint of Eq. 1; The constraint makes sure that none of flow is starved and is being offered some minimum portion of bandwidth. The weights in WFQ algorithm remain static and this weight updating process is made dynamic by fuzzy based method. On receiving new bandwidth request, fuzzy inference algorithm is called by base station. The fuzzy system reads values of latency and throughput, fuzzification of

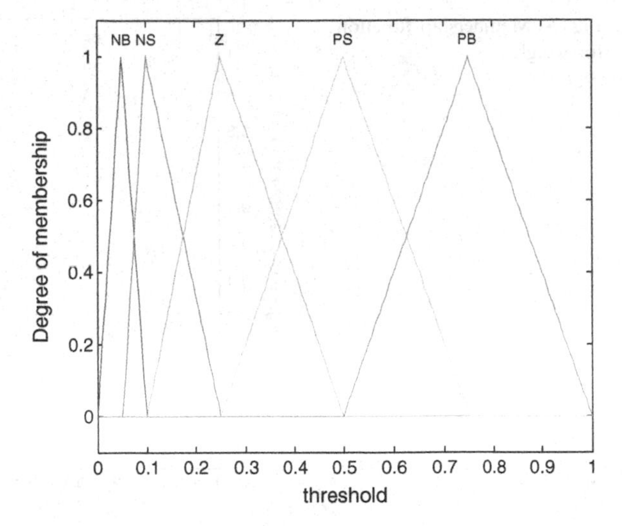

Fig. 3 Membership function for throughput

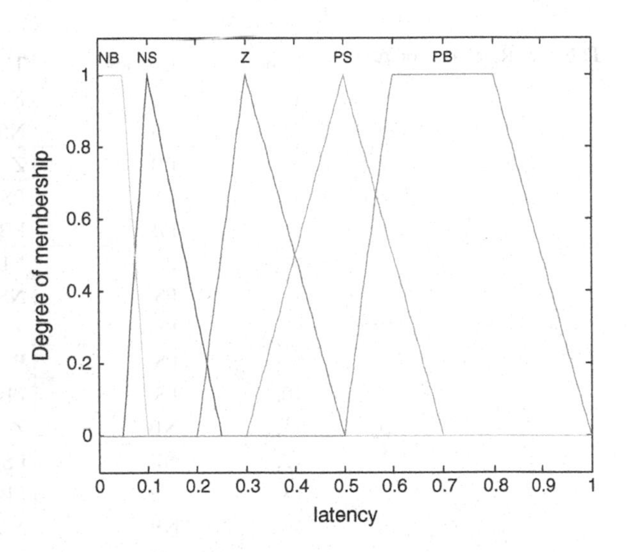

Fig. 4 Membership function for latency

these values is done and fuzzy reasoning is applied to get value of weight in fuzzy form. The said value is defuzzified to get a crisp weight value which is utilized as weights for queues of real time traffic classes. More widely accepted Mamdami's inference has been utilized and center of gravity method is used for de-fuzzification. Allocation of bandwidth to different flows using Eq. 2.

$$\text{Allocated Bandwidth} = R_{max} \times \frac{w_i}{\sum_{i=0}^{n} w_i} \tag{2}$$

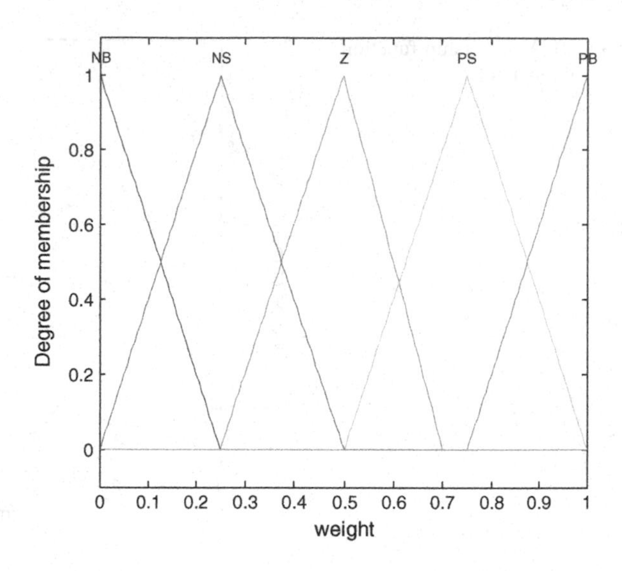

Fig. 5 Membership function for weight

Table 1 Rulebase for fuzzy system

S. no	Latency	Throughput	Weight
1.	PB	NB	NB
2.	PB	NS	NB
3.	PB	Z	NB
4.	PB	PS	NB
5.	PB	PB	NB
6.	PS	NB	NS
7.	PS	NS	NS
8.	PS	Z	NS
9.	PS	PS	NS
10.	PS	PB	NS
11.	NB	Z	PB
12.	NB	PS	PB
13.	NB	PB	PB
14.	NB	NB	PB
15.	NB	NS	PB
16.	Z	NB	Z
17.	Z	NS	Z
18.	Z	Z	Z
19.	Z	PS	Z
20.	Z	PB	Z
21.	NS	NB	PS
22.	NS	NS	PS
23.	NS	Z	PS
24.	NS	PS	PS
25.	NS	PB	PS

4 Results and Discussion

The proposed scheme is implemented in proprietary simulator written in C ++. Conducted experiments are aimed to evaluate if designed fuzzy inference system can meet quality of service levels of all types of classes in WiMAX. The main aim of scheme is to improve performance offered to various service classes in WiMAX specially non real classes. Evaluation of performance has been done by increasing number of real time connections and measuring delay and throughput parameters. Conducted experiments are discussed below.

Experiment 1 In this experiment number of connections for other real time classes has been kept fixed while UGS connections are increases significantly. Number of rtPS,ertPS and nrtPS connections have been kept as 10,10,25 respectively while UGS connections were varied between 10–40. Figures 6, 7 plots average delay and Figs. 8, 9 plots throughput performance of all classes as amount of UGS traffic is increased. Variations in delay for real time classes are minimum as supported by IEEE 802.16 standard. The delay of non real time classes increases a bit with high number of real time connections but still these values are considerably manageable considering amount of traffic in network. Throughput plots for non real time classes shows small updowns with nrtPS graph having better projections as compared to BE graph because BE does not provide any QOS requirements.

Experiment 2 This experiment studies effects of increase in extent of rtPS service class on other classes. rtPS can severely effect performance of other classes since it is bursty in nature. rtPS connections were taken in range of 5–18 and nrtPS and BE connections were taken as 25 and 40. For performance analysis, comparison delay and throughput has been measured for real and non real time classes respectively. The corresponding plots are presented in Figs. 10, 11. The increase in traffic was able to influence performance of these classes up to only a limited extend.

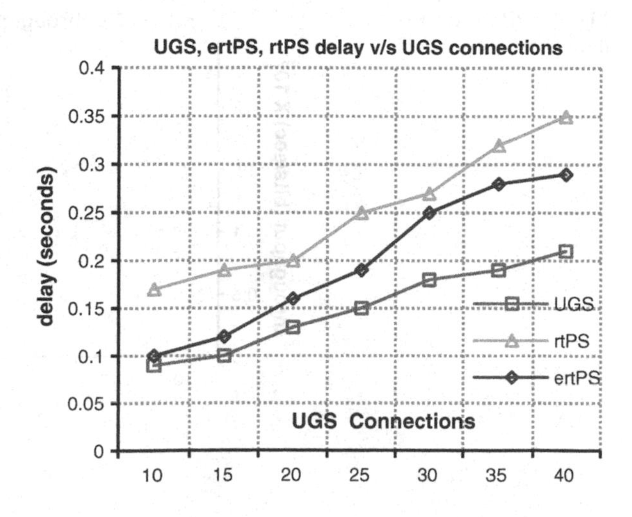

Fig. 6 UGS, ertPS, rtPS delay versus UGS

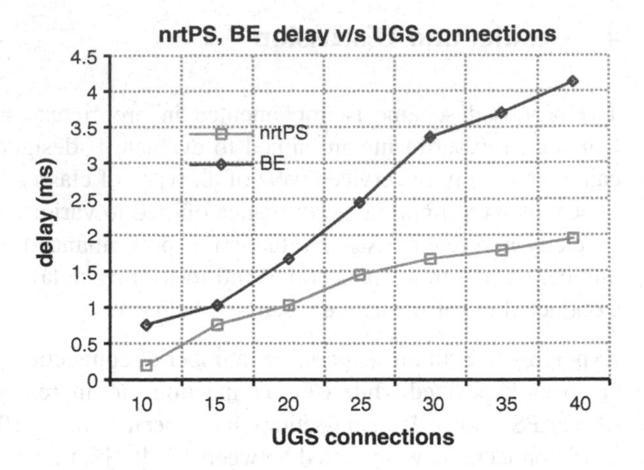

Fig. 7 nrtPS and BE delay versus UGS

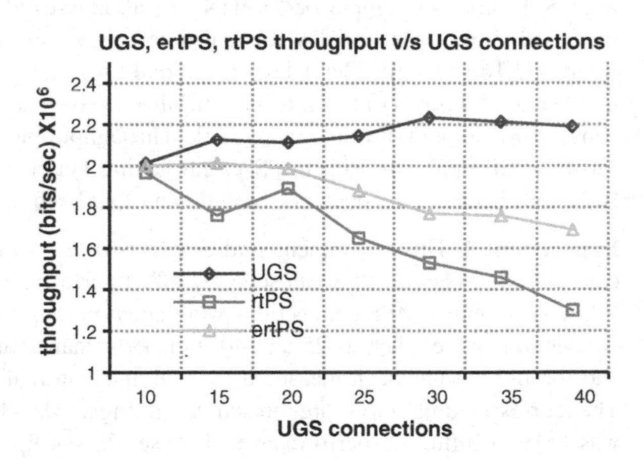

Fig. 8 UGS, ertPS, rtPS throughput versus UGS

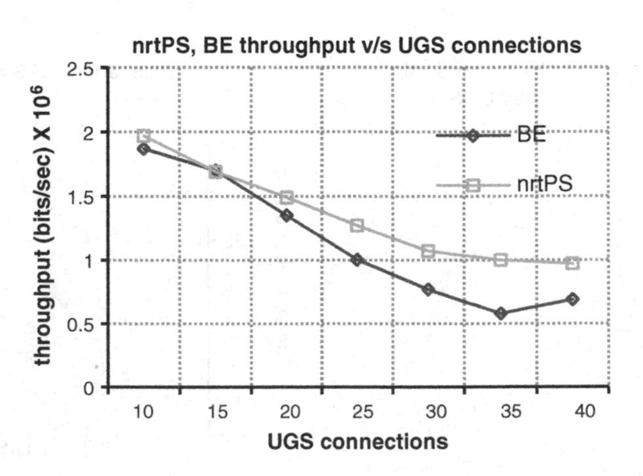

Fig. 9 nrtPS and BE throughput versus UGS

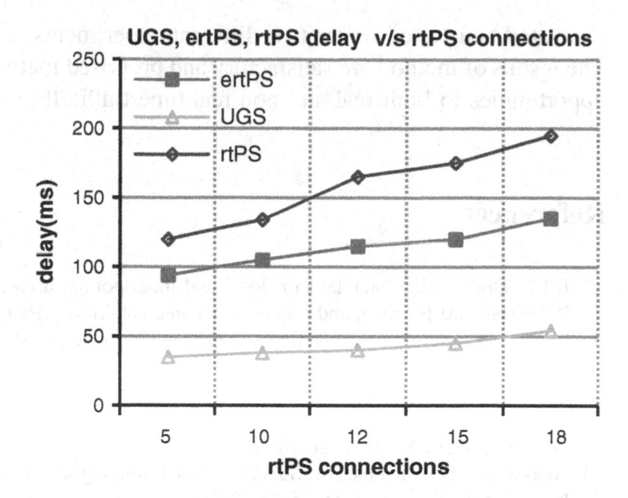

Fig. 10 UGS, ertPS, rtPS delay versus rtPS

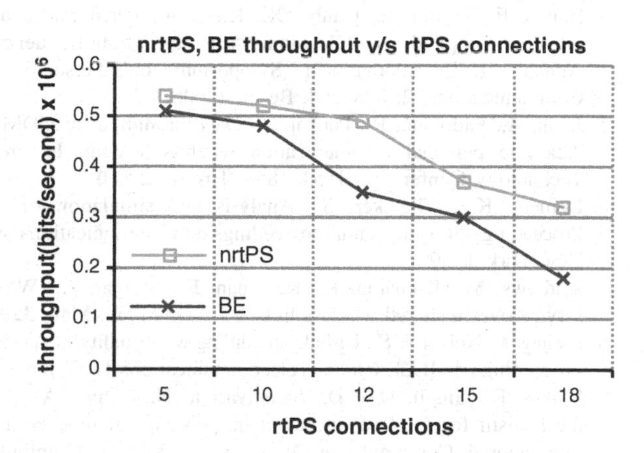

Fig. 11 nrtPS and BE throughput versus rtPS

The performance of BE classes was influenced to greater extent since BE class does not specify any quality of service requirements. Fuzzy inference system shows quick adaptive response towards changes in traffic flow for these classes and offer appropriate slots for transmission.

5 Conclusion

Proposed work in this paper utilizes fuzzy logic techniques to implement an automatic and intelligent version of WFQ algorithm. The dynamism is embedded into system by utilizing values of latency and throughput from incoming traffic to modify weights of queues in WFQ algorithm. The proposed algorithm has been simulated and tested under two different experiments for performance evaluation. The results of method are satisfactory and proposed method was able to provide fair opportunities to both real and non real time traffic flows.

References

1. IEEE, Draft.: IEEE standard for local and metropolitan area networks. 727 Corrigendum to IEEE standard for local and metropolitan area networks—Part 16: 728 Air interface for fixed broadband wireless access systems (Corrigendum to IEEE Std 729 802.16– 2004). IEEE Std P80216/Cor1/D2. 730. (2005)
2. IEEE, Draft.: IEEE standard for local and metropolitan area networks. 731 Corrigendum to IEEE standard for local and metropolitan area networks—732 Advanced air interface. IEEE P80216 m/D10, 1–1132. (2010)
3. Akashdeep, K.S., Kumar, H.: Survey of scheduling algorithms in IEEE 802.16 PMP networks. Int. Egyp. Inform. J. **15**(1), 25–36 (2014)
4. Ball, C.F., Tremil, F., Gaube, X., Klein, A.: Performance analysis of temporary removal scheduling applied to mobile wimax scenarios in tight frequency reuse, In Proceedings of 16th Annual IEEE International Symposium on Personal Indoor and Mobile Radio Communications, 2: 888–894. Berlin, (2006)
5. Jalali, A., Padovani, R., Pankaj, R.: Data throughput of CDMA-HDR a high efficiency high data rate personal communication wireless system. In Proceedings of IEEE Vehicular Technology Conference, 1854–1858. Tokyo, (2000)
6. Demers, K.S., Shenker, S.: Analysis and simulation of a fair queuing algorithm. In Proceedings of Symposium proceedings on Communications architectures & protocols, 1–12. New York, (1999)
7. Andrews, M., Kumaran, K., Ramanan, K., Stolyar, A., Whiting, P.: Providing quality of service over a shared wireless link. IEEE Commun. Mag. **32**(9), 150–154 (2001)
8. Freitag, J., Nelson L.S.: Uplink scheduling with quality of service in IEEE 802.16 networks. In Proceedings of IEEE Global Telecommunications Conference, 2503–2508 (2007)
9. Mukul, R., Singh, Das, D., Sreenivasulu, N., Vinay, A.: An adaptive bandwidth request mechanism for QoS enhancement in WiMax real time communication. In Proceedings of International Conference on Wireless and Optical Communications Networks, 102–108. Bangalore, India, (2006)
10. Pheng, Z., Guangxi, Z., Hongzhi, L., Haibin, S.: Adaptive scheduling strategy for WiMAX real-time communication. In Proceedings of International Symposium on Intelligent Signal Processing and Communication Systems, 718–721. Xiamen, (2008)
11. Hwang, J., Youngnam, H.: An adaptive traffic allocation scheduling for mobile Wimax. In Proceedings of IEEE 18th International Symposium on Personal, Indoor and Mobile Radio Communications,1–5. Athens, (2007)
12. Fathi, M., Rashidi-Bajgan, S., Khalilzadeh, A., Taheri H.: A dynamic joint scheduling and call admission control scheme for 802.16 networks. Int. J. of Telecommun. Syst. 52:195–202. DOI 10.1007/s11235-011-9555-8 http://www.springerlink.com/content/h81w80515trt0843 (2013)

A Comparative Study of Secure Hash Algorithms

Smriti Gupta, Sandeep Kumar Yadav, Alok Pratap Singh and Krishna C. Maurya

Abstract Important responsibility of every organization is to provide an adequate security and confidentiality to its electronic data system. Data should be protected during transmission or while in storage it is necessary to maintain the integrity and confidentiality of the information represented by the original data. Message digests are the condensed representation of a message after processing that message through some mathematical algorithm which is iterative in nature and one way. Integrity of message is ensured by these algorithms: any change to the message, with a very high probability, results in a different message digest. This paper explains the implementation of all available secure hash algorithms (SHA) and their performance evaluation. Performance evaluation has been done in terms of security. Randomness of all the hash algorithms have been evaluated with K-S, Chi Square and Autocorrelation tests. It is observed that all available SHAs are very useful in their area of applications because they produce totally random output even if their input is highly correlated. Hence it is found that a very high level of security attained using hash algorithms. It has been observed that SHA-512 provides a very high level of security because it has 512 bits inside which are randomly distributed but it is a bit lengthy and time taking.

Keywords Message digest (MD'S) · Secure hash algorithm (SHA) · Kolmogorov smirnov test · Chi square test

S. Gupta (✉) · S.K. Yadav
Government Mahila Engineering College, Ajmer, India
e-mail: smritigupta225@gmail.com

S.K. Yadav
e-mail: Sandeep.y9@gmail.com

A.P. Singh
I.I.T, (B.H.U), Varanasi, India
e-mail: alokpratap1@gmail.com

K.C. Maurya
NIT Kurukshetra, Kurukshetra, Haryana, India
e-mail: kcmaurya90@gmail.com

S.C. Satapathy and S. Das (eds.), *Proceedings of First International Conference on Information and Communication Technology for Intelligent Systems: Volume 2*, Smart Innovation, Systems and Technologies 51, DOI 10.1007/978-3-319-30927-9_13

1 Introduction

Cryptography is a technique for secure communication in presence of third parties or unauthorized person in wireless networks. In other words this is a protocol or scheme of the user to communicate over a channel without compromising in privacy or authentication of the transmission. Now days, modern cryptography has many applications like Security data, Voting process, Electronic payment, ATM protecting password, In Cryptography there are some specific requirements including authentication, privacy and confidentiality to be maintained of the secure data or message, integrity of the data, also the non repudiation characteristic should also be there. To achieve all these qualities there are basically three types of cryptography which are used to achieve these attributes which are Public Key Cryptography, Secure Key Cryptography and Has Function with various types which exhibits different properties.

In every scenario the unencrypted data which is not yet encrypted, the original data initially is called as plaintext. It is encrypted into cipher text, which will in the last be again decrypted into usable plaintext. Various types of Cryptographic schemes are present that have different weakness and strength, typically they are classified into two parts-

(1) Those are strong, but slow to run.
(2) Those that are weak but quick to run but less secure. SKC uses only one key for both the encryption and decryption process. This Cryptography is the example of traditional form of Cryptography. One key is used for encryption and other key is used for decryption in PKC that means two different keys are used one for encryption other for decryption.

Cryptographic Hash Functions uses a mathematical transformation to irreversibly encrypt information having some data of an arbitrary length (and possibly a key or password) and generate a fixed length hash based on this input and practically impossible to invert i.e. to regenerate the input data from its hash value alone.

This one way hash function has been called "the work horses of modern Cryptography". The input data is called message and hash value is often called message digest. It has no key for both the encryption and decryption process.

This paper is organized as follows:-

Section 2 describes different type of SHA on the basis of their parameters such as number of bits in input message block, number of iterations required, number of constants required.

In Sect. 4 various tests are performed to check the independence of output on input, results obtained are used to compare these algorithms are discussed in section second.

In Sect. 4 the simulation results are explained which are obtained with various tests

In Sect. 5 Conclusion is reported.

For the simulation of the described work, laptop with core-i5 with 64-bit microprocessor at 2.4 GHz, consisting of 4 GB RAM is used as machine, while the MATLAB 7.8 launched in February 2009, as a 64-bit software is employed. For the compilation of the report, Microsoft Office 2007 is used with their tools like Equation Editor, Visio and Picture Manager.

2 Sha

SHA-1 is the mostly and widespread used algorithm among all the available SHA algorithms and is been used in different applications and protocols. In 2005, some security related minor flaws in and mathematical weakness were identified in SHA-1, which arises the need of more powerful and strong Hash (SHA). SHA 2 is developed and introduced keeping the need of a more secure and strong hash function which is algorithmically same as SHA 1, But no successful reports of any security related attacks is reported till now, and so various efforts are taken to develop a improved and strong version.

A new hash function standard, SHA-3, is under development currently—NIST hash function competition was organized in 2012 with a winning function.

SHA-1 has only one algorithm SHA-160. SHA-2 has three algorithms SHA-256, SHA-384 and SHA-512. Input block sizes are also different. Block size of SHA-1and SHA-256 is 512 bits while block size of SHA-384 is 1024 bits and SHA-512 has also block size of 1024 bits. Except SHA-256 all other hash algorithms uses 80 iterations to calculate hash value. SHA-256 uses only 64 iterations. SHA-1 and SHA-256 can generate a hash code of a message length of maximum 264 bits. SHA-384 and SHA-512 can generate a hash code of a message length of maximum 2128 bits [1].

Comparison between all four SHA's summarized in the following table in terms of their message digest size, message size, block size, word size and number of steps required to perform the operation (See Table 1).

Table 1 Comparison between all four SHA's

	SHA-1	SHA-256	SHA-384	SHA-512
Message digest size	160	256	384	512
Message size	$<2^{64}$	$<2^{64}$	$<2^{128}$	$<2^{128}$
Block size	512	512	1024	1024
Word size	32	32	64	64
No. of steps	80	64	80	80

3 Performance Parameter

All secure hash algorithms must have uncorrelated and random output as if input is correlated data or random data. Some of the randomness tests have been performed on these SHA algorithms. Also a correlation factor of output is observed by providing a highly correlated sequence. The randomness tests are as follows [2]:

3.1 *Kolmogorov Smirnov Test*

The K-S test represents the largest absolute deviation between uniformly distributed continuous CDF and empirical CDF. In this process the input data is ranked in ascending order and maximum values are observed in particular interval.

$$D^{+} = \max_{1 \le i \le N} \left\{ \frac{i}{N} - R(i) \right\}$$

$$D^{-} = \max_{1 \le i \le N} \left\{ R(i) - \frac{i-1}{N} \right\}$$

$$D = max(D^{+}, D^{-})$$

Where *R (i)* is ranked input data, *N* is total number of data values and i is the index.

The maximum value obtained is compared with critical value and if obtained value is greater than critical value then hypothesis is rejected.

3.2 *Chi-Square Test*

It uses the technique of sample statistic as shown below

$$X_0^2 = \sum_{i=1}^{n} \left(\frac{(O(i) - E(i))^2}{E(i)} \right)$$

where O (i) represents the observed number in ith class.

E (i) denotes the expected number in ith class and n is the number of classes.

The expected number is given by E (i) = N/n;

where N represents total number of observations. Output of the test or results are compared with critical value to satisfy the hypothesis.

3.3 Autocorrelation Test

Autocorrelation represents a statistical test which is used to determine that whether independent random numbers are being generated in a sequence by a random number generator. The test requires computation of autocorrelation between every m numbers starting from *i*th number. So, autocorrelation factor of the in any sequence follows few numbers in interest:

R(i),R(i + 2 m),R(i + 3 m),…,R(i + (M + 1)m). The static is given as

$$\rho_{im} = 1/(M+1)\left[\sum_{K=0}^{M} R(i+Km)R(i+(K+1)m)\right] - 0.25$$

$$\sigma_{\rho im} = \frac{\sqrt{13M+7}}{12(M+1)}$$

$$Z_o = \frac{\rho_{im}}{\sigma_{\rho im}}$$

Where m stands for the lag, the space between the numbers being tested.
i denotes the index, or the number in the sequence that we start with.
N represents the number of numbers generated in a sequence.
M denotes the largest integer such that i + (M + 1)m < N.
The Z0 is compared with critical value to satisfy null hypothesis.

4 Result and Disscussion

All the results have been simulated on a machine with core i-5 (2nd generation) 64-bit processor with CPU speed 2.4 GHz and 4 GB RAM. As a simulation tool MATLAB 7.8, 64 bit, launched in February 2009, is used.

All secure hash algorithms with sizes 160, 256, 384 and 512 have been implemented and their outputs are crosschecked from FIPS website itself. As mentioned earlier, all secure hash algorithms must result to uncorrelated and random output as if input is correlated data or random data. Some of the randomness tests have been performed to check the results that it is random or not. The simulation results after applying the tests on all the SHA's are tabulated here.

4.1 Analysis by K-S Test

The bar diagrams for test have drawn below. In Fig. 1 K-S test values have drawn versus optimum K-S test value. As for all the SHA's obtained values are less than the optimum value means all the SHA's passed this test (See Tables 2, 3 and Fig. 2).

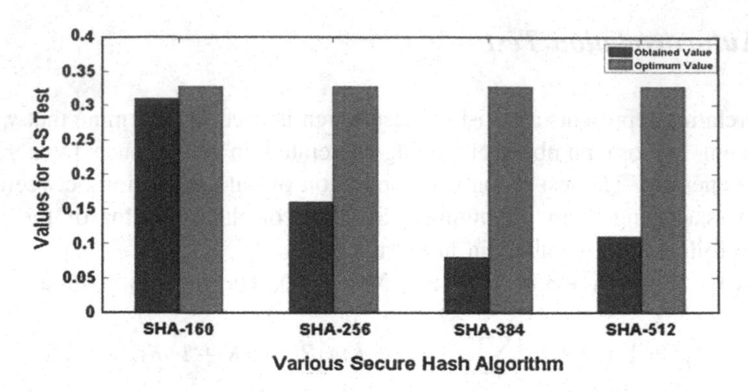

Fig. 1 Results obtained by K-S test

Table 2 Results obtained by K-S test

	Optimum value	SHA-1	SHA-256	SHA-384	SHA-512
K-S test	0.328	0.31	0.16	0.08	0.11

Table 3 Results obtained by Chi-square test

	Optimum values	SHA 1	SHA 256	SHA 384	SHA 512
Chi square	16.9	8.0	9.25	2.41	12.8

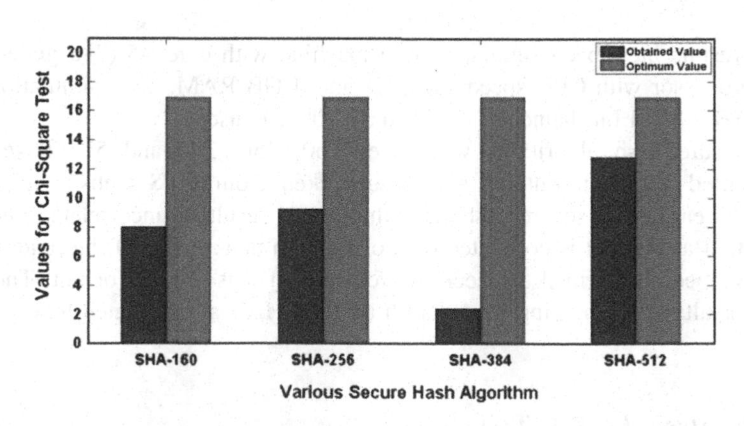

Fig. 2 Results obtained by Chi square test

4.2 Analysis by Chi Square Test

The bar diagrams for test have drawn Fig. 3 show us that Chi square test values have drawn versus optimum Chi square test value. As the obtained values of all the SHA's are less than the optimum value means all the SHA's passed this test.

4.3 Analysis by Autocorrelation Test

Figure 3 shows results of Auto correlation test and we can see SHA's passed this test too (See Table 4).

4.4 Analysis by Comparing Number of Cycles

As we can see in table below that runs of functions in SHA-1 and SHA-256 are almost same but in comparison of SHA-384 or SHA-512 are too low. Because of this reason cycles for SHA-384 and SHA-512 are has a very high value than SHA-1 and SHA-384. We can see number of cycles for each algorithm below in Fig. 4 (See Table 5).

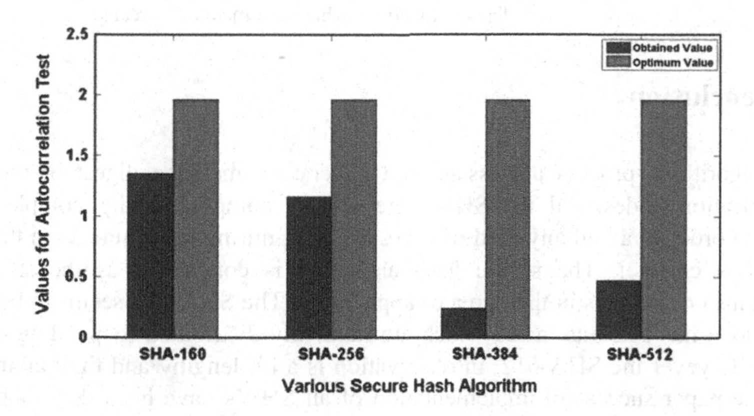

Fig. 3 Results obtained by autocorrelation test

Table 4 Results obtained by auto correlation test

	Optimum values	SHA 1	SHA 256	SHA 384	SHA 512
Auto-correlation test	1.96	1.35	1.16	0.24	0.47

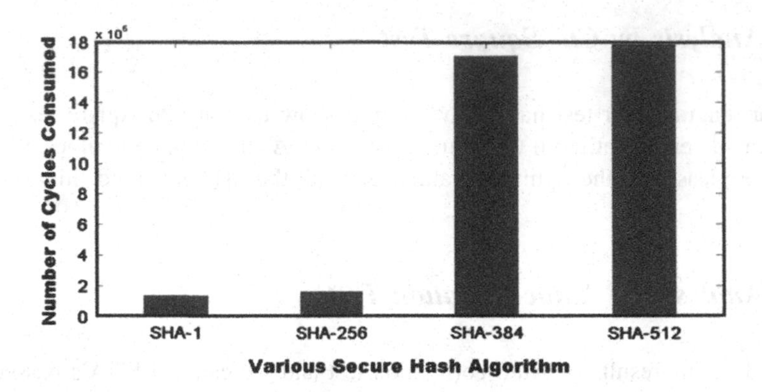

Fig. 4 Cycles consumed by various SHA's

Table 5 Runs of main functions used in algorithm

	SHA 1	SHA 256	SHA 384	SHA 512
strcat	502	502	48123	48125
num2str	502	502	48118	48125
int2str	500	500	1012	1012
bit_rotate*	224	576	736	736
bin2dec	16	16	1504	1504
dec2bin	2	2	1474	1474
main_add*	85	312	758	760
bitnot*	20	64	80	80

* denotes functions which are manually created

5 Conclusion

Hash algorithms process a message and generate a message digest or message authentication codes. All the SHA's are strong, computationally complex and secure in order to avoid any kind of eavesdropping during communication through a wireless channel. The secure hash algorithm is considered as the strongest algorithms now-a-days in their area of application. The SHA-512 seems to be most secure as it has 512 bits inside which are randomly distributed (depending on the input). However the SHA-512, in observation is a bit lengthy and time taking.

In the paper successful implementation of all SHA's have been done and analyzed. The basis of analysis completely depends on level of security which was measured in terms of randomness of the output of particular algorithm. After seeing the results we can say that all the SHA's produce totally random results as they passed all the randomness tests as mentioned above.

References

1. FIPS 180–3.: Secure Hash Standards, NIST, Gaithersburg, (2008)
2. E-mail from William E. Burr, Manager, Cryptographic Technology Group, NIST, 9 December 2010

Enhancing Data Security Using AES Encryption Algorithm in Cloud Computing

Snehal Rajput, J.S. Dhobi and Lata J. Gadhavi

Abstract Cloud Computing offers computation as per utility of the customer and hence it is known as utility computing. This model is attractive mainly for business oriented people because it reduces total cost of operation, maintainance cost, increases return of investment. But the only thing that is impeding popularity of cloud computing is security issues. This paper discuss about AES encryption algorithm (RIJNDAEL) that secure data stored on cloud. This method is more efficient than DES which is symmetry based algorithm and offers 56 bits key size whereas RIJNDAEL algorithm is asymmetry and offers 128 bits key and 128 bits blocks. RIJNDAEL is block cipher algorithm and is secure against cryptanalytic attacks. It is versatile, which means it can be implemented on different working environment efficiently, its key agility is good which means setup time of key is less and this algorithm is easy to be understood and using it. Rijndael requires less memory and hence make it is well suited for environments which have less space such as 8 bit micro-processor, also it shows marvellous performance in terms of software and hardware implementation.

Keywords AES encryption · RSA encyption · Blowfish · DES encryption · ISO · IETF · NIST · IEEE

S. Rajput (✉) · J.S. Dhobi
Government Engineering College, Gandhinagar, India
e-mail: snehalrajput89@gmail.com

J.S. Dhobi
e-mail: jsdhobi@gecg28.ac.in

L.J. Gadhavi
Saffrony Institute of Technology, Mehsana, India
e-mail: lata.gadhvi@saffrony.ac.in

S.C. Satapathy and S. Das (eds.), *Proceedings of First International Conference on Information and Communication Technology for Intelligent Systems: Volume 2*, Smart Innovation, Systems and Technologies 51, DOI 10.1007/978-3-319-30927-9_14

1 Introduction

Clouds are large pool of virtualized resources which comprises of hardware, platforms, and software. These resources are dynamically allocated among user to balance the load and to utilise resources optimally [1]. It offers remarkable features such as an illusion of infinite computing resources, pay- as-much-use model, elimination of any frontal commitment, self service interface, and elastic capacity, resources are abstracted and virtualised [2]. Cloud computing services are of three types:

Infrastructure as a Services (IaaS), Platform as a Services (PaaS), Software as a Service (SaaS) [3]. Infrastructure as a services provides various infrastructure to the user such as data server, storage, firewall, network and hardware. Various example are Amazon EC2, Simple Queue Service, Simple Storage Service, VPC Service etc. Platform as Services provides framework, platforms to develop our own applications such as Microsoft Azure, GoogleApp Engine, Force.com. Software as a services provides application as service to user such as video conferencing, office suits, social networking. Various examples are Salesforce, SQLAzure, GoogleApp etc. Cloud computing deployment models are: public cloud, private cloud, and hybrid cloud and community cloud.

Public Cloud: It is available to all end users developed by Cloud service provider. Services are charged as per usage basis.

Private Cloud: It is developed by particular Organisation for their personal use.

Hybrid Cloud: It is use to make certain organisation scalable. Here any private cloud can be extend to public cloud.

Community Cloud: Here numerous organisation combine to form their private cloud called community Cloud.

The biggest advantage of cloud is cost saving which pulled business people toward cloud computing and the biggest disadvantage of cloud is security. Cloud is booming technology because it contain more positive aspects compare to negative aspects. Some of the advantages are: User whenever needed can extend the resources as per their requirement, the maintenance cost is almost negligible, it is scalable. Cloud offers auto scaling and load balancing of workload. Cloud service provider guarantees QoS to the user through Service Legal Agreement(SLA). It is fault tolerance as replica of servers are installed at various site. Some of the challenges in cloud computing is to secure data stored on cloud, network latency, statelessness. The worrisome challenges is security of data stored on cloud, which need to be protected against various attack using some algorithm. Once stored data user himself do not know where his data is stored. Some cloud service provider uses one time authentication method i.e. username, password to authenticate user; some uses two way authentication or multiway authentication and some uses encryption techniques for server security. Using cryptographic technique, user can protect their

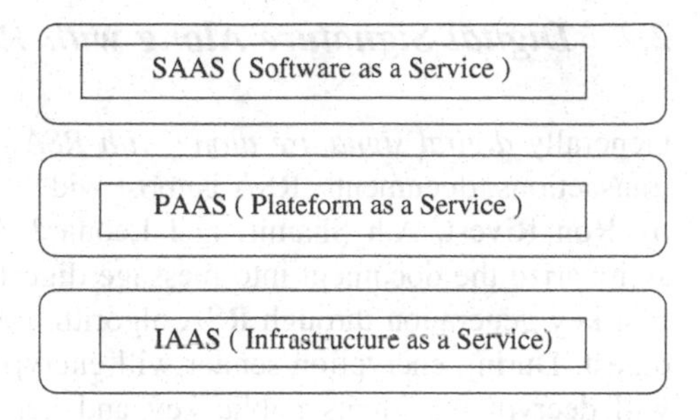

Fig. 1 Cloud deployment model

information from unauthorized access. Rijndael algorithm when implemented on cloud computing will make data stored on cloud more secure and speedy to access.

In 1997 several international organisations (ISO, IEEE and IETF) called for the implementation of AES algorithm. Later on, in 1999 NIST published five finalist: RC6, Rijndael, Serpent MARS, and Twofish. After second round Rijndael came out best among them when compare in terms on security, performance [2], cost, simplicity, versatility, key agility and hardware implementation. NIST concluded Rijndael with the following statement: Rijndael consistently showed remarkable performance both in terms of hardware and software on various computing environments. Its key setup time and key agility is excellent. As it requires very low memory, it is very much suitable for restricted space environments,without affecting its performance (Fig. 1).

2 Related Work

Cloud computing focuses on utility, and deployes computing resources on demand. Cloud provider must ensure that the all the resources provided by them i.e. infrastructure, platform, software, as well as client information is secure and customer must ensure that cloud provider has implemented proper algorithm which guarantees data security on cloud. This paper focus on securing data from threats without hindering performance. Also this paper shows the comparison of various well known algorithms.

Encryption is the technique to secure sensible data from the threats while transferring data onto cloud. Encryption technique can be asymmetry or symmetry.

In Asymmetry algorithm, different key are required during encryption and decryption whereas in symmetry algorithm same key is used during encryption and decryption time. Due to generation of different key at encryption and decryption time, asymmetric type of algorithm are much slower in compare to symmetric type of algorithms. Again, symmetric algorithm can be of block cipher or stream cipher.

In block cipher, ciphering of data occur blockwise while in stream cipher, ciphering of data occur bit wise.

2.1 Digital Signature Along with RSA Encryption

Generally *digital signature along with RSA encryption* is used to secure financial transactions, documents. RSA is most widely used asymmetric algorithm proposed by Ron Rivest, Adi Shamir, and Leonard Adleman. Here digital signature will summarize the document into message digest. This algorithm goes into two steps, first key generation through RSA algorithm and in second step preparing message digest. During encryption sender will encrypt it using its private key and receiver will decrypt it with its public key and get the message digest and again using hashing function plaintext can be generated. The main drawback of RSA algorithm is that, it is too slow.

2.2 DES Algorithm

It is a symmetry key Block cipher algorithm. At encryption site, it use 56 key size with plaintext of 64 bit and generates 64 bit ciphertext, at the decryption side it uses 64 bit ciphertext and generate 64 bit plaintext using same 56 bit of key size. The encryption process consists of two permutation viz. initial permutation and final permutation and 16 Fiestel cipher round. Here while encryption each block is XORed with previous block. The first block XORed with 64 bit vector called as *Initialisation vector*. This method is not viable now-a-days and is not secure due to 56 bit key size. It uses iterative round key having 16 round but since the same round transformation is used in each and every round, we can conclude that the DES algorithm has only one round transformation, which is easy to crack. Hence we can say that DES algorithm is not secure [4].

2.3 Single Sign-on Algorithm

Single sign-on algorithm provides a single interface to the users to access a group of software or resources. Here SSL or SAML can be used [5].

(i) Password Manager Agent (PMA): is installed on browsers of client as an extension. The main function is to communicating with the single interface onto cloud server for simultaneously sign on various SaaS.

(ii) Single Sign On SaaS Application (SSOSA): It is an application that manages usernames and passwords of users, encrypts those data and stores into PCS (password cloud server), stores keys into key cloud server, decrypts usernames and passwords, connects to various cloud computing application and serve request of users.

(iii) SSL: Secure socket layer to data transmission between PMA and SSOSA

(iv) Advanced Encryption Standard (AES) is used by SSOSA for encrypting information and storing it onto password cloud server.

2.4 *Elliptical Curve Cryptography*

It is a public key cryptography Algorithm, first proposed by Neal Koblitz and Victor Miller. Here each user has its own pair of private and public key. In any plane, an Elliptic curve on a field say 'f' consists of set of points (Xi, Yi).

The Standard equation of ECC is given by

$Y^2 = x^3 + ax + b$ where a and b are parameter [6].

The crux of ECC Algorithm is the of computation of new points on the curve and then its encryption to be share among users as information. Group Operator is used to find P which is one of the point on the curve. The computation goes as P+P, P+P+P, …………This method can be use for authentication of users using key agreements.

Public key of (Y_a) User A: $X_a + P$ (where X_a is private key of A and P is some point) $X_a = k1$

Public key of (Y_b) User B: $X_b + P$ (where $X_b = k2$ is private key of B and P is some point). Now both will exchange their keys. A calculates the session key by $K_a = X_a \text{ x } Y_b = k1 \text{ x } k2 \text{ x } P$.

B calculates the session key by $K_b = X_b \text{ x } Y_a = K2 \text{ x } kl \text{ x } P$ which means both KA and KB are same.

Other numeral encryption techniques are double DES, triple DES, MD, MD-5, blowfish, SHA, Elliptic Curve Cryptography (ECC) system.

3 Proposed Algorithm

AES algorithm can be classified into two type: First,128-bit plain text block paired with 128-bits key block and secondly,128-bit plain text block paired with 256-bit key block. Mostly 128-bit plain text block paired with 128-bits key block are widely used we will examine such case; The minimum number of round and the maximum round are 10, 14 respectively. This algorithm make uses of key alternative block cipher. Rijndael is similar to DES but only thing differs is that latter involves only bits in operation against entire block. The only difference between AES and Rijndael is that AES uses fix 128 bit block sizes, and the key lengths differs to 128, 192 or 256 bits whereas in Rijndael algorithm the block size in bits and the key size can be any multiple of 32 bits, with a minimum length of 128 bits and a maximum length of 256 bits.

3.1 Procedure [7]

The round transformation consists of four different transformations. The C code is as follow:

```
Round(State,RoundKey)
{
ByteSub(State);
ShiftRow(State);
MixColumn(State);
AddRoundKey(State,RoundKey);
}
```

The final round differs slightly as,

```
FinalRound (State,RoundKey)
{
ByteSub(State);
ShiftRow(State);
AddRoundKey(State,RoundKey); }
```

In Rijndael algorithm, *State* is defined as any intermediate round result. The State can be defined as a array of bytes which have four rows and number of columns (Nc) is equal to the (block length divide by 32). The Cipher Key is similarly defined as a array consists of four rows and number of columns of the Cipher Key (Nk) is equal to the (key length divide by 32).

The *ByteSub Transformation* functionoperates on each state independently and is a non-linear byte substitution where multiplicative inverse is applied first and affine function is done on it. Here each state is mapped to S-box and required array is obtained.

The *ShiftRow transformation* function shifts each row by some offset. First row is not shifted, second row is shifted by s1 byte and third row by s2 byte…

For example:

State array	New state array
W 1 Q 4	W 1 Q 4
A B C D	B C D A
0 P Q R	Q R 0 P

The MixColumn transformation: here state as considered as polynomial and multiplied with certain fix polynomial. Let C(x) = B(x) XOR A(x), where B(x) if fix matrix (Table 1).

Table 1 Time required in an AMD K7-700 processor (Per round, Using 100 k rounds) [10]

	DES (64, 64)	DES (64, 128)	Rijndael (128, 128)
Ciphering	3.4 μs	6.9 μs	35.8 μs
De-ciphering	3.5 μs	7.0 μs	36.0 μs

$$\begin{pmatrix} C0 \\ C1 \\ C2 \\ C3 \end{pmatrix} = \begin{pmatrix} 1 & 2 & 3 & 4 \\ 2 & 1 & 4 & 3 \\ 3 & 4 & 2 & 1 \\ 4 & 3 & 1 & 2 \end{pmatrix} \text{XOR} \begin{pmatrix} A1 \\ A2 \\ A3 \\ A4 \end{pmatrix}$$

The Round Key addition: Here XOR operation is performed between state and key blocks.

4 Limitation

Speed is less due to s-box generation [8]. Also there is few limitations of the ciphering technique with its inverse such as:

- When comparing Encryption with the decryption. Decryption is not implemented much on a smart cards because it requires more code and more CPU cycles for execution (Still it is faster than other algorithms.)
- While implementing this algorithm in software, the encryption and its decryption implementation codes are different.
- While implementing this algorithm on hardware, the decryption technique partially re-use the encryption circuitry, hence this will add further cost (Table 2).

Table 2 Time required in 8051 Micro-controller (Per round, Using 100 rounds) [10]

	DES (64, 64)	DES (64, 128)	Rijndael (128, 128)
Ciphering	2.8 ms	6.1 ms	28.8 ms
De-ciphering	2.7 ms	6.0 ms	28.0 ms

5 Comparison Between AES, DES and RSA [9]

Factors	AES (Rinjdael)	DES	RSA
Key size	128, 192, 256 bits	56 bits	1024–4096 bits
Block size	Any multiple of 32 bits	64 bits	512 bits or larger
Ciphering and deciphering key	Same key	Same key	Same key
Encryption and decryption	Faster	Moderate	Slower
Power consumption	Less	Less	High
Security	Excellent	Moderately secure	Least secure
Hardware and software algorithm	Faster compare to DES, RSA	Faster	Slow
Attacks vulnerabilities	Brute force attack	Brute force attack, Linear and differential cryptanalysis attack	Brute force attack
No. of rounds	10 or 12 or 14	16 Rounds	1 Rounds

6 Conclusion

We have seen numerous ciphering algorithm and have compared Rijndael algorithm performance with them and have found that Rijndael algorithm can be implemented in small space environment with remarkable performance, also it is secure against Numerous attack.

References

1. IBMCloudArchitectureforCrete052011b.pdf
2. The Rijndael's Algorithm. http://www.esat.kuleuven.ac.be/~rijmen/rijndael/
3. A secure data access control method using AES for P2P storage cloud. IEEE Sponsored 2nd International Conference on Innovations in Information, Embedded and Communication Systems (ICJJECS) 2015
4. http://csrc.nist.gov/archive/aes/rijndael/Rijndael-ammended.pdf
5. Shimbre, N., Deshpande, P.: Enhancing distributed data storage security for cloud computing using TPA and AES algorithm
6. Daernen, J., Rijrnen, V.: The design of Rijndael algorithm

7. Biham, E.: A note on comparing the AES candidates. In: Proceedings of the 2nd AES Candidate Conference, Rome, pp. 85–92, 22–23 March 1999
8. Web & security. Glob. J. Comput. Sci. Technol. Netw. Online ISSN: 0975–4172 & Print ISSN: 0975-4350
9. Penchalaiah, N. et al.: Int. J. Comput. Sci. Eng. (IJCSE) **02**(05), 1641–1645 (2010)
10. Daemen, J., Rijmen, V.: AES submission document on Rijndael. Version 2, http://csrc.nist.gov/CryptoToolkit/aes/rijndael/Rijndael.pdf. Sept 1999

Investigating the Performance of Energy Efficient Routing Protocol for MANET Under Pareto Traffic

Dhiraj Nitnaware

Abstract Mobile Ad Hoc Network (MANET) is a multi-hop, infrastructureless wireless network with limited bandwidth and battery power. The investigation of energy efficient protocols under Pareto traffic is being carried out in this paper. The routing protocols taken for analysis are ECG_AODV [1, 2], ECNC_AODV [3], EBG_AODV [4, 5] and Energy plus Node Cache plus Gossip (E+NC+G) and compare it with AODV. The behavior of all these algorithms under CBR traffic is already studied in [6]. Here we have focused on Stochastic (Pareto) traffic source. Based on the simulation results, we observed that there is reduction in energy and overhead up to 10–30 % with 2–12 % deprivation in the delivery ratio for all protocol as compared to AODV.

Keywords Pareto traffic · Energy consumption · Energy efficient protocol · Routing overhead · NS-2 simulator

1 Introduction

MANET is a network that is deployed anywhere without any infrastructure. MANET reactive protocols such as AODV practices flooding method during route discovery process which increases number of RREQ (route request) and RERR (route error) packets. That results in energy consumption as well as routing overhead incremental of each node.

Previously, we have proposed various energy efficient routing protocols. They are *Energy Constraint Gossip* (ECG_AODV) protocol where intermediate nodes send RREQ packets with possibility k that depend on the energy eminence of the each node [1, 2]. In [3], we have proposed *Energy Constraint Node Cache* (ECNC_AODV) protocol where the intermediate nodes forward the control packets

D. Nitnaware (✉)
Electronics and Telecommunication Department, Institute of Engineering and Technology, DAVV, Indore, Madhya Pradesh, India
e-mail: dhiirajnitnawwre@gmail.com

S.C. Satapathy and S. Das (eds.), *Proceedings of First International Conference on Information and Communication Technology for Intelligent Systems: Volume 2*, Smart Innovation, Systems and Technologies 51, DOI 10.1007/978-3-319-30927-9_15

that depend on *node cache* and energy eminence. Intermediate nodes will forward control packet with some probability that depends on the existing energy status is proposed Energy Based Gossip (EBG_AODV) [4, 5].

In the related work, Haas et al. have proposed Gossip concept to increase the MANET performance [7]. There selected parameters are packet delivery fraction, average end to end delay and normalized routing overhead. There is unnecessary increase in control packets for on-demand protocols like AODV [8] and DSR [9] due to use of flooding strategy. Energy constraint routing algorithm is proposed by Frikha et al. where the control packets are transmitted depending upon the energy status of the node [10].

In this paper, we have analysis of all these energy efficient algorithms under Pareto traffic. Energy consumption, routing overhead and throughput against speed, nodes, area size and packet transfer rate are considered as performance and variable parameters respectively.

2 MANET Routing Protocol

2.1 Working of AODV

AODV protocol falls under reactive category that finds route towards destination as demanded [8]. Every AODV node is having routing table that gives the details of sequence number and next hops. Hello packet helps in maintaining the node connectivity. It consists of two methods: namely Route Discovery and Route Maintenance. The path between source and destination node is established by sending RREQ and RREP packet in route discovery process. While error in the link is notified to the source using route error (RERR) message. The major drawback of AODV is use of flooding method during route discovery that results in excessive control packet which in turn have energy consumption and routing overhead incremental of each node.

2.2 ECNC_AODV

Node Cache is that node that has previously taken part in data transfer. An RREQ packet is send by the intermediate node if it is *node cache*. The working at intermediate node is explained below:

- Intermediate Routine: *n*th node will advance RREQ packet if current energy (E_n) is more than the set energy threshold (E_{th}) and transmitted time of data packet *T (N)* is more than T-τ. Else drop the packet. Where current time is T and τ is small set time threshold to decide the node's memory that is 30 s in this method.

2.3 ECG_AODV

In all the proposed protocol, different methods are used instead of flooding during route discovery process. Here, the intermediate nodes that are having energy more than the set threshold will be taken in the route discovery process which helps to reduce the routing packets [1, 2]. The steps involved are:

- **Source Routine**: After generating RREQ packet, the source node broadcast to its neighbors.
- **Intermediate Routine**: *n*th node will send request packet with *k* probability if current energy (E_n) is more than the set energy threshold (E_{th}), provided that its neighbor is more than one else with probability 1. Node will drop the packet if this condition is not satisfied.
- **Destination Routine**: Reply packet will be send by the destination node after receiving Request packet.

2.4 EBG_AODV

The request packet will be forwarded by intermediate nodes with *k* probability that depends on the status of its current energy [4, 5]. The working used in this protocol is as follows:

- **Intermediate Routine**: If the residual energy of intermediate node is 80 % of the original energy than the value of *k* will be 0.8, if 75 % then *k* will be 0.75 and so on. This scheme is used in forwarding the request packet.

2.5 Energy Plus Node Cache Plus Gossip Based AODV (E+NC+G)

The working of this protocol is as follows:

- **At intermediate node**: If conditions $E_n > E_{th}$, and $T\text{-}\tau \leq T(N)$ are satisfied than only node will send the RREQ packet with *k* probability else packet will be drop by the node.

3 Simulation Environment

There are 50 nodes which are randomly distributed in a region of 1000 m × 1000 m with 30 numbers of source-destination pair are taken for simulation. Each node has 200 J of primary energy with 250 m of communication range and data rate of

Table 1 Traffic model parameters

Parameter	Value
Nodes	50
Area size	1000 m × 1000 m
Communication range	250 m
Pause time	100 s
Speed	5 m/s
Probability (k)	0.75
Transmission rate	64 packets/s
Time threshold (τ)	30 s
Packet size	512 bytes
Idle time	1.0 s
Burst time	2.5 s
Shape	2.5
Simulation time	900 s

2 Mbps. Power consumed during transmission is 1.65 W while during reception it is 1.1 W. Stochastic (Pareto) traffic is used as traffic model that have various values given in Table 1. NS-2 version 2.31 is used for simulation [11] and script cbreng.tcl [12] for traffic model generation.

Stochastic (Pareto) traffic can be generated using three fields which are given below.

(a) ***Burst Time***: It represent the ON time of the generator during which the packets are generated.
(b) ***Idle Time***: It denotes the OFF time of the generator where no packet is generated.
(c) ***Traffic Shape***: It signifies the shape of the traffic with Poisson's distribution.

4 Simulation Results

For investigation of the above protocols are Energy consumption, overhead and Delivery ratio is taken as performance parameter against varying speed, nodes, area size and transmission rate.

4.1 Energy Consumed by All Nodes

The total energy consumed in Joules is shown in Fig. 1a–d by varying speed, nodes, grid area size and packet transmission rate. We observed that E+NC+G consume 30 % less energy as compared to other. ECNC_AODV consumes second least energy of 25 % under Pareto traffic as compared to AODV.

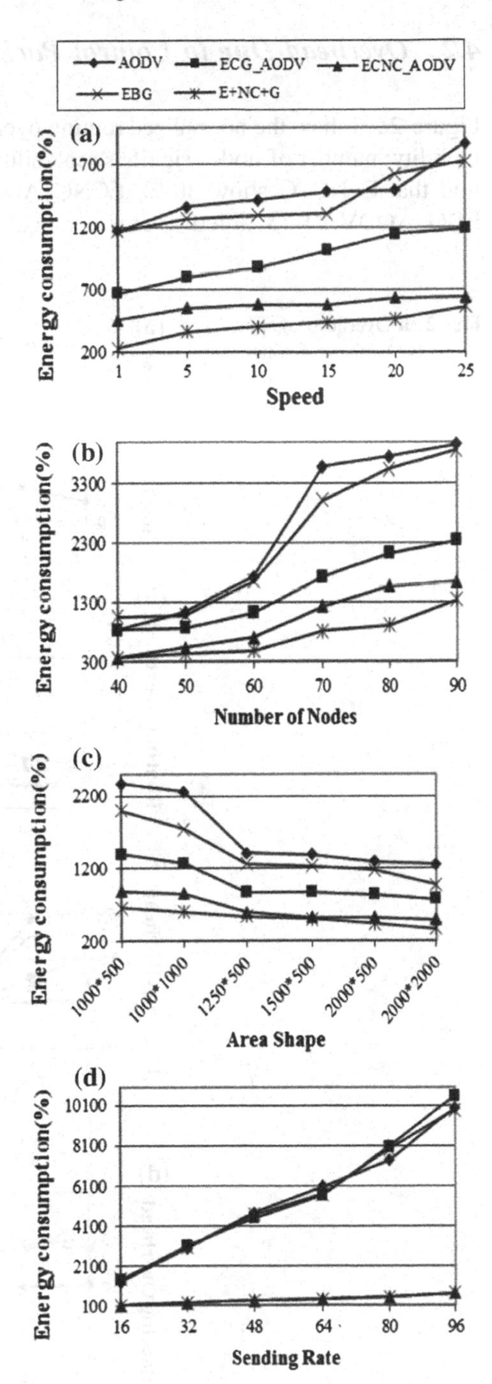

Fig. 1 **a** Energy consumption against speed **b** Energy consumption against nodes **c** Energy consumption against area size **d** Energy consumption against sending rate

4.2 Overhead Due to Control Packet

Figure 2a–d show the normalized routing overhead in terms of speed which signifies mobility; number of nodes signifies scalability, area grid and sending rate. Here we find that E+NC+G show 30 %, ECNC_AODV and ECG_AODV 20–25 %, and EBG_AODV 10–15 % less overhead reduction as compared to AODV.

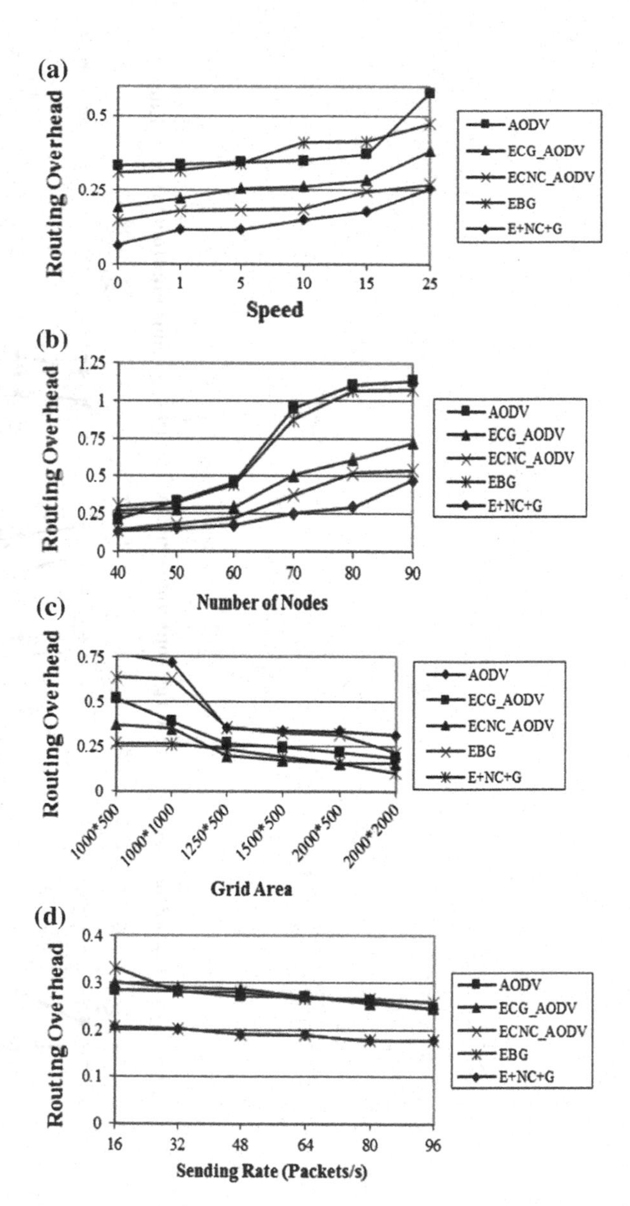

Fig. 2 **a** Overhead against speed **b** Overhead against nodes **c** Overhead against area size **d** Overhead against sending rate

4.3 Delivery Ratio

Figure 3a–d gives the throughput or delivery ratio against speed, nodes, area size and transmission rate. The throughput of E+NC+G is 10–12 % less, ECNC_AODV and ECG_AODV is 8–10 % less while EBG_AODV is having same 1–2 % less throughput as compared to AODV.

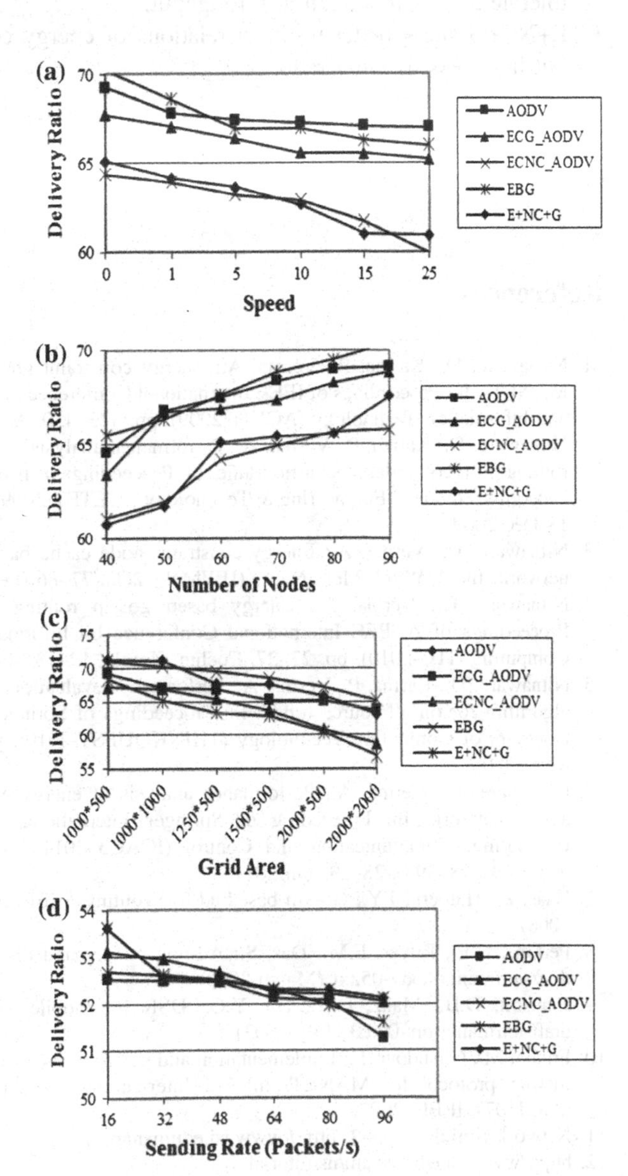

Fig. 3 **a** Delivery ratio against speed **b** Delivery ratio against nodes **c** Delivery ratio against area size **d** Delivery ratio against sending rate

5 Conclusion Remark

Following conclusion is drawn based on the simulation results:

- Overall performance of EBG protocol is better in terms of high mobility, network size and scalability as its throughput is only 1–2 %.
- ECNC_AODV and ECG_AODV protocol can be preferred if network can tolerate 5–8 % reduction in throughput.
- E+NC+G show better result in relations of energy consumption and overhead but have less delivery ratio.

Finding the optimum value τ, E_{th} and k can increase the throughput of the above protocol. Thus in future, we want to find the optimum value as well the end to end delay of various protocols.

References

1. Nitnaware, D., Karma, P., Verma, A.: Energy constraint gossip based routing protocol for MANETs. In: Proceedings of IEEE International Conference on Advances in Computer Vision and Information Technology (ACVIT-2009), pp. 423–430, Aurangabad, 16–19 Dec 2009
2. Nitnaware, D., Karma, P., Verma, A.: Performance analysis of energy constraint gossip based routing protocol under stochastic traffic. In: Proceedings of IEEE International Conference on Emerging Trends in Engineering & Technology (ICETET-2009), pp. 1110–1114, Nagpur, 16–18 Dec 2009
3. Nitnaware, D., Verma, A.: Energy constraint node cache based routing protocol for adhoc network. Int. J. Wirel. Mob. Netw. (IJWMN), **2**(1), 77–86 (February 2010)
4. Nitnaware, D., Verma, A.: Energy based gossip routing algorithm for MANETs. In: Proceedings of ACEEE International Conference on Information, Telecommunication and Computing (ITC-2010), pp. 23–27, Cochin, Kerala, 12–13 Mar 2010
5. Nitnaware, D., Karma, P., Verma, A.: Performance evaluation of energy based gossip routing algorithm for on-off source traffic. In: Proceedings of Springer International Conference on Contours of Computing Technology (THINKQUEST-2010), pp. 327–331, Mumbai, 13–14 Mar 2010
6. Nitnaware, D., Verma, A.: Performance analysis of energy efficient routing algorithms for adhoc network. In: Proceeding of Springer International Conference on Advances in Computing, Communication and Control (ICAC3–2011), pp. 222–230, Mumbai, ISBN: 978-3-642-18439-0, 28–29, Jan 2011
7. Haas, Z., Halpern, J.Y.: Gossip based ad hoc routing. IEEE Trans. Networking, **14**(3) (June 2006)
8. Perkins, C.E., Royer, E.M., Das, S.: Ad-hoc on demand distance vector routing (AODV). draft-ietfmanet-aodv-05.txt (March 2000)
9. Johnson, D.B., Maltz, D.A., Hu, Y.C.: DSR for mobile ad hoc network, internet-draft. draft-ietfmanet-drs-09.txt (July 2003)
10. Frikha, M., Ghandour, F.: Implementation and performance evaluation of an energy constraint routing protocol for MANET. In: 3rd International Conference of Telecommunications (AICT'07), IEEE, 2007
11. Network Simulator, ns-2. http://www.isi.edu/nsnam/-ns/
12. http://www.isi.edu/nsnam/ns/tutorial/

Implementing and Evaluating Collaborative Filtering (CF) Using Clustering

Sachin S. Agrawal and Ganjendra R. Bamnote

Abstract A tremendous increase has taken place in the amount of online content. As a result, by using traditional approaches, service-relevant data becomes too big to be effectively processed. In order to solve this problem, an approach called clustering based collaborative filtering (CF) is proposed in this paper. Its objective is to recommend services collaboratively in the same clusters. It is a very successful approach in such settings where interaction can be done between data analysis and querying. However the large systems which have large data and users, the collaboration are many times delayed due to unrealistic runtimes. The proposed approach works in two stages. First, the services which are available are divided into small clusters for processing and then collaborative filtering algorithm is used in second stage on one of the clusters. It is estimated to decrease the online execution time of collaborative filtering algorithm because the number of the services in a cluster is much less than the entire services available on the web.

Keywords Collaborative filtering · Clustering · Big data

1 Introduction

A large number of applications and websites are built on Internet which target on data and user interaction. It is assumed that the users of these systems will be introduced to the new content by recommendations given by their friends and can also submit feedback to make these recommendations better. Data collection has grown extremely and is in front of the ability of frequently used tools to capture, manage, and process by software's [1]. For large data sets, the term known as Big

S.S. Agrawal
College of Engineering and Technology, Akola, India
e-mail: sachin.s.agrawal@gmail.com

G.R. Bamnote (✉)
Prof Ram Meghe Institute of Technology and Research, Badnera, Amravati, India
e-mail: grbamnote@rediffmail.com

S.C. Satapathy and S. Das (eds.), *Proceedings of First International Conference on Information and Communication Technology for Intelligent Systems: Volume 2*,
Smart Innovation, Systems and Technologies 51, DOI 10.1007/978-3-319-30927-9_16

data is used. These data sets are very large or difficult that old data processing applications are insufficient. The main challenges that are accompanied with such data sets are storage, data creation, analysis, sharing, capture, search, transfer, information privacy and visualization. Thus to find the very large volumes of data and find useful information for future actions is the main challenge for Big Data applications [2]. These demands are met by using user data or preferences and algorithms to advise novel items that will help them in decision making process through the techniques called Recommender systems (RS).

A RS has many components like items, preferences, users, neighborhoods and ratings. Collaborative filtering (CF) such as item and user-based methods are the major techniques applied in RSs [3]. The objects or things that are recommended to a user are called items. Items, for example can be news articles, reviews, games, movies or songs. The characterization of these items can be by their specific metadata which comprise of relevant titles, tags or keywords. Users are the people who are being recommended. Many times they require guidance or assistance in preferring an item in an application [4].

The large number of services that need to processed in real time need to be decreased. The techniques which can decrease the volume of data by a huge factor by grouping similar services together are known as clustering. Thus, we propose a collaborative filtering method by using clustering. The proposed approach consists of two stages, i.e., clustering and collaborative filtering. The first step i.e., clustering, is a preprocessing step to separate big data into usable parts [5]. A cluster which contains some similar services like a movie contains some like-minded actors. The computation time of CF algorithm can be reduced significantly because in a cluster the total number of services is less than the overall number of services. Also the ratings of similar services within a cluster are more related than that of dissimilar services [6], thus the recommendation accuracy based on users ratings can be enhanced. Clustering technique is a promising way to improve the scalability of collaborative filtering by reducing the quest for neighborhoods between clusters instead of using complete data set. It recommends accurate and better recommendations to users. So for each specific user a user model can be made for enhanced recommendations. These user model works as profiles where actions and preferences are encoded. It also represents the history of a user along with their interactions with items in the recommender system and such interactions are called as preferences. Many a times preferences are classified as ratings if a recommender system gives a media to rate items [7]. The user view of an item in a recommender system can be understood as preferences and it can be both implicit/explicit. A group of similar users will be represented by a neighborhood that reports users and their preferences [8]. Classification of Collaborative filtering can be done in two classes. These classes are model-based methods in which the total user-item rating dataset is used to make predictions and neighborhood-based (memory based) methods.

2 Basic Knowledge

The metadata from the web service description language files was investigated by Liu [9] to compute the similarity amongst web services and defined a web service as =(*P*, *Q*, *R*, *S*), where *P* is the web service name, *Q* is the group of messages exchanged, *R* is the group of data type, and *S* is the group of operations provided by the web service. Li [10] defined a web service for evaluating reputation of service as *WS*(*A*, *R*, *B*, *C*, *D*, *E*) where *A* is identity, *B* is classification, *R* is description of text, *C* is transaction volume, *E* is degree of reputation and *D* is review group. A service proposed by Zielinnsk [11] can be defined as a conceptual specification of information technology functions that are business-aligned. While the definitions of service are different and application-specific, they have common elements which essentially include service descriptions and service functionalities. Another important user activity that reflects their opinions on services is rating. Service rating is an important element mainly in application of service recommendation. As a number of services are emerging on the Internet, such large volume of service-related elements are generated and distributed across the network, which cannot be effectively accessed by common database management system. To tackle this problem, Hadoop is used to store services. It uses a disseminated user level file system across the cluster to handle storage sources. Hadoop [12] is an open source achievement of Map-Reduce technique for large datasets analysis. A file system called HDFS (Hadoop distributed file system) is used in Hadoop. The HDFS provides the foundation in storing files in a storage node. Task trackers and job trackers are provided by Mapreduce. It has been successfully used by Google to process big datasets. A distributed file system is needed by Mapreduce and also an engine which monitors, coordinate, collect and distribute the consequences. HDFS contains nodes and name node and is a master and slaver framework. The name node manages namespace in the file system and is a center server and data node handles the data stored [13].

2.1 Architecture of HDFS

On the compute nodes, the HDFS stores data and also provides large average bandwidth amongst the cluster. The mechanism consists of master and slave nodes. The master node consists of single name node and the slave node consists of number of data nodes. In the cluster, the nodes are distributed one data node/machine, whose purpose is to handle data block attached to the machines. The operations on file system namespace are executed by the name node. It also maps data blocks to data nodes. For serving write and read requests data node are responsible from client and execute block procedures upon instructions [14]. For higher performance, resiliency and load-balancing, HDFS divide data into large pieces and on the server it makes multiple copies. By means of data copies at

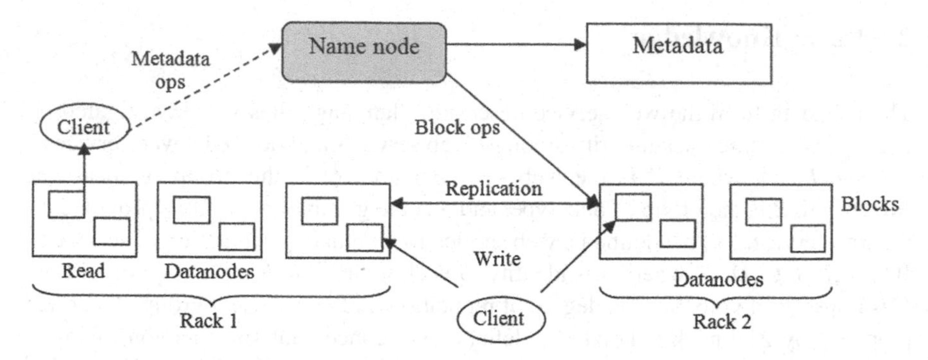

Fig. 1 HDFS architecture

multiple servers, whichever server at any time can access in computation, writing and reading of a data. Figure 1 shows the HDFS architecture. All decisions concerning block replication are made by name node. For a large cluster it cannot be realistic to join the nodes in flat topology.

The general method is dividing the nodes in many racks. The switches in a rack are joined by core switches and a switch is shared by a rack share. Nodes in multiple racks undergo number of switches.

2.2 *Mapreduce*

Its main objective is to give facilities which permit execution and development of jobs which are processed form big scale data. Its goal is to use the processing capacity in a much more efficient way which is given by processing cluster. Simultaneously it gives a logical model which makes the advancement of distributed applications simpler. It is made in such a way that it is flexible to machine crash failures. Map-reduce is used by Google to handle large data sets whose size can exceed in terabytes. It is motivated to achieve this through the thought of functions which are of high order.

3 Deployment of Clustering and Collaborative Filtering

The proposed approach works in two stages. First, the services which are available are divided into small clusters for processing and collaborative filtering algorithm is used in second stage on one of the clusters.

3.1 Calculate Description and Functionality Similarity

By Jaccard similarity coefficient, similarities are calculated which measure similarity amongst sample sets statistically. The coefficient of similarity can be defined as the ratio of the cardinality of their conjunction to the cardinality of the union for 2 sets. Mathematically, similarity of description between A and B is computed by

$$D\_Similarity(A,B) = \frac{|D'_A \cap D'_B|}{|D'_A \cup D'_B|} \quad (1)$$

From the above formula, the larger $|D'_A \cap D'_B|$ is, the more similar the two services are. Dividing by $|D'_A \cup D'_B|$ is the scaling factor which ensures that description similarity is between 0 and 1. Mathematically, functionality similarity between A and B is computed by

$$F\_Similarity(A,B) = \frac{|F'_A \cap F'_B|}{|F'_A \cup F'_B|} \quad (2)$$

3.2 Calculate Characteristic Similarity

The characteristic similarity between A and B is computed by weighted sum of description similarity and functionality similarity. Mathematically,

$$C\_Similarity(A,B) = \alpha \times D\_Similarity(A,B) + \beta \times F\_Similarity(A,B) \quad (3)$$

In above formula,$\alpha \in [0, 1]$ is the weight of description similarity, $\beta \in [0, 1]$ is the weight of functionality similarity and $\alpha + \beta = 1$. The relative importance between these two can be expressed by the weights.

3.3 Cluster Services

Clustering techniques are widely used in RS to divide object sets into clusters. The objects in the identical cluster are similar to each other than objects in diverse clusters. Usually, algorithms of cluster analysis are being employed where data storage is large. The algorithms used for clustering may be partitional and hierarchical. Some standard partitional approaches (e.g., K-means) suffer from several limitations such as their result depends on the accurate value of K which is originally unidentified, number of clusters K choices and size of cluster is not considered while executing the algorithm K-means, a number of clusters can turn out to be unfilled and can result is early completion of the algorithm. Also algorithms converge to a local minimum [15]. The hierarchical clustering methods are grouped into agglomerative or divisive on the basic of whether the hierarchy formation is from top-down fashion or bottom-up [16]. Many current modern clustering systems use a clustering strategy as agglomerative hierarchical clustering, because of its simple structure of processing and adequate performance level [17]. The AHC algorithm for service clustering is as follows:

```
Input:A set of items S = {I_1 ,............I_n}, a characteristic
      similarity matrix D = [d_{i,j}]_{n×n} the number of required
      clusters k.
Output: Dendrogram for k = 1 to | I |
Algorithm:
      1. c_i = {I_i},∀i.
      2. d_{ci,cj} = d_{i,j} ∀i,j
      3. for k = |I| down to K
      4. Dendogram_k = {c_1,.......c_k};
      5. l,m = argmax_{i,j} d_{ci,cj}
      6.c_l = Join (c_l ,c_m);
      7. for each c_h ∈ I
      8. if c_h ≠ c_l and c_h ≠ c_m.
      9. d_{cl,ch} = Average(d_{cl,ch}, d_{cm,ck});
      10. end if
      11. end for
      12. I = I - {c_m};
      13.end for
```

Algorithm: AHC algorithm

3.4 Compute Rating Similarity

The rating similarity computation between items is time consuming but important step in item-based CF algorithms. The cosine similarity between ratings vectors and the Pearson correlation coefficient (PCC) are the common rating similarity measures included [18, 19]. By measuring the resemblance amongst two users or items or measuring the character of two series to move together in a linear or proportional manner the PCC calculates the similarity between two items or users. Preferences calculation is given by:

$$PCC(a,b) = \frac{\sum_i (w_{a,i} - \bar{w}_a)(w_{b,i} - \bar{w}_b)}{\sqrt{\sum_i (w_{a,i} - \bar{w}_a)^2 \sum_i (w_{b,i} - \bar{w}_b)^2}} \quad (4)$$

where a and b are users/items, i is item, $w_{a,i}$ and $w_{b,i}$ are ratings from a and b for i, $\bar{w}_x$ and $\bar{w}_y$ are mean ratings for user a and b.

3.5 Select Neighbors

Based on the rating similarities between services, the neighbors of a target service A are determined as

$$N(A) = \{B | R\_Similarity(A, B) > \gamma, A \neq B\} \tag{5}$$

here $R\_similarity(A, B)$ is the enhanced rating similarity between service A and B,γ is a rating similarity threshold. The larger value of γ is, the chosen number of neighbors will moderately less but they can be more similar to the target service, hence the coverage of collaborative filtering will reduce but the accuracy can increase [20]. In contrast, the smaller value of γ is, the more neighbors are selected but some of them can be only somewhat similar to the target service, thus the coverage of CF will increase but the accuracy would decrease.

3.6 Compute Predicted Rating

For an active user U_A for whom predictions are being made, whether a target service A is worth recommending depends on its predicted rating. If $N(A) \neq \phi$, similar to the computation formula proposed by Wu et al. [20], the predicted rating $P(U_A, A)$ in an item-based CF is computed as:

$$P(U_A, A) = \bar{r}_A + \frac{\sum B \in N(A)(r(U_A, B) - \bar{r}_B) \times R\_Similarity(A, B)}{\sum B \in N(A) R\_Similarity(A, B)} \tag{6}$$

here, $\bar{r}_B$ is the mean rating of B, $N(A)$ is neighbor set of B, $A \in N(B)$ denotes A is a neighbor of the target service B, $r(U_A, B)$ is rating that an active user U_A give to B, $\bar{r}_B$ is average rating of B, and $R\_similarity(A, B)$, is enhanced rating similarity between service A and B [21].

4 Experiments and Results

4.1 Data Sets Used

To verify the proposed approach, the Movie-Lens and the Group-Lens data sets are used. The data sets include data about movies, movie ratings and users. Different sizes of data sets are used in experiments. Some of these are the Movie-Lens ML-100-K which has around one lakh ratings obtained from about one thousand users for about one thousand seven hundred movies, the Movie-Lens ML-1M data set which has about ten lakhs obtained from six thousand users for four thousand movies. The Movie-Lens ML-10-M data set which has about 1 crore ratings and one lakh tags obtained from ten thousand movies rated by about seventy two thousand users [22].

4.2 Evaluation of Similarity Algorithms

The main objective of this approach is to calculate the accurateness of a recommender system when there is increase in number of users, user preferences and items. By using the Movie-Lens data sets with different users preferences and feedback, we have evaluated description similarity and functional similarity by using the clustering algorithm. The results are shown in Figs. 2 and 3.

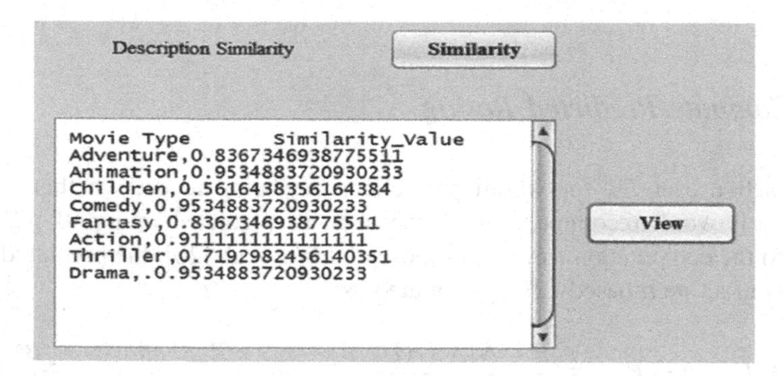

Fig. 2 Description similarity

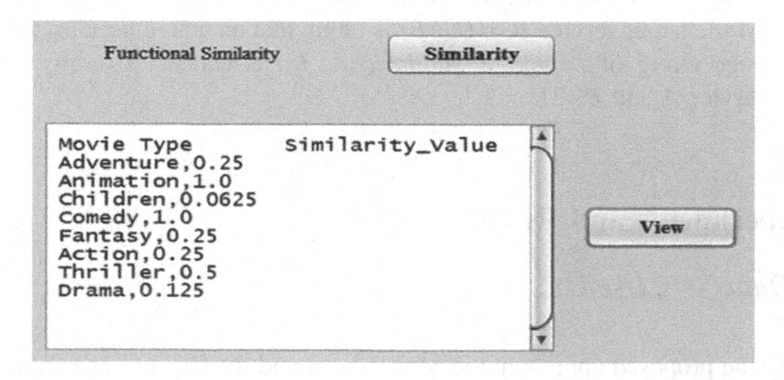

Fig. 3 Functional similarity

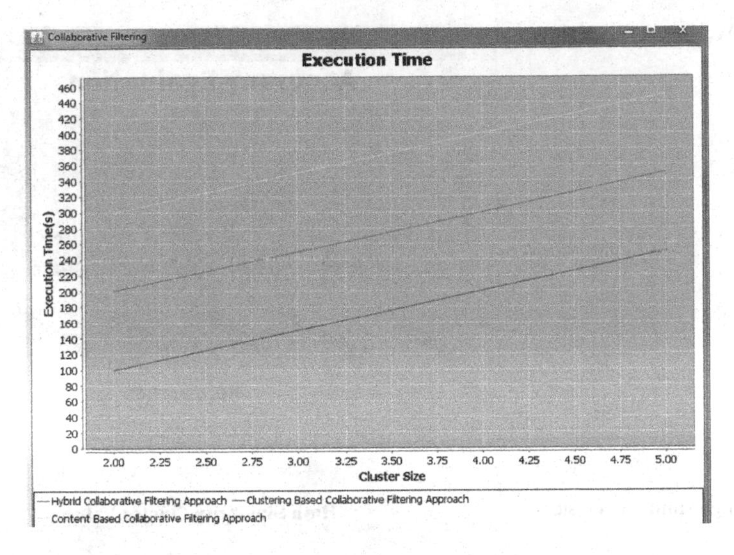

Fig. 4 Execution time

Figure 4 and 5 shows the graph of execution time (s) versus cluster size and graph of efficiency (similarity level) versus the cluster size for three different approaches i.e. hybrid, content and clustering based. From the graph we can see that the execution time is minimum by using the proposed approach and the efficiency of the clustering based approach is more as compared to other approaches.

The accuracy comparison of different approaches are shown in Fig. 6, from the results we can observe that the accuracy of clustering based approach is 80 % more

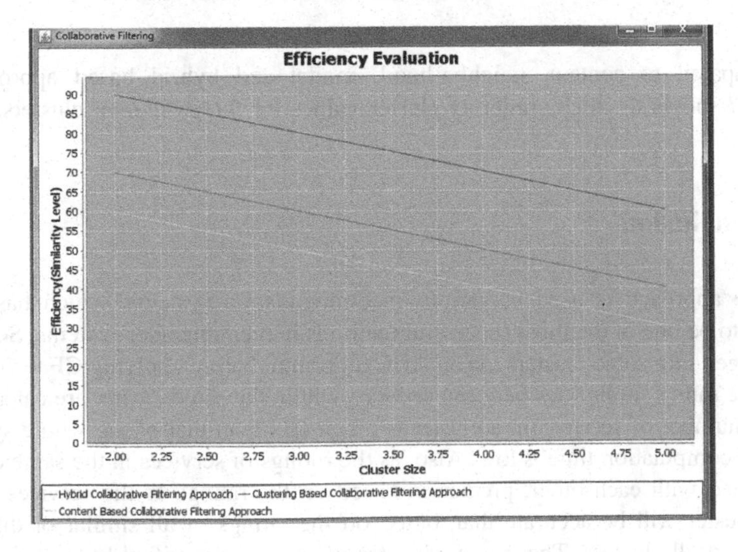

Fig. 5 Efficiency evaluation

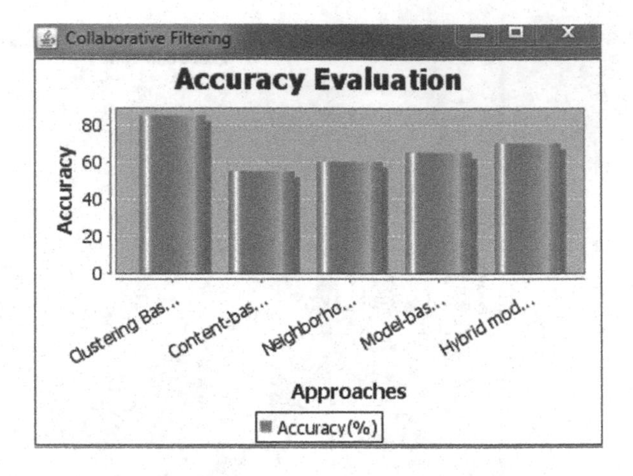

Fig. 6 Accuracy evaluation

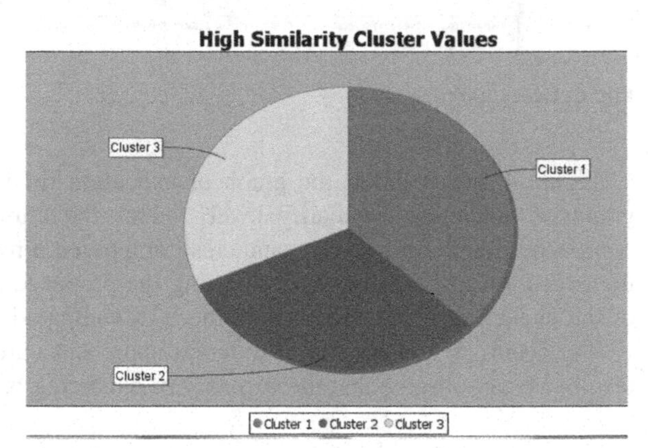

Fig. 7 High similarity cluster values

as compared to content, neighborhood, model and hybrid based approaches. Figure 7 shows the high similarity cluster values for three different clusters.

5 Conclusion

The CF approach is used in various recommendation systems, which has been proved to be one of the most successful methods in recommender systems. Services are merged into some clusters via an AHC algorithm before applying CF technique. Then the rating similarities between services within the same cluster are calculated. As the number of services in a cluster is much less than that of the whole system, costs of computation time is low. Also, as the ratings of services in the same cluster are related with each other, prediction based on the ratings of the services in the same cluster will be accurate than based on the ratings of all similar or different services in all clusters. These two advantages have been verified by experiments.

References

1. Wu, X., Wu, G., Zhu, X.: Data mining with big data. IEEE Trans. Knowl. Data Eng. **26**(1), 97–107 (January 2014)
2. Rajaraman, A., Ullman, J.D.: Mining of Massive Datasets. University Press of Cambridge
3. Bellogín, A., Díez, F., Cantador, I.: An empirical comparison of social, collaborative filtering (CF), hybrid recommender. ACM Trans. Intell. Syst. Tech. **4**(1), 1–37 (January 2013)
4. Zeng, W., Zhang, Q., Shang, M.: Can dissimilar users contribute to accuracy and diversity of personalized recommendation? IJMPC **21**(10), 1217–1227 (June 2010)
5. Havens, T.C., Hall, L.O., Leckie, C., Palaniswami, M.: Fuzzy c-means algorithm for very large data. IEEE Trans. Fuzzy Syst. **20**(6), 1130–1146 (December 2012)
6. Liu, Z., Zheng, Y., Li, P.: Clustering to find exemplar terms for key-phrase extraction. In: Proceedings of Conference on Empirical Methods in Natural Language Processing, pp. 257–266, May 2009
7. Rodriguez, A., Chaovalitwongse, W., Zhe, L.: Master defect record retrieval using network based feature ass. IEEE Trans. Syst. Man Cybern. App. Rev. **40**(3), 319–329 (October 2010)
8. Adomavicis, G., Zhang, J.: Stability of recommendation algorithms. ACM Trans. Inf. Syst. **30** (4), 23:1–23:31 (August 2012)
9. Liu, X., Mei, H., Huang, G.: Discovering homogeneous web services community in the user centric web environment. IEEE Trans. Serv. Comput. **2**(2), 167–181
10. Li, H.H., Tian, X., Du, X.Y.: A review based reputation evaluation approach for web services. Int. J. Comput. Sci. Tech. **249**(5), 893–900 (Sep 2009)
11. Zielinnski, K., Szydlo, T., Szymacha, R.: Adaptive soa solution stack. IEEE Trans. Serv. Comput. **5**(2), 149–163 (April-June 2012)
12. Shafer, J., Rixner, S.T., Cox, A.: The hadoop distributed file system (HDFS): balancing portability and performance. IEEE Int. Symp. Perform. Anal. Syst. S\W. doi: 10.1109/ISPASS.2010.5452045. pp. 122–133, 28–30 March 2010
13. Kirankumar, R., Vijayakumari, R., Gangadhara, R.K.: Comparative analysis of google file system and hadoop distributed file system. IJAT CSE. **3**(1), 24–25 (Feb 2014)
14. HDFS Guide [Online]. http://hadoop.apache.org/common/doc/current/hdfs_user_guide
15. Li, M.J., Cheung, Y., Ng, M.: Agglomerative fuzzy k means clustering algorithm with selection of number of clusters. IEEE Trans. Knowl. Data Eng. **20**(11), 1519–1534 (November 2008)
16. Zhao, Y., Fayyad, U., Karypis, G.: Hierarchical clustering algorithms for document datasets. Data Min. Knowl. Discov. **10**(2), 141–168 (November 2005)
17. Platzer, C., Dustdar, S., Rosenberg, F.: Web service clustering using multi-dimensional angle as proximity measures. ACM Trans. Internet Tech. **9**(3), 11:1–11:26 (July 2009)
18. Taherian, T.F., Niknam, T., Pourjafarian, N.: An efficient algorithm based on modified imperialist competitive algorithm & K means for data clustering. Eng. App. Artif. Intell. **24**(2), 306–317 (March 2011)
19. Thilagavathi, G., Aparna, N., Srivaishnavi, D.: A survey on efficient hierarchical algorithm used in clustering. Int. J. Eng. **2**(9), 306–317 (Sep 2013)
20. Julie, D., Kumar, K.: Optimal web service selection scheme with dynamic QoS property assignment. IJART. **2**(2), 69–75 (May 2012)
21. Wu, J., Chen, L., Feng, Y.: Predicting quality of service for selection by neighborhood based collaborative filtering (CF). IEEE Trans. Syst. Man Cybern. Syst. **43**(2), 428–439 (March 2013)
22. Lens G MovieLens. Available [Online] http://grouplens.org/datasets/movielens.html

Double Fully Homomorphic Encryption for Tamper Detection in Incremental Documents

Vishnu Kumar, Bala Buksh and Iti Sharma

Abstract The famous scheme of Van Dijk, Gentry, Halevi and Vaikuntanathan (DGHV) style of Fully Homomorphic Encryption (FHE) is simple to implement. This paper proposes how this scheme can be modified and used for double encryption to secure files stored at third party like Cloud service provider (CSP). The files belong to an incremental text model which is followed by majority of confidential documents. Encryption is similar to a homomorphic hash function. A protocol is also proposed to detect if files have been tampered, and the tampering can be located up to single file level or up to a word of file level.

Keywords Incremental documents · Homomorphic encryption · Tamper detection

1 Introduction

Confidential documents mostly follow an incremental structure, where files arrive like chunks of a large file. Say files have arrived in sequence $f_1, f_2, f_3, \ldots, f_k$, then each of f_i is a file in its own, and any sequence from beginning $f_1 f_2 f_3 \ldots f_k$ is also a file at the moment file f_k arrives. Tamper detection commonly relies on hashing and message authenticators (MAC) code computations. In case, these incremental documents are stored with a third party, like a cloud service provider, the files need to be stored in encrypted form. This encryption secures the crucial information stored in files from being leaked. At the same time, now files cannot be used, not even concatenated without decrypting them back. Considering the amount of data,

V. Kumar (✉) · B. Buksh
R.N. Modi Engineering College, Kota, Rajasthan, India
e-mail: vishnumkota@gmail.com

B. Buksh
e-mail: balabuksh@gmail.com

I. Sharma
Rajasthan Technical University, Kota, Rajasthan, India
e-mail: itisharma.uce@gmail.com

S.C. Satapathy and S. Das (eds.), *Proceedings of First International Conference on Information and Communication Technology for Intelligent Systems: Volume 2*,
Smart Innovation, Systems and Technologies 51, DOI 10.1007/978-3-319-30927-9_17

downloading the files back from cloud to the user site, decrypting and then concatenating, then uploading to the cloud is an impractical solution. If this stupendous task is outsourced to the cloud, there will be a need to share the encryption key; losing all the point of encryption in the first place. Here we assume that cloud might be honest but curious. So, the first requirement is that CSP should be able to concatenate files in encrypted form only. This can be simply achieved if the files are encrypted at word level, that is, each word of the file is encrypted separately and concatenated to form a file. Though it can be taken to bit-level, it would be very expensive. Once this has been done, one can implement any message digest or hash computation function homomorphically over the files for the purpose of tamper detection. The real problem starts when the files are a part of an incremental document. The message digest or hash has to be now computed for $f_1, f_1 f_2, f_1 f_2 f_3$ and so on. This requires a homomorphic hash as has been suggested in this paper.

2 Incremental Text Model

Documents like appeals, applications, project reports, annual reports etc. are crucial and large sized. They are incremental in both use and storage. Though so common in practice, the similar model for the soft copies of incremental documents has yet not been formalized. We present here an elementary idea how such documents can be stored and viewed as soft copies. Let every individual part of the document be called a file, denoted by f. At any point, the document is a sequential concatenation of files generated till now, $D_k = \langle f_1 f_2 .. f_k \rangle$. Since, the files are generated and maintained chronologically, for any file f_i and file f_j, $i < j$ suggests that file f_i has been generated before f_j. Files either have independent existence or the entire document has. No other sequence or combination of file can be retrieved or has any meaningful interpretation. Thus, at any point of time k, D_k is sequence of k files beginning at f_1 till f_k.

3 DGHV Style FHE Scheme

Homomorphic encryption is a technique which allows operations to be performed over encrypted data, and the results when decrypted give the same output as would have been produced by actual desired operation on plaintexts. Thus, the operation 'g' performed over cipher texts is homomorphic to desired function 'f' over plaintexts. If the operations 'f' contain any arbitrary circuit and homomorphic 'g' exist for any such 'f', the scheme is called fully homomorphic. The idea was asked for by Rivest et al. [1], and the first solution arrived in 2009 through Gentry [2]. The idea has been developed from bits to integers, based on several hard problems. The recent works based on ring learning with errors [3, 4] are efficient and secure enough. Yet, practical issues like number and size of keys involved, noise growth etc. make implementing FHE debatable.

In this section we first discuss the FHE scheme to be used, based on those suggested in [5], which in turn are symmetric versions of DGHV [6] style FHE systems. Any file is assumed to be divided into equal sized words, which may be 1, 2 or 4 bytes long or more as per the computation power available at the data owner. The size of word affects the runtime cost of encryption. File is to be encrypted before storage to ensure security. The key used remains secret with the data owner. Let the word size be w (256, 65,536 or 4,294,967,296), and length of key η bits. The file to be encrypted consists of 'B' words, and each word $b_1, b_2, \ldots, b_B$ is encrypted as

$$x_j = b_j + p * r_j \tag{1}$$

where p is the secret key, an odd number of length η bits. Each r_i is a random odd number of length η bits, $r_j \neq p, \forall j$. Thus, file $\langle b_1, b_2, \ldots, b_B \rangle$ is now encrypted and stored in Cloud as $\langle x_1, x_2, \ldots, x_B \rangle$. The length of each encrypted word, now onwards called cipherword, is of $O(\eta)$.

Each file is associated with a hash value, which is equivalent to bitwise XOR of all words of the file, producing a hash word of length $O(\lambda)$. Since the encryption is fully homomorphic, instead of computing XOR of all words and then encrypting it under key p, the encrypted hash value, y, can be directly computed as modulo sum of all cipher words. Here, we require a homomorphic hash, so the encrypted hash is re-encrypted under a new key q,

$$y = (\Sigma x_j)w \tag{2}$$

$$z = y + qs \tag{3}$$

where, q is a public key, an odd number of λ bits and s is a random odd number of λ bits, $s \neq q$. This produces the final hash value of a file to be stored as z, a number of length $O(\lambda)$ bits.

The fully homomorphic properties of the suggested scheme can be observed directly, as proved in [5].

Note on parameter selection: The word size, w, makes it mandatory that $\eta > \log_2 w$, so that modular arithmetic remains correct. Moreover, $\lambda > \eta$, for hash computations. Our experiments suggest that minimum difference of 2 be maintained.

4 Proposed Protocol

Let the incremental document be a chronological sequence of n files, $D = \langle f_1 f_2 \ldots f_n \rangle$. The user generates any file f_i, encrypts it into a sequence of cipherwords under key p using Eq. (1), and uploads it to the third-party storage (TPS). TPS computes the hash value of file as z_i using Eqs. (2) and (3) under key q. At a subsequent instant, file f_{i+1} is generated, and similar procedure is followed. At any arbitrary instant k, the document D_k is concatenation of k files $\langle f_1 f_2 .. f_k \rangle$, it is required to have hash values associated with documents too. The hash values of documents are computed as

$$Hash(D_1) = Hash(f_1) = z_1$$

$$Hash(D_2) = Hash(f_1 f_2) = z_1 + z_2$$

Thus,

$$Hash(D_k) = Hash(f_1, f_2 .. f_k) = \sum z_k.$$

We proceed to discuss how hash value of a document D_k can be verified for correctness.

ALGORITHM Verification
INPUT: Document D_k
OUTPUT: YES/NO
Step 1: For each file f_i in D_k, compute the sum of cipherwords as

$$A = \sum_{i=1}^{k} \left(\sum x_j\right)_{\forall x_j\, in f_i}$$

Step 2: Let hash value of D_k be z_k, then

$$B = z_k \bmod q$$

Step 3: If $A \equiv B \bmod w$, return YES else return NO

The steps 1 till 4 of the protocol are repeated many times at arbitrary instances. Files are concatenated in chronological sequence. At any point k, the command is D_k and user may want to perform a tamper check. The request of tamper check is handled by sending hash value of current document. It is verified using verification algorithm. If verification is positive, protocol stops In case verification is negative, tamper is detected and user now proceeds to locate the tamper. This is done by running verification algorithm over all documents beginning at D_1. But now the hash values have to be recomputed by TPS under a new key sent by user. Let the point at which verification fails is t. This implies that file f_t is tampered. The error can further be located by checking file f_t. The cipher word which does not decrypt correct is tampered.

When the verification process returns a negative reply, a tamper in one or more files is expected. The protocol for tamper detection can be summarized as shown in Fig. 1.

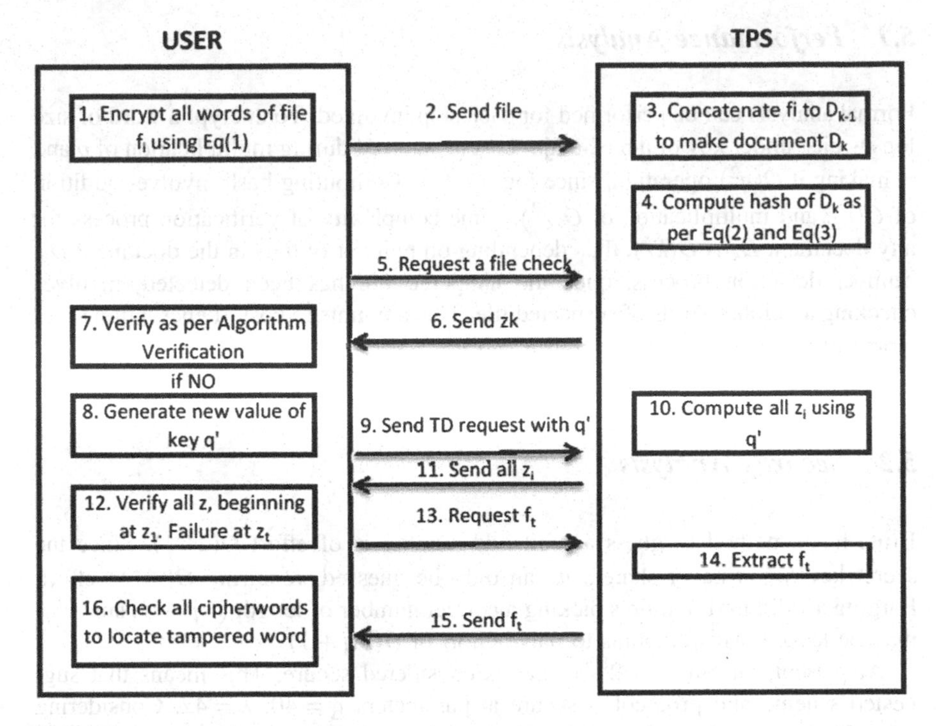

Fig. 1 Tamper detection protocol

5 Analysis of Proposal

The proposal was implemented as a Java program and growth of runtime recorded for different values of η, λ and w. The extensive results cannot be reported due to limitation of space. Figure 2 is a plot of encryption time with different combinations of η and λ, keeping $w = 256$. It can be seen that growth is quadratic as the value of parameters increase. This is apt even for a lightweight client.

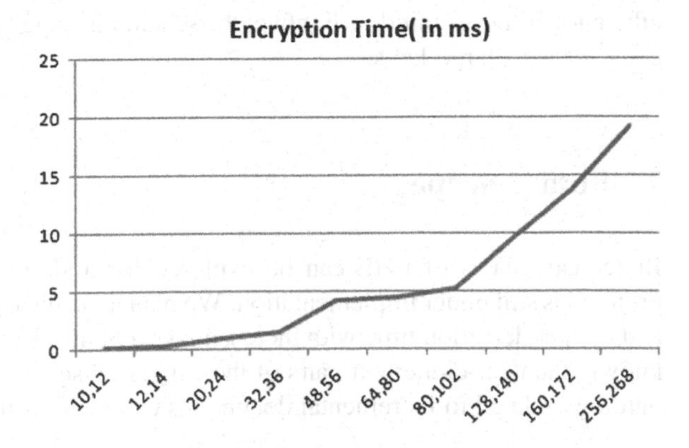

Fig. 2 Growth of encryption time with increasing values of parameters

5.1 Performance Analysis

Formal analysis can be performed for each step involved. To encrypt a word of size $\log_2 w$ bits, under key of η bits, major cost is incurred during multiplication of p and r, making it $O(\eta^2)$ operation, since $\log_2 w \ll \eta$. Computing hash involves addition of $O(\eta)$, and multiplication of $O(\lambda^2)$. Time complexity of verification process for any document D_k is $O(k\lambda)$, thus depending on number of files in the document D_k. Tamper detection process, once the tampered file has been detected, involves checking all cipherwords of suspected file. This amounts to cost of encryption of all words of file that depends on both n and η^2.

5.2 Security Analysis

Brute-force method to guess a b-bit odd number is of effort $O(2^{b-1})$. Since the secret key 'p' is never shared, it can only be guessed, requiring $O(2^{\eta-1})$ effort. Forging a valid hash requires picking a correct number of length $(2\eta + 1)$, since 'q' may be leaked. This amounts to naive effort of $O(2\eta + 1)$.

At present, an effort of 2^{80} cycles is considered secure. This means that suggested scheme and protocol is secure at parameters $\eta = 40$, $\lambda = 42$. Considering other kind of attacks, a key of length 128 bits is considered secure enough. The parameter values $\eta = 128$, $\lambda = 140$, when experimented incur encryption time of less than 10 ms.

6 Conclusion

The security and tamper detection problem for incremental documents has not been considered till now. We present a simple solution based on DGHV style FHE. The protocol for tamper detection and location is simple yet powerful. Keeping the efficiency issue in mind, a lightweight scheme is suggested, which is fast enough even for very large keys.

7 Future Scope

Better candidates of FHE can be explored for a similar application. The overall protocol is still under implementation. We plan to analyze growth of verification time and tamper detection time with increasing size of file. Security analysis in context of known plaintext-ciphertext pairs of the suggested scheme is yet to be done. Formal ontology related to incremental documents can be a further research area.

References

1. Rivest, R.L., Adleman, L., Dertouzos, M.L.: On data banks and privacy homomorphisms. Found. Secure Comput. **4**(11), 169–180 (1978)
2. Gentry, C.: Fully homomorphic encryption scheme. In: Diss. Stanford University (2009)
3. Brakerski, Z., Vaikuntanathan, V.: Fully homomorphic encryption from ring-LWE and security for key dependent messages. In: Advances in Cryptology–CRYPTO, pp 505–524. Springer, Berlin (2011)
4. Brakerski, Z., Gentry, C., Vaikuntanathan, V.: Fully homomorphic encryption without bootstrapping. In: Cryptology ePrint Archive, Report 2011/277
5. Aggarwal, N., Gupta, C., Sharma, I.: Fully homomorphic symmetric scheme without bootstrapping. In: International Conference on Cloud Computing and Internet of Things (CCIOT), Chengdu, China (2014)
6. Van-Dijk, M., Gentry, C., Halevi, S., Vaikuntanathan, V.: Fully homomorphic encryption over the integers. In: Advances in Cryptology–EUROCRYPT, pp 24–43. Springer, Berlin (2010)

Part II
Intelligent Systems for Health and IT Applications

Enhancement of GSM Stream Cipher Security Using Variable Taps Mechanism and Nonlinear Combination Functions on Linear Feedback Shift Registers

Darshana Upadhyay, Priyanka Sharma and Srinivas Sampalli

Abstract With the advance wireless communication, data security became a significant concern. The GSM standard hardware level encryption technique uses A5/1 algorithm circuit which embedded in the Mobile Equipment. A5/1 algorithm uses Linear Feedback Shift Register (LFSR) to produce a key streams for encode the information sent between the mobile station and the base station. It is a secure cipher among all the versions of ciphers using in GSM. However, latest research studies demonstrate that A5/1 can be subjected to several attacks owing to feeble clocking mechanism which results in a low rate of linear complexity. To overcome from these issues, we introduce a feedback tap mechanism enhanced by variable taps and four nonlinear combination functions. Analysis shows that the proposed method has a high algebraic degree of correlation immunity against basic correlation attack, mathematical attack, linear estimate attack and Berlekamp-Massey attack.

Keywords A5/1 algorithm · Linear feedback shift register (LFSR) · Correlation attack · Linear estimate attack · Nonlinear combination function · Polynomial primitives

D. Upadhyay (✉) · P. Sharma
Department of Computer Science and Engineering, Nirma University, Ahmadabad, Gujarat, India
e-mail: darshana.upadhyay@nirmauni.ac.in

P. Sharma
e-mail: priyanka.sharma@nirmauni.ac.in

S. Sampalli
Department of Computer Science, Dalhousie University, Halifax, NS, Canada
e-mail: srini@cs.dal.ca

S.C. Satapathy and S. Das (eds.), *Proceedings of First International Conference on Information and Communication Technology for Intelligent Systems: Volume 2*, Smart Innovation, Systems and Technologies 51, DOI 10.1007/978-3-319-30927-9_18

1 Introduction

People converse over distances by wireless communication. Since huge exposure of wireless network, it's vulnerable by an eavesdropper. In the GSM communications security has offered by A5/1 stream cipher. Initially the cipher was kept undisclosed, but through leaks and reverse engineering it became public. The number of severe limitations in the cipher has been identified [1]. The A5/1 stream cipher designed using three Linear Feedback Shift Registers, length of 19, 22, and 23 bits respectively. The output of this these Linear feedback shift register is combined using XOR gate to generate the key stream for secure communication in GSM Technology.

Recent research, analysis gives you an idea about limitations of GSM cipher due to which it is vulnerable to a number of attacks [2, 3]. GSM cipher was first broken by Golic and a rough sketch of A5/1 was disclosed. After A5/1 was inverse plotted, it was investigated by Biryukov et al. [4], Dunkelman and Biham [5], Johansson and Ekdahl [6], Johansson et al. [7], and freshly by Biham and Barkan [8]. The GSM stream cipher have been poorly broken down using a range of attacks like faster time-memory trade off attack have need of some pre working out, basic correlation attack, mathematical attack, linear estimate attack, Bereleykamp-Massey attack, general inversion attack and also the brute force attack requiring no pre computation. A suitable preference of merging nonlinear function significantly advances the performance of the cipher from the security aspects. A combination function has to be impartial and nonlinear in nature; it should have high statistical degree and correlation immunity against attacks [1, 9]. Thus to carry out modifications by considering above points on the existing A5/1 algorithm to make it more robust and non-linear.

The rest of the paper is planned as follow. Section 2 stretches the comprehensive information on the A5/1 algorithm. In Sect. 3, the improved scheme of proposed A5/1 architecture is discussed. Section 4 explained the mathematical proof of proposed algorithm to enhance security of GSM stream cipher, finally conclude in Sect. 5.

2 GSM Stream Cipher—A5/1 Algorithm

The GSM stream cipher is a part of SIM card which provides security during GSM communication between Mobile station and Base station. Before start the communication the Mobile Equipment requires to acquire authentication on the network. The authentication procedure is carried out by an A3 algorithm using challenge- response mechanism by Subscriber Authentication Key Ki and a 128 bits nonce called RAND. After the authentication process A8 algorithm is used to generates the 64 bits session key Kc using Ki and RAND [10]. The Mobile station and Base station uses same 64 bits of session key to initialize the three LFSRs.

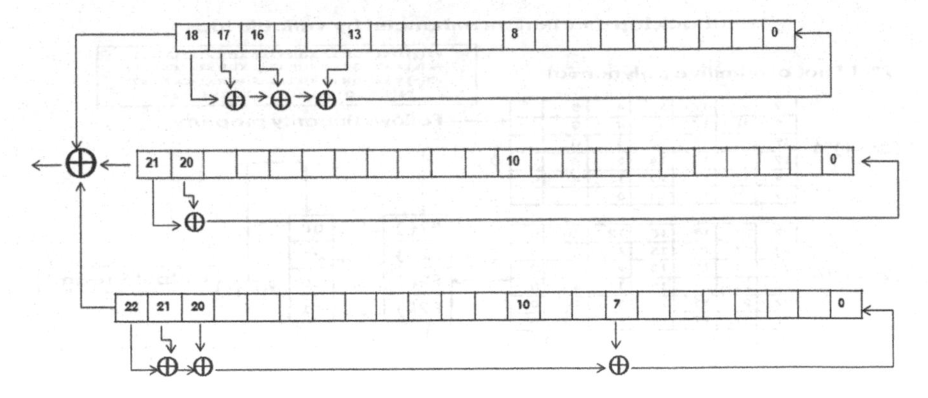

Fig. 1 Black-box view of conventional GSM stream cipher

The 64 bits of session key requires 64 clock cycles to load all the three registers. The GSM stream cipher as shown in Fig. 1 uses three LFSRs. The polynomial primitives of 19 bits, 22 bits and 23 bits are $x^{19} + x^{18} + x^{17} + x^{14} + 1$, $x^{22} + x^{21} + 1$ and $x^{23} + x^{22} + x^{21} + x^{8} + 1$ which is derived using Galois field [3]. After that 22 bits of frame counter (Fn) value is also loaded into the three LFSRs in the similar manner using 22 clock cycle. Subsequently the LFSR are irregularly clocked for 100 times using the majority rule. According to the majority rule if two or more LFSR's clocking bits are enable, then those LFSRs has been consider for that round and other become disable. Thus minimum two LFSRs has been enable in the particular round. To apply majority rule position 8 of 19 bit LFSR, position 10 for 22 bit of LFSR, and position 10 for 23 bit of LFSR is taken into consideration. These clocking bits are most irregular in nature and hence consider in majority rule. For this 100 clock cycles output bits are discarded. After that LFSRs are clocked for 228 times to generate 228 bits key stream where 114 bits are for uplink communication and 114 bits are downlink communication. This entire cycle repeats by incrementing the value of frame counter by one for a single session of communication in GSM Technology [5.10].

3 Proposed Algorithm of GSM Stream Cipher

The prototype of the suggested stream cipher involves primary modifications in the improvement in feedback tapping units as well in combining function of conventional A5/1 shown in Fig. 2. The feedback taps mechanism enhanced by six polynomial primitive for each LFSR and four nonlinear combination functions are introduced in the A5/1 stream cipher to make it more robust and protected. To rise the randomness of the output stream; instead of one polynomial primitive, to design LFSR, six polynomial primitives are used with the same degree of GF (2). Also, to

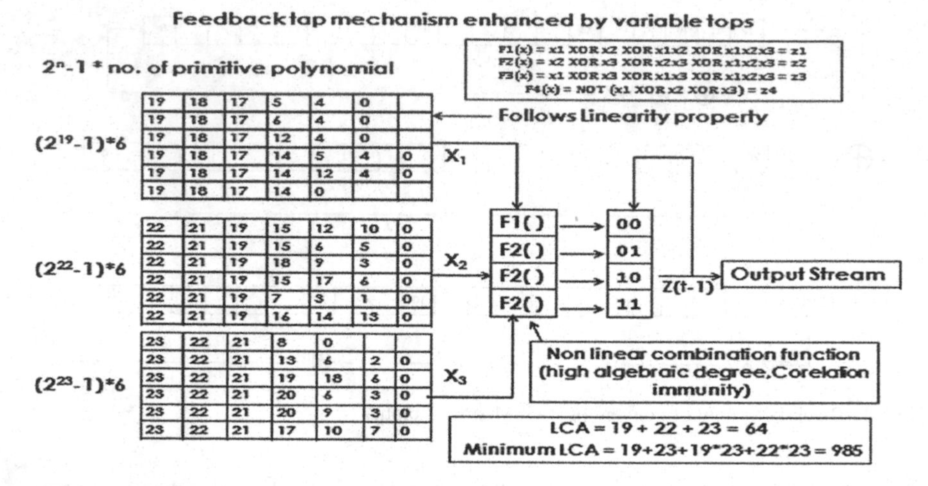

Fig. 2 Black-box view of proposed A5/1 stream cipher

decrease hardware complication, the polynomial primitives with minor distinction in tap positions is identified. Six polynomial primitives of each LFSR and access hardware requirement is also computed and state in Table 1. The key stream is produced based on four non liner function, which is known as the nonlinear combination function. Moreover, the combiner perhaps has flip-flops to store previous output key stream to compute next state of the key stream.

Table 1 Analysis of access hardware requirement

Degree of LFSR	Polynomial primitives	Access components required	
Degree of LFSR 1: 19	19, 18, 17, 14, 4, 5, 0 19, 18, 17, 14, 4, 12, 0	2 × 1 MUX	5—XOR gates
	19, 18, 17, 4, 5, 0 19, 18, 17, 4, 6, 0 19, 18, 17, 4, 12, 0	4 × 1 MUX	
	19, 18, 17, 14, 0		
Degree of LFSR 2: 22	22, 21, 19, 15, 5, 6, 0 22, 21, 19, 15, 5, 7, 0	2 × 1 MUX	10—XOR gates
	22, 21, 19, 15, 12, 10, 0 22, 21, 19, 15, 17, 6, 0	2(2 × 1 MUX)	
	22, 21, 19, 7, 3, 1, 0 22, 21, 19, 16, 15, 14, 0	3(2 × 1 MUX)	
Degree of LFSR 3: 23	23, 22, 21, 20, 3, 6, 0 23, 22, 21, 20, 3, 7, 0	2 × 1 MUX	7—XOR gates
	23, 22, 21, 8, 0 23, 22, 21, 13, 6, 2, 0 23, 22, 21, 19, 18, 6, 0 23, 22, 21, 17, 10, 7, 0	3(4 × 1 MUX)	

To enhance the linear complexity of the GSM stream cipher and to make it more conquer, four cryptography improved nonlinear functions are employed [9], each is having nonlinear order of degree three. Furthermore the combining functions are dynamic in nature using one selection bits from the preceding state of key stream P (t − 1). The four nonlinear functions f1(.), f2(.), f3(.) and f4(.) are as under:

$$f1(x) \rightarrow \alpha 1 \oplus \alpha 2 \oplus \alpha 1 \alpha 2 \oplus \alpha 1 \wedge \alpha 2 \wedge \alpha 3 = P1 \quad (1)$$

$$f2(x) \rightarrow \alpha 2 \oplus \alpha 3 \oplus \alpha 2 \alpha 3 \oplus \alpha 1 \wedge \alpha 2 \wedge \alpha 3 = P2 \quad (2)$$

$$f3(x) \rightarrow \alpha 1 \oplus \alpha 3 \oplus \alpha 1 \alpha 3 \oplus \alpha 1 \wedge \alpha 2 \wedge \alpha 3 = P3 \quad (3)$$

$$f4(x) \rightarrow \alpha 1 \oplus \alpha 2 \oplus \alpha 3 \quad P1 \quad (4)$$

4 Security Analysis of Proposed Algorithm

4.1 Linear Approximation Attack

It is easy to approximate linear function compare to nonlinear function by linear approximation attack. In the proposed scheme we convert linear function into nonlinear function to prevent key stream estimation [11] by linear approximation attack. Maximum linear complexity of proposed algorithm compare to original A5/1 algorithm is $LC \rightarrow 22 + 23 + 22 * 23 + 19 * 22 * 23 = 10165$ while conventional algorithm having linear complexity $LC \rightarrow 64$, hence proposed mechanism is robust and resistive to this attack.

4.2 Correlation Attack

Firstly is intending at the nonlinear combiners, Siegen haler primary pioneered the correlation attack in the middle of the 1980s [12]. Correlation attack discovers the flaw in the combination function of given stream cipher which has numerous LFSRs series inputs and identify the relationship amongst input literals and output literals of combination function and then apply methodology to taking out information about the correlated input literals. Truth Table of input streams $\alpha\mathbf{1}$, $\alpha\mathbf{2}$, $\alpha\mathbf{3}$ (output of each LFSR respectively) and output of nonlinear combination function **p1**, **p2**, **p3** and **p4** is as in Table 2. To pass up this attack in LFSR based stream ciphers is to decide the correlation probabilities of the function must be constant. To make it feasible, choose nonlinear function dynamically by applying a variation in its function. Correlation probabilities of sequences $\alpha\mathbf{1}, \alpha\mathbf{2}, \alpha\mathbf{3}$ two key streams **p1 (t)**, **p2(t)**, **p3(t)**, **p4(t)** is as in Table 3.

Table 2 Input stream output of nonlinear combination function

α_1	α_2	α_3	p1	p2	p3	p4
0	0	0	0	0	0	0
0	0	1	0	1	1	1
0	1	0	1	1	0	1
0	1	1	1	1	1	0
1	0	0	1	0	1	1
1	0	1	1	1	1	0
1	1	0	1	1	1	0
1	1	1	0	0	0	1

Table 3 Correlation probabilities

prob(p1(t) = α1) → 5/8 prob(p1(t) = α2) → 5/8	prob(p2(t) = α1) → 3/8 prob(p2(t) = α2) → 5/8
prob(p3(t) = α1) → 5/8 prob(p3(t) = α2) → 3/8	prob(p4(t) = α1) → 4/8 prob(p4(t) = α2) → 4/8

When four nonlinear function **p1(t), p2(t), p3(t), p4(t)** selected alternatively then correlation probabilities are as under.

Correlation Probability (CP) of α1:

$$\frac{1}{4}(\text{prob}(\text{p1}(t) = \alpha 1) + \text{prob}(\text{p2}(t) = \alpha 1) + \text{prob}(\text{p3}(t) = \alpha 1)$$
$$+ \text{prob}(\text{p4}(t) = \alpha 1)) \underset{\text{CP}(\alpha 1)}{\Rightarrow} 1/4\left[\sum_{i=1}^{4} \text{prob}(\text{pi}(t)\right] = \frac{5}{8} + \frac{3}{8} + \frac{5}{8} + \frac{4}{8} = 0.53 \quad (5)$$

Correlation Probability (CP) of α2:

$$\frac{1}{4}(\text{prob}(\text{p1}(t) = \alpha 2) + \text{prob}(\text{p2}(t) = \alpha 2) + \text{prob}(\text{p3}(t) = \alpha 2)$$
$$+ \text{prob}(\text{p4}(t) = \alpha 2)) \underset{\text{CP}(\alpha 2)}{\Rightarrow} 1/4\left[\sum_{i=1}^{4} \text{prob}(\text{pi}(t)\right] = \frac{5}{8} + \frac{5}{8} + \frac{3}{8} + \frac{4}{8} = 0.53 \quad (6)$$

Correlation Probability (CP) of α3:

$$\frac{1}{4}(\text{prob}(\text{p1}(t) = \alpha 3) + \text{prob}(\text{p2}(t) = \alpha 3) + \text{prob}(\text{p3}(t) = \alpha 3) +$$
$$+ \text{prob}(\text{p4}(t) = \alpha 3)) \underset{\text{CP}(\alpha 3)}{\Rightarrow} 1/4\left[\sum_{i=1}^{4} \text{prob}(\text{pi}(t)\right] = \frac{3}{8} + \frac{5}{8} + \frac{5}{8} + \frac{4}{8} = 0.53 \quad (7)$$

Therefore the correlation probability of output sequences $\boldsymbol{\alpha_i}$ of LFSRs and key stream **p(t)** can be removed as it is constant.

4.3 Algebraic Attack

The algebraic attack [11, 13–15] is reasonably fresh in the research literature but has so many reflexion [9]. The LFSR-based ciphers are susceptible against this attack and it has been successfully proved that the algebraic attack against a various stream ciphers is applied and well-organized [13–16]. To resist this attack, notion of algebraic degree is applied. Algebraic degree determined by the maximum number of variables employed to describe part of the function. In convention cipher an algebraic degree of the blend function is 1 instead in the proposed algorithm the algebraic degree of the blend function is 3. Hence proposed scheme offers more resistance to algebraic attack.

4.4 Berlekamp-Massey Attack

The notable Berlekamp-Massey algorithm is a very effective algorithm to determine the linear complexity of a finite binary series of bit length n within $O(n^2)$ bit operations [6, 17]. As greater the linear complexity avoids this attack. It identifies the shortest length of LFSR used in stream ciphers. This attack required twice of LC consecutive bits of the series generated by stream cipher in order to design LFSR of length LC which generates the same output key stream [1].

The change in polynomial primitive at Time instant t is

$$\text{19 bits LFSR: } 39 + 129 * t \ (38 + 44 + 46 = 128 \text{ where } t = 0 \text{ to } 5) \tag{8}$$

$$\text{22 bits LFSR: } 83 + 129 * t \ (38 + 44 + 46 = 128 \text{ where } t = 0 \text{ to } 5) \tag{9}$$

$$\text{23 bits LFSR: } 129 + 129 * t \ (38 + 44 + 46 = 128 \text{ where } t = 0 \text{ to } 5) \tag{10}$$

Thus a variable taps mechanism provides more prevention against Berlekamp-Massey attack.

5 Conclusion

This paper is attempted to upgrade security on GSM stream cipher using consolidating methodology applying on a linear feedback shift register using variable tap mechanism and nonlinear combination functions. Proposed algorithm improves keystrems in terms of randomness and offering more security. A5/1 has weak linear complexity and output keystream generation of A5/1 has a low rate of unpredictability. To defeat these issues we present variable tap system improved by six variable taps for every LFSR and four nonlinear combination functions. It has been mathematically inclined that proposed calculation is having high algebraic degree correlation immunity against correlation attack, linear approximation attack, algebraic attack and Berlekamp-Massey attack because of nonlinear combination generator on account of the nonlinear blending generator.

6 Future Work

Further work of this paper is to design existing algorithm as well as proposed algorithm using VHDL language, simulate using ISIM simulator and deploy it on FPGA-SPARTAN 6 Xilinx 12.4 ISE toolkit. NIST Statistical test suite is use to measure direct unpredictability of output key stream and compare it with an original A5/1 algorithm. Further extensions to this project is to build a generic framework of the pseudo random number generator.

Acknowledgment The authors would like to thank Nirma University and Dalhousie University for providing common platform for research collaboration. This work has been funded by Shastri research Grant—Canada. The authors would also like to thank program and member relations officer, Shastri Indo Canadian Institute for support and guidance related to project grant.

References

1. Shrestha, R., Paily, R.: Design and implementation of a linear feedback shift register interleaver for turbo decoding. In: VDAT'12 Proceedings of the 16th International Conference on Progress in VLSI Design and Test, Heidelberg (2012)
2. Sugimura, T., Shibata, K., Fujita, Y.: A method for deriving tap polynomials of LFSR generating syndromes by utilizing a matrix-reduction algorithm. Electron. Commun. Japan (Part III: Fundamental Electronic Science) **90**(1), 30–45 (2007)
3. Upadhyay, D.P., Sharma, P., Valiveti, S.: Randomness analysis of A5/1 stream cipher for secure mobile communication. Int. J. Comput. Sci. Commun. **3**, 95–100 (2014)
4. Biryukov, A., Shamir, A., Wagner, D.: Real time cryptanalysis of A5/1 on a PC. In: Advances in Cryptology, Proceedings of Fast Software Encryption'00, LNCS, pp. 1–18. Springer-Verlag (2001)
5. Biham, E., Dunkelman, O.: Cryptanalysis of the A5/1 GSM stream cipher. In: Progress in Cryptology, Proceedings of INDOCRYPT'00, LNCS, pp. 43–51. Springer-Verlag (2000)
6. Johanson, T., Ekdahl, P.: Another attack on A5/1. IEEE Trans. Inf. Theory **49**, 284–289 (2003)
7. Maximov, A., Johansson, T., Babbage, S.: An improved correlation attack on A5/1. In Proceedings of SAC 2004, LNCS, vol. 3357, pp. 1–18. Springer-Verlag (2005)
8. Barkan, E., Biham, E.: Conditional estimators: an effective attack on A5/1. In: Proceedings of SAC 2005, LNCS, vol. 3897, pp. 1–19. Springer-Verlag (2006)

9. Yamada, T., Nakajima, H.: Pseudorandom pattern built-in self-test for embedded rams. Syst. Comput. Japan **7**(12), 1–8 (2012)
10. Upadhyay, D.P., Shah, A., Sharma, P.R.: In: IEEE International Conference on Computational Intelligence and Communication Networks, Udaipur (2014)
11. Ahmad, M, Izharuddin.: Randomness evaluation of stream cipher for secure mobile communication. In: IEEE International Conference on Network Security (2010)
12. Courtois, N.T., Meier, W.: Algebraic attacks on stream ciphers with linear feedback. In: Lecture Notes in Computer Science, vol. 2656, pp. 345–359. Springer, Berlin (2003)
13. Ahmad, M, Izharuddin.: Enhanced A5/1 cipher with improved linear complexity. In: IEEE International Conference on Impact (2009)
14. Feregrino-Uribe, C., Kitsos, P., Cumplido, R., Morales-Sandoval, M.: Area/performance trade-off analysis of an FPGA digit-serial GF(2 m) Montgomery multiplier based on LFSR. Comput. Electr. Eng. **39**, 542–549 (2013)
15. Karpovsky, M., Wang, Z.: Design of strongly secure communication and computation channels by nonlinear error detecting codes. IEEE Trans. Comput. **63**(11), 2716–2728 (2014)
16. Hawkes, P., Rose, G.G.: The complexity of fast algebraic attacks on stream ciphers. In: Lecture Notes in Computer Science, Advances in Cryptology—CRYPTO2004, pp. 390–406. Springer, Berlin (2004)
17. Konheim, A.G.: Computer Security and Cryptography, p. 544. Wiley, California (2007)
18. Shah, T., Upadhyay, D.P., Sharma, P.: A comparative analysis of different LFSR based ciphers and parallel computing platforms for development of generic cipher compatible on both hardware and software platforms, Jaipur (2014)

Implementation of Taguchi Method in the Optimization of Roller Burnishing Process Parameter for Surface Roughness

Kiran A. Patel and Pragnesh K. Brahmbhatt

Abstract Need of industrial growth for developing country gives rapid acceleration in the field of technical research. Industries are very much aware of producing mechanical component with good surface quality without allowing a margin of error. Among the different challenges of industry, surface quality is the key factor now a day's which can be improved by a novel after machining process known as burnishing process. This paper is mainly concerned with the effect of different process parameter on the surface roughness of aluminum alloy and the optimization of response measure. To achieve the goal of proposed work first pilot experiment is intended to ascertain the range of different parameters required for the experimental design methodology. Analysis of variance and signal to noise ratio are applied as statistical analysis to find out the significant control factor and optimize the level. The result shows the optimum set of process parameter having a value of 850 RPM spindle speed, 8 mm interference, 0.024 mm/rev feed and 4 no. of tool pass predict 0.010 μm surface roughness value which is having a greater agreement with the experimental value.

Keywords Burnishing · Design of experiment · Pilot experiment · Taguchi's orthogonal array

1 Introduction

Surface treatment is to be considered most important aspect for growing industrial manufacturing processes. To confer various physical and mechanical properties it has been utilized, for example, appearance, consumption, grinding, wear and

K.A. Patel (✉)
PAHER University, Udaipur, Rajasthan, India
e-mail: kpatelp@gmail.com

P.K. Brahmbhatt
L. D. College of Engineering, Ahmadabad, Gujarat, India
e-mail: pragneshbrahmbhatt@gmail.com

S.C. Satapathy and S. Das (eds.), *Proceedings of First International Conference on Information and Communication Technology for Intelligent Systems: Volume 2*,
Smart Innovation, Systems and Technologies 51, DOI 10.1007/978-3-319-30927-9_19

weariness resistance [15]. The execution of a machined part, for example, weariness quality, burden bearing limit, erosion, and so on depends to a vast degree at first glance as geography, hardness, nature of anxiety and strain impelled at the first glance area [5]. Roller Burnishing is one of the most important finishing processes which produce plastic deformation of surface in addition with compressive residual stresses and thereby it helps in improvement of fatigue resistance [16, 19]. Material from the peak is cold flow into valley under the action of plastic deformation formed with the help of roller burnishing. Result of this process gives a mirror-like finish with a work hardened, tough and wears as well as corrosion resistant surface which is key objective of the proposed study [17].

This research paper investigates the surface roughness for roller burnishing process (RBP) for aluminum alloy 6061. Pilot experiment is proposed to determine the range of different parameters required for roller burnishing process. Taguchi's L25 orthogonal array is used to analyze the parametric combination of four process parameters namely speed, Interference, feed and no. of tool pass [8]. Analysis of variance (ANOVA) and signal to noise ratio are employed to identify the significant process parameter and optimum level [4].

1.1 Objectives of the Present Investigation

In this section, an effect of roller burnishing process parameter is examined for better surface finish on Al alloy 6061 with following different modules.

- To identify the levels and working range of Roller burnishing process parameters by conducting Pilot experiment
- To examine the effects of different process parameters on measured surface roughness in Roller burnishing process from measured experimental results.
- To optimize the surface roughness using Taguchi method
- To validate the results by conducting confirmation experiments.

1.2 Material and Experimental Work

1.2.1 Work-Piece Detail

The chemical composition of aluminum alloy 6061 material used for the proposed work is given in Table 1. A specimen of diameter 40 mm was taken in roller burnishing process.

Table 1 Machine specification [1]

Constituent	Actual	Standard
Al	97.5	Balance
Cr	0.1	0.04–0.35
Cu	0.17	0.15–0.40
Fe	0.48	0.7 max.
Mg	0.8	0.8–1.2
Mn	0.15	0.15 max.
Si	0.5	0.40–0.8
Ti	0.15	0.15 max.
Zn	0.15	0.25 max.

1.2.2 Experimental Condition

The experiments have to be accomplished on SPIN FLAT CNC lathe machine. The following steps are followed in the operation:

- The work piece has mounted and clamp on the work table.
- The attachments have made for the burnishing operation of the work piece and a work piece of 40 mm diameter is selected.

While performing various experiments, the following precautionary measures have taken:

- The order and replication of experiments has randomized to avoid bias, if any, in the results.
- Each set of experiments has to be performed at room temperature in a temperature range of (32 ± 2 °C).
- Before taking measurements of surface roughness, the work piece has cleaned with Acetone.

2 Pilot Experiments

The intention of the pilot experiments is to learn the variations of the Roller burnishing process parameters effect on surface roughness [18]. The effects of these input process parameters are examined on surface roughness using one factor at a time approach [7].

Apart from the parameters mentioned above following parameters were kept constant during the experiments:

- Work Material: Aluminum alloy 6061
- Burnishing Tool: Tungsten carbide roller

Table 2 Effect of process parameter on surface roughness

Exp. no.	N	Ra	I	Ra	F	Ra	n	Ra
1	50	0.120505	2	0.216772	0.024	0.086501	1	0.140683
2	250	0.11175	3.5	0.154996	0.044	0.09809	2	0.123197
3	450	0.11084	5	0.11084	0.064	0.11084	3	0.11084
4	650	0.117774	6.5	0.084306	0.084	0.124752	4	0.103611
5	850	0.132554	8	0.075392	0.104	0.139826	5	0.101509

2.1 *Effect of Process Variable on Surface Roughness*

The observed data of experiment for the surface roughness with different values of process parameters are given in Table 2. The spindle speed is varied from 50 to 850 RPM. The values of the other parameters are kept constant and their values are given as Interference = 5 mm; Feed = 0.064 mm\rev and No. of Tool Passes = 3.

The Interference is varied from 2 to 8 mm. The values of the other parameters are kept constant and their values are given as Spindle speed = 450 RPM; Feed = 0.064 mm\rev and No. of Tool Passes = 3. The spindle speed is varied from 0.024 to 0.104 mm/rev. The values of the other parameters are kept constant and their values are given as Spindle speed = 450 RPM; Interference = 5 mm and No. of Tool Passes = 3. The No. of Tool Passes is varied from 1 to 5. The values of the other parameters are kept constant and their values are given as Interference = 5 mm; Feed = 0.064 mm\rev and No. of Tool Passes = 3 μm.

Figure 1 shows that the value of surface roughness decreases with the increase in spindle speed up to a certain limit. The value of surface roughness then starts to increase rapidly with a little wavy pattern. The surface roughness decreases with an increase in the Interference and shows a decreasing trend continuously. The surface roughness increases with the increase in the feed rate in a straight line fashion. The values of the surface roughness show a decreasing pattern with increase in No. of Tool Passes.

2.2 *Selection of Parameter Range Based on Pilot Investigation*

The pilot experiments were carried by varying the process parameters, e.g. spindle speed, Interference, Feed and No. of tool to study their effect on surface roughness. The ranges of these process parameters are given in Table 3.

From these ranges of the process parameters, different levels of process parameters would be selected for the Taguchi experimental design and experimental design methodology using response surface methodology.

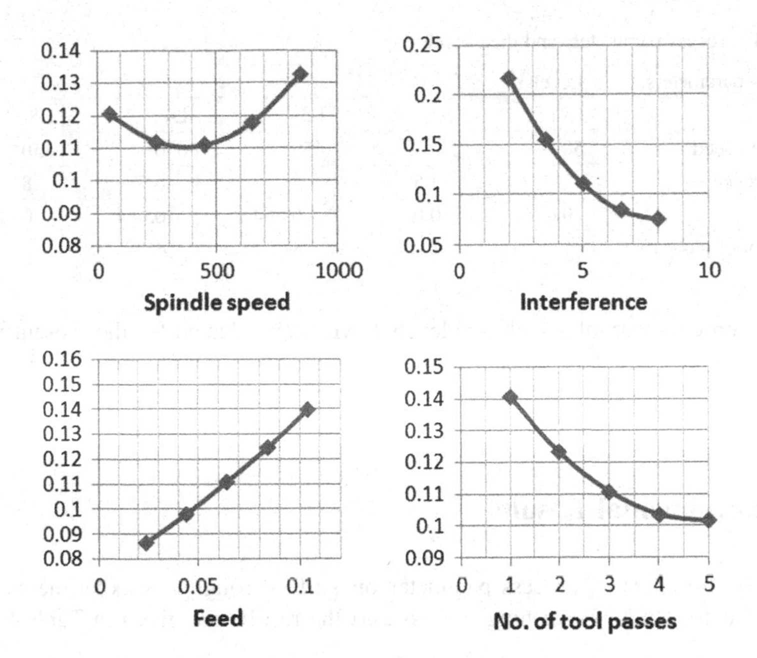

Fig. 1 Scatter plots of parameters versus surface roughness

Table 3 Process parameters with their ranges

Process parameters	Symbol	Units	Range
Spindle speed	N	RPM	50–850
Interference	I	mm	2–8
Feed	F	mm\rev	0.024–0.104
No. of tool passes	N	–	1–5

3 Taguchi Method of Optimization

The present work gives the application of Taguchi experimental method for the purpose of optimizing the real engineering problem. To find out the effect of parameters involved in process on the surface roughness the design of experiment was selected and accordingly the experiments were conducted [13]. The experimental results are discussed subsequently in the following sections.

3.1 Selection of Orthogonal Array

Generally experimental procedures are expensive and time consuming. So the prime requirement of experimental procedures is to fulfill the design objectives with the least number of experiments [14].

Table 4 Process parameters and their levels

Process parameters	Levels				
	L1	L2	L3	L4	L5
Spindle speed	50	250	450	650	850
Interference	2	3.5	5	6.5	8
Feed	0.024	0.044	0.064	0.084	0.104
No. of tool passes	1	2	3	4	5

Four process variables with five levels have been selected for the present work. Process variable are denoted and mentioned in the column of factors. The levels of the individual process parameters/factors are given in Table 4.

4 Experimental Results

To study the effect of process parameter on surface roughness experiments were conducted for Roller burnishing process and the results are given in Table 5.

4.1 Analysis and Discussion of Results

From the experimental data S/N ratio of the surface roughness for each variable at various levels were obtained. The main effects of process variables on surface roughness were plotted. The response curves are used for finding out the parametric effects on the response characteristics [9]. The purpose of analysis of variance (ANOVA) is to identify the variables having significant effect and to find their effects on the surface roughness. Optimal settings of process variables in terms of mean response characteristics are established by analyzing the response curves and the ANOVA tables [11].

4.2 Effect of Process Variable on Surface Roughness

Experiments were conducted using L27 orthogonal array to check the effects of process parameters on the surface roughness [6]. Figure 2 with main effects plot for means shows an average value for each reading of a particular parameter. From the graph, the mean value is a maximum at 650 RPM and a minimum at 850 RPM. The Mean value is highest at first level and observed lowest at the final level of Interference. Mean value is minimum for 0.024 mm\rev and observed higher at 0.084 mm\rev feed rate. Better surface property is achieved at level 4 of No. of tool pass and higher value is observed at level 1.

Table 5 Taguchi's L27 standard orthogonal array

Exp. no.	Spindle speed	Interference	Feed	No. of tool passes	Surface roughness
1	50	2	0.024	1	0.210
2	50	3.5	0.044	2	0.159
3	50	5	0.064	3	0.120
4	50	6.5	0.084	4	0.093
5	50	8	0.104	5	0.077
6	250	2	0.044	3	0.191
7	250	3.5	0.064	4	0.142
8	250	5	0.084	5	0.104
9	250	6.5	0.104	1	0.133
10	250	8	0.024	2	0.079
11	450	2	0.064	5	0.185
12	450	3.5	0.084	1	0.226
13	450	5	0.104	2	0.161
14	450	6.5	0.024	3	0.073
15	450	8	0.044	4	0.084
16	650	2	0.084	2	0.286
17	650	3.5	0.104	3	0.212
18	650	5	0.024	4	0.089
19	650	6.5	0.044	5	0.090
20	650	8	0.064	1	0.088
21	850	2	0.104	4	0.097
22	850	3.5	0.024	5	0.126
23	850	5	0.044	1	0.136
24	850	6.5	0.064	2	0.110
25	850	8	0.084	3	0.097

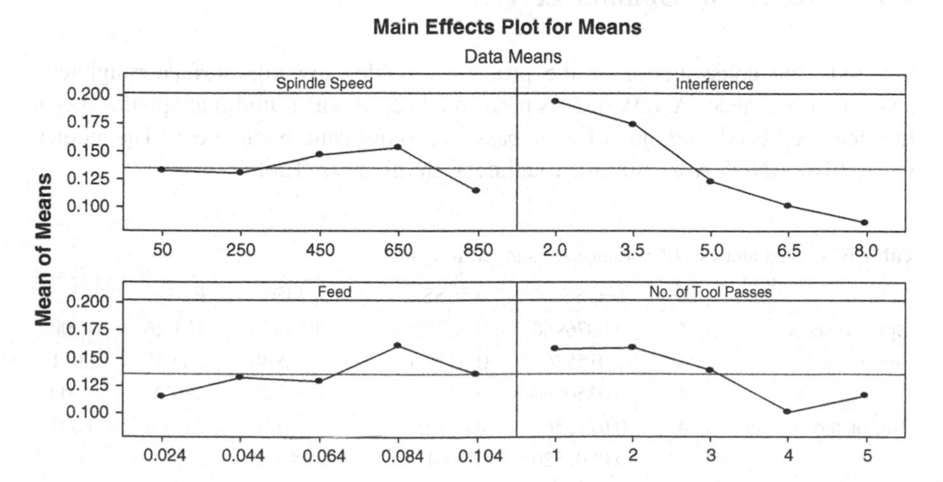

Fig. 2 Effects of process parameters on surface roughness (means)

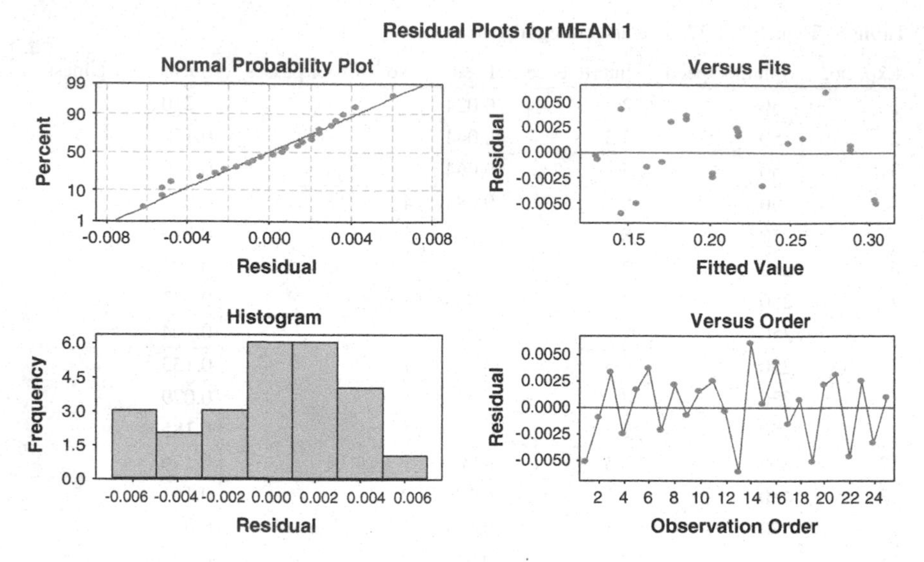

Fig. 3 Residual plots for surface roughness

To estimate the statistics for the problems like non normality, non random variation, non constant variance, higher-order relationships, and outliers' generally residual plots are used. It can be seen from Fig. 3 that the residuals trail a straight line in the normal probability plot and the symmetric nature of histogram indicates that the residuals are distributed normally. Residuals possess constant variance as they are scattered randomly around zero in residuals versus the fitted values.

4.3 Selection of Optimal Levels

To study the consequence of the process variables towards surface roughness, analysis of variance (ANOVA) was performed [2]. It was found that spindle speed, Interference, Feed and no. of tool pass are significant parameters. The pooled versions of ANOVA for surface roughness are given in Table 6.

Table 6 Pooled analysis of variance for surface roughness

Source	DF	Seq SS	Adj SS	Adj MS	F	P
Spindle speed	4	0.0076960	0.0076960	0.0019240	101.26	0.000
Interference	4	0.0155360	0.0155360	0.0038840	204.42	0.011
Feed	4	0.0250960	0.0250960	0.0062740	330.21	0.000
No. of tool passes	4	0.0212560	0.0212560	0.0053140	279.68	0.004
Error	8	0.0001520	0.0001520	0.0000190		
Total	24	0.0697360				

Table 7 Response table for surface roughness

Level	Spindle speed	Interference	Feed	No. of tool passes
1	14.69	15.09	12.09	15.64
2	14.20	14.69	12.84	14.68
3	13.95	13.71	13.49	13.57
4	13.35	13.20	14.67	12.87
5	12.66	12.16	15.75	12.07
Delta	2.02	2.93	3.66	3.57
Rank	4	3	1	2

Table 8 Factor levels for predictions

Factor	Spindle speed	Interference	Feed	No. of tool passes
Levels	850	8	0.024	4

From this table, it is clear that all four process variables are significantly affecting the surface roughness. The response table (Table 7) shows the average of surface roughness for each level of each factor. The tables consist of ranks based on delta information, which compare the relative degree of effects. The delta statistic is the highest minus the lowest average for each factor [10].

The ranks and the delta values show that Feed have the greatest effect on surface roughness and is followed by no. of tool pass, interference and spindle speed in that order [20]. As surface roughness is the "smaller the better" type quality characteristic, the fifth level of spindle speed, the fifth level of interference, the first level of feed and fourth level of no. of tool pass provides better value of surface roughness as given in Table 8.

4.4 Predict Performance at the Optimum Setting

Using an optimum set of parameters achieved by response curve analysis, predict the value of S/N ratio and surface roughness in roller burnishing process [3] as shown in Table 9.

Table 9 Predicted performance

S/N ratio	Surface roughness
25.7655	0.010

Table 10 Comparison between predicated value and experimental value

Surface roughness	
Predicated value	Experimental value
0.010	0.013

4.5 Confirmation Experiment

In this step confirmation experiment was carried out to evaluate the predictive capability of the Taguchi method for surface roughness using roller burnishing process. The optimum parameters were settled and performance was measured for that set of parameter. Table 10 shows performance was compared with predicated performance and was found that the experimental value was nearer to the predicated value [12].

5 Conclusion

The viability of using Taguchi method to optimize preferred roller burnishing process parameter for highest performance was investigated using CNC Spin Flat Lathe machine. In this paper, the effect and optimization of four process variables on surface roughness has been evaluated by applying L25 orthogonal array. For surface roughness, the most significant factor observed from the data is feed. The optimum results of the Taguchi experiment identify that 850 RPM spindle speed, 8 mm interference, 0.024 mm/rev feed and 4 no. of tool pass predict 0.010 μm surface roughness value which is having a greater agreement with the experimental value. Response characteristic is mostly influenced by Feed and is least influenced by spindle speed. Confirmation experiment was done using an optimum combination show the closeness of the prediction capability of Taguchi method.

References

1. ASTM Stand. B 209-96, In 1996 Annual Book of ASTM Standards, vol. 03.03, ASTM, West Conshohocken, PA (1996)
2. Benardos, P.G., Vosniakos, G.C.: Prediction of surface roughness in CNC face milling using neural networks and Taguchi's design of experiments. Robot. Comput. Integr. Manuf. **18**(5), 343–354 (2002)
3. Benardos, P.G., Vosniakos, G.C.: Predicting surface roughness in machining: a review. Int. J. Mach. Tools Manuf. **43**(8), 833–844 (2003)
4. Chen, C.H., Shiou, F.J.: Determination of optimal ball-burnishing parameters for plastic injection moulding steel. Int. J. Adv. Manuf. Technol. **21**(3), 177–185 (2003)
5. El-Axir, M.H.: An investigation into roller burnishing. Int. J. Mach. Tools Manuf. **40**(11), 1603–1617 (2000)

6. El-Axir, M.H., El-Khabeery, M.M.: Influence of orthogonal burnishing parameters on surface characteristics for various materials. J. Mater. Process. Technol. **132**(1), 82–89 (2003)
7. El-Khabeery, M.M., El-Axir, M.H.: Experimental techniques for studying the effects of milling roller-burnishing parameters on surface integrity. Int. J. Mach. Tools Manuf. **41**(12), 1705–1719 (2001)
8. El-Taweel, T.A., El-Axir, M.H.: Analysis and optimization of the ball burnishing process through the Taguchi technique. Int. J. Adv. Manuf. Technol. **41**(3–4), 301–310 (2009)
9. Garg, R., Singh, H.: Optimisation of process parameters for gap current in wire electrical discharge machining. Int. J. Manuf. Technol. Manage. **25**(1), 161–175 (2012)
10. Gharbi, F., Sghaier, S., Al-Fadhalah, K.J., Benameur, T.: Effect of ball burnishing process on the surface quality and microstructure properties of AISI 1010 steel plates. J. Mater. Eng. Perform. **20**(6), 903–910 (2001)
11. Kiran, A., Nikunj, K., Disha, B.: Parametric optimization of process parameter for roller burnishing process. Int. J. Sci. Res. Dev. **1**(4), 806–811 (2014)
12. Li, F.L., Xia, W., Zhou, Z.Y., Zhao, J., Tang, Z.Q.: Analytical prediction and experimental verification of surface roughness during the burnishing process. Int. J. Mach. Tools Manuf. **62**, 67–75 (2012)
13. Lin, Y.C., Yan, B.H., Huang, F.Y.: Surface improvement using a combination of electrical discharge machining with ball burnish machining based on the Taguchi method. Int. J. Adv. Manuf. Technol. **18**(9), 673–682 (2001)
14. Lin, C.L.: Use of the Taguchi method and grey relational analysis to optimize turning operations with multiple performance characteristics. Mater. Manuf. Process. **19**(2), 209–220 (2004)
15. Nemat, M., Lyons, A.C.: An investigation of the surface topography of ball burnished mild steel and aluminium. Int. J. Adv. Manuf. Technol. **16**(7), 469–473 (2000)
16. Randjelovic, S., Tadic, B., Todorovic, P.M., Vukelic, D., Miloradovic, D., Radenkovic, M., Tsiafis, C.: Modelling of the ball burnishing process with a high-stiffness tool. Int. J. Adv. Manuf. Technol., 1–10 (2015)
17. Rao, D.S., Hebbar, H.S., Komaraiah, M., Kempaiah, U.N.: Investigations on the effect of ball burnishing parameters on surface hardness and wear resistance of HSLA dual-phase steels. Mater. Manuf. Process. **23**(3), 295–302 (2008)
18. Sarıkaya, M., Güllü, A.: Taguchi design and response surface methodology based analysis of machining parameters in CNC turning under MQL. J. Clean. Prod. **65**, 604–616 (2014)
19. Sastry, M.N.P., Devi, K.D., Reddy, K.M.: Analysis and optimization of machining process parameters using design of experiments. Ind. Eng. Lett. **2**(9), 23–32 (2012)
20. Sathiya, P., Aravindan, S., Haq, A.N.: Optimization for friction welding parameters with multiple performance characteristics. Int. J. Mech. Mater. Des. **3**(4), 309–318 (2006)

Relationship Strength Based Access Control in Online Social Networks

Abhinav Kumar and Nemi Chandra Rathore

Abstract Online social networking site is a platform, where a person can communicate with others by creating his own profile. Lots of contents (e.g., photos, videos, etc.) are being generated every minute, because of its popularity. Generally, users do not want to share their all information with everyone as some of them are very sensitive, this raises privacy and confidentiality issues. So, Online Social Network (OSN) requires an effective and reliable access control mechanism to protect the users' contents. This paper presents a novel Relationship Strength Based Access Control (RSBAC) mechanism which allow users to provide the access to their contents based on intimacy degree or closeness with respect to their friends. A reliable intimacy degree is calculated by considering the social activities and profile similarities between the friends. The content owner can make the access control policy by assigning different range of intimacy degrees. Users whose intimacy degrees are in the acceptable range, can access the contents.

Keywords Online social networks · Access control · Access policies

1 Introduction

Online social networks are web-based services that provide a platform to people to share their information and communicate with other members of the network through links. Sharing objects, organizing online events, creating groups etc. are the various type of social communications [1]. With the development of Web 2.0 technologies in the last few years, the number of users on OSN has increased exponentially [2]. Fostering relationships and sharing data, OSN attract up to 4

A. Kumar (✉) · N.C. Rathore
Department of Computer Science, Central University of South Bihar,
BIT Campus, Patna 800014, India
e-mail: abhinavanand05@gmail.com

N.C. Rathore
e-mail: nemichandra@cub.ac.in

S.C. Satapathy and S. Das (eds.), *Proceedings of First International Conference on Information and Communication Technology for Intelligent Systems: Volume 2*,
Smart Innovation, Systems and Technologies 51, DOI 10.1007/978-3-319-30927-9_20

users among each 5 Internet users [3]. So, social networks have become the most successful service on the web. Now-a-days, most developed social networks only provide the basic access control policies like publicly available, private or accessible to direct contacts only [2]. It is either too prohibitive or too loose because it limits too much information sharing i.e., private or grant access to all users i.e., public. Currently, there are many underlying problems in access control mechanisms: (1) only a small percentage of users change the default access control settings to define their own access control policies [4], (2) when these access control mechanisms are used, they fail to address the required fine-grained control to avoid privacy violations [5]. Studies have showed that due to lack of privacy control on MySpace, users have abandoned this OSN and have migrated to other OSNs for their better privacy-preserving means [6]. Almost all the previous mechanisms like, Role-based, Rule-based and Reachability-based ignore the fact that for a client, even two of his friends having the same role, fit in the same rule and can be reached by the same way, but they still can have different intimacy degrees with respect to the client. Thus, to protect sensitive personal data in OSNs, it requires a high level of protection by means of appropriate access control mechanisms.

2 Related Works

Privacy and security in OSNs are currently challenging areas for researchers. Several studies have been done in past few years to resolve these issues. Wang et al. [7] proposed an access model which uses intimacy degree between friends for access control. Li et al. [8] discussed an access model in which, permissions are associated with roles and users are assigned to appropriate roles for access control. Carminati et al. [9] gave a Rule-based access control model that allows the specification of access rules for online resources where authorized subjects are denoted in terms of the relationship type, depth and trust level existing between users in the network. Abdessalem et al. [2] proposed a Reachability-based access control where, access policies are specified in terms of constraints on the type, direction, depth of relationship and trust level between users, as well as on the users properties.

3 Proposed Methodology

3.1 System Model

In a typical online social networking site, $U = \{u_1, u_2, u_3, \ldots, u_n\}$ represents the set of users of OSN, who communicate and interact with each other via different type of activities (e.g., sharing resources, commenting on resources etc.). We can model the whole social networking site as a directed random graph $G = (V, B)$, where

Table 1 List of social activities of users

ID	social activities
t_1	**Leaving a message on a user's Timeline** Example: Alice leaves a message on Bob's timeline
t_2	**Commenting a resource that is owned by a user** Example: Alice comments on one of the Bob's photo
t_3	**Sharing a resource owned by a user** Example: Alice share a photo of Bob
t_4	**Tagging a user** Example: Alice tag Bob in her one of the photos
t_5	**Visiting user's Timeline** Example: Alice visit Bob's Timeline for searching Bob's latest update

$V = \{v_1, v_2, v_3, \ldots, v_n\}$ is a set of user-resource-attribute tuple. Each user-resource-attribute tuple is denoted as $v_i = \langle u_i, R_{u_i}, A_{u_i} \rangle$ where R_{u_i} and A_{u_i} respectively denotes the set of resources (e.g., photos, videos etc.) and set of attributes (e.g., hometown, school name, gender etc.) of user $u_i \cdot R_{u_i} = \{r_{u_i,k} | u_i \in U, k = \{1, 2, 3, \ldots, n\}\}$, $A_{u_i} = \{a_{u_i,k} | u_i \in U, k = \{1, 2, 3, \ldots, n\}\}$. Each resource has its own identifier and it can be represented as '*rid*'. Here we introduce sensitivity level (S_{level}) for each of the resources. More sensitive information is assigned higher value of $S_{level} \in [0, 1]$. Each resource with its sensitivity level is represented by a tuple: $\langle rid, S_{level} \rangle \cdot \mathrm{B} = \{b_1, b_2, \ldots, b_n\}$ is the set of the user's activities on OSN, where b_i is the set of directed random edges from user u_i to resources belonging to user u_j. For a typical user u_i, $b_i \in B$ as the set $b_i = \{e_{i,1}, e_{i,2}, \ldots, e_{i,n}\}$, where each $e_{i,x}$ is represented by a tuple: $\langle u_i, r_{u_j,k}, t_l \rangle$, where t_l (see Table 1) denotes one of the u_i's activities. The tuple means a user u_i does a activity of type t_l to the resource $r_{u_j,k}$ which belongs to another user u_j. If *Alice* leaves a comment (*Alice's* activity) on a photo in one of the user *Bob's* albums, the activity is denoted as $\langle Alice, r_{Bob,k}, t_2 \rangle$.

3.2 *Intimacy Degree Model*

In our daily lives, we do friendship with number of peoples, our communication with some of the people is quit often. If two people belong to the same city and if they also share the same school for their education and if there is some communication between them, then those people may have more closeness between them with respect to other. The situation on online social networking is quite similar. If two person on OSN have some profile similarity (in terms of attribute) and if both of them communicate with each other very frequently, then they may have more closeness with each other in comparison to people that have no profile similarity. We can represent intimacy degree model as: *DEG* = (*V*, *InDeg*), Where *V* is the set of user-resource-attribute tuple and *InDeg* is the set of single intimacy degree description which can be represented as:

$$InDeg = \{<u_i, u_j, I_{Deg}> | u_i, u_j \in U, 0 \leq I_{Deg} \leq 1\} \quad (1)$$

which means the intimacy degree of u_i with respect to u_j equals to I_{Deg}. If value of I_{Deg} is higher, means more intimacy or closeness between the users.

3.3 Social Access Control Model

A good access control model should have the ability to adapt the dynamic changes of closeness between a client and his friends, because relationship on the OSNs might change with time [7]. Current mechanisms like role based, rule based, reachability based don't adapt the dynamic social relation between users [7]. So, here we introduce RSBAC model which might be very useful to overcome such type of problem because, we are using different types of social activities to classify the friends and we know that the social activities are dynamic in nature, so intimacy degree will also be dynamic which depends on the amount of social activities done by client's friends. In our RSBAC model, we provide access based on intimacy degree which is dynamic, so users don't need to change policies with changing interaction distribution. Figure 1 shows the access control model where, $\phi_{I_{Deg}} \in [0, 1]$ denotes the threshold value defined by the owner. For higher level of security owner have to choose greater threshold value, because if an owner put very less value for threshold then large number of his friends can access his all the contents. If the intimacy degree of requester with owner is greater than equal to $\phi_{I_{Deg}}$ or S_{level} (sensitivity level), access will be granted.

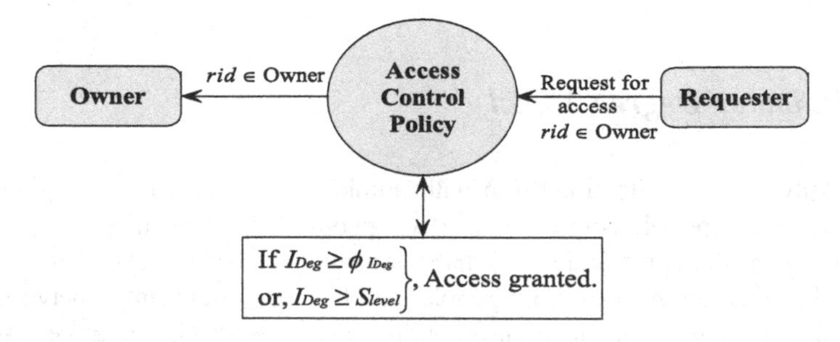

Fig. 1 Access control model

3.4 Access Control Policies

Our access control policy is represented by a tuple:

$$P = \langle owner, \langle rid, S_{level} \rangle, accessor, access\ condition, action \rangle, \quad (2)$$

where

- *Owner: The contents generated by a user on OSN in the form of videos, images,* etc. *are known as the resources of that user and that user is known as the owner of those resources.*
- *rid represents the identifier of a resource and S_{level} is the sensitivity level of that resource belonging to the owner.*
- *Accessor: Accessor is any set of potential audiences or users to whom the authorization is granted, which is defined in Definition 1.*
- *Access condition: It will decide whether access of resource will be granted or denied.*
- *Action ∈ {grant, deny} is the authorization effect of the access control policy.*

Definition 1 (*Accessor Specification*): Let $ac \in U \cup RT \cup G$ be a user $u \in U$, a relationship type $rt \in RT$ or a group $g \in G$, where *RT* is the set of relationship type supported by an OSN and *G* is the set on groups on OSN. Let $at \in \{UN, RN, GN\}$ be a type of the accessor specification (user name, relationship type, group name respectively). The accessor specification is defined as a set, accessors $= \{as_1, as_2, \ldots as_n\}$, where each element is a tuple $\langle ac, at \rangle$.

Access Condition: Access will only be granted if the accessor satisfy at least one of the access conditions:

$$AC_1 : I_{Deg} \geq \phi_{I_{Deg}} \quad or, \quad AC_2 : I_{Deg} \geq S_{level} \quad (3)$$

- *I_{Deg} represents the intimacy degree between owner and accessor.*
- *$\phi_{I_{Deg}} \in [0, 1]$ represents a threshold value defined by the owner.*

Access Policies: *Alice* makes an access policy for her friends' for her one of the resource *holiday.jpg* ($S_{level} = 0.6$) with the access condition AC_1, where she defined $\phi_{I_{Deg}} = 0.8$

$$P_1 = \langle Alice, \langle holiday.jpg, 0.6 \rangle, \langle friends\ of, RN \rangle, I_{Deg} \geq 0.8, grant \rangle \quad (4)$$

Alice makes an access policy for her friends' for her one of the resource *holiday.jpg* ($S_{level} = 0.6$) with the access condition AC_2:

$$P_2 = \langle Alice, \langle holiday.jpg, 0.6\rangle, \langle friends\ of, RN\rangle, I_{Deg} \geq 0.6, grant\rangle \tag{5}$$

Algorithm 1 Access Control Protocol

```
 1: procedure ACCESS CONTROL ENFORCEMENT
 2:     P ← getPolicies(rid)                    ▷ P= set of access policies
 3:     if P = φ then
 4:         P ← default access policy
 5:     else
 6:         if I_Deg ≥ φI_Deg then
 7:             Access granted
 8:         else if I_Deg ≥ S_level then
 9:             Access granted
10:         else
11:             Access deny
12:         end if
13:     end if
14: end procedure
```

Suppose one of the *Alice's* friends, *Bob* wants to access resource *holiday.jpg*, where intimacy degree of *Alice* with *Bob* is 0.7. The access control to a user will be granted on the basis of Algorithm 1. According to Algorithm 1 access is granted to *Bob* for resource *holiday.jpg* because, access condition $I_{Deg} \ngeq \phi_{I_{Deg}}$ but access condition $I_{Deg} \geq S_{level}$.

3.5 User Activities Model

Activity is an action which is done by a user on the resources. We are considering five (M = 5) (frequently done social activities on Facebook) types of social activities, listed in Table 1 which can be represented as: $T = \{t_i | 1 \leq i \leq M\}$. The set of social activities for a social networking site may change depending on functionality and applications which changes over time.

3.6 Friends Classification Definition

We classify friends into three classes cls_1, cls_2, cls_3 depending on ranking and trust level, as shown in Table 2. $cls_1 > cls_2 > cls_3$ is the classification order. For a user u, social activities of user u over his friend f_i is defined by the tuple:

Table 2 Classification of friends

CID. Classification and description	
cls_1	**Close friends**: Friends with C communicates frequently
cls_2	**Normal friends**: Friends that C communicates occasionally
cls_3	**Casual friends**: Friends with C communicate rarely

$B_{f_i} = \{<t_j, N_{t_j}> | 1 \leq j \leq M, N_{t_j} \neq 0\}$, where $f_i = i^{th}$ friend of user $u, f_i \subseteq U, t_j =$ a type of user u's social activity with respect to f_i's resources, $t_j \in T, N_{t_j} =$ the number of social activities (whose type are t_j) that user u does to all of f_i's resources.

4 Intimacy Degree Calculation Process

4.1 *Naive Bayes Based Classification of Friends*

Naive Bayes classifier [10, 11] has been used for the classification of friends under the categories Close, Normal and Casual. In the experiment, first the user is asked to manually pick some of the friends (normally 3–5) for each of the classes cls_1, cls_2, cls_3. Now we know the classification of each picked friend and by considering the social activities of those picked friend, we make a training set for Naive Bayes classifier and calculate the probability of residence of each unclassified friend in each classes. For an unclassified friend of the client, the maximum probability determines its classification, i.e., the one in which he get the maximum probability.

4.2 *Profile Similarity*

Similarity of profile [12] gives an idea of closeness between the friends. We have considered different attributes (1) Mutual Friends, (2) Home Town, (3) Age Group, (4) Gender, (5) Language, (6) School Name, (7) Graduate University/College, (8) Post Graduate University/College, (9) Employment (Professional Experience), (10) Religious Views, for profile matching to find profile similarity score. There are some challenges in profile matching: how to determine if two attribute values are equal, when the same conceptual value may be represented differently. This is commonly referred as the data disambiguation problem [13]. To resolve this problem, we make a dictionary of common variations of attribute like school names, graduation school names etc. as a lookup table. After that, we process each value using the Levenshtein Algorithm [14] to find the matching between the attributes.

4.3 Calculation of Profile Similarity Score

Different weights are assigned to different profile attributes. Binary weights are assigned to every attribute which is in the form of string. We use Levenshtein Algorithm [14] to process each of the attribute to find the matching. We assign 1 to matched and 0 to non-matched attribute fields. For our implementation, we use age difference up to "*3 years*" for the "Age group". It means, if we consider a person whose age is "*23 years*" then we assign weight as 1 to only those who belong to age group "*(23–26) years*" and rest as 0. Now, cosine similarity [15, 16] is calculated for all those attributes whose weight assignment has been done by the use of Levenshtein Algorithm. For our work, we add weight of mutual friends in the cosine similarity score to calculate the profile similarity score. Weight for mutual friends is calculated by the expression [17]:

$$W_{M_{f_i}} = \left[\frac{\frac{Mutual\,friends(u_j, f_i)}{Total\,number\,of\,u_j} \times 100}{WAF} \right] \tag{6}$$

where, u_j = Base profile, f_i = Friends profiles, $W_{M_{f_i}}$ = Mutual friends weight between (u_j, f_i), WAF = Weight Adjustment Factor.

$$\text{Profile Similarity Score} = cosine\ similarity\ score + W_{M_{f_i}} \tag{7}$$

4.4 Ranking of Friends

Based on profile similarity score of each client's friend, sorting (decreasing order based on profile similarity score) of friends is done inside each of the classifications cls_1 (close), cls_2 (normal), cls_3 (casual). Now, by taking classification order into account $(cls_1 > cls_2 > cls_3)$, final ranking of all the friends is done. After calculating the final ranking of all the client's friends, the ranking is bounded between 0 and 1, which is the intimacy degree scale for all the client's friends. Suppose a client has total 100 friends and a friend *Akash* have intimacy degree (I_{Deg}) = 0.9 with him, it means *Akash* is at 90th position in ranking from the last.

5 Result Discussion

We have used *Give Me My Data* [18] as Facebook application for collecting users' public informations like user name, gender, educational and other information. However availability of such informations are subjected to the privacy setting of an individual. For our experiment, we require data related to the social activities and

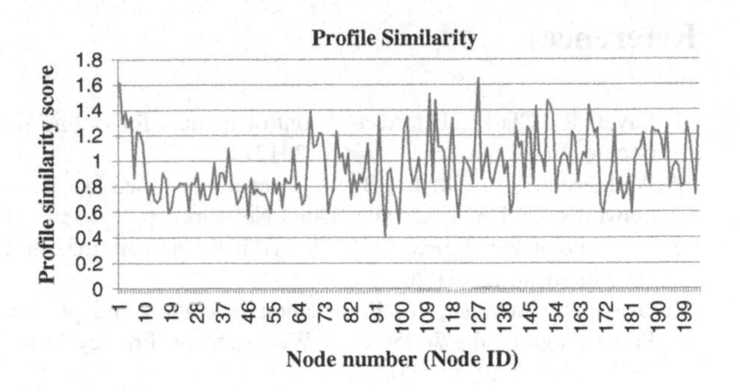

Fig. 2 Profile similarity between friends

profile attribute, but we are not able to find all such data, so a synthesized data set is created by using Python language. The data contains 204 friends from which 15 friends are taken for training, 5 friends from each of the classes viz. close (Class label = 1), normal (class label = 2) and casual (class label = 3). 32, 70 and 102 friends are respectively classified under Close, Normal and Casual classes after Naive Bayes classification and then profile similarity score has been calculated. Profile similarity score of each friend with respect to client is shown in the Fig. 2. It can be seen from the Fig. 2 node number 127 has the highest profile similarity with client. Now, ranking will be done inside each of the classification (cls_1, cls_2, cls_3) by sorting friends on the basis of profile similarity score (friend whose profile similarity is higher, is at the top). For our data set, a client who chooses the threshold value as 0.90 and the sensitivity of a resource belonging to client being 0.95, then only top 20 (top 10 % (approx) of the total friends) friends of the client get access.

6 Conclusion and Future Work

In this work, a novel access control model RSBAC for OSNs has been discussed, in which intimacy degree or closeness between a client and his friends is calculated by considering the social activities and profile similarities of his friends with client. Our ranking put such friends at the top whose social activities and profile similarity score with the client is higher with respect to all other friends. One of the limitations of this model is, if the attacker (mostly, profile similarity score = 0) does huge amount of social activities with respect to client, he is still unable to access all the sensitive resource of client, but he can rose to higher levels in ranking. In future, we are planning to consider bidirectional social activities (in terms of response on activities) which will be very useful to reduce attack from anomalous users. Bidirectional activities means, if *Alice* comments on *Bob* photo, *Alice* gets response on that comment from *Bob* too.

References

1. Sayaf, R., Clarke, D.: Access control models for online social networks. Soc. Netw. Eng. Secure Web Data Serv., 32–65 (2012)
2. Abdessalem, T., Dhia, I.B.: A reachability-based access control model for online social networks. In: Databases and Social Networks, pp. 31–36. ACM, New York (2011)
3. The state of social media 2011: Social is the new normal: http://www.briansolis.com/2011/10/state-of-social-media-2011/
4. Gross, R., Acquisti, A.: Information revelation and privacy in online social networks. In: Proceedings of the 2005 ACM Workshop on Privacy in the Electronic Society, pp. 71–80. ACM, New York (2005)
5. Masoumzadeh, A., Joshi, J.: Osnac: an ontology-based access control model for social networking systems. In: IEEE Second International Conference on Social Computing (SocialCom), pp. 751–759 (2010)
6. Baracaldo, N., López, C., Anwar, M., Lewis, M.: Simulating the effect of privacy concerns in online social networks. In: IEEE International Conference on Information Reuse and Integration (IRI), pp. 519–524 (2011)
7. Wang, Y., Zhai, E., Lua, E.K., Hu, J., Chen, Z.: isac: intimacy based access control for social network sites. In: Ubiquitous Intelligence & Computing and 9th International Conference on Autonomic & Trusted Computing (UIC/ATC), 2012 9th International Conference on, IEEE, pp. 517–524 (2012)
8. Li, J., Tang, Y., Mao, C., Lai, H., Zhu, J.: Role based access control for social network sites. In: IEEE Joint Conferences on Pervasive Computing (JCPC), pp. 389–394 (2009)
9. Carminati, B., Ferrari, E., Perego, A.: Rule-based access control for social networks. In: On the Move to Meaningful Internet Systems 2006: OTM 2006 Workshops, pp. 1734–1744. Springer, Heidelberg (2006)
10. Lewis, D.D.: Naive (bayes) at forty: the independence assumption in information retrieval. In: Machine Learning: ECML-98, pp. 4–15. Springer, Heidelberg (1998)
11. Rish, I.: An empirical study of the naive bayes classifier. In: IJCAI 2001 Workshop on Empirical Methods in Artificial Intelligence, vol. 3, pp. 41–46. IBM, New York (2001)
12. Jain, P., Kumaraguru, P., Joshi, A.: @ i seek'fb. me': identifying users across multiple online social networks. In: Proceedings of the 22nd International Conference on World Wide Web Companion, International World Wide Web Conferences Steering Committee, pp. 1259–1268 (2013)
13. Becker, J.L., Chen, H.: Measuring privacy risk in online social networks. PhD thesis, University of California, Davis (2009)
14. Su, Z., Ahn, B.R., Eom, K.Y., Kang, M.K., Kim, J.P., Kim, M.K.: Plagiarism detection using the levenshtein distance and smith-waterman algorithm. In: IEEE 3rd International Conference on Innovative Computing Information and Control, ICICIC'08, pp. 569–569 (2008)
15. Huang, A.: Similarity measures for text document clustering. In: Proceedings of the Sixth New Zealand Computer Science Research Student Conference (NZCSRSC2008), pp. 49–56. Christchurch, New Zealand (2008)
16. Tata, S., Patel, J.M.: Estimating the selectivity of tf-idf based cosine similarity predicates. ACM Sigmod Rec. **36**(2), 7–12 (2007)
17. Singh, R.R., Tomar, D.S.: Approaches for user profile investigation in orkut social network. arXiv preprint arXiv:0912.1008 (2009)
18. Facebook: https://apps.facebook.com/give_me_my_data/

Enhancement for Power Transfer Loadability in Long Transmission Line by Reactive Power Compensation

Manisha Jaswani and Satyadharma Bharti

Abstract This paper presents the reactive power management for 765 kV Extra High Voltage (EHV) transmission system. Reactive power compensation is used to control the power system voltage stability and increase the power transfer capability of the transmission line. The proposed 765 kV transmission line between Sipat—Seoni in India is simulated on PSCAD Library. The shunt compensation has been stimulated in the transmission line to maintain the system voltage. The effect of voltage, active power and reactive power has been studied at different transmission length and verifying the optimization transmission length for high power transfer. In order to increase the power transfer capability of the transmission system the series compensation is used. The work presented here also compares the shunt and series compensation using PSCAD simulation for better voltage control and its performance for power transfer capability.

Keywords Reactive power · Compensation · 765 kV line · Voltage stability · EHV

1 Introduction

With the rapid growth of our national Economy, the demand for power have been increased. As the demand of power increases, the power system develops large capacity, long distance and ultra high voltage, so that the stability of transmission system also increased. But one of the most important problem in the control of stability of the transmission system is the reactive power compensation [1].

M. Jaswani (✉) · S. Bharti
Department of Electrical Engineering, Rungta College of Engineering and Technology, Bhilai, India
e-mail: manishajaswani@gmail.com

S. Bharti
e-mail: sbharti@rungta.ac.in

S.C. Satapathy and S. Das (eds.), *Proceedings of First International Conference on Information and Communication Technology for Intelligent Systems: Volume 2*, Smart Innovation, Systems and Technologies 51, DOI 10.1007/978-3-319-30927-9_21

Compensation of transmission lines is meant the use of electrical circuits to modify the electrical characteristics of the lines within the prescribed limits.

The line requires the compensation of transmission line due to reduction of power transfer capability of lines which reduces the margin between the stable and unstable operation of the system. Reactive power is used to provide the voltage levels necessary for active power to do useful work [2]. Reactive power compensation is used as the most effective method for better efficiency of power generation, transmission, and distribution, voltage stability, increasing the load capability, reducing the system losses etc.

In order to compensate the reactive power, different types of reactive compensation devices are used [3]. Reactive compensation devices is the device which is connected in series or parallel with load and which is capable of supplying or absorbing reactive power demanded by load, by the help of proper reactive power compensation devices, the power limit of the transmission system will be improved [4]. Therefore, both the shunt and series compensation devices are to be placed one by one in the transmission system so to verify which reactive power compensation device is suitable for the transmission system [5, 6].

2 Reactive Power Management

Under no load condition, in any high voltage transmission line the receiving end voltage is higher than sending end voltage because of Ferranti effect. When there is a sudden load throw off then steady state over voltage occurs [7–9]. The easiest way to take these over voltages is to use line reactors during line charging. Under heavy load conditions, the loading capability of the line reduce due to the fixed reactors which may cause voltage collapse [10]. The line reactors capacity depends on the length of the line and the operating voltage of the line. Normally, for 400 kV transmission lines 60 % of line (MVAR) is usually compensated by the help of shunt reactors to control the voltages [11]. But this is not sufficient for full compensation, some other reactive power compensation devices should also be used for the compensation. Other devices used to manage the reactive power in the line is series compensation, shunt compensation or the combination of both the device i.e. series-shunt compensation [12].

2.1 Principle of Series Compensation

Series Compensation is basically a powerful tool to improve the performance of EHV lines. It consists of capacitors connected in series with the line at suitable locations.

Power transfer capacity of a line with and without compensation is given by,

$$P_1 = \frac{V_1 V_2}{X_L} \sin \delta. \tag{1}$$

$$P_2 = \frac{V_1 V_2}{(X_L - X_c)} \sin \delta. \tag{2}$$

$$\frac{P_2}{P_1} = \frac{X_L}{(X_L - X_c)}. \tag{3}$$

$$\frac{P_2}{P_1} = \frac{1}{(1 - \frac{X_c}{X_L})}. \tag{4}$$

$$\frac{P_2}{P_1} = \frac{1}{(1 - k)}. \tag{5}$$

where,

V_1 sending end voltage
V_2 receiving end voltage
X reactance of line
δ phase angle or load angle between V_1 & V_2
k degree of compensation

2.2 *Principle of Shunt Compensation*

Shunt capacitor are connected in parallel in the system are used mainly for power factor improvement and in harmonic filters. It also boost the voltage of bus. To improve the voltage at the receiving end shunt capacitors may be connected at the receiving end to generate and feed the reactive power to the load so that reactive power flow through the line and consequently the voltage drop in the line is reduced. Shunt compensation is also called load compensation.

3 System Description

The transmission line model is situated between Sipat (C.G) to Seoni (M.P) in India. The 3phase 765 kV, 675 km long transmission line having a load of 2000 MW is considered in Fig. 1.

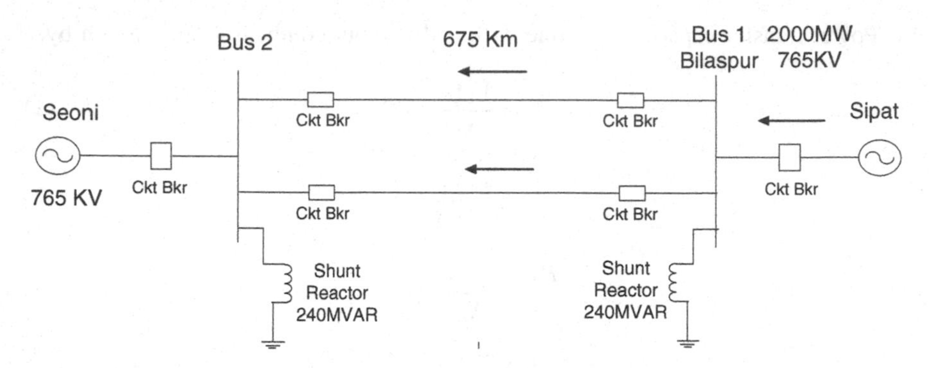

Fig. 1 Schematic wired diagram of transmission line

The 765 kV transmission line between Sipat and Seoni is represented in Schematic diagram as shown in Fig. 1. The line is double circuit transmission line and is the frequency dependent phase model. The length of the transmission line between Sipat and Seoni is 675 km. Equivalent sources at Sipat-Seoni is represented by 3-phase 50 Hz, 765 kV by infinite source. At both the end of bus, shunt reactors of 240 MVAR is connected in order to manage the reactive power. The circuit breakers is also connected in the double circuit transmission line at both the ends.

4 Simulation Model

The 3 phase, 765 kV double circuit transmission line between Sipat and Seoni is represented in PSCAD with using the standard block of library. PSCAD is used for simulation of the given power system and for analysis of the active and reactive power with and without compensation.

4.1 Simulation During Normal Condition

Under normal condition different output results for Sipat side. The system is simulated and loaded without any reactive power compensation. Active power and Reactive power flow are plotted. The simulation time is taken as 5 s.

The Active power and Reactive power at the source side i.e. at Sipat side is shown in Fig. 2a, b. The instantaneous value of active power from Sipat grid side is 210 MW and the instantaneous value of reactive power at grid side is—100 MVAR.

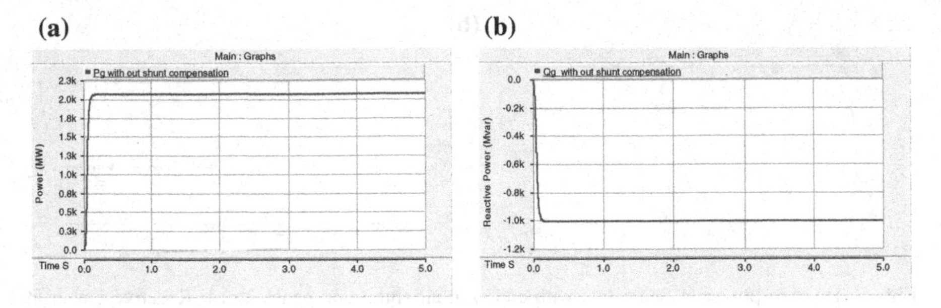

Fig. 2 **a** Active power at Sipat bus without compensation. **b** Reactive power at Sipat bus without compensation

4.2 Simulation with Shunt Compensation

At source side i.e. at Sipat the active power and reactive power is shown in Fig. 3a, b and also at load side i.e. at Seoni side the active power and reactive power is shown in Fig. 4a, b.

The instantaneous value of active power at load side with and without compensation is 200 and 199.5 MW respectively. The instantaneous value of reactive power at Seoni side with and without compensation is 190 and 650 MVAR respectively.

Thus, by comparing the graph of active power with and without compensation at load side, the value differs slightly but at the same time while comparing the values of reactive power with and without compensation at the load side has significant difference is observed.

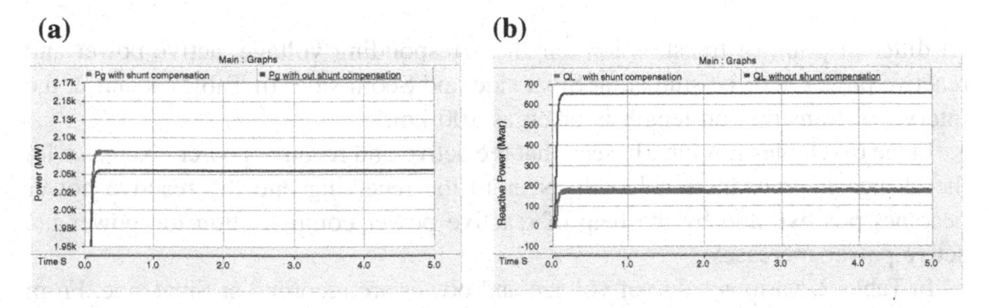

Fig. 3 **a** Active power at Sipat bus with and without shunt compensation. **b** Reactive power at Sipat bus with and without shunt compensation

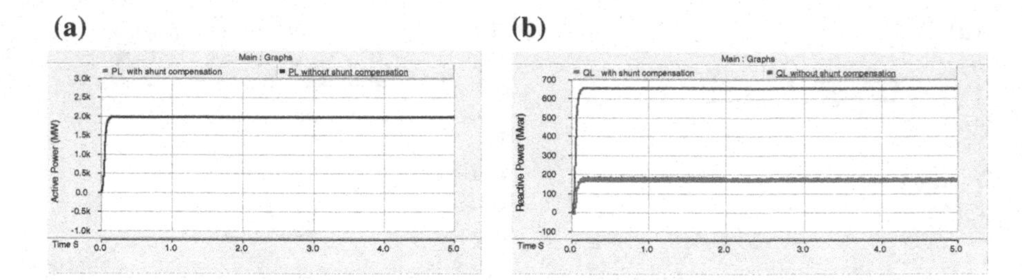

Fig. 4 **a** Active power at Seoni bus with and without shunt compensation. **b** Reactive power at Seoni bus with and without shunt compensation

Table 1 Measured value at Seoni side with and without compensation at different transmission length

Transmission length (km)	Without compensation			With compensation		
	Voltage (kV)	Active power (MW)	Reactive power (MW)	Voltage (kV)	Active power (MW)	Reactive power (MW)
100	765.3	11,900	−4203	765.3	11,930	−4707
200	765.3	6074	−1696	765.3	6086	−2193
300	765.3	4105	−804	765.3	4114	−1299
400	765.3	3125	−279.5	765.3	3132	−728.2
500	765.3	2545	−108.1	765.3	2551	−383.2
600	765.3	2166	431.8	765.3	2172	−58.82
700	765.3	1903	723.6	765.3	1908	232.5

4.3 *Transmission Length Optimization for Reactive Power Compensation*

At different point of transmission length corresponding voltage, active power and reactive power is to be studied at Sipat side and Seoni side. In Tables 1 and 2, the interval of transmission length is taken as 100 km.

In the two tables it is clearly seen that the active and reactive power changes with the change in transmission length. Near to the receiving end the reactive power becomes positive and by the help of reactive power compensation the power i.e. active power increases.

In Table 2, various value of voltage and power are recorded at Sipat side. From the table, it is clear that active power is decreasing, current is also decreasing so the voltage drop also reduces and hence, the terminal voltage increase.

At different values of transmission length, the voltage graph is plotted and shown in Fig. 5 and at the same time the values of active power and reactive power is also plotted at Sipat side shown in Fig. 6. Studies shows that by increasing the distance

Table 2 Measured value at Sipat side with and without compensation at different transmission length

Transmission length (km)		Without compensation		With compensation		
	Voltage (kV)	Active power (MW)	Power (MW)	Voltage (kV)	Active power (MW)	Reactive power (MW)
100	749.4	12,390	637.8	749.3	12,410	1103
200	757.7	6321	232.6	757.6	6330	706.8
300	760.4	4237	−96.77	760.4	4280	379.7
400	761.8	3256	−368	761.8	3260	110.7
500	762.7	2655	−610.9	762.6	2658	−131.6
600	763.2	2264	−841.4	763.1	2266	−361.5
700	763.6	1994	−1068	763.5	1997	−587.4

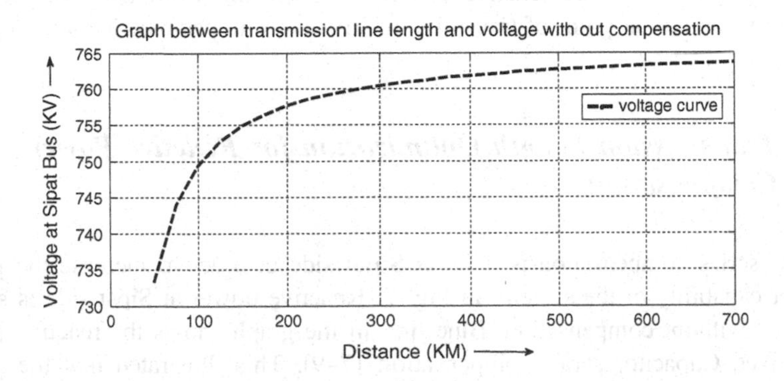

Fig. 5 Transmission length and voltage at Sipat bus

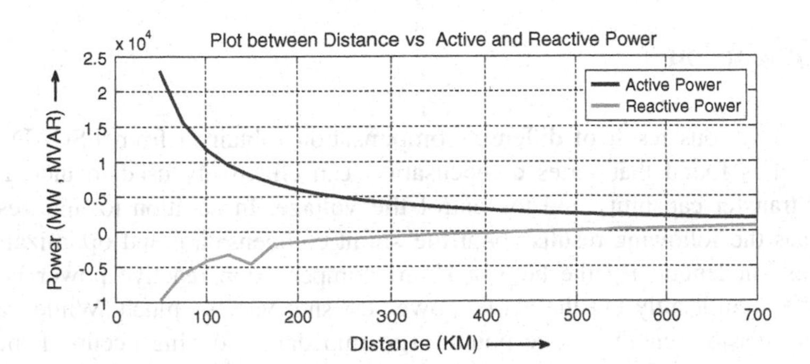

Fig. 6 Transmission length versus active and reactive power

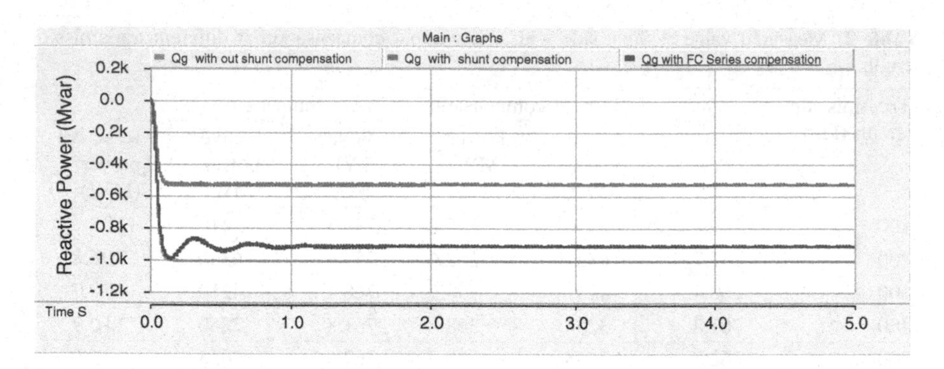

Fig. 7 Reactive power with shunt and series compensation

of transmission line, the active power is decreasing and the terminal voltage increasing gradually. The reactive power is increasing from negative value to positive value and at point of transmission line surge impedance loading occurs.

4.4 Transmission Length Optimization for Reactive Power Compensation

Finally, series compensation is done at Sipat side in order to increase the power transfer capability of the system. In Fig. 7, Reactive power at Sipat side is shown with and without compensation. Blue line in the graph shows the reactive power with Fixed Capacitor series compensation [7–9]. This illustrated that the power transfer capability increases. This shows that the reactive power is compensated by the help of Fixed Capacitor series compensation.

5 Conclusion

From the various result of different compensation (obtained from PSCAD simulation); it is found that series compensation can effectively used to increase the power transfer capability and to control the voltage. In addition to this, research work has the following results about the shunt compensation and optimization of transmission length. By the help of shunt compensation reactive power is compensated significantly but the active power doesnot vary so much. While varying the transmission length, at one point surge impedance loading occurs, hence the optimized transmission length is obtained. By adding the FACT devices with the compensation techniques the power transfer capability can also be enhanced, but in this work only simple series and shunt compensation techniques are used. By the help of FACT devices the reactive power is compensated but at the same time the cost of installation and maintenance is also increased. Finally, by using the Fixed Capacitor series compensation the power transfer capability increased.

References

1. Katal, G.: A survey paper on extra high voltage AC transmission line. Int. J. Eng. Res. Technol. **2** (2013)
2. Hingorani, N.G., Gyugi, L.: Understanding Facts: Concepts and Technology of Flexible AC Transmission Systems, pp. 155–190. IEEE press, NJ, USA (2001)
3. Chopade, P., Bikdash, M., Kateeb, I., Kelkar, A.D.: Reactive Power Management and Voltage Control of Large Transmission System Using SVC, vol. 2. IEEE press (2011)
4. Sundar, S.V.N.J., Vaishnavi, G.: Performance of a continuously controlled shunt reactor for bus voltage management in EHV systems. In: International Conference on Power systems transients Lyon, pp. 4–7, France (2007)
5. Song, Y.H., Johns, A.T.: Flexible AC Transmission Systems. IEEE press, London (1999)
6. Dantas, K.M.C.: An approach for controlled reclosing of shunt compensated transmission lines. IEEE Trans. Power Deliv. **29**, 1203–1211 (2014)
7. Daneshpooy, A., Gole, A.M.: Frequency response of the thyristor controlled series capacitor. IEEE Trans. Power Deliv. **16**, 53–58 (2001)
8. Matsuki, J., Hasegawa, S., Abe, M.: Synchronization schemes for a thyristor controlled series capacitor. In: Proceedings of IEEE International Conference on Industrial Technology, vol. 1, pp 536–541. IEEE press (2000)
9. Kiranmai, S.A., Manjula, M., Sarma, A.V.R.S.: Mitigation of various power quality problems using unified series shunt compensator in PSCAD/EMTDC. In: 16th National power systems conference (2010)
10. Hannan, M.A., Mohamed, A.: PSCAD/EMTDC simulation of unified series shunt compensator for power quality improvement. IEEE Trans. Power Deliv. **20**, 1650–1656 (2005) (IEEE press)
11. Hannan, M.A. Mohammed, A., Hussian, A., Dabbay, M.: Development of the unified series-shunt compensator for power quality mitigation. Am. J. Appl. Sci. pp. 978–986 (2009) (Science publication)
12. Rahmani, S., Hamid, A.: A combination of shunt hybrid power filter and thyristor controlled reactor for power quality. IEEE Trans. Indus. Electron. **61**, 2152–2164 (2014) IEEE press

A Contextual Approach for Modeling Activity Recognition

Megha Sharma, Bela Joglekar and Parag Kulkarni

Abstract In this paper, we propose a contextual approach for modeling human activity recognition. Activity recognition is performed using motion estimation based on context. Here contextual information is derived from motion, which is predicted from previous frames. This greatly enhances the process of activity recognition, by setting up a particular scenario which helps in constructing the activity. Context is acquired with the help of external inputs which surround an activity and help towards accurate reasoning about that activity. Context Modeling for any object can be done in terms of its relationship to other objects, called as contextual associations that lead towards accurate estimate of object position and presence. Here our focus is on vision based activity recognition. This process involves efficient feature extraction and subsequent classification for image representations. Classification accuracy is enhanced through Support Vector Machine (SVM) classifier, used along with Principle Component Analysis.

Keywords Activity recognition · Context · Principle component analysis (PCA)

1 Introduction

Activity recognition is the process of identifying actions of one or more agents based on series of observations of the agents' actions and environmental conditions. Human Action recognition which focuses on understanding and recognizing human

M. Sharma · B. Joglekar (✉)
MIT, Pune, India
e-mail: bela.joglekar@mitpune.edu.in

M. Sharma
e-mail: meghasharma1104@gmail.com

B. Joglekar
BVDUCOE, Pune, India

P. Kulkarni
Anomaly Solutions Pvt. Ltd, Pune, India

S.C. Satapathy and S. Das (eds.), *Proceedings of First International Conference on Information and Communication Technology for Intelligent Systems: Volume 2*, Smart Innovation, Systems and Technologies 51, DOI 10.1007/978-3-319-30927-9_22

activities in a video have gained substantial attention in the fields like medical imaging, video surveillance, video browsing, human computer interaction etc. Providing extra information for recognizing the activity is called as context.

Context can be categorized as inter-frame and intra-frame context. All the sources of information other than the region of interest from within the frame of a video are termed as intra-frame context and all the sources of information from outside the frame i.e. past or future frames of a video are termed as inter-frame context. Activity Recognition can be classified as Sensor based and Vision based. Sensor based activity recognition uses on body sensors or accelerator sensors to track human and identify the motion [1].

In our paper we have used vision based activity recognition. It is the process of identifying activity from a video which is captured from camera. Different methods such as Optical Flow, Hidden Markov models, etc., have been tried by many researchers that tend to use different setups such as single camera, stereo, etc. Here, we have considered a case study to recognize the activity of a single person performing action with respect to a car. Consider two humans having same position. Through this it is not always clear of what action is being performed as both frames appear visually similar. Once we take context into account, we obtain other additional information previously not known which will make more clear of what action is performed [2].

2 Related Work

Various motion and context features can be used to model related activities in a video, instead of modeling each activity individually [3]. Spatial and temporal relationships between activity classes can enhance recognition accuracy. This is achieved by labeling the activities from the learned model using greedy search method [3]. Promising directions in the field of context such as current trends, common datasets used, image representations in the form of global/local representations are presented by Poppe [2]. The reason for using Contextual approach is that, context provides an accurate estimation about the object's position and scale with respect to its presence [4]. Context has its effects at three levels, namely, semantic, spatial and pose. The percentage accuracy of recognition of an object/activity significantly depends on whether the object appears, or, is presented with respect to a familiar background/consistent scene, in contrast with an object presented against an unfamiliar background/inconsistent scene. Recognizing activities based on group, was found to be more efficient rather single human activity recognition, as the Action Context descriptor proposed by Lan et al. [5] encoded information about the action of the individual, as well as the behavior of other people nearby. Anomaly detection in video is addressed in [6], where motion and context patterns which deviate from the learned patterns of the training dataset are detected. Instances of normal activities are taken and labeled. The test instance labels which deviated from normal activities were detected as anomalies. Feature extraction and Optimization

Techniques for detecting Region Of Interest (ROI) is addressed in [7]. Human Action Recognition using Hidden Markov Model is discussed in [8].

3 System Architecture and Implementation

Figure 1 represents the system architecture of our system. The system has three sections as shown in the figure i.e. pre-processing in which degradation, noise etc. is removed, Feature Extraction in which some information from the image is extracted and Classification to classify and recognize the action.

Figure 2 represents how the inputs are acquired and processed to the next level in order to achieve the recognized activity. Initially we load the frame as an input on which the background subtraction is performed. After pre-processing, we obtain ROI to avoid unnecessary computation from which the features are then extracted using color and PCA extraction with the help of context to identify the motion region and extract features only for that region forming a feature vector. Once the features are extracted it is further classified into a class using Support Vector Machine which uses a trained dataset for learning and training of the classifier to recognize the activity.

Some problems that we are trying to minimize using our system are high probability of false classification by the classifier which in turn leads to less accuracy of the system. For this we use inter-frame context i.e. motion information

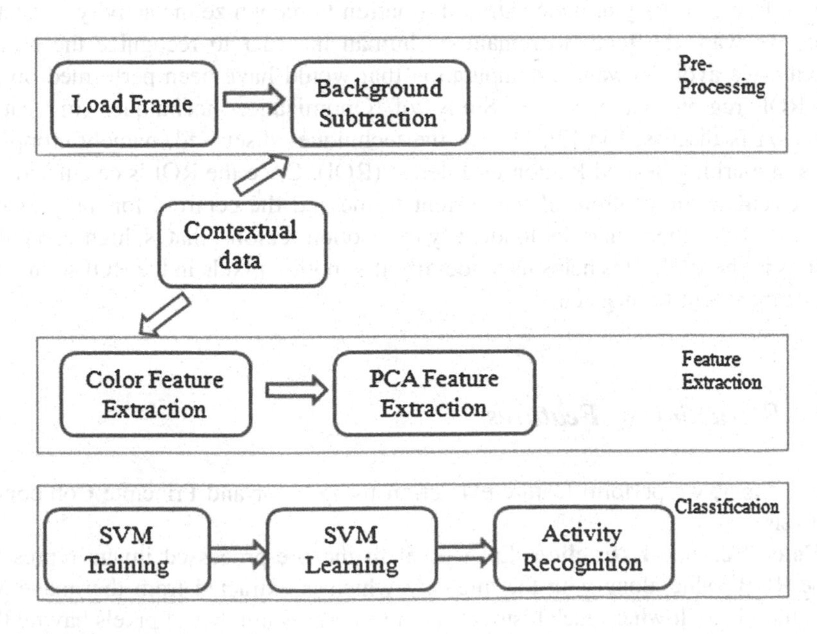

Fig. 1 System architecture

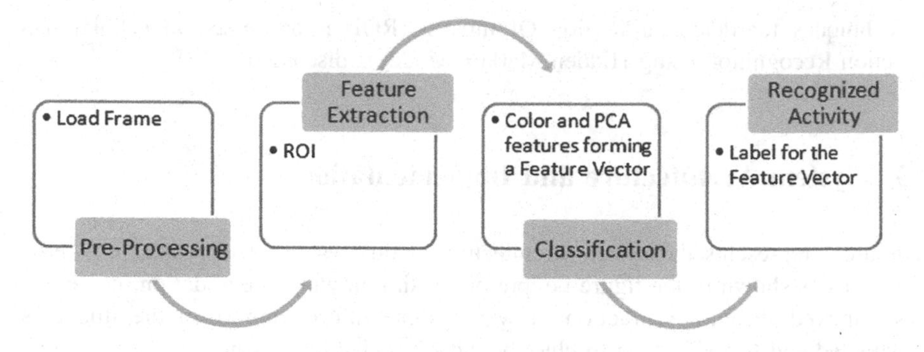

Fig. 2 System implementation

from the past frames as context. We consider time frame 't' as the current frame from which the action needs to be recognized and the past frames serve to act as context to it. This is called as temporal context.

3.1 Background Subtraction

By calculating variations in the sequential consecutive frames, we identify the motion region, as the pixels having motion will give variation in their values. Thus the previous frames used to identify the motion region acts as a context for the current frame, as they provide extra information to recognize the activity correctly. Since we want the focus to remain on human in order to recognize the human activity, we avoid unwanted computation that would have been performed on the non-ROI region. Comparative Study of Optimization techniques in feature extracton is discussed in [7]. One of the techniques discussed, namely cropping, helps in marking desired Region Of Interest (ROI). Once the ROI is calculated, we then calculate the centroid of the current frame and the centroid for the previous ones to overlap them, in order to identify the motion regions, that is, high variability regions in the ROI. This helps us to identify the motion pixels in the ROI so that the rest of them can be negated.

3.2 Extraction of Features

In this paper, we perform feature extraction using color and Principle Component Analysis.

Color Feature Extraction. Each pixel in the pre-processed image represents some RGB value along with the intensity which is extracted from the image and histogram is built where each histogram bin represents number of pixels having that

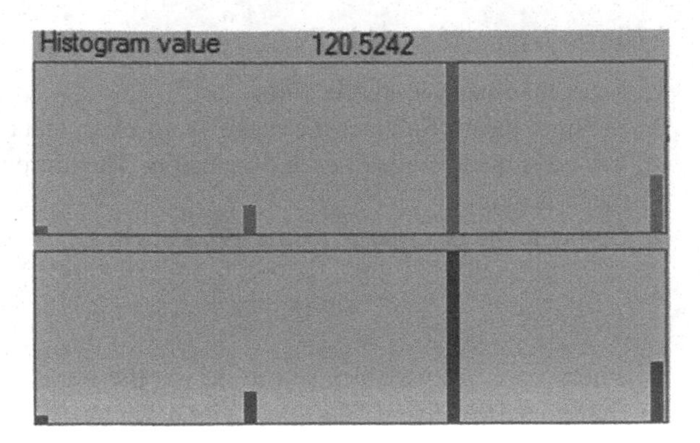

Fig. 3 Histogram for current and previous frame

color. This histogram information from the back-projected image i.e. previous frame histogram acts as a context for our active frame. The steps are as follows:

1. Load the image.
2. Convert it into the required form i.e. pre-process the image.
3. Decide the number of bins in the histogram formation.
4. Calculate the histogram for the image i.e. back projection.
5. Using back projection as reference to form histogram for the required image.

Back projection simply creates an image of the same size as that of our input image i.e. the previous frame. We first create the histogram for our previous frame which contains our object of interest i.e. human. Then this histogram is back-projected to our test image where we need to find the object. Figure 3 represents the histograms of previous and current frame respectively with the histogram value for the current frame.

Principle Component Analysis (PCA) Feature Extraction. PCA identifies and highlights important image representations, that is, features from the image, ignoring less important ones. It gives the principle components which lead toward direction of maximum variability. All these components are set in a growing order.

We perform feature extraction using PCA on the pre-processed frame obtained from the previous stage as shown in Fig. 4. It shows the frame of "open dikki" as input and the next picture box shows the pre-processed image with PCA performed. The two lines indicate the two components showing the directions of variability.

Fig. 4 Principle component analysis on pre-processed image

Steps to be performed in PCA:

1. Input the image in matrix form.
2. Subtract mean: Subtract the mean from every data dimension. The mean subtracted is the average of each dimension. Therefore we acquire a data set whose mean is zero.
3. Compute the covariance matrix [9] as below:

$$\mathrm{cov}(x_i, x_j) = \langle (x_i - \mu_i)\,(x_i - \mu_j) \rangle \tag{1}$$

   where x_i, x_j are variables and μ_i, μ_j are the mean.
4. Determine the eigen values and eigen vectors of the covariance matrix: More prominently, they deliver us with information about the patterns in data. Hence, by this process of extracting the Eigen vectors of the covariance matrix, we ought to extract the lines that characterize data.
5. Choosing components to form a feature vector: The Eigen vector with the uppermost Eigen value is the principal component of the data set. In general, once Eigen vectors are originated from the covariance matrix, the subsequent step is to organize them by their Eigen values, highest to lowest. As a result we attain the components in the order of their significance. Now we can ignore the components of lesser significance.

3.3 Support Vector Machine

After all the features are extracted by means of Principle Component Analysis (PCA), along with histogram features and is given as input to Support Vector Machine (SVM). The contextual approach and high level of feature extraction is to upsurge the classification and recognition accuracy of the classifier. Support Vector Machine scrutinizes data and isolates patterns that are used for classification. Figure 5 shows that to identify the action for our test data, we use a labelled i.e. known training dataset for leaning and training the classifier to correctly classify the activity. Based on the decision rule, it predicts a label for the test set.

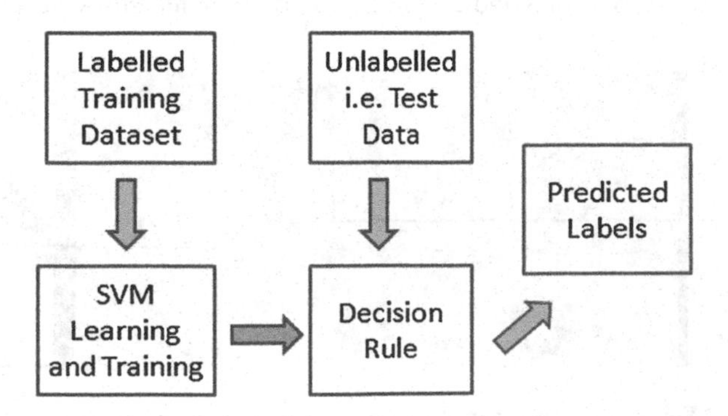

Fig. 5 Learning through support vector machine classifier

This classifier classifies the data by means of hyperplane separation which divides the space into different classes. On the two sides of hyperplane, two additional hyperplanes are plotted built on the criteria that the separating hyperplane must maximize the distance amongst the two hyperplanes. The two hyperplanes are drawn using the data from the space i.e. the points closest to separating hyperplane. Given a training dataset where each point has a label i.e. the class to which it belongs, SVM builds a model that assigns the test data to a class using the decision rule. Let the training dataset be D with a set of n points as shown in equation [10] below:

$$D = \{(x_i, y_i) | x_i \in R^P, y_i \in \{-1, 1\}\} \quad \text{for } i = 1\ldots n \tag{2}$$

where $y_i = 1$ or -1, describing the class to which x_i will belong. x_i is a vector. Hyperplane is described [11] as:

$$w \cdot x - b = 0 \tag{3}$$

where '·' represents dot product and 'w' is a normal vector to the hyperplane.

In our method, we perform linear separation of the hyperplane as shown in Fig. 6.

Figure 6 demonstrates that the problem is in a finite space and is linearly separable. Accordingly by using SVM for training and classification, activity recognition is achieved and its accuracy is calculated in terms of their precision and recall values.

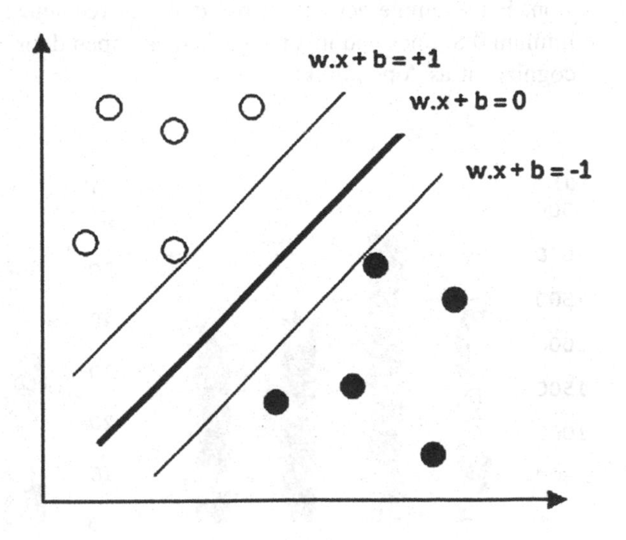

Fig. 6 Linear separation of hyperplanes

4 Experimental Results

Figure 7a shows the graph obtained from our system which represents the process time i.e. the time it takes to perform the pre-processing and the feature extraction. X axis: Time in ms Y axis: Actions Performed.

Lesser is the process time, more fast will be the system.

Figure 7b shows the recognition time of the system for different activities i.e. the time taken by the classifier to classify the activity into a particular class. If there are many number of known examples in the training set, the classifier becomes slow and takes more time for recognition. We have kept the required number of data in the training set and to avoid misclassification, more context information is added which helps to increase the accuracy of the system and to increase speed of the classifier.

We performed processing on four actions, but out of them our system recognizes three correctly whereas it has a high misclassification rate for the fourth activity. So the classification rate of the system is 3 out of 4.

Figure 8 shows the precision and recall obtained for different activities in the system. Precision is the fraction of retrieved instances that are relevant and Recall is the fraction of relevant instances that are retrieved.

Precision = tp/(tp + fp)
Recall = tp/(tp + fn)

where tp = true positive, fp = false positive and fn = false negative

Figure 9 shows the matrix which tells the minimum accuracy rate for each action. For example activity "close door" is recognized correctly as "close door" minimum 0.8 times and it is recognized as "open door" 0.2 times where as it never recognizes it as "open dikki".

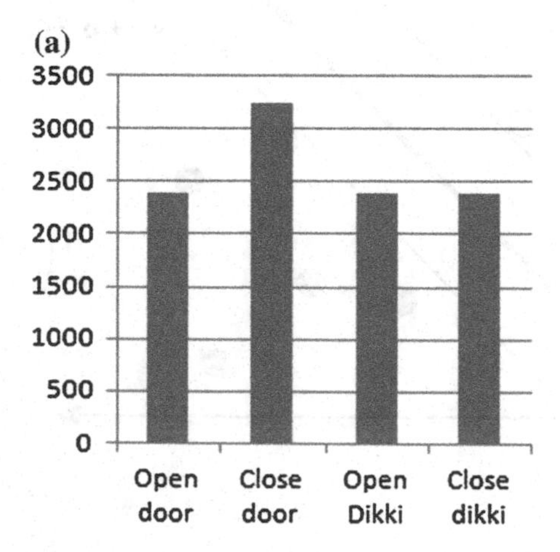

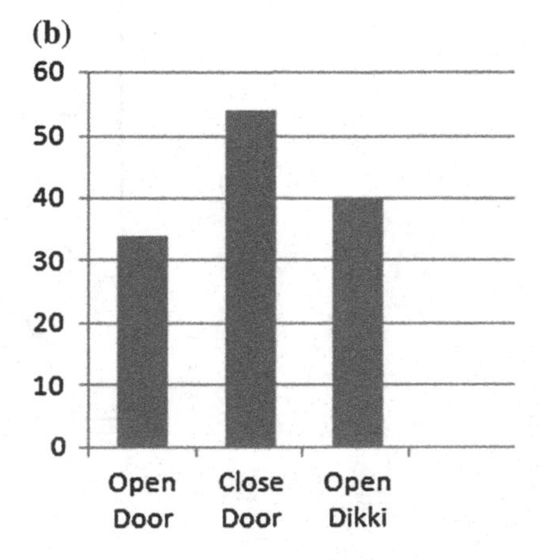

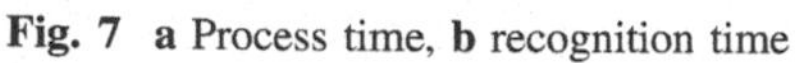
Fig. 7 **a** Process time, **b** recognition time

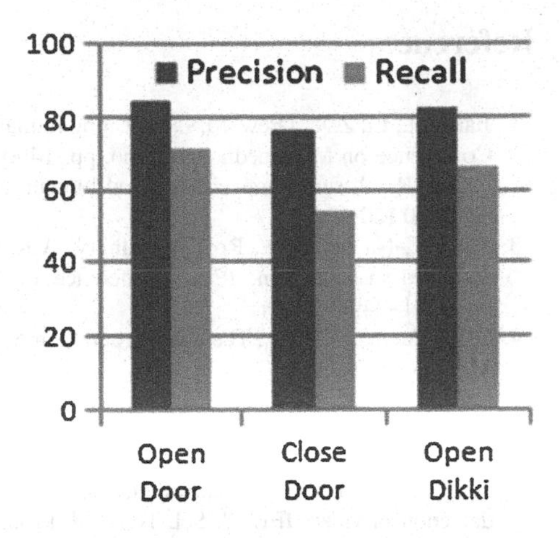

Fig. 8 Precision and recall

Fig. 9 Confusion matrix

Actions	Open Dikki	Close Door	Open Door
Open Dikki	0.8	0.1	0.1
Close Door	0	0.8	0.2
Open Door	0.1	0.1	0.8

5 Conclusion

The proposed method describes activity recognition using context for extracting useful information from any image. Principle Component Analysis (PCA) approach is used for capturing contextual information, thereby increasing visual accuracy of the system. The context presented to the user best represents the interaction of images which is faster to localize and recognize visual activities. In the absence of enough local evidence about an object's identity, prior knowledge of scene structure, additional features etc. provide additional information needed for localizing and recognizing an object through context. Classification accuracy is enhanced through Support Vector Machine (SVM) classifier, used along with Principle Component Analysis.

References

1. Jiang, H., Li, Z.N., Drew, M.S.: Detecting human action in active video. In: IEEE International Conference on Multimedia and Expo, pp. 1490–1500 (2006)
2. Poppe, R.: A survey on vision-based human action recognition. In: J. Image Vis. Comput. 976–990 (2010)
3. Zhu, Y., Nayak, N.M., Roy-Chowdhury, A.K.: Context-aware modeling and recognition of activities in video. In: IEEE Conference on Computer Vision and Pattern Recognition, pp. 2491–2498 (2013)
4. Oliva, A., Torralba, A.: The role of context in object recognition. Sci. Dir. J. Trends Cogn. Sci. **11**(12) (2007)
5. Lan, T., Wang, Y., Yang, W., Robinovitch, S.N., Mori, G.: Discriminative latent models for recognizing contextual group activities. IEEE Trans. Pattern Anal. Mach. Intell. **34**(8), 1549–1562 (2012)
6. Zhu, Y., Nayak, N., Roy-Chowdhury, A.K.: Context aware activity recognition and anomaly detection in video. IEEE J. Sel. Top. Sig. Process. **7**(1), 91–101 (2013)
7. Chadha, A., Mallik, S., Johar, R.: Comparative study and optimization of feature-extraction techniques for content based image retrieval. Int. J. Comput. Appl. **52** (2012)
8. Yamato, J., Ohya, J., Ishii, K.: Recognizing human action in time sequential images using Hidden Markov Model. In: IEEE Computer Society Conference on Computer vision and Pattern Recognition, pp. 379–385 (1992)
9. Mudrova, M., Prochazka, A.: Principal component analysis in image processing. Department of Computing and Control Engineering (2005)
10. Lovell, B.C., Walder, C.J.: Support vector machines for business applications. The University of Queensland and Max Planck Institute, Tübingen (2006)
11. Schuldt, C., Laptev, I., Caputo, B.: Recognizing Human Actions: a Local SVM approach. In: IEEE International Conference on Pattern Recognition, Vol. 3, pp. 32–36 (2004)

A Fuzzy Trust Enhanced Collaborative Filtering for Effective Context-Aware Recommender Systems

Sonal Linda and Kamal K. Bharadwaj

Abstract The Recommender systems (RSs) are well-established techniques for providing personalized recommendations to users by successfully handling information overload due to unprecedented growth of the web. Context-aware RSs (CARSs) have proved to be reliable for providing more relevant and accurate predictions by incorporating contextual situations of the user. Although, collaborative filtering (CF) is the widely used and most successful technique for CARSs but it suffers from sparsity problem. In this paper, we attempt toward introducing fuzzy trust into CARSs to address the problem of sparsity while maintaining the quality of recommendations. Our contribution is twofold. Firstly, we exploit fuzzy trust among users through fuzzy computational model of trust and incorporate it into context-aware CF (CACF) technique for better recommendations. Secondly, we use fuzzy trust propagation for alleviating sparsity problem to further improve recommendations quality. The experimental results on two real world datasets clearly demonstrate the effectiveness of our proposed schemes.

Keywords Recommender systems · Context-aware recommender systems · Context-aware collaborative filtering · Fuzzy trust · Fuzzy trust propagation

1 Introduction

The increasingly common appearance of e-commerce websites are taking advantage of recommender systems (RSs) that have been proven effective tool for non-experienced users facing decision-making problem. Context-aware recommender systems (CARSs) have played a leading role in mobile devices such as

S. Linda (✉) · K.K. Bharadwaj
School of Computer and Systems Sciences, Jawaharlal Nehru University, New Delhi, India
e-mail: lindasonal@gmail.com

K.K. Bharadwaj
e-mail: kbharadwaj@gmail.com

S.C. Satapathy and S. Das (eds.), *Proceedings of First International Conference on Information and Communication Technology for Intelligent Systems: Volume 2*, Smart Innovation, Systems and Technologies 51, DOI 10.1007/978-3-319-30927-9_23

Smartphone and Tablet computers for providing appropriate online recommendations from last few years. Researchers in academia as well as industries have recognized the importance of contextual information in several areas including e-commerce personalization, information retrieval, ubiquitous and mobile computing, data mining, marketing, and management. The area of CARSs brings together many researchers with wide-ranging backgrounds to identify important research challenges and innovations [1].

The ultimate goal of CARSs is to avoid information explosion on web and provide high quality personalized recommendations. The contextual recommendations affect the behaviour of customers in real-life scenario and establish their trust in the provided recommendations [2]. Trust is a key for strengthening the recommendations and it has a positive impact on sales of e-commerce society. Collaborative filtering (CF) is widely used and the most successful recommendation technique, which provides personalized and useful recommendations to users by utilizing their past transactions and opinion of like-minded users [3]. A context-aware collaborative filtering (CACF) based system predicts user's preference not only from opinions of like-minded users but also from feedback of other users in a context similar to that of the active user by leveraging the pervasive context information [4]. The collaborative approach raises the issue of quality assessment due to vast amount of data, whereas "web of trust", the elicitation of trust values among users, allows enhancement of CARSs [5].

In this paper, we propose a fuzzy trust enhanced CF using fuzzy computational model for effective CARSs that suggests items to users based on trustworthy neighbourhood formation. Our main contributions are as follows:

- Exploitation of context features from users' profile to find similarity among them and applying suitable similarity measures to generate neighbourhood set.
- Incorporation of fuzzy trust to capture trustworthiness among users through fuzzy computational model [6] to enhance predictive accuracy.
- Use of fuzzy trust propagation technique towards alleviating the problem of sparsity in CACF based CARSs.

The paper is outlined as follows: Sect. 2 discusses the related studies of trust-aware RSs and CF based CARSs, and also highlights the associated challenges. Section 3 presents our proposed framework that is based on fuzzy trust CACF and at the end methodology used to handle sparsity problem based on fuzzy trust propagation. The experimental evaluation of various schemes CACF, FT-CACF and FTprop-CACF is presented in Sect. 4. Finally, Sect. 5 concludes with future directions.

2 Related Work

There is a wide range of uses and meaning attached to trust on the web including how web site design influences trust on content and content providers, capturing ratings from users about the quality of information, propagation of trust over links, etc. [7]. It has been proved that trust matrices used in trust-aware RSs increase the coverage and decrease the prediction error when compared with traditional CF [5]. It is also effective for the problem of new users who have rated few items. Similarly, context has a multifaceted nature and its general concept is very broad. It has been clearly established, through empirical study that in terms of predictive accuracy and users' satisfaction with recommendations context-aware approach significantly outperforms the corresponding non-contextual approach [1]. Most CARSs perform adaptation of recommendations to the context as well as user preferences [8]. There are three general approaches for implementing CARSs:

- Contextual pre-filtering: Data is filtered using specified set of contextual attributes before applying recommendation technique.
- Contextual post-filtering: Data is filtered using specified set of contextual attributes after applying recommendation technique to adjust its ranking in recommendation list.
- Contextual modeling: A specified set of contextual attributes is directly incorporated into recommendation function in addition to the users and items data.

Sometimes, using exactly specified set of contextual attributes may not have enough data for accurate predictions. Therefore, we need to choose optimal set of contextual attributes to generate neighbourhood set that is suitable for recommendations.

The key challenge particular to the CARSs that affect the quality of recommendations is sparsity [9]. Scarcity of data points required to describe the exact context raises the issue of sparsity. Recent research has shown that trust aware RSs generate better recommendations and are more robust against sparsity. People naturally use linguistic expression rather than numeric values to describe their trust, especially in contradictory situations, therefore crisp modelling of trust is not enough for inferring accurate information. Thus, Fuzzy sets seem to be an ideal choice for trust modelling [10, 11]. This is the main motivation to use fuzzy trust in CACF for our proposed CARS framework.

3 Proposed Fuzzy Trust Based CACF for CARS Framework

In CARSs, contextual information is an essential key factor to generate relevant recommendations. There are several approaches (contextual pre-filtering, contextual post-filtering and contextual modeling) for incorporating contextual information in the recommendation process. We use hybrid approach in CACF i.e. combination of contextual pre-filtering and modeling into CF. The main focus of this work is to incorporate fuzzy trust through fuzzy computational model [6], to enhance capability of CARSs. In the following sections, details of proposed hybrid CACF, fuzzy trust based CACF (FT-CACF) and fuzzy trust propagated CACF (FTprop-CACF) are discussed.

3.1 Context Similarity and Users' Correlation

Displayed Context similarity metric for CARSs can improve the prediction accuracy by aggregating of similar context based on past experience of other users which reduces the quantum of irrelevant information. We use Cosine metric cosine $Sim(c,d)$ for context similarity between two vectors of contextual attributes c and d, as given in Eq. 1.

$$\text{cosine}\, Sim(c,d) = \frac{\sum_{j=1}^{n} c_j \times d_j}{\sqrt{\sum_{j=1}^{n} (c_j)^2 \times \sqrt{\sum_{j=1}^{n} (d_j)^2}}} \quad (1)$$

To identify users with similar inclination which leads to the formation of neighbourhood set, we use modified Pearson's correlation coefficient, as given in Eqs. 2 and 3.

$$I_1 = \{\langle i, c_a, c_j \rangle : \exists r_{u_a,i,c_j} \Lambda \text{cosine}\, Sim(c_a, c_j) > \mu_1\} \quad (2)$$

$$user\, Sim(u_a, u_j) = \frac{\sum_{(i,c_a,c_j)\in I_1} (r_{u_a,i,c_a} - \bar{r}_{u_a})(r_{u_j,i,c_j} - \bar{r}_{u_j}) \text{cosine}\, Sim(c_a, c_j)}{\sqrt{\sum_{(i,c_a,c_j)\in I_1} (r_{u_a,i,c_a} - \bar{r}_{u_a})^2 \sum_{(i,c_a,c_j)\in I_1} (r_{u_j,i,c_j} - \bar{r}_{u_j})^2 \sum_{(i,c_a,c_j)\in I_1} \text{cosine}\, Sim(c_a, c_j)^2}}. \quad (3)$$

Here, the set I_1 is obtained by collecting all co-rated items i and a pair of contexts c_a and c_j for users u_a and u_j respectively, such that each has rated i in that context and cosine $Sim(c_a, c_j) > \mu_1$.

3.2 Hybrid CACF and Fuzzy Trust Based CACF (FT-CACF)

A CACF based system predicts a user's preference in any context by leveraging past experiences of like-minded users in similar context. We use hybrid CACF i.e. combination of contextual pre-filtering and modeling into CF, where the current context is used for selecting only the relevant set of data, and ratings are predicted using multidimensional recommendation technique on the selected data. Suppose there is a list of n users $U = u_1, u_2, \ldots, u_n$, a list of m items $I = i_1, i_2, \ldots, i_m$, and a list of k context attributes $C = c_1, c_2, \ldots, c_k$. Each user u has a list of items consumed in certain context. Additionally, we need to obtain user-user-trust rating matrix based on following assumption that trust established among users when users have similar preferences on co-rated items and then incorporate fuzzy trust into CACF to enhance the predictive accuracy.

Following [6], we use fuzzy computational model for trust computation which defines two fuzzy subsets on each user's ratings namely satisfied and unsatisfied, and membership values of each subset always sum up to one. Moreover, we need each user to express his opinion for other users in terms of reciprocity (mutual favour or revenge). It is obtained by agreement and disagreement between two users. The agreement between two users implies that either both individuals are satisfied or unsatisfied in co-rated items. The disagreement between users implies that only one of them is unsatisfied in co-rated items. Reciprocity can be expressed mathematically in terms of *agr* (u_a, u_j) and *disagr* (u_a, u_j) as follows:

$$rec(u_a, u_j) = (1 - disagr(u_a, u_j))agr(u_a, u_j) \quad (4)$$

The formulae are opted for agreement and disagreement measures from fuzzy computational model [6]. Four values for any pair of users are defined based on four possible combinations of satisfied and unsatisfied fuzzy subsets. Since the computed value of reciprocity is not always reliable, therefore reliability measure needs to be incorporated into reciprocity to measure how much uncertainty is there in the system. The value of reliability is multiplied to reciprocity value which needs a new frame of discernment to cover the uncertainty for both individuals. Let, a proposition A defines good and bad ratings user u_a has given to user u_j, so the uncertainty u of proposition A is width of the Shafer interval $p(A) - s(A)$, where $p(A)$ is plausibility of A and $S(A)$ is support of A i.e. $u(A) = p(A) - s(A)$. The reliability measure $reliab_{u_a}(u_j)$ of u_j as seen by u_a is defined as:

$$reliab_{u_a}(u_j) = 1 - u(A) \quad (5)$$

To establish trust from active user u_a to other user u_j there could be experience of the u_j on purchasing items and u_a has confidence on that experience. Thus, trust is a function of reciprocity, experience and confidence in Eq. (6).

$$trust_{u_a}(u_j) = \frac{2 \times conf_{u_a}(u_j) \times ex(u_j) \times recip(u_a, u_j)}{(conf_{u_a}(u_j) \times ex(u_j)) + recip(u_a, u_j)} \quad (6)$$

Finally, for fuzzy trust based recommendations, predicted rating P_{u_a,i,c_a} for an item for an active user in current context with neighbourhood set N_{u_a,μ_1} is obtained by adapted Resnick's formula (8) as shown below.

$$N_{u_a,\mu_1} = \left\{ u : \max_{r_u,i,c_j}(\text{cosine } Sim(c_a, c_j) > \mu_1) \right\} \quad (7)$$

$$P_{u_a,i,c_a} = \bar{r}_{u_a,c_a} + \frac{\sum_{u_j \in N_{u_a,\mu_1}} ((r_{u_j,i,c_j} - \bar{r}_{u_a,c_a}) \times user\ Sim(u_a, u_j) \times trust_{u_a}(u_j))}{\sum_{u_j \in N_{u_a,\mu_1}} user\ Sim(u_a, u_j) \times trust_{u_a}(u_j)} \quad (8)$$

3.3 Fuzzy Trust Propagated CACF (FTprop-CACF)

In this section, we present further enhancement of proposed FT-CACF (Sect. 3.2) by alleviating the problem of sparsity through fuzzy trust propagation. The resulting scheme is referred as fuzzy trust propagated CACF (FTprop-CACF). Since the number of ratings that exist in CARSs, underlying user-item-context graph (Fig. 1) is very sparse, there could be possibility of getting no appropriate set of neighbours and rendering CACF useless [12]. We employ fuzzy trust propagation for alleviating sparsity to provide high quality recommendations even when datasets are sparse. According to this process trust is propagated in the graph through Maximin Distance approach [13] and association between users are built, even if they have no co-rated items. It incorporates transitive interactions and recognizes the indirect associative relationship between them. But in real life trust is not always transitive because it could not completely transfer without decay through distance. Since, fuzzy trust values reduce as path length increases, so propagation distance should be a constraint in transitive relationship [14].

Depending on the number of associations, a fuzzy trust path from a source user to a target user can be of variable length. The fuzzy trust value in direct associations is computed by Eq. (6) and for indirect associations we need to apply Eq. (9) that facilitates fuzzy trust value between the source user and the target user.

$$T(u_1,u_3)=\max(\min(T(u_1,u_2),T(u_2,u_3)),\min(T(u_1,u_2),T(u_2,u_4),T(u_4,u_3)),\min(T(u_1,u_2),T(u_2,u_4),T(u_4,u_5),T(u_5,u_3))) \tag{9}$$

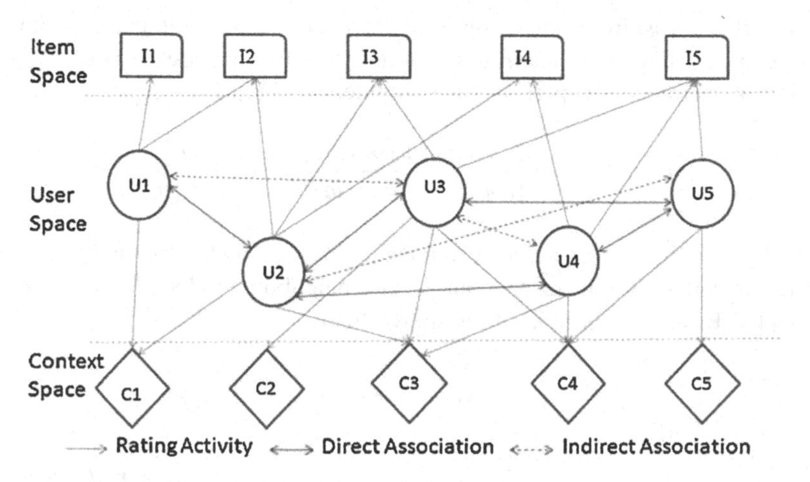

Fig. 1 Underlying user-item-context graph in CARSs

The main steps of proposed fuzzy trust propagated CACF (FTprop-CACF) schemes for CARS framework are summarized as:

Step 1: Compute similarity between users as $user\,Sim(u_a, u_j)$ and contextual attributes as cosine $Sim(c, d)$.
Step 2: Use Fuzzy trust computation to exploit trust values $trust_{u_a}(u_j)$ among users.
Step 3: Generate enhanced neighbourhood sets based on fuzzy trust propagation path.
Step 4: Make recommendations using FTprop-CACF technique with predicted rating P_{u_a,i,c_a}.

4 Experimental Evaluation and Results

We evaluate our proposed schemes for CARSs on two datasets, Movie and Restaurant-Customer belonging to two different domains. In the first experiment, the neighbourhood set of users is formed by collecting neighbours based on their contexts similarity and users' correlation. Then fuzzy trust values are computed to prune the neighbourhood set to get highly trusted recommenders for enhanced recommendations. In the next experiment, we evaluate the impact of variation of users' trust (low or high) towards sparsity and quality of recommendations based on the underlying user-item-context graph of direct and inferred associations. We measure the effectiveness of proposed CARS by two evaluation metrics: accuracy and coverage.

Mean Absolute Error (MAE) is employed as statistical accuracy measure, defined as the average absolute difference between the predicted ratings P_{u_j} and the actual Ratings r_{u_j} of active users from testing set $|S_t|$ i.e.

$$MAE = \frac{1}{|S_t|}\sum_{j=1}^{|S_t|} \left|P_{u_j,i,c_j} - r_{u_j,i,c_j}\right| \tag{10}$$

Coverage is computed as the number of items for which the CARS can generate correct predicted items I_{cp} for an active user over the total number of unseen items $|I_t|$ i.e.

$$Coverage = \frac{|I_{cp}|}{|I_t|} \tag{11}$$

4.1 Datasets Description

We use subsets of two real world datasets: LDOS-CoMoDa provided by Prof. Kosir [15] and Restaurant and consumer data from Machine Learning repository with rating scales 1–5 and 1–3 respectively to conduct the experimental evaluation of proposed schemes. The dataset extracted from LDOS-CoMoDa named as Movie dataset contains 81 ratings by 15 users for 10 movies in 10 various kind of contextual attributes (Time, Daytype, Season, Location, Weather, Social, Mood, Physical, Decision and Interaction). Similarly, the dataset extracted from Restaurant and consumer data named as Restaurant-Customer contains 92 ratings by 15 users for 10 movies in 10 contextual attributes (R-alcohol, R-accessible, R-price, R-ambience, R-area, C-ambience, C-marital_status, C-transport, C-interest, C-personality). The actual ratings of an active user are randomly divided into two sets: testing set and training set. Training set is used to construct neighbourhood set for rating prediction, whereas testing set is used to evaluate the quality of recommendations.

4.2 Results

Figure 2 illustrates the comparison between MAE for hybrid CACF (based on pre-filtering and modeling) and MAE for fuzzy trust based CACF (FT-CACF) on datasets with five random sample sets for active users 3, 5 and 10. The results show that incorporation of fuzzy trust into hybrid CACF enhanced the accuracy of predictions.

The variation in MAE on different testing sample sets of active users (S1, S2, S3, S4, S5, and S6) with fuzzy trust parameter t for various size of neighbourhood set which in turn can affect the quality of neighbours is depicted in Fig. 3. It is clear that increasing the value of t for improving the prediction results might add some irrelevant neighbours, which lead to drastic error in predictions. The results summarized in Table 1, present the MAE comparison of various CARS schemes with appropriate choice of parameter value t. It is clearly showed that fuzzy trust propagated CACF (FTprop-CACF) consistently outperforms the CACF as well as FT-CACF.

The percentage of correct predictions obtained from CACF, FT-CACF and FTprop-CACF for the active user is also given in Fig. 4. The higher number of correct predictions by FTprop-CACF illustrates that the accuracy gets enhanced through the better set of neighbours generated.

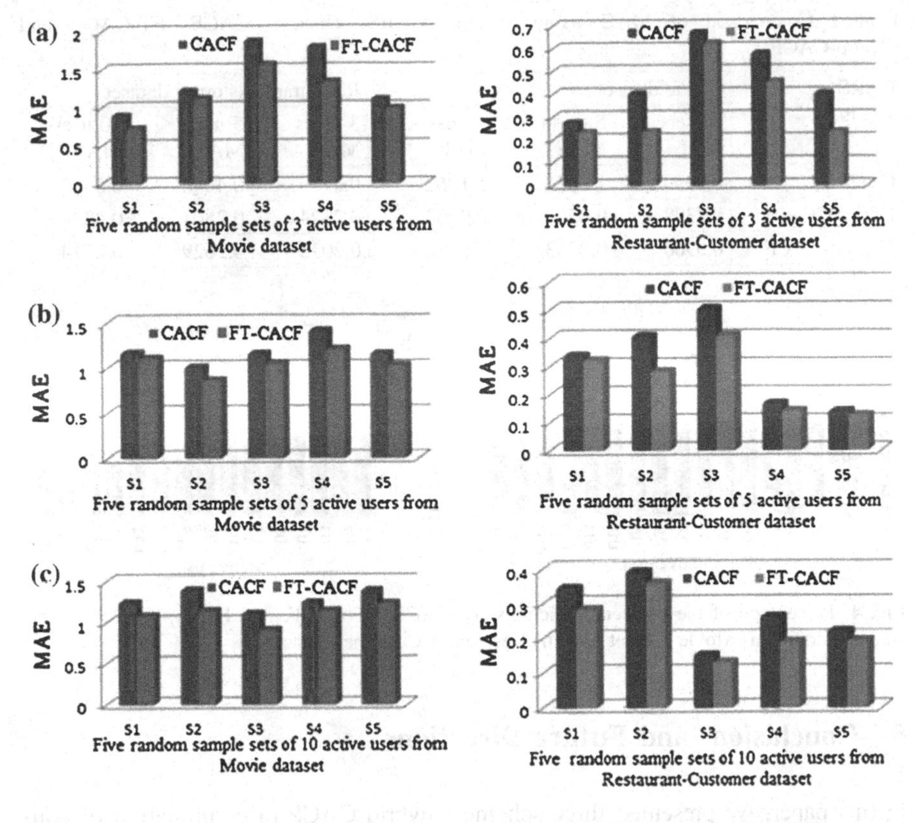

Fig. 2 Average MAE for hybrid context-aware collaborative filtering (CACF) and Fuzzy trust based context-aware collaborative filtering (FT-CACF)

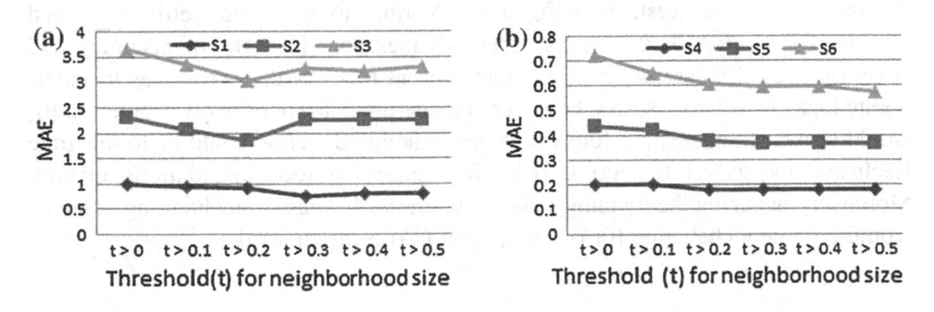

Fig. 3 Comparing MAE based on different thresholds for neighbourhood size using (**a**) Movie dataset (**b**) Restaurant-customer dataset

Table 1 Comparison of MAE using various CARS schemes: CACF, FT-CACF and FTprop-CACF

CARS schemes	Movie dataset			Restaurant-customer dataset		
	3 users MAE	5 users MAE	10 users MAE	3 users MAE	5 users MAE	10 users MAE
CACF	0.8990	1.0033	1.1062	0.2750	0.4048	0.3442
FT-CACF	0.7328	0.8590	0.9063	0.2324	0.2801	0.2838
FTprop-CACF	0.6900	0.8333	0.8928	0.2038	0.2629	0.2514

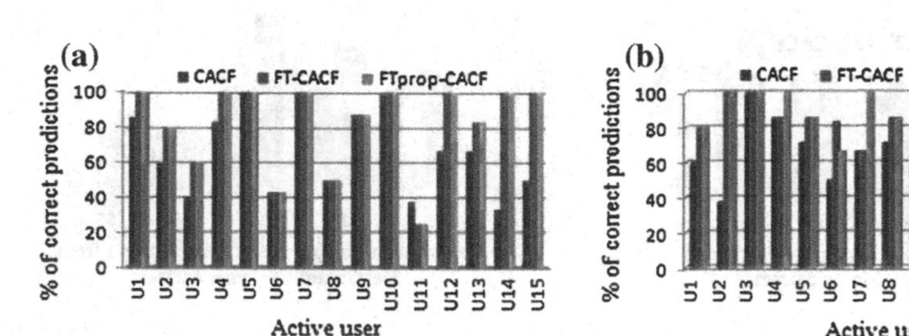

Fig. 4 Percentage of the correct predictions by CACF, FT-CACF, and FTprop-CACF for the active users of (**a**) Movie dataset and (**b**) Restaurant-Customer dataset

5 Conclusions and Future Directions

In this paper, we presented three schemes: hybrid CACF (a combination of contextual pre-filtering and modeling approaches), fuzzy trust based CACF (FT-CACF) and fuzzy trust propagated CACF (FTprop-CACF) to generate contextual recommendations. We compared these schemes across two real world datasets: Movie and Restaurant-Customer. Various experimental settings showed that the FTprop-CACF dominates other schemes both in terms of accuracy and coverage even if datasets are sparse. Our current focus in this work was to incorporate fuzzy trust into CARSs, however considering concept of fuzzy reputation [6] would also be an interesting future research. Another direction could be to integrate Reclusive methods [16] into CARSs for enhancing recommendations quality. Moreover, managing the dynamics in user interests [17] based on changing contexts would be a new challenge for our proposed CARSs framework.

References

1. Adomavicius, G., Tuzhilin, A.: Context-aware recommender systems. In: Recommender Systems Handbook, pp. 217–253. Springer, Heidelberg (2011)
2. Gorgoglione, M., Panniello, U., Tuzhilin, A.: The effect of context-aware recommendations on customer purchasing behavior and trust. In: RecSys'11, Chicago, USA (2011)

3. Goldberg, D., Nichols, D., Oki, B.M., Terry, D.: Using collaborative filtering to weave an information tapestry. Commun. ACM **35**(12), 61–70 (1992). (Magazine—Special issue on information filtering)
4. Chen, A.: Context-aware collaborative filtering system: predicting the user's preference in the ubiquitous computing environment. In: Location and Context-Awareness. LNCS, vol. 3479, pp. 244–253. Springer, Heidelberg (2005)
5. Massa, P., Avesani, P.: Trust-aware collaborative filtering for recommender systems. In: On the Move to Meaningful Internet Systems 2004, CoopIS, DOA, and ODBASE. LNCS, vol. 3290, pp. 492–508. Springer, Heidelberg (2004)
6. Bharadwaj, K.K., Al-Shamri, M.Y.H.: Fuzzy computational models for trust and reputation systems. Electron. Commer. Res. Appl. **8**(1), 37–47 (2009)
7. Artz, D., Gil, Y.: A survey of trust in computer science and the semantic web. Web Semant. Sci. Serv. Agents World Wide Web **5**(2), 58–71 (2007)
8. Adomavicius, G., Mobasher, B., Ricci, F., Tuzhilin, A.: Context-aware recommender systems. AI Mag. **32**(3), 67–80 (2011)
9. Abbas, A., Zhang, L., Khan, S.U.: A survey on context-aware recommender systems based on computational intelligence techniques. Computing **97**(7), 667–690 (2015)
10. Kant, V., Bharadwaj, K.K.: Incorporating Fuzzy trust in collaborative filtering based recommender systems. In: Swarm, Evolutionary, and Memetic Computing. LNCS 7076, pp. 433–440. Springer, Heidelberg (2011)
11. Al-Shamri, M.Y.H., Bharadwaj, K.K.: Fuzzy-genetic approach to recommender systems based on a novel hybrid user model. Expert Syst. Appl. **35**(3), 1386–1399 (2008)
12. Papagelis, M., Plexousakis, D., Kutsuras, T.: Alleviating the sparsity problem of collaborative filtering using trust inferences. In: Third International Conference on iTrust. LNCS, vol. 3477, pp. 224–239. Springer, Heidelberg (2005)
13. Luo, H., Niu, C., Shen, R., Ullrich, C.: A collaborative filtering framework based on both local user similarity and global user similarity. Mach. Learn. **72**(3), 231–245 (2008)
14. Agarwal, V., Bharadwaj, K.K.: Trust-enhanced recommendation of friends in web based social networks using genetic algorithms to learn user preferences. In: First International Conference on Computer Science, Engineering and Information Technology (CCSEIT 2011), CCIS, vol. 204, pp. 476–485 (2011)
15. Odic, A., Tkalcic, M., Kosir, A., Tasic, J.F.: Relevant context in a movie recommender system: users' opinion vs. statistical detection. In: CARS 2012. Dublin, Ireland (2012)
16. Kant, V., Bharadwaj, K.K.: Integrating collaborative and reclusive methods for effective recommendations: a Fuzzy Bayesian approach. Int. J. Intell. Syst. **28**(11), 1099–1123 (2013)
17. Ma, S., Li, X., Ding, Y., Orlowska, M.E.: A recommender system with interest-drifting. In: Web Information Systems Engineering–WISE 2007. LNCS, vol. 4831, pp. 633–642, Springer, Heidelberg (2007)

An Optimal LEACH-Based Routing Protocol to Balance the Power Consumption and Extending the Lifetime of the WSNs

Yahya Kord Tamandani, Mohammad Ubaidullah Bokhari and Qahtan Makki

Abstract Wireless sensor networks (WSNs) are made of tiny sensor nodes (SNs) and a base station (BS). As these tiny sensor nodes run on non- rechargeable batteries due to their wide distribution the lifetime of the network is of great importance. LEACH routing protocol that is the earliest hierarchical routing protocol suffers from some drawbacks such as not considering the remaining energy of SNs and the clusters' size to prolong the network life time to a higher level. In this paper we propose an efficient LEACH-based routing protocol in an effort to prolong the lifetime of WSNs. The proposed protocol consider residual energy of sensor nodes, their distance from base station and the size of each cluster to prolong the network life to a higher level. The result of our simulation indicates that our improved protocol works significantly better than existing LEACH its improved versions K-LEACH and T-LEACH in case of balancing the energy consumption of the network.

Keywords Wireless sensor networks · Hierarchical routing protocols · Energy efficiency · Clusters

1 Introduction

Wireless sensor networks (WSNs) are constructed of several tiny nodes, capable of sensing, collecting and transmitting of data which has been depicted in Fig. 1. Typical sensor nodes (SNs) are composed of four main units [1]. Figure 1 depicts a

Y.K. Tamandani (✉) · M.U. Bokhari · Q. Makki
Department of Computer Science, Aligarh Muslim University, Aligarh, India
e-mail: Yahya.kord@gmail.com

M.U. Bokhari
e-mail: mubokhari.cs@amu.ac.in

Q. Makki
e-mail: qahtan.mekki@yahoo.com

S.C. Satapathy and S. Das (eds.), *Proceedings of First International Conference on Information and Communication Technology for Intelligent Systems: Volume 2*,
Smart Innovation, Systems and Technologies 51, DOI 10.1007/978-3-319-30927-9_24

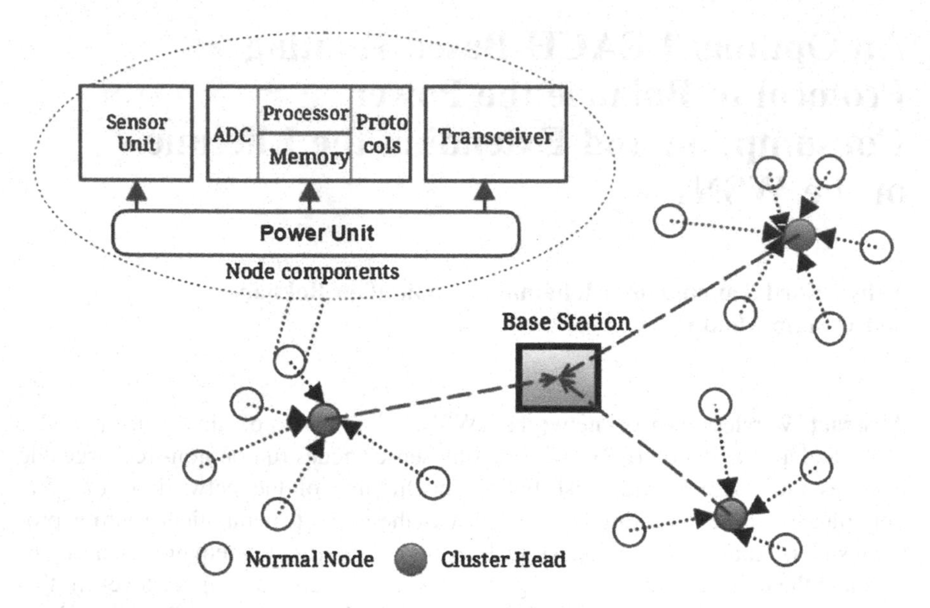

Fig. 1 A typical WSN along with the main components of a SN

WSN along with the main components of a SN. The power unit of a SN, which typically has limited amount of energy, is required for supplying power into the other components [1]. WSNs due to their wide applications [2–4] have gained an increasing attention over the past decade. However WSNs do have some issues. Limited energy supply, the replacement of the energy source might not be feasible due to arbitrary distribution of sensor nodes of wireless network. Data gathering and transmitting are known as the major energy consuming procedures and they are managed via the network layer, therefore the routing protocol has an extremely significant role in network lifetime optimization. Many routing techniques [5–8] has been enhanced and designed in order to prolong the life time and overall performance of the network. Power efficient routing protocols can potentially decrease the number of sent packets and also enhance the choice of traces and nodes for data transmission. Hierarchical routing protocol is regarded as the most energy efficient routing protocols for WSNs.

The LEACH [9] which is the earliest hierarchical protocol, reduces the energy consumption by choosing the CHs in a random fashion to distribute energy load evenly to each and every node. However the LEACH protocol is dependent upon the period of time to re-establish new clusters among the entire network without ever taking into consideration the remaining energy of the various clusters or their distance from the base station (BS). In addition the protocol does not have a numerical method to specify the optimal number of CHs as either small or large number of CHs will lead in energy dissipation of the network quickly. The paper proposes an efficient LEACH-based routing protocol to resolve the above issues.

2 Related Work

LEACH protocol, has been introduced by Heinzelman in 2000. In LEACH protocol to become a CH, all sensor nodes produce a number randomly. In case the number is lower than the threshold T(n), the node becomes a CH for the current round, the threshold is given in Eq. (1) [9].

$$T(n) = \begin{cases} \frac{p}{1-P*\left(r \bmod \frac{1}{p}\right)} & if\ n \in M \\ 0 & otherwise \end{cases} \quad (1)$$

A message will be broadcasted by nodes which are selected as cluster heads, thereafter other nodes which have not been chosen as cluster heads collect the information of all CHs. Based on the intensity of signal of each CH, nodes which have the lowest cost of communication can join the cluster [10].

Most of proposed and enhanced LEACH-based algorithms has been focusing on the threshold T(n) by taking into account different factors such as distance and residual energy of the nodes in an effort to select the most suitable nodes as CHs, e.g., LEACH-SWDN [11], K-LEACH [12], T-LEACH [13], MAP-LEACH [14], IBLEACH [15].

For instance, in [16] an enhancement scheme regarding LEACH protocol has been proposed, by taking into consideration the selection probability according to the increase probability for being CH. In [16] authors used Eq. (2) to calculate the threshold T(n).

$$T(n) = \begin{cases} c \times \left[\frac{E_{residual}}{E_{initial}} + \left(r_s\, div \frac{1}{p}\right) \times \left(1 - \frac{E_{residual}}{E_{initial}}\right)\right], & if\ n \in G \\ 0 & otherwise \end{cases} \quad (2)$$

$$c = \frac{p}{1 - p \times \left(r \bmod \frac{1}{p}\right)} \quad (3)$$

where *Eresidual* indicates the remaining energy of node (i); *Einitial* presents the initial energy of a given node (i); *rs* shows the number of successive rounds, during which node (i) is not selected as a CH.

K-LEACH [12] implemented the optimal number of clusters, which uses the consumed transmission energy in multi-path ($\varepsilon_{multi-path}$), free-space ($\varepsilon_{free-space}$) as well as distance to base station. Equation (4) is used to compute the threshold in K-LEACH.

$$T(n) = \begin{cases} \frac{p}{1-p\times\left(r \bmod \frac{1}{p}\right)} \times \frac{E_{residual}}{E_{initial}} \times K_{optimal}, & if\ n \in G \\ 0 & otherwise \end{cases} \quad (4)$$

Equation (5) is to find the optimal number of clusters.

$$K_{optimal} = \frac{\sqrt{n}}{\sqrt{2\pi}} \frac{\sqrt{\varepsilon_{free-space}}}{\sqrt{\varepsilon_{multi-path}}} \frac{M}{d_{bs}^2} \tag{5}$$

In [13] an algorithm T-LEACH has been proposed which considers the residual energy of sensor nodes in an effort to balance the load of the network and changing the round time by utilizing a new probability (*Phead*). Distance of a node to the base station and the area coverage (M) are taken into consideration by this probability. Equation (6) indicates how threshold T(t) s computed.

$$T(n) = \begin{cases} \frac{p_{head}}{1 - p_{head} \times \left(r \, mod \frac{1}{p_{head}} \right)} \times \frac{E(t)}{E_{total}(t)}, & if \; n \in G \\ 0 & otherwise \end{cases} \tag{6}$$

The probability *Phead* is calculated by Eq. (7) as follows:

$$p_{head} = \frac{\sqrt{n}}{\sqrt{2\pi}} \frac{\sqrt{\varepsilon_{free-space}}}{\sqrt{\varepsilon_{multi-path}}} \frac{M}{d_{bs}^2 \times N} \tag{7}$$

3 Proposed Routing Protocol

In this part we introduce our enhanced routing protocol which will reduce the energy consumption of SNs in order to improve the life time of the WSNs. First we make few assumption in our network model which will be used in our experimental section. These assumptions are as follows:

- All the SNs, deployed in WSN are homogenous and their position is fixed
- All the SNs have battery-limited power and not rechargeable
- SNs know their location.

3.1 Energy Consuming Model of the Improved Protocol

In this section the energy consumption model of the proposed protocol for radio communication has been given. The energy consumption for transmission of m-bit of data from one point to another would be calculated as:

$$E_{Tx}(m, d) = \begin{cases} m \times E_{elec} + m \times E_{fs} \times d^2, & if \; d < d_t \\ m \times E_{elec} + m \times E_{amp} \times d^4, & if \; d \geq d_t \end{cases} \tag{8}$$

where E_{elec} represents the base energy needed by either receiver or transmitter electronics; d represents the distance of transmission. d_t symbolises the threshold distance. E_{amp} and E_{fs} are standing for the unit energy needed for the transmitter amplifier which depends on the propagation model and the distance to approximate the power loss (d^2 for free space, d^4 for multipath fading). Therefore the energy consumption for receiving m-bit of data is:

$$E_{Rx}(m) = m \times E_{elec} \tag{9}$$

The energy consumption of data fusion for m-bit data is:

$$E_{Rx}(m) = m \times E_{DA} \tag{10}$$

where E_{DA} is the energy consumption for data fusion for each bit of data.

3.2 Cluster Heads Election and Clusters Formation

The proposed algorithm selects cluster heads by considering the distance and residual energy of each and every node. The geographical position as well as remaining energy of all nodes are transferred to BS. Average energy is calculated by BS and then nodes with more energy than average are added to the candidate CH set. Equation (11) is used to select the CHs:

$$P_i(t) = \frac{E_{r(i)}}{E_{av}} \times \frac{|D_{bs\_av} - D_{bs}(i)|}{D_{bs\_av}} \tag{11}$$

where $E_{r(i)}$ would be the residual energy of a node i for the current round; Eav signifies the average energy of all live nodes. D_{bs_a} represents the average distance of the all nodes from the BS and $D_{bs}(i)$ is for the geographic distance of a node (i) from the BS.

Eav can be calculated by the BS, after receiving the residual energy of all the sensor nodes as follows:

$$E_{av} = \frac{\sum_{i=1}^{n} E_{r(i)}}{n} \tag{12}$$

Similarly Eq. (13) is used to compute the average distance of all SNs from the base station.

$$D_{bs\_av} = \frac{\sum_{i=1}^{n} D_{bs}(i)}{n} \tag{13}$$

3.3 Determining the Ideal Number of CHs

Either having a small or large number of cluster heads will lead to inefficient energy consumption. In the proposed protocol an explicit method to find out the ideal number of CHs to maximize the lifetime of the network has been employed. From Eqs. (8) and (10) the total energy for data transmission can be written as:

$$E_{t \approx c}\left(m \times E_{elec}\, N/c + m \times E_{amp} \times d_{bs}^{4} + m \times E_{DA}\, N/c\right) + N\left(m \times E_{elec + m \times E_{fs} S^2/(2\pi c)}\right) \quad (14)$$

where n indicates the entire number of nodes; S shows the side length of the network which is a square; c stands for the number of CHs; d_{bs} is the distance from BS to CH. Furthermore from Eq. (14), we can obtain the optimum number of CHs as follow:

$$c_{O} = \frac{\sqrt{N}}{\sqrt{2\pi}} \sqrt{\frac{E_{fs}}{E_{amp}} \times \frac{S^2}{d_{bs}^{4}}} \quad (15)$$

3.4 Considering the Size of Clusters

Considering the following equation we can clearly see that clusters having less data member have to send more data than others.

$$F_r = \frac{T_r}{(t_s \times n)} \quad (16)$$

where the F_r is the data frames per round, T_r is the Time frame per round, t_s is the time slot per round and n number of members in a cluster.

In the proposed protocol to balance the energy variable time slots are used for data transmission in which the time slot is decreased if cluster has many members and increased if it has small number of members. Information about each cluster is being sent to the BS and then variable time is allocated for each cluster to transmit data.

4 Simulation and Results

To find out the efficiency and performance of the algorithms, simulation tests has been done using MATLAB (2015) [17]. Simulation parameters has been given in Table 1.

Table 1 Simulation parameters

Sr. no.	Parameters	Value
1	Environment size	100 × 100
2	Number of nodes	100
3	BS position	(52,174)
4	Packet size	6400 bits
5	Initial energy per node (E_0)	0.1 J
6	$E_{elec} = E_{bit}$	50 nJ/bit
7	E_{fs}	10 pJ/bit/m^2
8	E_{mp}	0.0013 pJ/bit/m^4
9	E_{DA}	5 nJ/bit

4.1 Network Lifetime

Figure 2 illustrates the total of alive nodes in different rounds. Considering Table 2, and Fig. 3 it is observed that the death of last node have been prolonged by 137.43, 112.53 and 74.58 % in the proposed algorithm compared with LEACH, K-LEACH and T-LEACH respectively.

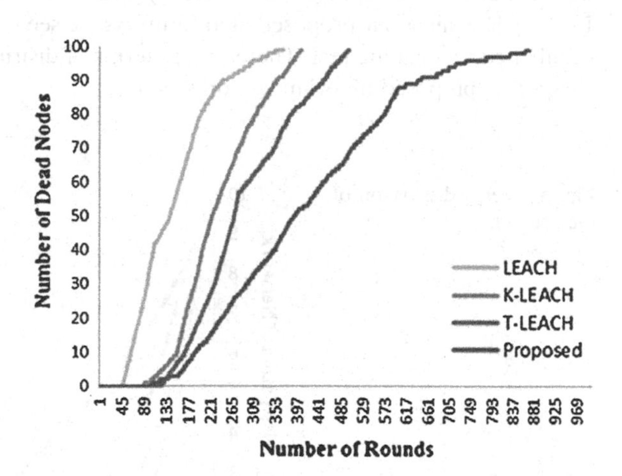

Fig. 2 Number of alive nodes

Table 2 Comparison of result

Protocols	Fist node died	Half nodes died	Last node died
LEACH	46	139	350
K-LEACH	85	226	391
T-LEACH	114	275	476
Proposed	93	395	831

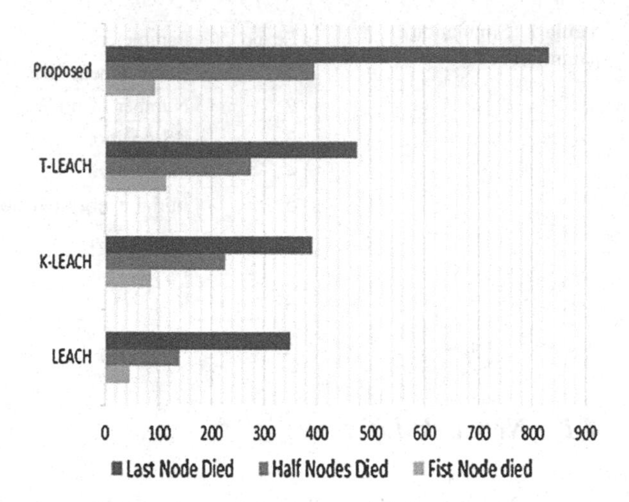

Fig. 3 Comparison of different algorithms

4.2 Network Residual Energy

Figure 4 shows the comparison of energy dissipation of LEACH, K-LEACH and T-LEACH against our proposed algorithm. As we see the proposed algorithm has a smother curve than the rest. Therefore, in terms of distributing and balancing of the energy the proposed algorithms works better.

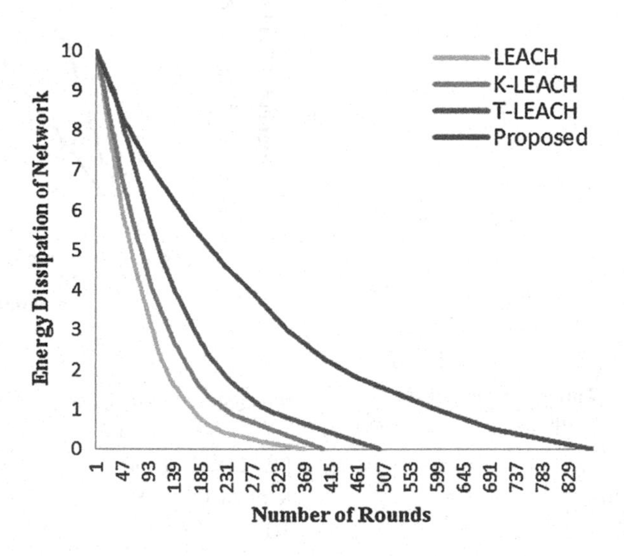

Fig. 4 Energy dissipation of the network

5 Conclusion

The proposed routing protocol takes advantages of the vital information such as nodes remaining energy and their distance from the base station to build the optimal clusters. In addition it uses different time slots for each clusters considering number of its members while data transmission in order to balance the overall energy consumption of the network. The result of our simulation indicates that proposed algorithm prolonged the life of the network by 137.43, 112.53 and 74.58 % compared with LEACH, K-LEACH and T-LEACH respectively. Therefore the proposed algorithm uses a better strategy to balance the energy of the network and improve its lifetime.

References

1. Elshakankiri, N.M., Moustafa, N.M., Dakroury, Y.H.: Energy efficient routing protocol for wireless sensor network. In IEEE International Conference on Intelligent Sensors, Sensor Networks and Information Processing, pp. 393–398, Dec 2008
2. Al-Anbagi, I., Erol Kantarci, M., Mouftah, H.: A survey on cross-layer quality of service approaches in WSNs for delay and reliability aware applications. IEEE Commun. Surv. Tutorials 1–1 (2014)
3. Navarro, K., Lawrence, E.: WSN applications in personal healthcare monitoring systems: a heterogeneous framework. In: 2010 Second International Conference on eHealth, Telemedicine, and Social Medicine (2010)
4. Mouradian, A., Nguyen, X., Auge-Blum, I.: Preventing alarm storms in WSNs anomaly detection applications. In: 2014 IEEE 25th Annual International Symposium on Personal, Indoor, and Mobile Radio Communication (PIMRC) (2014)
5. Ya, L., Pengjun, W., Rong, L., Huazhong, Y., Wei, L.: Reliable energy-aware routing protocol for heterogeneous WSN based on beaconing. In: 16th International Conference on Advanced Communication Technology (2014)
6. Tamandani, Y., Bokhari, M.: SEPFL routing protocol based on fuzzy logic control to extend the lifetime and throughput of the wireless sensor network. Wireless Netw. (2015)
7. Gupta, D., Verma, R.: An enhanced cluster-head selection scheme for distributed heterogeneous wireless sensor network. In: 2014 International Conference on Advances in Computing, Communications and Informatics (ICACCI) (2014)
8. Heinzelman, W.B., Chandrakasan, A.P., Balakrishnan, H.: An application-specific protocol architecture for wireless micro sensor networks. IEEE Trans. Wireless Commun. **1**(4), 660 (2002)
9. Heinzelman, W., Chandrakasan, A., Balakrishnan, H.: Energy efficient communication protocols for wireless sensor networks. In: IEEE Proceedings of the Hawaii International Conference System Sciences, pp. 3005–3014 (2000)
10. Xubo, L., Na, L., Liang, C.: An improved LEACH for clustering protocols in wireless sensor networks. In: 2010 International Conference on Measuring Technology and Mechatronics Automation (2010)
11. Udompongsuk, K., So-In, C., Phaudphut, C., Rujirakul, K., Soomlek, C., Waikham, B.: MAP: an optimized energy-efficient cluster header selection technique for wireless sensor networks. Adv. Comput. Sci. Appl. **279**, 191–199 (2014)
12. Wang, A., Yang, D., Sun, D.: A clustering algorithm based on energy information and cluster heads expectation for wireless sensor networks. Comput. Electr. Eng. J. 662–671 (2012)

13. Hou, R., Ren, W., Zhang, Y.: A wireless sensor network clustering algorithm based on energy and distance. In: Proceedings of Work on Computer Science and Engineering, pp. 439–442 (2009)
14. Xiangning, F., Yulin, S.: The improvement of LEACH protocol in WSN. In: Proceedings of International Conference on Computer Science and Network Technology, pp. 1345–1348 (2011)
15. Salim, A., Osamy, W., Khedr, A.M.: IBLEACH: intra-balanced LEACH protocol for wireless sensor networks. Wireless Netw. J. 111 (2014)
16. Handy, M.J., Haase, M., Timmermann, D.: Low energy adaptive clustering hierarchy with deterministic cluster-head selection. In: Proceedings of IEEE Conference on Mobile and Wireless Communication Networks, pp. 368–372 (2002)
17. In.mathworks.com, MATLAB—The Language of Technical Computing, 2015. [Online]. Available: http://in.mathworks.com/products/matlab/. Accessed 03 Jul 2015

Social Spider Algorithm Employed Multi-level Thresholding Segmentation Approach

Prateek Agarwal, Rahul Singh, Sandeep Kumar and Mahua Bhattacharya

Abstract Multi-level based thresholding is one of the most imperative techniques to realize image segmentation. In order to determine the threshold values automatically, approaches based on histogram are commonly employed. We have deployed histogram based bi-modal and multi-modal thresholding for gray image using social spider algorithm (SSA). We have employed Kapur's and Otsu's functions and in order to maximize its value, we have employed social spider algorithm (SSA). We have used the standard pre-tested images. Results have shown that the social spider algorithm has out-performed the results obtained by Particle Swarm Optimization (PSO) as far as optimal threshold values and computational time are concerned.

Keywords Social spider algorithm · Multi-level thresholding · Otsu's methodology and Kapur's methodology

1 Introduction

Typically almost every computer vision application and analysis requires an image segmentation pre-processing [1]. The primary objective of segmentation is region independent segregation of an image into separate regions which are visually distinct with respect to certain features [2]. Image segmentation is most basic pre-processing

P. Agarwal (✉) · R. Singh · S. Kumar · M. Bhattacharya
ABV-Indian Institute of Information Technology and Management,
Gwalior, Madhya Pradesh, India
e-mail: prateekagarwal1211@gmail.com

R. Singh
e-mail: rhlsingh720@gmail.com

S. Kumar
e-mail: sandeep2006iiitm@gmail.com

M. Bhattacharya
e-mail: mahuabhatta@gmail.com

S.C. Satapathy and S. Das (eds.), *Proceedings of First International Conference on Information and Communication Technology for Intelligent Systems: Volume 2*, Smart Innovation, Systems and Technologies 51, DOI 10.1007/978-3-319-30927-9_25

in any digital image processing. Image segmentation is the pre-process that divides the images into feature defined regions. Image segmentation can be categorized based of two characteristics that is discontinuity and homogeneity. Techniques exploiting discontinuities in the image are called as boundary based methods and the techniques exploiting homogeneity are called region based methods. The prime concern of image segmentation is to partition the pixels and intensity contained in the image into its salient featured regions. There are far too many applications of image segmentation. Some of them are: machine vision, object recognition, object detection, medical imaging application [3] and the list goes on and on.

Although there are many image segmentation techniques, but the following image segmentation approaches are quite popular: thresholding, region-based approaches, and edge-based techniques. Amongst them, image thresholding is commonly used technique because of its simplicity, accuracy and robustness [4]. In the last few decades, many thresholding techniques have been proposed. The detailed survey of thresholding [5, 6] presented various types of thresholding techniques which includes global and local multi-level thresholding. Amongst which, the global thresholding is the one which is widely used [7]. There has been a comparative in depth analysis of global multi-level thresholding techniques and advanced various useful criteria for thresholding performance evaluation. Previous papers also defined several evaluation criteria. Sahoo et al. analyzed nine thresholding techniques and illustrated comparative performances [8]. Glasbey pointed the relationships and performance differences between eleven histogram-based algorithms depending upon an extensive statistical study [9]. Threshold based technique can be mainly divided into bi-modal and multi-modal thresholding. For bi-modal threshold segmentation, a threshold value is chosen which divides the entire histogram into two classes, one representing the object and other representing background depending upon the image illumination. As far as multi-modal threshold segmentation is concerned, there are several classes corresponding to the objects in the image. The pixels pertaining to the same class are having grey levels within a specified range between two threshold values. But the key problem with this approach is to the positions of different valley in the histogram. Hence finding the multi-modal thresholding is still an interesting area to be explored by researchers. Despite the fact multi-level thresholding are easy to implement, [10, 11] their computational complexity increases manifolds with each new threshold being introduced [7, 12].

In this paper, we have used Otsu's and Kapur's method to realize our objective. Otsu's method [13] optimizes the threshold value by maximizing the variance between threshold classes. On the other hand, Kapur's methodology chooses the best possible threshold values by maximizing the entropy of the histogram [14]. Both Otsu's and Kapur's methodologies can be extended to multi-level thresholding problems. However, it fails to work as best to multi-level thresholding techniques owing to their exponential growth as far as computational time is concerned. In order to improve their performance, many techniques have been put forward to resolve this issue. To curtail the time of computation, evolutionary algorithms are used for choosing multi-modal thresholds values like PSO [15].

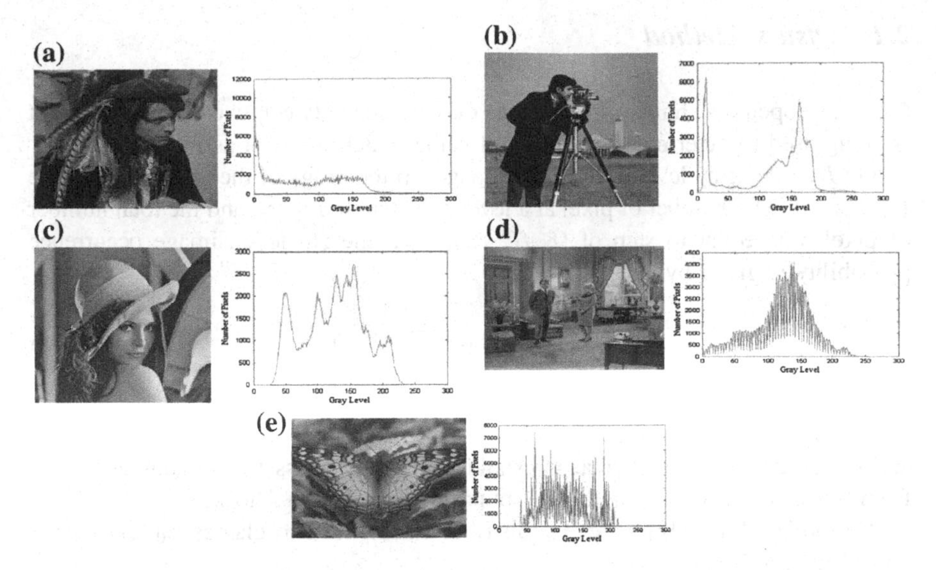

Fig. 1 Standard images and corresponding histograms (**a**–**e**)

We have employed social spider algorithm to perform optimization. Initially a pre-defined number of spiders are put on the web. In social spider, a spider will generate a vibration when it reaches the next position different from other. The quality of vibration is correlated with the fitness of that position. The vibration will flow over the web and others can sense it. Vibration play important role in SSA, this distinguish the SSA from other algorithms. The spiders can share their personal information with others spiders to gain the social knowledge. We have used this social spider technique to find the optimum threshold value in case of multi-level segmentation. We have compared the results of this approach to particle swarm optimization (PSO) technique. We have used standard images: hunter, camera man, Lena, living room and butterfly having their histograms (Fig. 1) for multilevel thresholding.

2 Problem Formulation

The best in class multi-level thresholding techniques select valley point values such that all classes of the histogram satisfy the specific characteristics and have same features and is carried out by considering an objective function. We have either maximize or minimize this function which is dependent on threshold value. In this paper, we have used two thresholding strategies. First methodology is given by Otsu which basically is a gray level histogram based approach. And the second one is Kapur's methodology which is based entropy of histogram.

2.1 Otsu's Method

Otsu developed a technique which relies on variance between two classes [3] and it is being used to determine the threshold value which is given as follows.

Let L be the gray level of the given image, so the range of the gray level will be $\{1, 2, \ldots L\}$. The number of pixel at a level z, is denoted by s_z, and the total number of pixel N is equal to sum of $(f_1, f_2, \ldots, f_L)$. At the zth level, image occurrence probability is given by:

$$p_z = \frac{s_z}{N}, p_z \geq 0, \sum_{z=1}^{L} p_z = 1. \tag{1}$$

If image is bi-modal, the histogram of that image is divide into two classes, C_0 and C_1. And at some particular threshold level s, the class C_0 contains the levels from 0 to s and class C_1 contain the other levels with s + 1 to L.

The probabilities (W_0 *sand* W_1) distribution for the two classes C_0 and C_1 are described as follows:

$$C_0 : \frac{p_1}{w_0(s)}, \ldots \frac{p_s}{w_0(s)} \quad \text{and} \quad C_1 : \frac{p_{s+1}}{w_1(s)}, \ldots \frac{p_L}{w_1(s)} \tag{2}$$

where $w_0(s) = \sum_{z=1}^{s} p_z$ and $w_1(s) = \sum_{z=s+1}^{L} p_z$.

The mean values of C_0 and C_1 classes are τ_0 and τ_1 respectively and describe as follows:

$$\tau_0 = \sum_{z=1}^{s} \frac{zp_z}{w_0(s)} \, and \, \tau_1 = \sum_{z=s+1}^{L} \frac{zp_z}{w_1(s)}. \tag{3}$$

Suppose the mean intensity value of the whole image is μ_S and then it is easy to show that $w_0\tau_0 + w_1\tau_1 = \tau_S$, and $w_0 + w_1 = 1$.

The total variance between any two levels is given as:

$$\sigma_{BC}^2 = \sigma_0 + \sigma_1 \tag{4}$$

where $\sigma_0 = w_0(\tau_0 - \tau_S)^2$ and $\sigma_1 = w_1(\tau_1 - \tau_S)^2$.

The Otsu's objective function is given as:

$$Maximize\, \mathrm{F}(s) = \sigma_{BC}^2 = \sigma_0 + \sigma_1, \tag{5}$$

The above analysis can be expended for multi-modal thresholding analysis [3] and can be described as:

N-dimensional improvement drawback, for obtaining n best thresholds for the given image $[s_1, s_2, \ldots, s_n]$, which divide the initial image into n- categories such as C_0 for $[0, \ldots, s_1 - 1]$, C_1 for $[s_1, \ldots, s_2 - 1]\ldots$ and C_n for $[s_n, \ldots, L - 1]$ and the threshold valley point values are obtained by maximizing the following equation

$$Maximize\,\mathrm{F}(s) = \sigma_0 + \sigma_1 + \sigma_2 \cdots + \sigma_n, \tag{6}$$

where

$\sigma_0 = w_0(\tau_0 - \tau_s)^2, \sigma_1 = w_1(\tau_1 - \tau_s)^2, \sigma_2 = w_2(\tau_2 - \tau_s)^2$ and so on
$\sigma_n = w_n(\tau_n - \tau_s)^2$.

2.2 *Kapur's Method*

Kapur has developed a method for bimodal [2] or bi-level thresholding which can be described as follows: Let L be the gray within the given image and these gray level ranges between {0, 1, 2,... $(L - 1)$}. Then, prevalence probability of gray level z is given by:

$$P_z = \frac{k(z)}{N}; \quad \text{for } (0 \leq z \leq (L-1)). \tag{7}$$

where k (z) signifies the number of pixel corresponding to each gray level, L and the total number of pixels are represented by N which is denoted as:

$$N = \sum_{z=0}^{L-1} k(z). \tag{8}$$

The objective is maximizing the fitness function as follows:

$$f(s) = H_0 + H_1, \tag{9}$$

where

$$H_0 = -\sum_{z=0}^{s-1} \frac{P_z}{w_0} \ln \frac{P_z}{w_0}, \; w_0 = \sum_{z=0}^{s-1} P_z \text{ and } H_1 = -\sum_{z=s}^{s-1} \frac{P_z}{w_1} \ln \frac{P_z}{w_1}, \quad w_0 = \sum_{z=s}^{L-1} P_z$$

The above method can be used to evaluate multi-level threshold valley points and is described as: N-dimensional optimization problem, for choosing n threshold values for a given image $[s_1, s_2, \ldots, s_n]$.

Here also our aim is to maximize of the objective function:

$$\mathrm{f}([s_1, s_2, s_3, \ldots\ldots, s_n]) = H_0 + H_1 + H_2 + \cdots + H_n \tag{10}$$

where

$$H_0 = -\sum_{z=0}^{s_1-1} \frac{P_z}{w_0} \ln \frac{P_z}{w_0}, \quad w_0 = \sum_{z=0}^{s-1} P_z, \quad H_1 = -\sum_{z=s1}^{s_2-1} \frac{P_z}{w_0} \ln \frac{P_z}{w_0}, \quad w_0 = \sum_{z=s1}^{s_2-1} P_z$$

$$H_2 = -\sum_{z=s2}^{s_3-1} \frac{P_z}{w_0} \ln \frac{P_z}{w_0}, \quad w_0 = \sum_{z=s2}^{s_3-1} P_z, \quad H_3 = -\sum_{z=s3}^{s_4-1} \frac{P_z}{w_0} \ln \frac{P_z}{w_0}, \quad w_0 = \sum_{z=s3}^{s_4-1} P_z$$

$$H_n = -\sum_{z=sn}^{L-1} \frac{P_z}{w_0} \sum_{z=sn}^{L-1} \ln \frac{P_z}{w_0}, \quad w_0 = \sum_{z=sn}^{L-1} P_z$$

2.3 *Social Spider*

In this SSA technique is employed to seek for the best possible threshold values by maximizing the target the objective of Kapur's methodology and Otsu's methodology for multi-modal thresholding.

2.3.1 Social Spider Algorithm

Social Spider is the latest optimization algorithm used for multi-level thresholding and basically for foraging behavior. In this algorithm process, each spider (s) will receive **|pop|** different vibrations generated by any other spiders where pop is the spider population.

The information which is received from spider vibrations includes the source position of the vibration and its attenuated intensity. We take V to represent above |pop| vibrations. To receive V, s will be select the highly vibration v_s^{best} from V and compared it's intensity with the target vibration intensity v_s^{tar} stored in its memory. S will store v_s^{best} as v_s^{tar} if the intensity of v_s^{best} is bigger and c_s, or the number of iterations since s has last changed then its target vibration, is reset; otherwise, the original v_{tar} is retained and c_s is increased by one. We use J_s^k and J_s^{tar} to represent the source positions of V and v_{tar}, respectively, and k = $\{1, 2, \ldots, |\text{pop}|\}$. The algorithm then manipulates s to perform a random walk towards v_s^{tar}. At this time we utilize a dimension mask to guide the movement. Each spider holds a dimension mask M, which is a (0-1) binary vector of length d and d = dimension of the optimization problem.

Initially all values in the mask are reset. In each iteration value, spiders has a probability of $1 - p_c^{C_s}$ to change its mask where $p_c \in (0, 1)$ is a user-defined attribute that describes the probability of changing the mask. If the mask is decided to be changed, then each bit of the vector has a probability of p_m to be assigned with a one, and $1 - p_m$ to be a zero. p_m is also a user controlled parameter defined in $(0, 1)$. Each bit of a mask is changed independently and does not have any correlation with the previous mask. In case all bits are zeros, one random value of the mask is changed to 1. Similarly, one random bit is assigned to zero if all values are ones. After the dimension mask is determined, a new following position J_s^{f0} is generated based on the mask for s. The value of *k*th dimension of the following position $j_{s,k}^{f0}$ is generated as follows.

$$J^{fo}_{s,k} = \begin{cases} J^{tar}_{s,k} M_{s,k} = 0 \\ J^{r}_{s,k} M_{s,k} = 1, \end{cases} \quad (11)$$

where is a random integer value generated in [1, |pop|], and $M_{s,k}$ stands for the kth dimension of the dimension mask M of spiders. Here the random number r for two different dimensions with $M_{s,k} = 1$ is generated independently. With the generated J^{f0}_{s}, s performed a random walk to that position. This random walk is conducted using the following equation.

$$J_s(t+1) = J_s + (J_s - J_s(t+1)) * r + \left(J^{fo}_{s} - J_s\right) \odot R \quad (12)$$

where $\odot$ denotes element wise multiplication and R is a vector of random float-point numbers generated from zero to one uniformly. Before following J^{f0}_{s}, s first moves along its previous direction, which is the direction of movement in the previous iteration. The distance along this direction is a random portion of the previous movement. Then s approaches J^{f0}_{s} along each dimension with random factors generated in (0, 1). This random factor for different dimensions is generated independently. After this random walk, s stores its movement in the current iteration for the next iteration. This ends the random walk sub step. The final step of the iteration phase is the constraint handling.

2.3.2 Algorithm Steps:-

1: Assign values to the parameters of SSA.

2: Create the population of spiders pop and assign memory for them.

3: Initialize v^{tar}_{s} for each spider.

4: **while** stopping criteria not met **do**

5: **for** each spider s in pop **do**

6: Evaluate the fitness value of s.

7: Generate a vibration at the position of s.

8: **end for**

9: **for** each spider s in pop **do**

10: Calculate the intensity of the vibrations V generated by all spiders.

11: Select the strongest vibration v^{best}_{s} from V.

12: **if** the intensity of v^{best}_{s} larger than v^{tar}_{s} Then

13: Store v^{best}_{s} as v^{tar}_{s}.

14: **end if**

15: Update cs.

16: Generate a random number r from [0, 1).

17: **if** r > p^{cs}_{c} **then**

18: Update the dimension mask $\boldsymbol{M_s}$.

19: **end if**

20: Generate J^{fo}_{s}.

21: Perform a random walk.

22: Address any violated constraints.

23: **end for**

24: **end while**

25: Output the best solution found.

In this we used fixed parameter which is helpful to test the performance of the social spider algorithm with other algorithm like PSO.

In SSA we used 3 user-controlled parameters which help us to searching the behavior:

- r_a: This parameter states that the rate of vibration attenuation when propagating over the spider web.
- p_c: This parameter controls the probability of the spiders which changing their dimension mask in the random walk step.
- p_m: This parameter states that the probability of each value in a dimension mask to be one.

3 Result

The applicability of the Social spider algorithm (SSA) for multi-level segmentation for can be extended to almost every possible with some constraints being taken into consideration. However for the sake of analysis, we have used five standard images:

Table 1 Comparison when Otsu's objective function is used

Image name	N	Objective values		Threshold values		CPU time	
		SSA	PSO	SSA	PSO	SSA	PSO
Hunter	3	3315.3230	3314.3967	36, 85, 128	38, 87, 114	3.9645	4.4282
	4	3390.4504	3257.1256	32, 80, 120, 151	35, 83, 131, 158	4.2682	4.9681
	5	3295.1229	3275.3808	31, 74, 108, 142, 178	36, 84, 126, 123, 178	4.8043	5.2372
Cameraman	3	3684.6623	3678.4006	62, 119, 153	71, 135, 127	3.7253	4.2123
	4	3838.7202	3725.5342	49, 105, 143, 170	65, 122, 148, 073	4.3255	4.8233
	5	3868.3239	3860.6515	39, 87, 126, 152, 075	36, 84, 126, 123, 178	4.8043	5.2372
Lena	3	2129.1786	2128.0007	78, 125, 170	79, 128, 170	3.7588	4.2887
	4	2190.2567	2181.6634	77, 117, 159, 183	78, 113, 135, 138	4.5286	4.7289
	5	2250.7022	2215.6425	66, 92, 123, 159, 083	79, 111, 141, 168, 189	4.9577	5.2533
Living room	3	1759.8257	1757.2367	75, 124, 164	80, 126, 166	3.8281	4.3456
	4	1836.3242	1822.8743	65, 103, 135, 171	68, 109, 142, 179	4.0132	4.7211
	5	1868.9966	1865.9812	57, 95, 126, 148, 182	57, 99, 129, 157, 191	5.1623	5.8134
Butterfly	3	1668.8189	1665.4364	79, 118, 166	78, 119, 161	3.8131	4.2035
	4	1708.5732	1703.5245	75, 106, 138, 068	80, 114, 146, 078	4.5813	4.8188
	5	1734.2212	1731.8467	73, 105, 128, 152, 182	74, 108, 130, 158, 081	5.1802	5.4224

hunter, camera man, Lena, living room and butterfly. We have done the analysis on Matlab. We have compared the results of SSA algorithm with that of PSO.

The proposed social spider algorithm (SSA) is applied on histogram's entropy based objective function which is Kapur's method and also on variance based technique which is Otsu's method. The results of both the function have out-performed the particle swarm optimization (PSO) approach. The same can also be inferred from Tables 1 and 2 showing the results for Otsu and Kapur's approaches respectively. Also it is noticed during the analysis that the level (quality) of processed image is much better when the value of n is high (we have taken as n = 5(max) during analysis). For both the techniques we have obtained better results in terms of optimum threshold values and computational time. So results clearly shows in Fig. 2 that our SSA is faster than PSO as far as computational time is concerned.

Table 2 Comparison when Kapur's objective function is used

Test image	N	Objective values		Threshold values		CPU time	
		SSA	PSO	SSA	PSO	SSA	PSO
Hunter	3	15.7002	15.1068	58, 105, 276	85, 127, 177	8.1813	8.8055
	4	18.9718	18.0002	50, 98, 138, 282	73, 132, 182, 200	8.8354	9.0025
	5	21.9525	20.2509	49, 95, 138, 278, 222	90, 120, 175, 191, 218	9.7252	10.1521
Cameraman	3	16.7409	15.1246	95, 150, 194	96, 139, 292	8.1267	9.1165
	4	19.1082	18.0000	41, 96, 140, 201	76, 116, 157, 202	8.6269	9.3525
	5	21.5912	19.6252	41, 85, 115, 751, 199	70, 96, 129, 255, 198	9.1262	10.1084
Lena	3	16.2706	15.9252	88, 147, 189	87, 150, 181	7.7334	8.0121
	4	17.9878	17.6388	73, 119, 153, 185	92, 130, 169, 191	8.5656	9.1734
	5	20.6709	20.1452	64, 95, 139, 192, 195	74, 114, 149, 170, 197	8.8845	9.4024
Living room	3	15.6055	15.1250	71, 125, 174	73, 159, 188	7.9905	8.2562
	4	18.9195	18.1410	60, 105, 148, 188	58, 125, 173, 203	8.2521	9.0000
	5	21.2298	20.6252	47, 94, 135, 168, 202	72, 97, 118, 159, 199	8.7563	9.5241
Butterfly	3	12.5846	12.2131	74, 178, 153	63, 136, 171	7.9905	8.2962
	4	14.9726	14.1571	71, 97, 128, 187	71, 113, 163, 185	8.2521	9.0800
	5	17.9292	16.0144	74, 97, 120, 144, 189	92, 116, 163, 157, 182	8.7563	9.5741

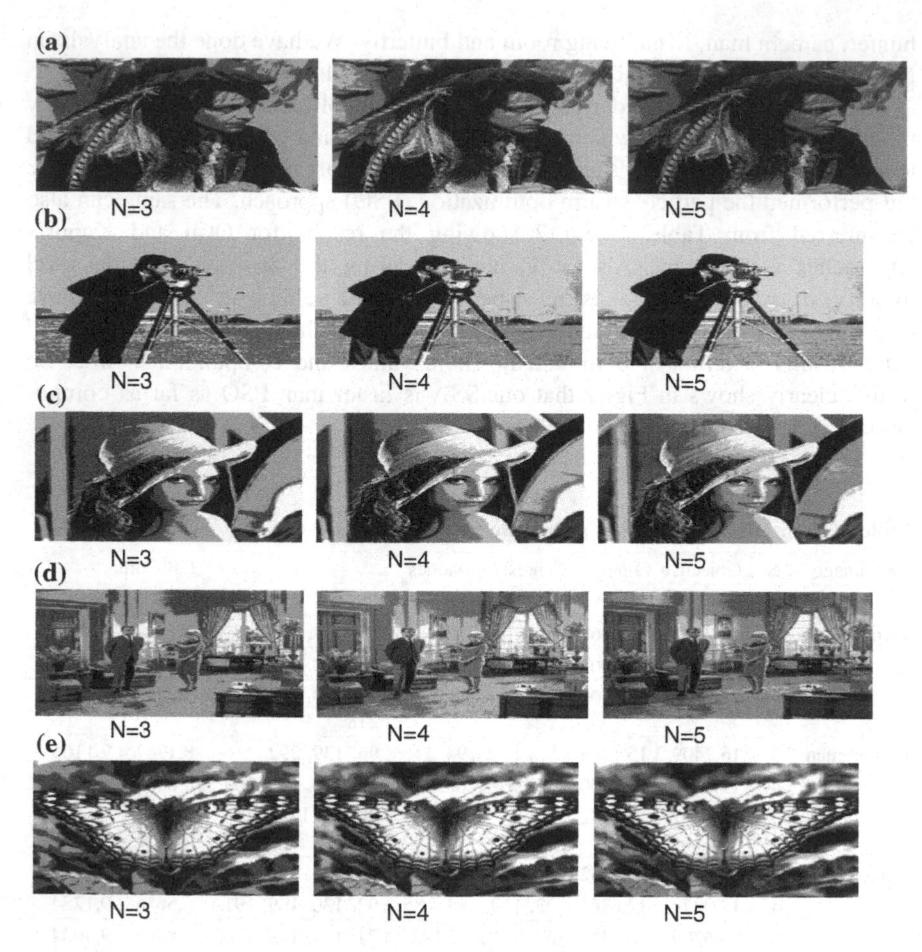

Fig. 2 Segmented images of SSA for different values of N (**a**–**e**)

4 Conclusion

We have proposed the social spider algorithm (SSA) which deals with the problem of solving thresholding in mutli-level segmentation. Verification and validation of effectiveness of our algorithm was examined on two widely known methods that is Kapur and Otsu. We carried the examination on standard tested images and compared the results with that of obtained by PSO. The observations revealed that SSA gives better results as compared to PSO as far as threshold valley values and computational time are concerned.

5 Future Scope

In this paper, we have compared social spider technique and particle swarm optimization (PSO). And we come up with a result that social spider technique is beter than PSO approach. Future work can be carried out to examine the feasibility of our technique with other mutli-level segmentation approaches also (other than PSO).

References

1. Tobias, O.J., Seara, R.: Image segmentation by histogram thresholding using fuzzy sets. IEEE Trans. Image Process., vol
2. Zuva, T., Olugbara, O.O., Ojo, O., Ngwira, S.M.: Image segmentation, available techniques, developments and open issues. Can. J. Image Process. Comput. Vis. **2**(3), 20–29 (2011)
3. Saha, S., Bandyopadhyay, S.: Automatic MR brain image segmentation using a multiseed based multiobjective clustering approach. Appl. Intell. **35**(3), 411–427 (2011)
4. Pal, N.R., Pal, S.K.: A review on image segmentation techniques. Pattern Recognit. **26**(9), 1277–1294 (1993)
5. Sezgin, M., Sankur, B.: Survey over image thresholding techniques and quantitative performance evaluation. J. Electron. Imaging **13**(1), 146–165 (2004)
6. Sankur, B., Sezgin, M.: Image thresholding techniques: a survey over categories. Pattern Recogn. **34**(2), 1573–1607 (2001)
7. Pal, S.S., et al.: Multi-level thresholding segmentation approach based on spider monkey optimization algorithm. In: Proceedings of the Second International Conference on Computer and Communication Technologies. Springer India (2016)
8. Sahoo, P.K., Soltani, S., Wong, A.K.C., Chen, Y.: A survey of thresholding techniques. Comput. Graph. Image Process. **41**, 233–260 (1988)
9. Glasbey, C.A.: An analysis of histogram-based thresholding algorithms. Graph. Models Image Process. **55**, 532–537 (1993)
10. Maitra, M., Chatterjee, A.: A hybrid cooperative–comprehensive learning based PSO algorithm for image segmentation using multilevel thresholding. Expert Syst. Appl. **34**(2), 1341–1350 (2008)
11. Bhandari, A.K., et al.: Cuckoo search algorithm and wind driven optimization based study of satellite image segmentation for multilevel thresholding using Kapur s entropy. Expert Syst. Appl. **41**(7), 3538–3560 (2014)
12. Hassanzadeh, T., Vojodi, H., Moghadam, A.M.E.: A multilevel thresholding approach based on levy-flight firefly algorithm. In: Machine Vision and Image Processing (MVIP), 2011 7th Iranian. IEEE (2011)
13. Otsu, N.: A threshold selection method from gray-level histograms. Automatica **11**(285-296), 23–27 (1975)
14. Kapur, J.N., Sahoo, P.K., Wong, A.K.C.: A new method for gray-level picture thresholding using the entropy of the histogram. Comput. Vis. Graph. Image Process. **29**(3), 273–285 (1985)
15. Akay, B.: A study on particle swarm optimization and artificial bee colony algorithms for multilevel thresholding. Appl. Soft Comput. **13**(6), 3066–3091 (2013)

Simulation Based Comparison Between MOWL, AODV and DSDV Using NS2

Balram Swami and Ravindar Singh

Abstract Mobile Ad-hoc network is collection of mobile nodes which can move anywhere within the network and because of the mobility of node which have limited resources, routing faces challenges as maximize the delivery ratio and minimize the delay with less consumption of nodes resources. MANET does not have any central coordinator. AODV is the most popular MANET routing protocols as it works very well in the dynamic nature of MANET. AODV is based on BFS and high scalability. DSDV is proactive routing which is very efficient for the small scale network. It uses a routing table maintained by every node of the network which contain the information about the network and remaining nodes of the network. MOWL is the new enhancement of basic OWL which is based on HS. MOWL tries to achieve scalability of basic OWL by using multiple BFS. In this research we going to analysis and compare the simulation results of MOWL with well known MANET routing protocols.

Keywords MANET · AODV · DSDV · MOWL · BFS · DFS · HS · OWL

1 Introduction

Mobile ad-hoc network is self-adjustable network with dynamic topology. It is a collection of mobile nodes like laptops and mobile phone which have limited resources like battery power and bandwidth. It does not have central control/coordinator that can manage and take decisions about the network and routing. So routing faces challenges like maximize the delivery ratio and minimize the delay. AODV is a reactive routing protocol which discovers route/path from source to destination on demand. It uses three kinds of messages to establish

B. Swami (✉) · R. Singh
Computer Science Department, Government Engineering College, Ajmer, India
e-mail: mbswami8@gmail.com

R. Singh
e-mail: ravikaviya@gmail.com

S.C. Satapathy and S. Das (eds.), *Proceedings of First International Conference on Information and Communication Technology for Intelligent Systems: Volume 2*, Smart Innovation, Systems and Technologies 51, DOI 10.1007/978-3-319-30927-9_26

communication between the nodes of the network. DSDV is proactive routing protocol which stores the paths from source to destination in the routing tables which is maintained by every node of the network. MOWL is the enhancement of basic OWL which is based on hybrid searching i.e. combination of DSF and BFS. Different routing techniques have different working strategies and have different advantages and disadvantages. In this research we are going to introduce a new enhancement of basic OWL in which it tries to increase the scalability of basic OWL which was based on single DFS instead of BFS.

In this research we are going to analyze the performance of MOWL and compare with well known MANET routing protocols. Section 2 contains the overview of routing protocols. Section 3 contains the comparison between the AODV, MOWL and DSDV. Section 4 will contains the simulation results and Sect. 5 is the conclusion of the research and last one is the references.

2 Routing Protocols

Routing protocols are the set of rule for the routing strategies of the network. MANET support mobility of nodes within the network and it does not have any central control so this makes challenges for the routing protocols.

Mobile ad-hoc network routing Protocols can be divided in to three types on the basis of their working strategies [1, 2].

- Reactive Routing Protocols
- Proactive Routing Protocols
- Hybrid Routing Protocols

2.1 Reactive Routing Protocols

Reactive routing protocols find the route/path from source node to destination node on demand. Nodes of the network do not know about the network structure. Routing protocols have three steps Route discovery, Packet sending/receiving and Route maintenance

- **AODV**

It is the well known Reactive routing protocol which uses three kinds of messages to establish communication between the nodes of the network [3, 4].

- RREQ: it is generated by the source node when it want to communicate the other node called destination node. Same node can generate many RREQ for different destination and more than one source nodes can generate RREQ for same destination node. A RREQ contains the source id, Destination id, and

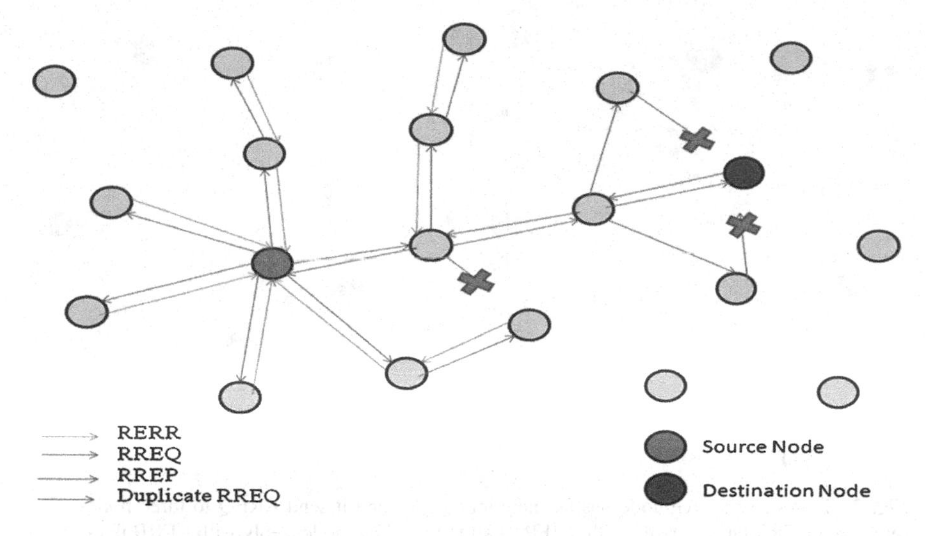

Fig. 1 Working scenario of AODV, whenever any node want to communicate it generate RREQ and broadcast to all of its neighbor nodes, neighbor nodes will forward to their neighbor nodes if RREQ's destination id is not match with its own id. This happens until all nodes get the RREQ. Only destination node can reply with RREP

RREQ sequence number etc. These information uniquely identified a RREQ. Source node broadcast the RREQ to all of its neighbor nodes.

- RERR: it is generated by the leaf nodes or the node which does not have any neighbor node to forward the RREQ than it generate RERR and send to that neighbor node from which RREQ was received and this will received by the source node at the last it will notify failure of last route request and start new search.
- RREP: it is generated by the destination node after receiving the RREQ and send back to that node from which destination node was received the RREQ (Fig. 1).

- **MOWL**

Multiple ordered walk is the enhancement of basic OWL [5, 6] in which it uses Hybrid Searching. HS is the combination of BFS and DFS. MOWL based on multiple DFS simultaneously at same time but the number of DFS if the half of the number of neighbor nodes i.g number of DFS is equal to the half of BFS/number of neighbor nodes. HS have less network overhead than BFS and it is faster than single DFS.

It is s reactive routing protocols and it is also uses the same three kinds of messages like AODV [7] i.e. RREQ, RERR, and RREP for the same purpose as they used in the AODV. MOWL is different in route discovery phase of routing, as it uses HS instead of BFS (Fig. 2).

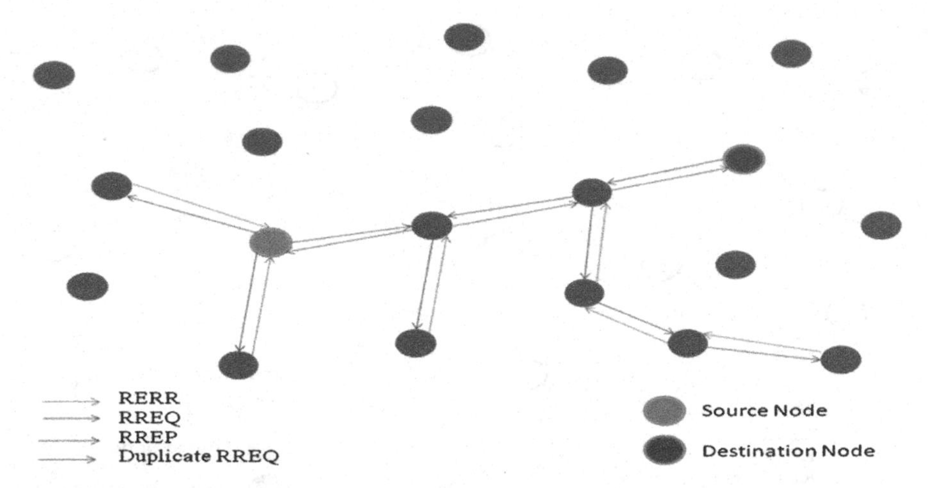

Fig. 2 Shows that source node has six neighbor nodes and it send RREQ to three nodes. Two reply with RERR but one reply with RREP. If all the neighbor nodes reply with RERR than source send RREQ to the remaining neighbor nodes

2.2 *Proactive Routing Protocols*

Every node in the Proactive routing protocols maintains a routing table which contains the routing information like list of neighbor nodes and route to the remaining nodes of the network. Every node of the network knows about the network structure.

- **DSDV**

It is also called table driven routing protocol in which every node of the network maintains routing table [8]. Whenever any node want to communicate than first it search in the routing table for the route to the destination node with number of hop counts then send data packets to the destination node and the last step route maintenance in which if there is any update to the routing table of any node than that update broadcast to the neighbor nodes of the node to indicate them about the update. The broadcasting of table or updates may be event based or time based in which updates are broadcast after time interval or after en event occurred (Fig. 3).

2.3 *Hybrid Routing Protocols*

It is based on the combination of both reactive routing protocol and proactive routing protocol. ZRP is the hybrid type routing protocol.

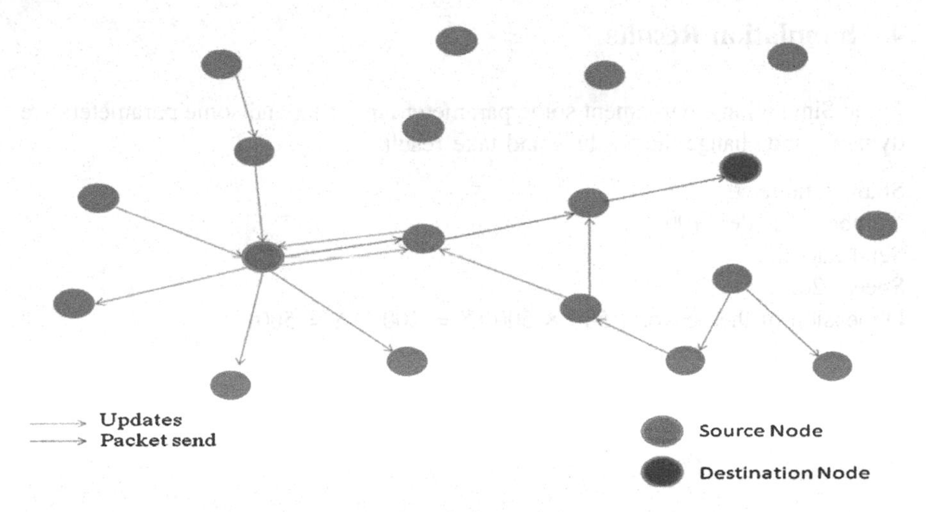

Fig. 3 Shows that source node search the route from the table and send data packet to the destination node and *red arrow* show the table update of the nodes of the network

3 Comparisons Between AODV, MOWL and DSDV

Parameter	AODV	MOWL	DSDV
Network overhead	Higher than MOWL	Lower than AODV	High overhead of tables and updates
Speed in small network and large network	Faster in large network and slow than MOWL in small network	Fast in small network and slow in large network, MOWL' worst performance equal to AODV' performance	Fastest in small network and slower in large network
Messages used	RREQ, RERR and RREP	RREQ, RERR and RREP	HELLO
Type	Reactive routing	Reactive routing	Proactive routing
Scalability	Highest	Medium	Lowest
Efficiency	Medium	Highest	Lowest
Routing table size	Only neighbor list	Same as AODV	Very large as neighbor list and path to remaining nodes
Communication	Via RREQ, RREP and RERR	Via RREQ, RERR and RREP	Broadcasting of tables or table updates
Based on	BFS	Hybrid searching	Pre-stored route in tables
Route discovery	On demand	On demand	Pre-stored in table

4 Simulation Results

In the Simulation environment some parameter are set fix and some parameters are dynamic and change their values and take results

Static parameters
Number of nodes: 100
Send rate: 0.25
Speed: 20
Dimension of the network: 500 × 500 (X = 500 × Y = 500)
And dynamic parameters are
Number of connections: {15, 30, 45}
Pause Time: {0, 20, 40, 60, 80, 100}

4.1 Delay

Delay is the total time taken by the packets to deliver at destination node. Table 1 contains the simulation results takes with dynamic connections.

Delay (ms) = (sum/recvnum)*1000;
Sum = total duration of packets;
Recvnum = Number of received packets; (Fig. 4).

If the network is large than AODV will work best and MOWL work average and DSDV will be the worst if network is too large.

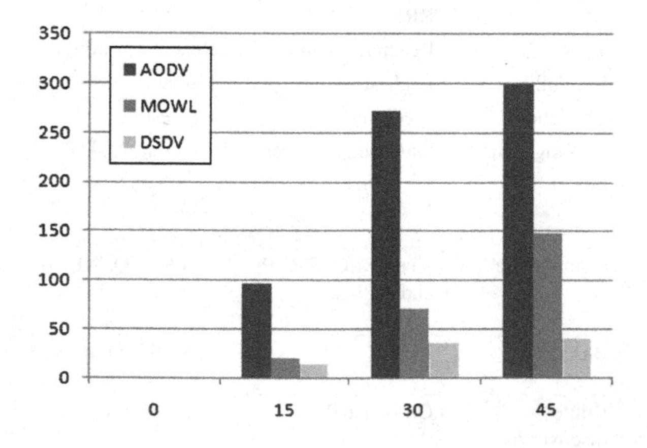

Fig. 4 Shows the Delay versus number of connections per node. Delay in all protocols increases as we increase the number of connections but DSDV has less delay than AODV and MOWL because more nodes direct connected to other nodes of the network and we have a small network of 100 nodes so DSDV work best in small network

Table 1 Delay comparison between AODV, MOWL and DSDV

Connections	AODV	MOWL	DSDV
0	0	0	0
15	97.6	20.57	15.26
30	272.50	71.93	36.70
45	299.87	149.44	40.45

4.2 *Packet Delivery Fraction/Packet Delivery Ratio*

PDF = (recvs/sends)*100;
Recvs = total number of received packets;
Sends = total number of sent packets; Fig. 5, Table 2.

4.3 *DPR*

DPR = (dropped Packets/sends)*100;
Dropped Packets = total number of dropped packet within the network.
Sends = total number of sent packets; Fig. 6, Table 3.

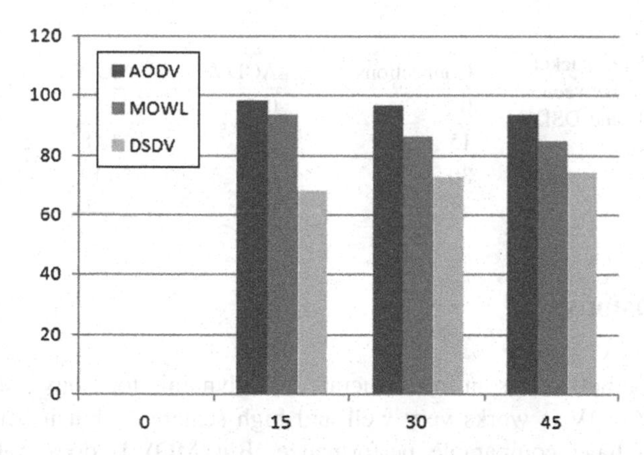

Fig. 5 Shows the packet delivery fraction versus number of connection per node of AODV, MOWL and DSDV. In both APDV and MOWL PDF decreases as we increases the number of connection because it will increases the time of route discovery but there are many alternatives route so the packets will be delivered successfully. But in DSDV PDF increases as we increases the number of connection because more nodes will connect directly with the node and we have small large table entry this will increase the network load so in DSDV PDF increases as we increase the connections per node

Table 2 Packet delivery ratio comparison between AODV, MOWL and DSDV

Connections	AODV	MOWL	DSDV
0	0	0	0
15	98.72	94.09	68.19
30	97.21	86.44	72.89
45	94.10	84.98	74.66

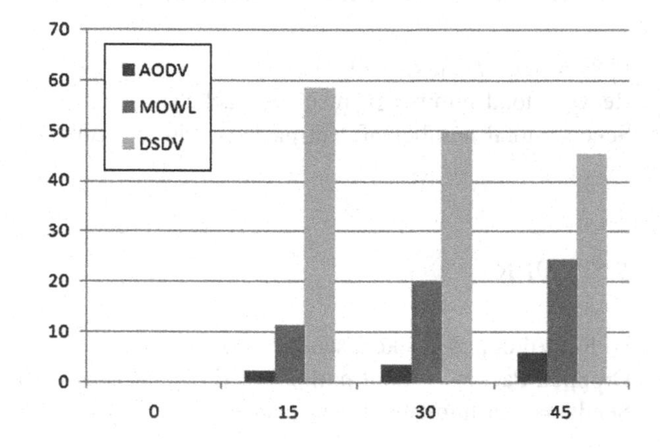

Fig. 6 Show delivered packet ratio of AODV, MOWL and DSDV. In both ADOV and MOWL works similar as both are Reactive routing but DSDV works opposite because it is proactive routing

Table 3 Delivered packet ratio comparison between AODV, MOWL and DSDV

Connections	AODV	MOWL	DSDV
0	0	0	0
15	2.35	11.68	58.77
30	3.74	20.19	49.91
45	6.12	24.55	45.72

5 Conclusions

Mobile ad-hoc network is an infrastructure less, dynamic topology without central coordinator AODV is works very well and high scalability but in small network MOWL also have comparable performance. But MOWL does not have high scalability like AODV but it has higher than DSDV and basic OWL protocols. DSDV is also good for very small network but it has very poor scalability and very less efficient than other AODV and MOWL. MOWL is very efficient and medium scalable routing technique for Mobile ad-hoc Network. But we can use past information and some kind of priority technique to improve the performance of MOWL.

References

1. Shenakaran, S., Parvathavarthim, A.: An overview of routing protocols in mobile Ad-hoc network. In: IJARCSSE, vol. 3, Issue 2 (2013)
2. Gill, A., Diwaker, C.: Comparative analysis of routing in MANET. In: IJARCSSE, vol. 2, Issue 7 (2012)
3. Perkins, C.E., Royer, E.M.: Ad-hoc on-demand distance vector routing. In: 2nd IEEE Workshop on Mobile Computing Systems and Applications, pp. 90–100, New Orleans (1999)
4. Perkins, C.E., Royer, E.M.: An implementation study of the AODV routing protocol, IEEE (2000)
5. Dabideen, S., Garcia-Luna-Aceves, J.J.: Efficient routing in MANETs using ordered walks. Springer Science Business Media, Heidelberg (2011)
6. Dabideen, S., Garcia-Luna-Aceves, J.J.: OWL: towards scalable routing in MANETs using depth-first search on demand. In: 6th IEEE International Conference on Mobile Ad-hoc and Sensor Systems (2009)
7. Perkins, C.E., Belding-Royer, E.: Ad hoc On-Demand Distance Vector (AODV) Routing (2003)
8. Perkins, C.E., Bhagwat, P.: Highly Dynamic Destination-Sequenced Distance-Vector Routing (DSDV) for Mobile Computers (1994)

PeTelCoDS—Personalized Television Content Delivery System: A Leap into the Set-Top Box Revolution

Mangesh Bedekar, Saniya Zahoor and Varad Vishwarupe

Abstract At home, on the television, the sheer number of Channels and the vast number of Programs on each Channel has itself made the task of identifying the "appropriate" program to watch difficult for the common user. There is a need of a system to generate suggestions/recommendation to the common user about which Programs to watch and when. In this paper, we propose a method and system which assists the user to choose which Programs on which Channels to watch without any inputs from the Viewer about his "Likes" or "Dislikes". It learns from the Viewer Implicitly over time and learns all the patterns that the Viewer exhibits over the course of Television watching.

Keywords STB · Personalization · Data mining · Pattern recognition · User profiling · Recommendation system

1 Introduction

The Television Content consumed today is sourced from the Set Top Box (STB), which in turn receives content from a roof top micro dish which in turn receives content broadcast by satellites. The content commonly is terms of multi-media content, primarily, Audio and Video, segregated in Frequency bands, called Channels which are uniquely identifiably by a number and the content is labeled as per the genre it relates to. Historically the schedule of television programs were

M. Bedekar (✉) · S. Zahoor
Department of Computer Engineering, MAEER's MIT, Pune, India
e-mail: mangesh.bedekar@gmail.com

S. Zahoor
e-mail: saniya.zahoor@yahoo.com

V. Vishwarupe
Department of Information Technology, MAEER's MIT College of Engineering, Pune, India
e-mail: varad44@gmail.com

S.C. Satapathy and S. Das (eds.), *Proceedings of First International Conference on Information and Communication Technology for Intelligent Systems: Volume 2*, Smart Innovation, Systems and Technologies 51, DOI 10.1007/978-3-319-30927-9_27

made available in a printed form as a grid listing the date and time, channel and program title through which the Viewer had to look up the program of their interest through these guides. Currently however, the date and time of availability of the content is published in the form of Electronic Program Guides (EPG). The EPG's today hold upwards of 500 channels with each channel having programs individually. EPG's definitely are helping the Viewers in better identifying the programs of their choice. As the number of these television channels has increased very rapidly, it has become difficult to correctly identify the program desired by a common Viewer even using the EPG's.

With the advent of the STB's the television channels have a small snippet explaining the content of the program, along with a scheduler, Viewers can look-up these programs on the television itself instead of printed copies, set alarms, manually, schedule their viewing patterns by specifying explicitly, what they wish to watch and when. Other variants/extensions of STB's can record programs on Solid State Data Storage Devices, as and when indicated by the Viewer or through a schedule recorded beforehand. The STB can also record programs simultaneously from different channels too.

Many television viewers have a particular preference from amongst the vast channels which are streamed to them. The number of channels that an average Viewer watches is a fraction and the programs that the Viewer watches on these channels are also very few. These channels and programs if identified can give a lot of clues of the likes of the Viewer. These programs watched on the respective channels if logged can reveal a pattern about the watching pattern of the Viewer. These patterns usually repeat over time and can to a very high probability deduce what the Viewer would watch next. In this paper, we propose a method and system that learns from the Viewer Implicitly over time and learns all the patterns that the Viewer exhibits over the course of Television watching.

2 Related Work

A number of tools and techniques have been proposed in the past to recommend television programs to the Viewer. Most of these systems require the Viewer to rate programs with indicators like "Thumbs Up" or "Thumbs Down", through the Remote of the STB. These indicators reveal whether the Viewer, "Likes" or "Dislikes" programs. After the Viewers likes and dislikes are asked, the recommender system compares it to the EPG to identify programs of "interest" to the Viewer on a scale of [1–5].

These systems rely completely on the Viewers program rating, always done explicitly with indicators like "Thumbs Up" for "Like" or "Thumbs Down" for "Dislikes". For the system to generate correct results the Viewer has to rate each and every program watched. This rating is Annoying for any user as it affects the primary task of the Viewer. Furthermore there are no intermediate indicators, like "Partially Like", etc. A need therefore exists for a Method and Apparatus to

Generate and Display Personalized Television Content Based only on the viewers viewing patterns learned implicitly.

3 System Overview

We propose a system that logs the Viewers watching patterns in a database in the Key-Value pair format, Date_and_Time, Channel _Number _Watched. Here the Date_and_Time is the system generated Timestamp, while the Channel_ Number_ Watched is initiated by the Viewer, and is a part of the EPG. The system also logs all the commands initiated by the Viewer through the Remote in a database, Channel_Number_Active–Current_Program_Active_ Command_Issued. Here the Channel_Number_Active, is the current program being displayed on the Television and the Command_Issued is one of the many commands possible from the Remote like, Volume_Increase, Volume_Decrease, Flip_to_another_Channel, Activate_EPG, etc. and is a part of the User Manual of the STB. The system "Learns Implicitly" from these indicators as exhibited the Viewer. All these indicators are the normal course of action of watching programs in Channels in the television. No rating questions of the type of "Thumbs Up" or "Thumbs Down" are asked at all.

Patterns are formulated from the logs and Viewers interests are determined automatically. As these Patterns get repeated, over the course of time of watching Television by the Viewer, the Confidence level of the system on these Patterns increases. Using these patterns, the system Clusters the Channels from the STB (part of the EPG) into Channels_Seen and Channels_Not_Seen. This is further sorted to Channels_ Seen_Frequently as well as along with those channels_Not_Seen_Frequenty Furthermore from amongst these sub-set of Channels the system identifies which Programs are watched by the Viewer on which Channels and how often. The Programs on these Channels are also sorted into Channel_Number_&_Program_Seen and the counts of how many times any Particular Program on a Particular Channel was seen by the Viewer. Each of these Channels has a Label which indicates its Genre and each of the Program has a snippet, both mentioned in the EPG. Thus the interest areas of the Viewer are determined.

A need also exists for a method to generate recommendations on content based on the method described above. If the Viewer still continues to watch the current Program on the television, other than the one that's recommended to the Viewer, the recommended program will be recorded on the STB automatically for future viewing. A message/intimation in that regard should be given to the Viewer.

Various components of the system are discussed below:

(a) [Hardware Component] denotes the Infrared Receiver of Signals/Commands from the Remote Control by the System and other hardware peripherals.

(b) [Storage Component] denotes the logging of the Commands from the Remote into a Table.
(c) [Software Component] denotes the co-relation of the Command for Channel from the Remote control to a Channel on the EPG.
(d) [System Component] denotes the identification of the current Program from the current System Date and Time.
(e) [Learning Software Component] denotes the Pattern generator engine which co-relates the Current Channel, Current Program, Current System Date, System Time.
(f) [System Component] denotes the Recommended Program message to be indicated to the Viewer.
(g) [Hardware Component] denotes the Infrared Transmitter of Signals/Commands to the Television.

4 Hardware Details and Mechanism

The hardware system contains following modules:

1. Microcontroller,
2. IR Receiver,
3. Data logging shield
4. Power source.

Depending on the context in which the system is used i.e. the amount of computation needed, the appropriate hardware can be used so as to cater to the relevant needs. In our system, we have used two microcontrollers. The first being an Arduino Uno which is based on the Atmega328 and has 14 digital I/O pins [6]. The second being, Raspberry Pi B+ that can be used in applications needing intensive computing [7]. We tested both of them for our system. The TSOP1738, an IR Receiver consists of a PIN diode and a pre amplifier all embedded on a board which captures the IR signals. The third hardware component includes a data logging shield having a Real Time Clock, which is used as a local database and the final module is a 9 V Battery which acts a power anchor for the system. The Real Time Clock module can be used as an integrated component of the data logging shield or as a separately assembled circuit.

5 Working

The system starts functioning as soon as the keys on the remote are pressed by the user. Kept atop the STB, the IR Receiver captures the keystrokes from the remote control. Thereafter, these keystrokes are sent to the serial monitor of Arduino Uno

Table 1 Mapping of remote codes with its applications

Prefix code	Actual code	Applications
1258	3044	Home
1258	3040	TV
1258	2924	Power off
1258	3116	Guide
1258	3037	Showcase
1258	3038	Organizer
1258	3157	Active
1258	2925	Mute
1258	3115	Information
1258	2928	Volume up
1258	2929	Volume down
1258	2944	Channel up
1258	2945	Channel down
1258	3000	Navigate up
1258	3001	Navigate down
1258	3002	Navigate left
1258	3003	Navigate right
1258	3004	Select
1258	3005	Back

which displays a unique number which is associated with every key pressed as shown in the Table 1. Apparently, there is no direct provision in Arduino to display the time and date while working as a standalone system. The *millis* function only displays the number of seconds elapsed, after the Arduino was turned on. So to overcome this hurdle, we used a Real Time Clock which is soldered with the data logging shield providing the much important insight into the date and time at which the keystrokes were captured. The programming was done in Embedded C and using Python scripts we were able to export the serial monitor data onto an SD Card fitted in the data logging shield. The TSOP1738 is an off the shelf IR receiver which aptly captures the keystrokes at a brisk pace [8]. In fact, while testing our system, we found the number of times a remote key-press signal is recorded, to be around 5–6 signals per second, which is a testimony to the fact that often, pressing the remote control keys continuously leads to remote control 'abusage'. Arduino Uno processes the IR signals and displays them on the serial monitor at a baud rate of 9600 Hz. This data is then stored in a .csv/.txt file which is then fed as an input to the learning component of the system. The system functions as a standalone unit owing to the 9 V Battery which powers the Microcontroller. A separate coin cell battery powers the RTC which provides a seamless capability for long time logging of data along with timestamps. A diagrammatic representation of the hardware system is as shown in the Fig. 1.

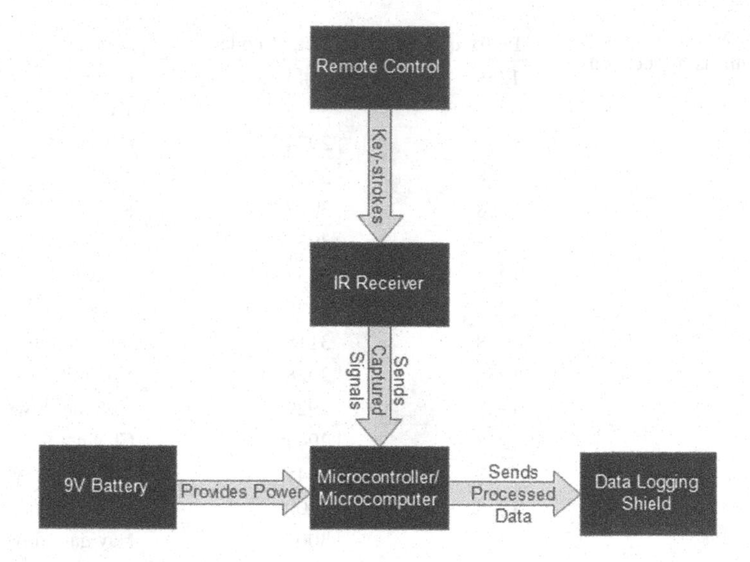

Fig. 1 Hardware and storage system

The logged data is furthermore mapped with the Actual Button pressed as shown below. We tested it using an STB remote which has a one-to-one relation with a function on the remote with the button code recorded. Every key press has a particular application associated with it. Prefix code is a unique code for every STB remote, which is a sort of passkey that ensures the use of only proprietary remotes for the STB. It is then followed by an actual code which denotes the operation it relates itself to. The relation is shown in Table 1.

6 Algorithm Design

Algorithm works on following Input and produces the desirous output.

Input: A continuous stream of Television data.
Output: A personalized Television viewing experience.
The system takes in the commands issued by the Viewer through the Remote destined for the STB. The commands indicate the Television functionality desired by the Viewer, namely, Sound Increase/Decrease, Channel Up/Down, Remainders, Schedule etc. All these commands are logged with reference to the current System Date and Time. The steps followed are as under:

The system starts with the data capture stage where the viewers TV watching pattern is recorded, the Channel number and the Program watches at a particular date and time is recorded. In the Data Interpretation stage, the frequency of watching particular programs and channels are identified. The sessions in which the

TV is watched is also identified. The more the occurrence of a particular pattern of Program or Channel watching the higher is the weight assigned to the pattern. These patterns are identified and Users Part Profiles are built which are based on Days watching or are based on Sessions. More the occurrence of a particular pattern more is the relevancy of the particular program/channel to the user. If a pattern does not repeat again, its weight reduces as per time (Time Decay) and hence unlearned over a period of time. These part profiles are then combined together to build User Profiles. Based on these User Profiles which contain the viewers TV viewing details, TV programs and channels can be recommended to the user based on the relevancy learned from the users TV watching patterns.

The TV viewing patterns when repeated for a sufficiently large number of times invokes the system to generate Recommendations which are sent as commands to the Set-Top-Box to switch to a particular Program on a particular Channel which is of Interest to the Viewer. If the generated Recommendation is over-ruled by the Viewer, the system learns from it too and reduces the confidence on the exhibited pattern automatically. The algorithmic steps are shown in the Fig. 2.

Fig. 2 Algorithmic steps

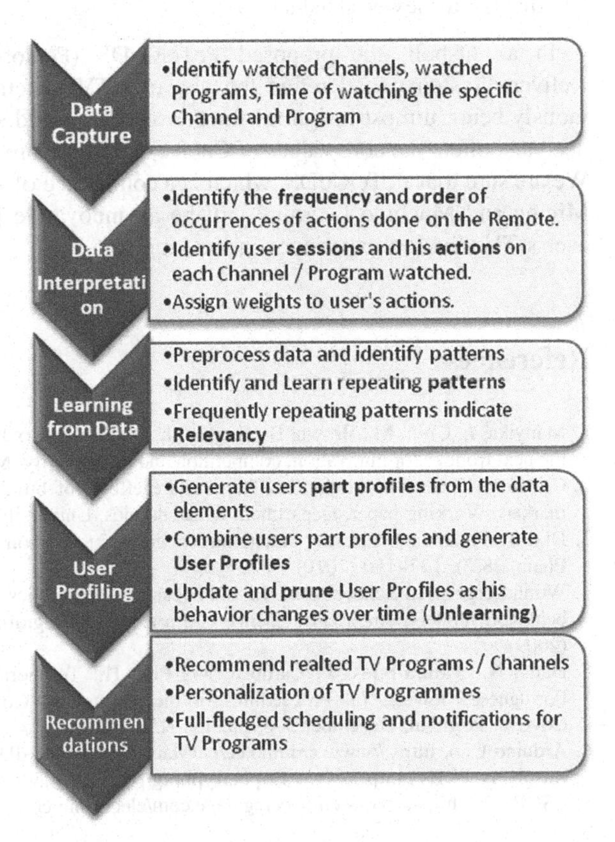

7 Conclusion

The proposed system replaces the monotonous TV viewing experience which is prevalent in the current scenario. By getting an insight into the user's likes and interests, we can make sure that the time which he/she spends watching the TV is very intuitive and engaging. The benefits, both for the user as well as for the STB service provider are enunciated as follows:

1. Improved user experience and better user engagement
2. Personalization of TV programs
3. User identification through user's usage patterns
4. Insight into the remote usage/abusage as a whole
5. A new USP for the STB service provider
6. Customized Subscription Recommendations
7. Enhanced and Accurate TRP Prediction
8. Customized advertisements
9. User data analysis for business models
10. Increased viewer satisfaction

In a nutshell, the proposed PeTelCoDS (Personalized Television Content Delivery System) shall refine the viewer's TV watching experience by synonymously being utmost insightful for the service provider. PeTelCoDS caters to this need and offers the core values of Choice, Control Convenience and Customization. We are sure that PeTelCoDS, which is a confluence of Hardware, Electronics, Data Mining and Machine Learning will be an innovative proponent for adorning the user's TV viewing experience in the long run.

References

1. Manyika, J., Chui, M., Brown, B., Bughin, J., Dobbs, R., Roxburgh, C., Byers, A.H.: Big data: the next frontier for innovation, competition and productivity. McKinsey Global Institute (2011)
2. Crawford, G.S., Yurukoglu, A.: The welfare effects of bundling in multi-channel television markets. Working paper, Department of Economics, University of Arizona, Tucson, AZ (2008)
3. Dix, S., Phau, I.: Measuring situational triggers of television channel switching. Mktg. Intel. Plann. **28**(2), 137–150 (2010)
4. Wonneberger, A., Schoenbach, K., van Meurs, L.: Dynamics of individual television viewing behavior: models, empirical evidence and a research program. Comm. Stud. **60**(3), 235–252 (2009)
5. Datia, N., Moura-Pires, J., Cardoso, M., Pita, H.: Temporal patterns of TV watching for Portuguese viewers. In: Proceedings of the Portuguese Conference Artificial Intelligence, Covilha, Portugal, December 5–8, pp. 151–158 (2005)
6. Arduino Uno, https://www.arduino.cc/en/Main/arduinoBoardUno
7. Raspberry Pi B+, https://www.raspberrypi.org/products/model-b-plus/
8. TSOP1738, https://www.engineersgarage.com/electronic-components/tsop1738-datasheet

Wave Front Method Based Path Planning Algorithm for Mobile Robots

Bhavya Ghai and Anupam Shukla

Abstract Path planning problem revolves around finding a path from start node to goal node without any collisions. This paper presents an improved version of Focused Wave Front Algorithm for mobile robot path planning in static 2D environment. Existing wave expansion algorithms either provide speed or optimality. We try to counter this problem by preventing the full expansion of the wave and expanding specific nodes such that optimality is retained. Our proposed algorithm 'Optimally Focused Wave Front algorithm' provides a very attractive package of speed and optimality. It allocates weight and cost to each node but it defines cost in a different fashion and employs diagonal distance instead of Euclidean distance. Finally, we compared our proposed algorithm with existing Wave Front Algorithms. We found that our proposed approach gave optimal results when compared with Focused Wave Front Algorithm and faster results when compared with Modified Wave Front Algorithm.

Keywords Wave front · Path planning · Static environment · Mobile robot

1 Introduction

Path Planning is one of the key research areas in Dynamic Robotics [1]. Although the field of robot path planning is more than 30 years old but still it is an active topic for research. In very raw form, Path planning is moving of the robot from the starting node to the target node without any collisions. The evaluation criteria for Path Planning Algorithms may vary from Completeness [2], Computational Complexity [2], optimality, etc. as per the application. Path Planning can be modeled as a multi objective optimization problem [3]. Objectives may be to reduce

B. Ghai (✉) · A. Shukla
ABV—Indian Institute of Information Technology and Management, Gwalior, India
e-mail: bhavyaghai@gmail.com

A. Shukla
e-mail: anupamshukla@iiitm.ac.in

S.C. Satapathy and S. Das (eds.), *Proceedings of First International Conference on Information and Communication Technology for Intelligent Systems: Volume 2*,
Smart Innovation, Systems and Technologies 51, DOI 10.1007/978-3-319-30927-9_28

the energy consumption, path length, execution time, communication delay, etc. [4] Mobile robots have wide domestic, military and industrial applications [4]. They may be used for cleaning where they navigate around the entire space [5]. They are widely used in dangerous environments which may be hazardous for humans such as aerospace research, mining industry, defense industry, nuclear industry, etc.

Path Planning algorithms may deal with known/unknown environments, static/dynamic [6] obstacles, single/multiple robots, 2d/3d space, etc. Numerous methods are employed to deal with all this problems such as heuristics, Genetic Algorithms, soft computing, statistical approaches, etc. In our case, we have dealt with static 2d environment. We have proposed a new approach based on wave front method. In wave front based methods, values are assigned to each node starting from target node. It is followed by traversal from start node to target node using the values assigned. Our major concern is to ensure optimal path length along with faster execution time. We tried to address this problem by preventing the full expansion of waves and used a new cost function so that optimality is not compromised. Finally, we compare our proposed approach with the existing wave front based path planning algorithms to verify the effectiveness of our proposed algorithm. We used Player Stage Simulator for testing. Player/Stage is a widely used open source multi robot simulator which is compatible with multiple platforms [7, 8].

This paper is organized as follows. Section 2 will discuss about the related work in this field. Section 3 will discuss about the major wave front based algorithms and also present our proposed approach (Optimally Focus Wave front Algorithm) to this problem. Section 4 will discuss about the assumptions and the comparison of our approach with existing algorithms. Sections 5 and 6 contain Conclusion and Future Work respectively.

2 Related Work

Numerous methods have been employed to solve different aspects of Path Planning Algorithm such as Heuristics [9], Wave Front Method [1, 2], Genetic Algorithms [10], Neural Network [11], etc. Some of the common examples include A*, artificial potential field, D*, etc. Environment for Path Planning algorithms can be modeled as grid of Polygons. Typically, it is modeled as a rectangular Grid but it also be modeled as a triangular grid so that the number of directions and hence path length can be further optimized [5]. Path planning algorithms are of two types based on data available about environment: static and dynamic. In static path planning, entire information about obstacles is known beforehand. We have the entire map of the environment at the beginning then we go for preprocessing based on the map and starting and goal node positions. The algorithm returns a path and then robot simply follow the co-ordinates of the path [1, 2]. In case of dynamic Path Planning,

Robot is dependent on its sensors. Only small fraction of information about obstacles is known in advance. The robot has to take navigation decisions while moving. As the robot moves and interacts with the environment, more information becomes available about obstacles. There are many algorithms for dealing with static environment effectively [2, 10–12]. In case of dynamic environment obstacles may change their position. This kind of situation is dealt using sensor information [4, 6, 9].

In case of Wave front algorithms, Robot moves from the source node to target node based on the waves emitted by the target [1]. In this paper we have purposed a modified version of Focused Wave Front Algorithm.

3 Proposed Approach

Wave Front algorithm uses breadth first search from the target node to the start node. In the wave front algorithm values are assigned to each node in increasing order from target node. The nodes in a wave i.e. nodes at equal distance from the target node are assigned the same value provided node is not an obstacle. The numbers assigned to cells in adjacent waves differ by 1. In our case, we have considered that robot can move in 8 directions so waves are square in shape. The following formula to assign value to each cell [1]:

$$\text{map}(i,j) = \begin{cases} \min(\text{neighborhood}(i,j)) + 1 & \text{Empty Cell} \\ \text{Nothing} & \text{Obstacle Cell} \end{cases}$$

Here i, j are the co-ordinates on the grid. Neighborhood (i, j) represents the cell adjacent to the cell (i, j). In our case, each node will have 8 neighbors. In every stage each cell who has not got any values will get values. This goes on until all the nodes in the map are assigned a value. After all nodes are assigned a value, Traversal from the start node begins towards the target node such that at each step it chooses the next node with minimum value. This algorithm always provides a path if it exists but it has two major drawbacks. Firstly, it is very time consuming and computationally expensive as it needs to explore all nodes i.e. it assigns value to each node. Secondly, it is possible that two or more nodes in the neighborhood have the same value. Hence, we have to choose the best path among different possible paths.

As the name suggests, Modified Wave Front Algorithm is an improved version of Wave Front Algorithm. The main advantage of MWF over wave front algorithm is that it returns the best optimal path. Like Wave Front Algorithm, MWF also explores all nodes and allocates value to each node in increasing order starting from the target node. The key difference lies in the way it allocates values to each node.

MWF differentiates between orthogonally adjacent and diagonally adjacent nodes. This is reflected in the following formula:

$$\text{map(i,j)} = \begin{cases} \text{min(neighborhood(i,j))} + 4 & \text{Empty Diagonal Cell} \\ \text{min(neighborhood(i,j))} + 3 & \text{Other Empty Cell} \\ \text{Nothing} & \text{Obstacle Cell} \end{cases}$$

The above formula describes the value allocated to the node with coordinates (i, j). This algorithm provides a solution if it exists(Completeness) and gives the optimal solution. The only drawback is that it is very slow as it explores all nodes. Specifically for bigger maps, it might take long to calculate optimal path.

Focused Wave Front Algorithm is a further modification to MWF. This algorithm is quite faster than previous algorithms because it explores only a limited number of nodes. Each node is allocated two values—weight and cost. Weight is the value assigned to node depending on its position. It is assigned in exactly same fashion as we allocate values in modified wave front algorithm. Weight can be understood as a measure of minimum path length of a node from the target node although it may be on a different scale. If we consider each node to be of unit length and we increment weight by 3 and 4 between adjacent nodes then weight of a node will be approximately 3 times of the path length from the target node. Cost of a node is its Euclidean distance from the start node.

Initially target node is assigned 0 weight. All its neighbors are assigned weight and cost value. The node with the minimum cost is expanded until source node is reached. This algorithm follows a greedy approach whereby it gives priority to those nodes which are near to start node. It reaches the start node quite swiftly. It is time efficient but not optimal. It will return a path if it exists although it may suggest a relatively longer route. It might not be suitable when movement cost is high and optimal path length is a priority.

Optimally Focused Wave Front Algorithm (OFWF) is our proposed approach. Optimal Path length is one of the most important properties sought in Path Planning Algorithms for a vast number of applications. OFWF is a further modification of FWF and it returns path with optimal path length. Like FWF, it explores only a limited number of nodes and hence is quite faster than MWF. FWF focuses on those nodes which are closer to source node irrespective of its distance from the target node. Hence, it doesn't provide optimal solution due to its greedy approach. On the other hand, OFWF weighs distance from the target node and approximate minimum distance from the start node equally. Weight is assigned in the same way as FWF. OFWF also expands nodes with minimum cost but cost is defined in a different fashion. Cost of a node with coordinates (i, j) is defined as follows:

$$\text{Cost(i, j)} = \text{Weight(i, j)} + \text{heuristic(i, j)}$$
$$\text{heuristic(i, j)} = 3 * ((\text{dx} + \text{dy}) + (\sqrt{2} - 2) * \text{Min(dx, dy)})$$

where

dx absolute difference of x coordinates of the given node and start node
dy absolute difference of y coordinates of the given node and start node
Min(dx, dy) returns minimum value between dx and dy

***Algorithm*:**

Step 1: Insert target node into priority queue
Step 2: c = Pop node from priority queue
Step 3: if c == start node goto Step 7
Step 4: Assign weight and cost to neighbors of c
Step 5: Insert neighbors of c to priority queue
Step 6: Goto Step 2
Step 7: Traverse from start to target node by choosing the node with least weight among neighborhood at each step

Internally, Priority queue will arrange the nodes in ascending order based on the cost of each node. Instead of using Euclidean distance for measuring approximate minimum distance from start node, we use a variant of diagonal distance [13] which is better suited in our case as our robot can move in only 8 directions. Since this algorithm focuses on optimal path, so we decided to call it Optimally Focused Wave Front Algorithm (OFWF). Firstly, we push target node into priority queue. Then we allocate weight and cost to its immediate neighbors and push them into priority queue as well. Then we pop a node from the priority queue and repeat this process until start node is popped out. Lastly, we traverse from start node to target node by moving to nodes with least weight among other neighbors.

4 Results

We simulated MWF, FWF, OFWF using Player 3.0.2 and Stage 3.2.2 on Ubuntu 12.04 Platform. We feed the starting and target locations along with the environment as the input to the algorithm. The algorithm returns the set of x, y co-ordinates of adjacent cells which will form the path. We have used gray image to represent the 2d environment where black pixels represent obstacles and white spaces represent the free region. An image of size $P \times Q$ pixels represent a map of $P \times Q$ cells.

We have assumed that robot can move in 8 directions (North, West, East, South, North-East, North-West, South-East, South-West) and number of obstacles are finite and static. We have assumed that robot can rotate in clockwise and anti-clockwise direction, hence robot can rotate 45°, 90°, 135° or 180 °. We have used 4 different maps for comparison. In each map the starting node will be the top left cell and the target node will be the bottom right cell. To measure the total angle turned, we have considered that in the beginning the robot faces towards north.

Number of explored nodes is the count of all nodes to which weight and cost has been assigned. We have considered each cell to be of 1 unit length. For every horizontal or vertical movement, Path length will be incremented by 1 and for each diagonal move path length will be incremented by $\sqrt{2}$. We will compare the performance of MWF, FWF and OFWF based on 6 constraints i.e. Number of Nodes Explored, Number of Steps, Path Length, Execution Time, Number of turns and total angle turned (Fig. 1, Tables 1, 2, 3 and 4).

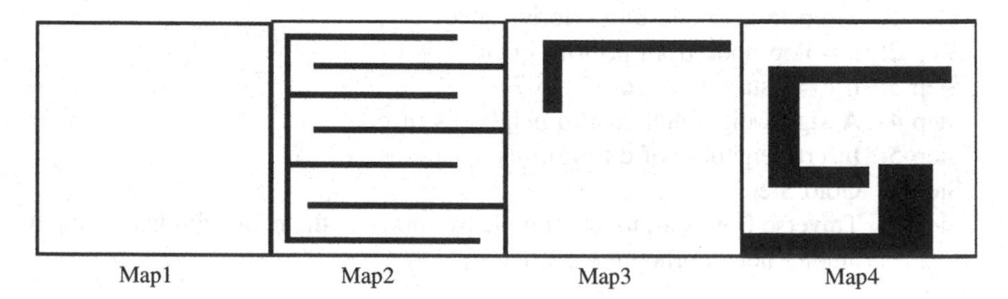

Fig. 1 Simulation environment

Table 1 Results for Map1 (200 × 200 pixels)

Parameters	MWF	FWF	OFWF
Nodes explored	40,000	989	994
Number of steps	199	199	199
Time (ms)	3895	49	53
Path length	281.428	281.428	281.428
Number of turns	1	1	1
Total angle turned (°)	135	135	135

Table 2 Results for Map2 (40 × 40 pixels)

Parameters	MWF	FWF	OFWF
Nodes explored	1344	938	405
Number of steps	75	197	75
Time (ms)	120	73	48
Path length	76.2426	213.154	76.2426
Number of turns	4	27	7
Total angle turned (°)	270	1530	450

Table 3 Results for Map3 (200 × 200 pixels)

Parameters	MWF	FWF	OFWF
Nodes explored	37,548	2389	11,820
Number of steps	247	247	247
Time (ms)	3550	173	1095
Path length	309.546	322.801	309.546
Number of turns	4	8	39
Total angle turned (°)	270	540	1890

Table 4 Results for Map4 (200 × 200 pixels)

Parameters	MWF	FWF	OFWF
Nodes explored	28,389	8625	14,153
Number of steps	342	373	342
Time (ms)	2724	807	1346
Path length	365.196	418.978	365.196
Number of turns	4	15	11
Total angle turned (°)	270	810	540

5 Conclusion

In uncluttered environment such as map1, all algorithms returned optimal results but MWF took considerably longer than FWF and OFWF. On observing the number of turns and Total angle turned for Map2, Map3 and Map4, we can deduce that MWF, FWF and OFWF propose 3 different paths for each case. Based on the observations, we can approximately compare the performance of OFWF with MWF and FWF. When we compare OFWF with MWF, we observe that path length and number of steps is same as both return optimal path length. Execution time and nodes explored is quite less for OFWF and number of turns and angle turned is better for MWF. When we compare OFWF with FWF, we observe that FWF execution time and number of nodes explored is better while path length and number of steps is better for OFWF.

Among the three, MWF provides the most optimal results in terms of path length, number of turns and total angle turned. However, it is computationally expensive and time consuming as it explores a relatively larger number of nodes. FWF is faster than MWF and OFWF but it compromises optimality for high speed. OFWF provides results with optimal path length and is a lot faster than the MWF. OFWF seems as a balanced algorithm which provides optimal path length with good execution time. If rotation cost is not a major concern, then OFWF may prove to be a good alternative among other path planning algorithms.

6 Future Work

In the future, Optimally Focused Wave front algorithm may be further modified so that apart from path length it may also optimize number of turns and total angle turned. The current algorithm may also be extended to work in 16 directions which will further optimize path length. The current algorithm might also be modified to work in unknown environment or with dynamic obstacles.

References

1. Nooraliei, A., Nooraliei, H.: Path planning using wave front's improvement methods. In: International Conference on Computer Technology and Development, ICCTD'09, IEEE, vol. 1, pp. 259–264 (2009)
2. Pal, A., Tiwari, R., Shukla, A.: A focused wave front algorithm for mobile robot path planning. In: Hybrid Artificial Intelligent Systems, pp. 190–197. Springer, Heidelberg (2011)
3. Liu, G., et al.: The ant algorithm for solving robot path planning problem. In: Third International Conference on Information Technology and Applications (ICITA), pp. 25–27 (2005)
4. Ganeshmurthy, M.S., Suresh, G.R.: Path planning algorithm for autonomous mobile robot in dynamic environment. In: 3rd International Conference on Signal Processing, Communication and Networking (ICSCN), IEEE, pp. 1–6 (2015)
5. Oh, J.S., Choi, Y.H., Park, J.B., Zheng, Y.F.: Complete coverage navigation of cleaning robots using triangular-cell-based map. IEEE Trans. Ind. Electron. **51**(3), 718–726 (2004)
6. Zelek, J.S.: Dynamic path planning. In: IEEE International Conference on Systems, Man and Cybernetics, 1995. Intelligent Systems for the 21st Century, vol. 2, pp. 1285–1290 (1995)
7. Biggs, G., et al.: All the robots merely players: history of player and stage software. IEEE Robot. Autom. Mag. **20**(3), 82–90 (2013)
8. Player/Stage Source Forge Homepage, http://playerstage.sourceforge.net10
9. Guo, X.: Coverage rolling path planning of unknown environments with dynamic heuristic searching. In: 2009 WRI World Congress on Computer Science and Information Engineering, IEEE, vol. 5, pp. 261–265 (2009)
10. Manikas, W., Ashenayi, K., Wainwright, R.: Genetic algorithms for autonomous robot navigation. IEEE Instrum. Meas. Mag. **10**(6), 26–31 (2007)
11. Du, X., Chen, H.-H., Gu, W.-K.: Neural network and genetic algorithm based global path planning in a static environment. J. Zhejiang Univ. Sci. **6**, 549–554 (2005)
12. Behnke, S.: Local multiresolution path planning. Preliminary version. In: Proceedings of 7th RoboCup International Symposium, Padua, Italy, pp. 332–343 (2003)
13. Diagonal Distance, http://theory.stanford.edu/~amitp/GameProgramming/Heuristics.html#diagonal-distance

Multimodal Database: Biometric Authentication for Unconstrained Samples

S. Poornima

Abstract Biometrics provides a reliable authentication of a human in a wide variety of applications such as security systems, surveillance and human-computer interaction. Biometric system was started with utilization of a single biometric feature referring as a unimodal biometric system, which is unable to fulfill the security needs extensively in a highly sensitive environment and hence multibiometrics has emerged gaining more importance in the research area. Though there is a shortage of publicly available multimodal databases acquired in real unconstrained environment, a multimodal biometric system can succeed with the assistance of suitable multiple sensors providing higher accuracy rate than that of unimodal biometrics, of course subject to cost, time and subject's acceptance This paper presents a new multimodal dataset which is developed using simple acquisition setup and devices to capture features belonging to the same person in uncontrolled scenarios. The dataset is composed of color images collected from 100 subjects (50 male and 50 female) under the age group 18–22. Totally 6 samples per trait were collected at different time internals between 2011 and 2014 with various occlusions. The dataset is also tested and analyzed by our developed biometric recognition system.

Keywords Biometrics · Multimodal · Segmentation · Noise removal · Feature extraction · Fusion

1 Introduction

Biometric serves society in various real time security applications, providing high recognition rates and low error rates. Unibiometric systems suffer from limitations such as non-universality, noisy sensor data, large intra user variations and susceptibility to spoof attacks [1], whereas multibiometrics [2, 3] addresses the above issues and provides improved performance with very low error rates. However,

S. Poornima (✉)
SSN College of Engineering, Chennai, India
e-mail: poornimas@ssn.edu.in

S.C. Satapathy and S. Das (eds.), *Proceedings of First International Conference on Information and Communication Technology for Intelligent Systems: Volume 2*,
Smart Innovation, Systems and Technologies 51, DOI 10.1007/978-3-319-30927-9_29

there is immense opportunity for betterment and it must be acceptable to general public for being user friendly, convenient and affordable for having satisfactory level of reliability and security. But researchers run the risk of high cost and high storage space when using multibiometric system. Consequently, there is a shortage of publicly available multimodal databases acquired in real unconstrained environment and most of the systems are experimented on virtually built multimodal databases from publicly available unimodal datasets either into signals or scores. As well as, when multiple traits are involved, acquisition of input samples requires different sensors like scanners, camera or printer and further, the samples of multiple features traits of the same person is to be acquired. Under unique circumstances, the user must not feel discomfort and annoyed during acquisition of inputs using different sensors and certain constraints like position, distance, and contact nature. Taking these into consideration along with two essential characteristics of a biometric system, namely collectability and acceptability, our objective is to promote higher collectability and choice of feature which helps continue its significant growth in security related applications. Hence the above things motivate us to create a new database using a cost effective, suitable acquisition device such as any digital or mobile camera to capture contactless features like iris, face, ear and and a web camera to acquire Dorsa palm vein of a person respectively. These features are chosen to carry out research work in the area of multibiometrics, which is suitable and adaptable for any real time biometric applications.

This paper explores the impact of various biometric features on our developed biometric systems. In Sect. 2, the proposed system that handles iris, face, ear and palm dorsal vein characteristics is presented. Section 3 describes the experimental analysis and results obtained by matching individual biometric traits and concluded in Sect. 4.

2 Description and Imaging Principle of Multimodal Dataset

The type of biometric features considered in this proposed system requires different and suitable acquisition devices. Conventionally, high quality scanner for iris, digital camera for face and ear and an infrared camera for vein are used to capture these features. Our intention is to introduce a reliable, versatile and economical acquisition setup to collect the above discussed biometric features and also suitable for small scale biometric applications to provide security access. The noisy samples are collected from 100 subjects (50 Males and 50 Females) from the age group of 18-30, totally 6 samples per trait, in different intervals of 2011–2014, to account for time variance. A digital/mobile camera of not less than 5 Megapixels is used to acquire iris, face and ear in an unconstrained environment. The distance of the camera is variable (20–50 cm) depending on the type of image sample to be acquired. The datasets contain various samples of normal condition and with

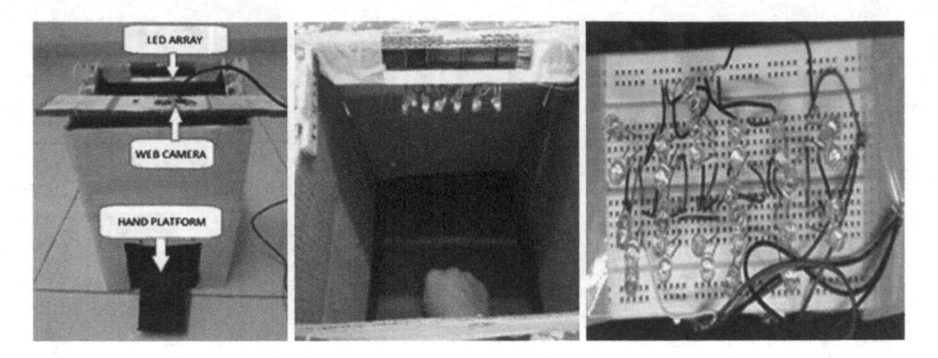

Fig. 1 Palm vein acquisition setup; *Top view*; Array of LEDs (*Left* to *right*)

occlusions such as eyelids drooped, half closed eye, reflections, shadows, wearing contact lens, scarf, sunglass, hazy samples, ornaments, covered with hair, illuminations, and blurred samples, etc.

An INTEX WEBCAM IT-LITE-VU is used as a Near Infrared (NIR) camera to capture Dorsa palm vein patterns of human and the setup is shown in Fig. 1 [4]. INTEX WEBCAM IT-LITE-VU has 1/7" CMOS sensor with a frame rate of 30fps and its focus distance ranges from 4 cm to infinity. The lens view angle is around 54°. The camera produces an image around 15 megapixels. This camera is designed for taking images in the visible spectra by blocking out the infrared light using an IR filter. A camera is converted into an IR camera by removing the IR filter and placing a filter for visible light. The best filter for visible light is to use a new negative photographic film, which blocks out visible light and allows Infrared light to pass through the camera. To view the vein patterns under a near infrared camera we need an infrared source emitting infrared rays in near-infrared region. This illuminates the underlying vein patterns and can be viewed under the near-infrared camera.

The vein patterns are viewed using 30 infrared LEDs connected serially in a breadboard powered by 18 V battery source with precision of 780 nm light source. A black background is chosen to improve the perspective of the acquired hand images. The camera is mounted horizontally parallel to the base on which the hand is placed at a height of 34 cm. The array of infrared LEDs emits infrared rays in all directions. In order to regulate the amount of light falling on the hand, a breadboard is placed in an angle of 60° to the platform on which the camera is mounted. A hand is placed on a slope at an angle of 50° to the base to provide acute focus on our region of interest which includes the knuckle tips and the surface of the dorsal palm. All acquired images are processed to fix to a black background for visual clarity of the feature [4]. Image samples of the database are shown in Fig. 2.

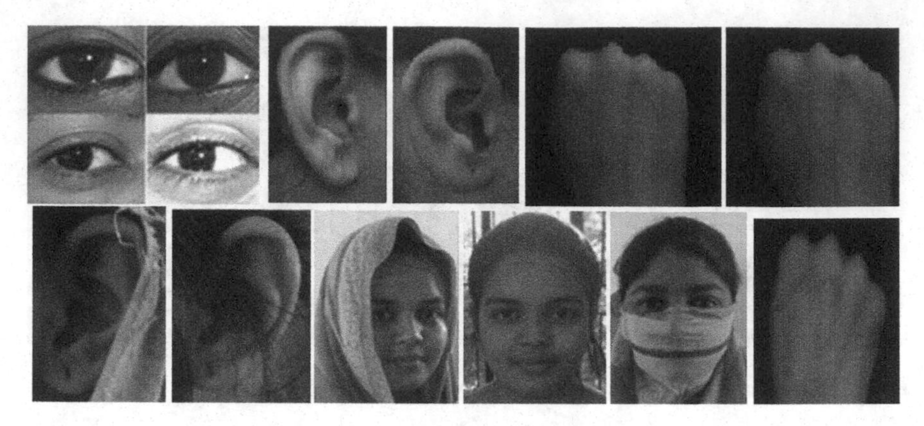

Fig. 2 Samples of biometric traits from multimodal database

3 Overview of Unimodal Biometric Systems

The proposed biometric system is developed for different physiological biometric features, namely, iris, face, ear and Dorsa palm vein. A biometric system consists of four main modules such as acquisition, preprocessing, feature extraction and matching. In the following subsections, we focus individually on the processing methods of each distinct feature, namely iris, face, ear and Dorsa palm vein of the same individual for the samples collected in the above multimodal database.

3.1 Processing of Iris

Images captured in an unconstrained environment contain various noises. Before starting the process of the collected features, the sample images are preprocessed using median filter and Gaussian filter to reduce or remove noise, according to the type of noise exists in the input sample. The main modules in the processing of iris feature are iris localization and feature extraction [5–8].

Since the iris region is surrounded by outer sclera and inner pupil boundaries, iris is segmented by detecting the edges of iris as well as pupil using Roberts and Canny edge detectors respectively [5]. Circular Hough transform helps to detect the centre and radii of both iris and pupil boundaries using the property of a circle. Followed by, unrelated parts like eyelid and eyelashes which acts as noise are also to be removed [5]. Iris normalization compensates the stretching of texture according to the changes in pupil size and maintains the same texture information, regardless of pupil dilation [6]. Normalization approach produces a 2D array using the pixel coordinates within the iris region which is performed using Daugman's rubber sheet model [6, 8] by mapping of each pixel in the iris region into rectangular region like unwrapping the image considering the size inconsistence and dilation of pupil and

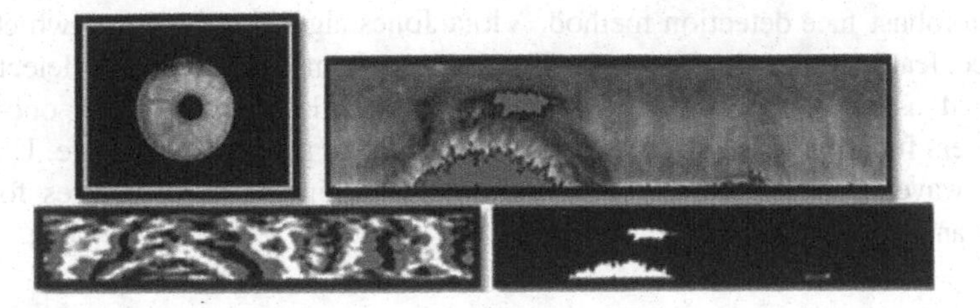

Fig. 3 Noise removed iris; Normalized representation; Feature template; Noise mask (*Left* to *right*)

generates a normalized representation of iris pattern. But the rotational inconsistencies are handled using shifting operation as in Daugman's system.

The discriminant information from the normalized iris are extracted using Log-Gabor filter, where zero DC components can be obtained for any bandwidth on a logarithmic scale [9]. Log-Gabor filter in terms of frequency response is given as,

$$G(f) = exp\left(\frac{-log(\frac{f}{f_0})^2}{2log(\frac{\sigma}{f_0})^2}\right). \quad (1)$$

where, f_0 is the centre frequency, and σ represents the bandwidth of the filter. Convolve the filter with the normalized iris image for the given the wavelength and bandwidth and produce the iris template and a noise mask relatively [5]. Sample outputs from iris processing is shown in Fig. 3.

3.2 *Processing of Face*

The input image is initially cropped to have focus. This image is preprocessed using Gaussian filter to remove any noise and ROI of input, i.e., face region is extracted

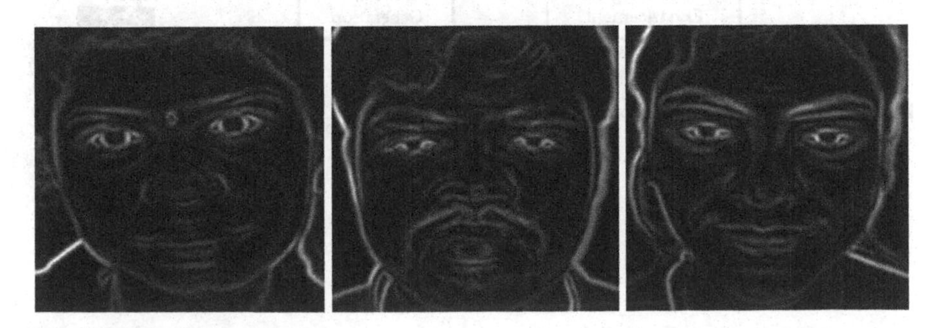

Fig. 4 Preprocessed face samples

using a robust face detection method, Viola-Jones algorithm [10] and which helps to detect feature points in the image. The features from the face region detection are extracted using a Gabor function similar to iris. Gabor function is one of the prominent functions used for extraction of the feature points in an image. Using the Gabor wavelets generated by this function we obtain the base features for comparing any input images (Fig. 4).

3.3 *Processing of Ear*

An ear is an external biometric feature which is easier to acquire and process in two steps namely ear detection and feature extraction. Noises in the image would degrade the edge detection result. Hence the noise effect was removed by convolving the image with Gaussian operator. Morphological operations [11] are used to analyze the shape of the image by choosing an appropriate structuring element. The primary morphological functions, dilation and erosion process are performed by laying the structuring element B on the image F and sliding it across the image in a manner similar to convolution [11]. The edges of the image are determined by the dilation residue edge detector. Dilation allows objects to expand, thus potentially filling in small holes and thus connecting the disjoint objects. Erosion shrinks objects by fetching away (eroding) their boundaries. This morphological operator helps to find the clean edges of the input ear using a sample as a structuring element. It is observed that inner ear region posses' a higher amount of information useful for recognition (Fig. 5). The essential features are extracted from the detected ear region using a shape descriptors.

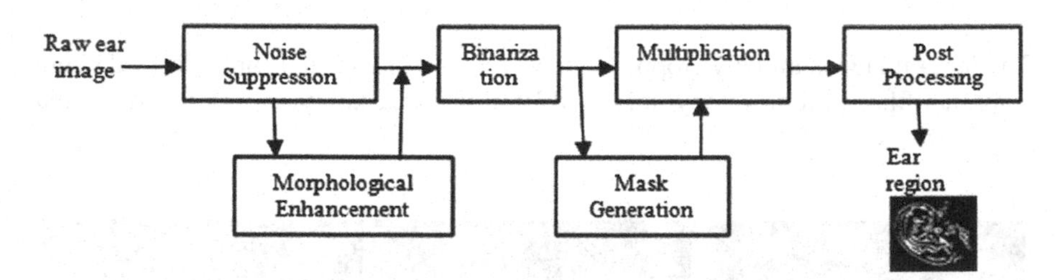

Fig. 5 Framework for ear detection

3.4 Processing of Palm Vein Pattern

The images acquired from the web camera are of dimensions 640 × 480. The image acquired is reduced to a dimension of 320 × 240 so as to improve the visibility of the vein patterns and is presented for further processing. The images are then passed through a median filter initially for removing noise, followed by a low pass and a Gaussian filter for image smoothing. Thereafter, improved the contrast of the image by Contrast-Limited Adaptive Histogram Equalization and binarized using Otsu threshold. On this binarized image, canny method is applied for detecting edges and the image is cropped to locate the region of interest. These images are again filtered and applied with adaptive threshold. Further, the obtained output sample is then thinned to provide the final output region of interest (Fig. 6) and features are extracted using Hierarchical Multiscale Local Binary Pattern (HMLBP) which helps to extract the non uniform patterns present in the ROI and enhances the depth of information through feature extraction.

$$H(k) = \sum_{i=1}^{N}\sum_{j=1}^{M} f(LBP_{P,R}(i,j),k), \quad k \in [0,K]. \tag{2}$$

Where,

$$f(x,y) = \begin{cases} 1, & x = y \\ 0, & otherwise \end{cases} \tag{3}$$

$$LBP_{P,R} = \sum_{P=0}^{p=1} s\ (g_P - g_c)2^P. \tag{4}$$

$$s(x) = \begin{cases} 1, & x \geq 0 \\ 0, & x < 0 \end{cases} \tag{5}$$

where K is the maximal LBP pattern value [12]. In Multi-scale LBP [12], histogram is built from bigger to smaller radius and grouping the patterns as, ‘uniform’ and

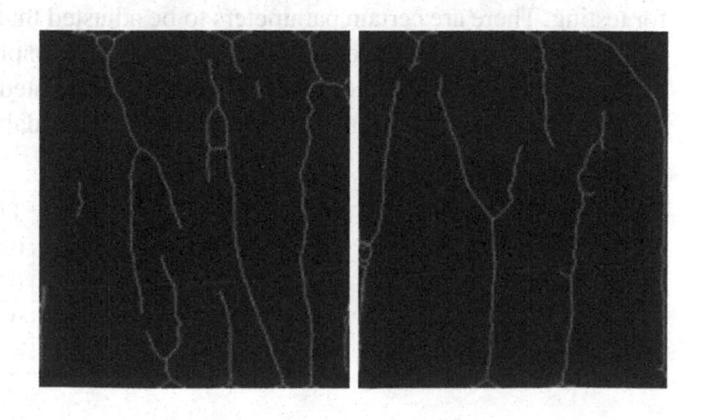

Fig. 6 Sample extraction of region of interest

'non-uniform' through mapping each point. Hence, a sub histogram is constructed from the uniform patterns and LBP patterns from non-uniform pattern are extracted starting smaller radius. Finally, these histograms are concatenated into one multi-scale histogram [13].

3.5 Distance Matching

Hamming Distance (HD) is used for the measurement of similarity distance between the test and training templates of iris, ear and vein. A decision can be made to determine the relation between the templates whether they belong to the same feature or different one. Sometimes there will be an overlap between inter-class and intra-class distributions which would result in higher false acceptances and rejections. Hence, two thresholds (T1, T2) have been set. One for intra and other for inter class comparisons respectively. An intra class comparison is determined as a match if the HD measure is ≤T1, and if <T2, but >T1, it determines that probability of having images belong to the same feature [5]. The modified HD is defined as,

$$HD = \frac{1}{\mathrm{N} - \sum_{k=1}^{N} \mathrm{Xnk\ (OR)Ynk}} \sum_{j=1}^{N} \mathrm{Xj(XOR)\ Yj(AND)\ Xn'j(AND)\ Yn'j}. \quad (6)$$

where X_j and Y_j are the two bit-wise templates, X_{nj} and Y_{nj} are the corresponding noise masks for X_j and Y_j, and N is the number of bits represented by each template [5, 7]. The similarity distances are measured for all three feature types by comparing their respective trained templates. The lowest score among each type gives the better match of it.

4 Experimental Setup and Results

Two samples from each trait have been trained and the rest of the samples are utilized for testing. There are certain parameters to be adjusted that influence the performance of the system. The parameters used in the normalization phase of iris feature and used in feature encoding phase of all traits had to be adjusted to give maximum performance. These parameters are selected using the decidability factor [7], which is a

Table 1 Thresholds for intra and inter class comparisons

Feature type	T1	T2
Iris	0.349	0.443
Ear	0.17	0.25
Palm (Dorsa) vein	0.277	0.391
Face	0.315	0.392

Table 2 Performance measures of multimodal dataset

Feature type	FAR	Accuracy
Iris	0.1	89
Ear	0.1	88
Palm vein	0.1	87.3
Face	0.1	82.3

function of mean and standard deviation of intra and inter class comparisons. The higher the decidability, the greater is the separation of intra-class and inter-class distributions [5]. The computed distance measures are analyzed based on thresholds T1 and T2 (Table 1) to find out the match decision. The overall performances obtained for multimodal dataset are given in Table 2. Individual measures of each trait at various thresholds are shown in Table 3 and Fig. 7. The proposed system is also experimented with few existing public dataset whose performances are given in Table 4.

The accuracy rates obtained for samples of multimodal dataset is moreover equivalent to the performance obtained for public dataset samples. When multiple features are considered for recognition, the features must belong to the same person for an exhausted analysis. Hence the fusion at score level is implemented by

Table 3 Performance measures of biometrics traits at various thresholds

Iris			Face			Ear			Palm (Dorsa) vein		
T1	T2	GAR	T1	T2	GAR	T1	T2	GAR	T1	T2	GAR
0.10	0.20	40.7	0.10	0.19	65.1	0.10	0.18	78.46	0.08	0.18	68.5
0.16	0.26	73	0.12	0.20	69.9	0.12	0.20	82.3	0.11	0.22	72.1
0.20	0.31	75.3	0.14	0.22	72.8	0.13	0.21	84.3	0.14	0.25	75.2
0.25	0.34	80	0.16	0.29	76.1	0.14	0.22	85.9	0.18	0.30	80.7
0.29	0.39	84.6	0.21	0.29	78.7	0.15	0.23	86.3	0.22	0.33	82.6
0.30	0.40	86.3	0.27	0.35	80.5	0.16	0.24	87.4	0.26	0.37	86.0
0.35	0.45	89.1	0.32	0.39	82.1	0.17	0.25	88.0	0.28	0.39	87.3

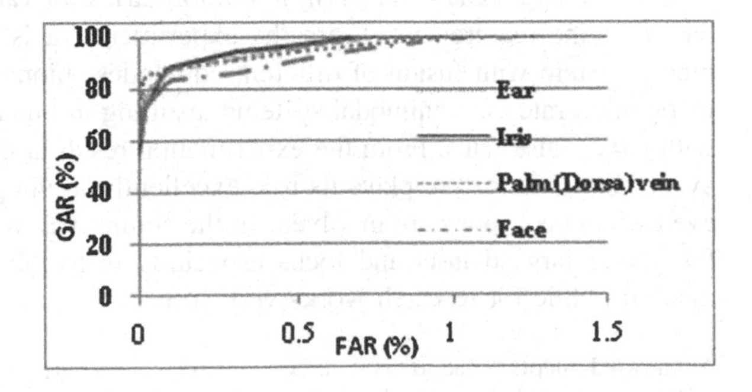

Fig. 7 Performance curves of unimodal biometrics

Table 4 Performance rates of public dataset samples measured at 0.1 % FAR

Feature type	Database	No. of Samples	Accuracy (%)
Iris	IR-TestV1	2000	92
Ear	IIT-Delhi	421	88
Palm vein	Dr. Badawi dataset	1000	88.2
Face	AR faces	1200	87

computing the total score from the similarity scores obtained in each modality. The person with fused score less than the threshold is determined as authenticated. This score level fusion is easier to implement than other fusion methods and doesn't involve any feature vectors directly. Hence the information loss is zero when compared to other fusion methods. This helps in reducing the false rate, in turn increasing the matching rate to greater than 94 % for any combination of traits. This is an improved performance when compared to unimodal system. This system is also tested with occluded samples such as reflections, contact lens and spectacles on iris, 3/4th close of eyelids, bunch of hairs covering the ear, poor visibility of palm veins and samples captured under different lighting conditions. The proposed system is responding to most of the occluded samples except iris sample with 3/4th close of eyelid, side faces and poor visibility of palm vein. Fusion gives a good performance even for occluded samples. These performances are not liable to compare with any literature works, since the features are acquired from the same person. No such public dataset with our chosen features. Hence we have tested our system with various combinations within our dataset and with the unimodal performances.

5 Conclusion

The focus of this work is to develop a versatile biometric system using cost effective acquisition devices and to emphasize the role of fusion in providing secure authentication when multiple features of an individual are handled. This work helps to learn the essentials in adapting multibiometrics for various practical applications based on the requirement. From the experiments, it is evident that the multibiometric system with fusion of different contactless biometric features improves the recognition rate than unimodal systems resulting in higher accuracy rate of 96.3 % with lower false rates. From the experimental result tables, it is shown clearly that every biometric feature plays its role excellently in the process of recognition and even when occlusions are involved. In the future, we would like to test the system for a very large dataset and focus especially in handling various occlusions and make it public for research works very soon.

Acknowledgments Glad to thank the persons who contributed their feature images to build this versatile multimodal dataset with great cooperation.

References

1. Jain, A.K., Ross, A., Prabhakar, S.: An introduction to biometric recognition. Circ. Syst. Video Technol. IEEE Trans. on **14**(1), 4–20 (2004)
2. Bigun, J., Fiérrez-Aguilar, J., Ortega-Garcia, J., Gonzalez-Rodriguez, J.: Combining biometric evidence for person authentication. In: Advanced Studies in Biometrics, pp. 1–18, Springer, Berlin Heidelberg (2005)
3. Ross, A., Jain, A.: Information fusion in biometrics. Pattern Recogn. Lett. **24**(13), 2115–2125 (2003)
4. Poornima, S., Nasreen, F., Prakash, A.D.S., Raghuraman, A.: Versatile and economical acquisition setup for dorsa palm vein authentication. Procedia Comput. Sci. **50**, 323–328 (2015)
5. Poornima, S, Subramanian, S.: Unconstrained iris authentication through fusion of RGB channel information. Int. J. Pattern Recogn. Artif. Intell. **28**(5), 1–18, 1456010 (2014)
6. Proença, H., Alexandre, L.: Iris recognition: an analysis of the aliasing problem in the iris normalization stage. Int. Conf. Comput. Intell. Secur. **2**, 1771–1774 (2006)
7. Daugman, J.: How iris recognition works. IEEE Trans. Circ. Syst. Video Technol. **14**(1), 21–30 (2004)
8. Masek, L.: Recognition of human iris patterns for biometric identification. Doctoral Dissertation, Master's Thesis, University of Western Australia (2003)
9. Wang, Q., Zhang, X., Li, M., Dong, X., Zhou, Q., Yin, Y.: Adaboost and multi-orientation 2D gabor-based noisy iris recognition. Pattern Recogn. Lett. **33**(8), 978–983 (2012)
10. Viola, P., Jones, M.J.: Robust real-time face detection. Int. J. Comput. Vis. **57**(2), 137–154 (2004)
11. Kumar, A., Wu, C.: Automated human identification using ear imaging. Pattern Recogn. **45**, 956–968 (2012)
12. Guo, Z., Zhang, L., Zhang, D., Mou, X.: Hierarchical multiscale LBP for face and palmprint recognition. In 17th IEEE International Conference on Image Processing (ICIP), pp. 4521–4524 (2010)
13. Poornima, S., Subramanian, S.: Personal authentication through dorsa palm vein patterns. Int. J. Appl. Eng. Res. **10**(34), 27286–27290 (2015)

Study and Comparison of Non-traditional Cloud Storage Services for High Load Text Data

L. Srinivasa Rao and I. Raviprakash Reddy

Abstract With the fastest growing nature of application migrating to cloud, the problem of choosing the best suitable cloud service for storage is always a challenge. The cloud storage services are majorly paid service and some of them are designed to deal best with specific kinds of data like read only or update only. Some of the services are designed to match small amount of data blocks and some of the services are designed to best match the larger blocks of data. Many of the application uses file storage, rather than RDBMS storage structure on cloud services. The storage hardware, replication and distribution policies are decided and controlled by cloud storage service providers. The storage options are converted and associated with manageable access portals. This access portal demonstrates the user interface view of the storage visualization and usability. In this paper we compare multiple cloud storage services on high textual data loads and implement an algorithm to choose the best cloud storage service based on cost and storage efficiency.

Keywords Cloud storage · Performance comparison · Evolution application · Response time · Comparison

1 Introduction

Files in the cloud storage services are usually stored in the container offered by third-party companies. Instead being provided by a single host, the containers are integrated and distributed through centralized management. A great set of works are

L.S. Rao (✉)
Mother Teresa Institute of Science and Technology, Sathupally
Telangana State, India
e-mail: srinucsea4@gmail.com

I.R. Reddy
G. Narayanamma Institute of Technology and Science, Hyderabad
Telangana State, India
e-mail: irpr@rediffmail.com

S.C. Satapathy and S. Das (eds.), *Proceedings of First International Conference on Information and Communication Technology for Intelligent Systems: Volume 2*,
Smart Innovation, Systems and Technologies 51, DOI 10.1007/978-3-319-30927-9_30

been conducted on the area of cloud storage services, however the comparative study for storage and retrieval on the loaded network needs to be studies in real time. Cloud computing services can be seen as either computing or storage offering [1–4].

As far as data storage is concerned, there are multiple schemes are available to improve file and data compression. The other most influencing parameters For instance, a data file that is uploaded and accessed on the server may seriously be effected by the network bandwidth as well as the server workload. This will degrade the efficiency. Moreover the cloud storage services deals with a great scope and domain of the data being storage and retrieved along with the frequency of access varying depending on the mode of the operation performed on the data [5–8]. Offering unlimited storage container space might cause a high economic drawback on the cloud storage provider and as well as the users due to inefficient storage. Hence, a technique or automation is needed to find the best suitable storage structure based on cost and other influencing factors [9–12]. There are many free offerings of the cloud storage services; however they may not suite the application requirement to the best always.

The work here is demonstrated as Sect. 2 explores and discuss the proficiency and constraints of multiple storage services with the cost variation based on the data size, in Sect. 3 we study the related works, in Sect. 4 we demonstrate the architecture of the application for choosing the best suitable storage service automatically, in Sect. 5 we compare the storage services based on the data storage and retrieval inclusive of network and server load parameters along with formulation of these parameters and in Sect. 6 we conclude.

2 Cloud Storage Services

As the choice of storage services from cloud is not limited and most of those are configured to give best advantages for specific type of data and operation, we compare most of the services here [13–15].

2.1 Dropbox

The Dropbox is a storage service which is available for client side access for Windows systems, Linux Systems, Macintosh systems, Blackberry mobile operating systems, Android mobile operation systems and finally the IPhone operating systems. The free Basic account comes with a paltry 2 GB of storage. For document based applications this is huge. The Storage service is good choice for applications using the container for read only data (Table 1).

Table 1 Cost comparison for dropbox

Data load	Cost
Load in giga bytes	Price in US Dollars
100	99 USD
200	99 USD
300	99 USD
400	499 USD
500	499 USD
1000	Not available
>1000	Not available

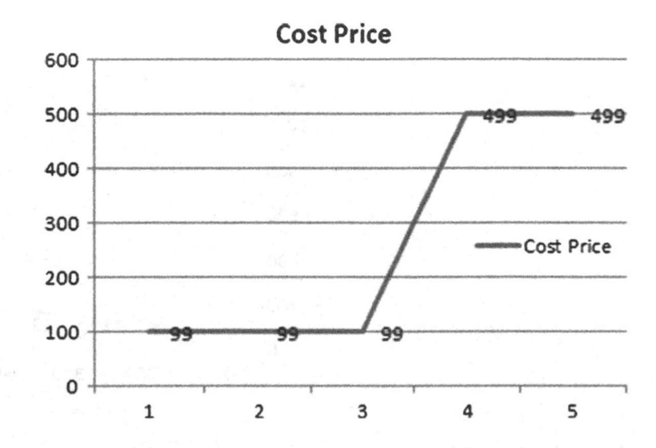

Fig. 1 Cost comparison for dropbox

Table 2 Support for mobile based cloud applications in dropbox

Client OS type	Support
Apple IPhone operating systems	Available
Android mobile operating systems	Available
Blackberry operating systems	Available
Microsoft mobile operating system	Available

Here we provide a graphical representation of the cost price comparison (Fig. 1 and Table 2).

2.2 *Google Drive*

The most popular cloud storage service is Drive storage from Google. The basic account comes with 15 GB of storage for a new customer account or an existing account created with Google Email. The highest rated benefit of the Google Drive is

Table 3 Cost comparison for Google drive

Data load	Cost
Load in giga bytes	Price in US dollars
100	60 USD
200	120 USD
300	120 USD
400	240 USD
500	240 USD
1000	600 USD
>1000	1200 to 9600 USD

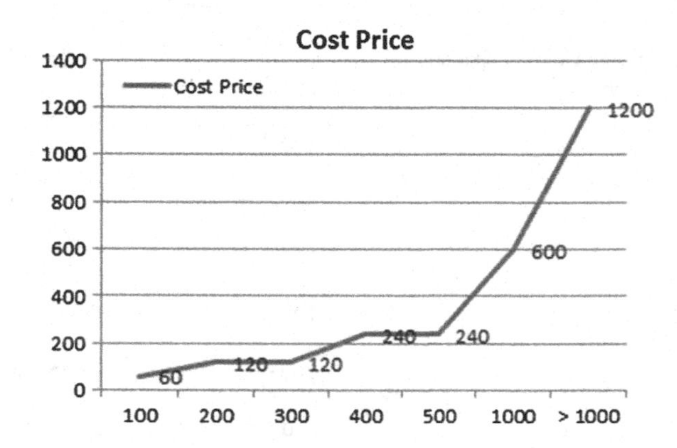

Fig. 2 Cost comparison for Google Drive

Table 4 Support for mobile based cloud applications in Google Drive

Client OS type	Support
Apple IPhone operating systems	Available
Android mobile operating systems	Available
Blackberry operating systems	Not available
Microsoft mobile operating system	Not available

the service can be also be integrated with other existing google services for storing various types of data from other services (Table 3).

Here we provide a graphical representation of the cost price comparison (Fig. 2 and Table 4).

2.3 *Hightail*

The previous version of business cloud storage of Hightail was popular by name of YouSendIt. The basic reason for creating the name was the core of the features that Hightail provides. Hightail is majorly known for sharing files, which can be digitally signed for verifications. The core technology behind this provider is link

Table 5 Cost comparison for hightail

Data load	**Cost**
Load in giga bytes	Price in US dollars
100	Free
200	Free
300	Free
400	Free
500	Free
1000	Free
>1000	195 USD

Table 6 Support for mobile based cloud applications in hightail

Client OS type	Support
Apple IPhone operating systems	Available
Android mobile operating systems	Not available
Blackberry operating systems	Not available
Microsoft mobile operating system	Not available

sharing, where the sender can upload a file and the link to that same file can be shared with the recipient. The recipient can click on the link to download the same. This service is popular for business users as it provides the private cloud storage and the desktop version of the client, which can be used for syncing local files to the cloud storage (Tables 5 and 6).

2.4 *OneDrive*

The OneDrive was previously popular as SkyDrive. The functionalities are mostly same as Dropbox. The most important factor for this storage service is that the client version is available for Windows systems, Linux Systems, Macintosh systems, Blackberry mobile operating systems, Android mobile operation systems and finally the IPhone operating systems. Moreover the supports for social media plug-ins are also available here. This feature makes the application more compatible with other applications to access data directly (Table 7).

Here we provide a graphical representation of the cost price comparison (Fig. 3 and Table 8).

2.5 *SugerSync*

The SugerSync is majorly popular among business users for its effective and fast online backup solutions. The service can also be used for complete folder and

Table 7 Cost comparison for onedrive

Data load	Cost
Load in giga bytes	Price in US dollars
100	50 USD
200	100 USD
300	Not available
400	Not available
500	Not available
1000	Not available
>1000	Not available

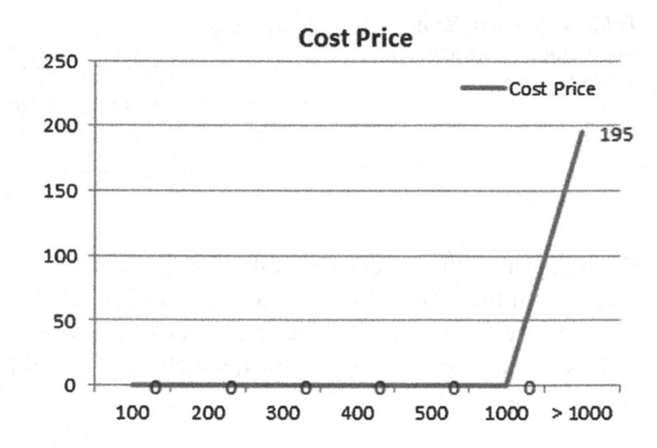

Fig. 3 Cost comparison one drive

Table 8 Support for mobile based cloud applications in onedrive

Client OS type	Support
Apple IPhone operating systems	Available
Android mobile operating systems	Available
Blackberry operating systems	Available
Microsoft mobile operating system	Available

individual file syncingwith multiple applications and multiple users. Moreover the service provides a unique function to share the stored content over multiple devices at same point of time but with different permission levels. The most important factor for this storage service is that the client version is available for Android mobile operation systems and also the IPhone operating systems (Fig. 4 and Table 9).

Here we provide a graphical representation of the cost price comparison (Table 10).

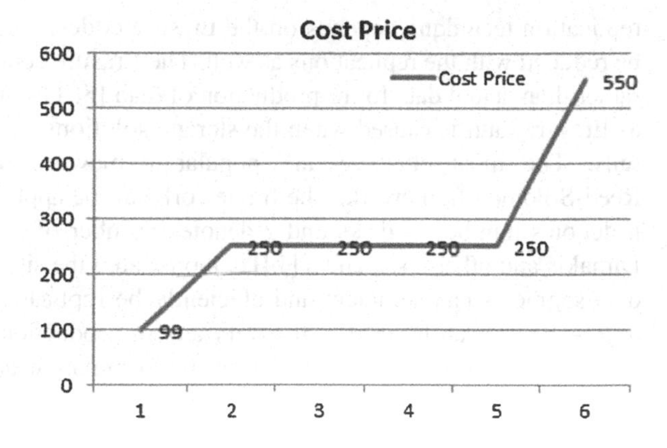

Fig. 4 Cost comparison for SugerSync

Table 9 Cost comparison for sugersync

Data load	Cost
Load in giga bytes	Price in US Dollars
100	99 USD
200	250 USD
300	250 USD
400	250 USD
500	250 USD
1000	550 USD
>1000	Pay per use

Table 10 Support for mobile based cloud applications in sugersync

Client OS type	Support
Apple IPhone operating systems	Available
Android mobile operating systems	Available
Blackberry operating systems	Available
Microsoft mobile operating system	Available

3 Parallel Research Outcomes

Here we analyze the outcomes in parallel research carried out in the area distributed storage solutions. In the early researches we understand the use of Network Attached Storage or NAS [8] devices and as well as the Network File Systems [9] are the only source to provide the additional spaces for business application requirements. In the next decade the storage solutions had a tremendous growth in terms of latency efficiency, security, power consumption and robustness [1, 2, 10]. An additional central control point can leverage more scalability to the architecture.

The next advancement was on the replication methodologies, where the data is replicated over multiple other servers rather than the production server. The replication technique depends on the Erasure codes, where the cost of storage can be reduced with the replications as well. The Erasure codes involve storage of codes rather than actual data for re-production of data [5, 11–14]. During a failure, known as Erasure fault is caused when the storage solutions on the cloud service provider fails. The most effective and popular framework under Erasure Coding is Reed-Solomon framework. The framework can be applied in case of $n \leq 2^z$, where n denotes number of disks and z denotes number of customer data. The work of Dimakis and other researchers [14] demonstrated the situation where allocation of n data segments can sufficient and efficiently be replicated with a likelihood of $1 - n/p - o(1)$ for an faster data recovery and re-production. Where m = b log n with m is a collection of constant of Erasure formulation and b is always five times greater than "a". Here the cost reduction is notable for the replication and ensures high data availability. However the security can be compromised in these architectures as now the data is available over multiple storage servers which can be attacked by the intruders. Hence the next set of researches focus on encryption of the replicated data. The notable research by Lin and Tzeng demonstrates [5] the use of a secure and robust solution for distributed storage solutions powered by Erasure codes [16–19].

Hence in the storage solutions the need for a central system for storage of codes over distributed systems. To retrieve any communication the client nodes request the central security server for the key. Upon receiving the code, the likelihood of the request must be checked. Upon confirm the above mentioned likelihood; the data can be recovered with high efficiency.

The results of the previous research shows that for "m" storage containers, the factor "v" is equal to sqrt (n) log n. Hence the efficiency of this framework stands on $1 - n/p - o(1)$.

4 Performance Evaluation Application

The following application is created to demonstrate the load versus response time comparison for the tested cloud service providers (Fig. 5).

4.1 Local Storage Container

The local storage container stores the load data in form of textual file to be uploaded to multiple cloud storage services for testing. The textual data or the normalized data is considered as the data which cannot be formulized in data base or any other structured data types. However, it is important to understand that the unstructured data can be only text based or rich text including other media data. Most of the time, the unstructured data is collected from multiple applications used for business communication as emails, power point presentations, documents containing images

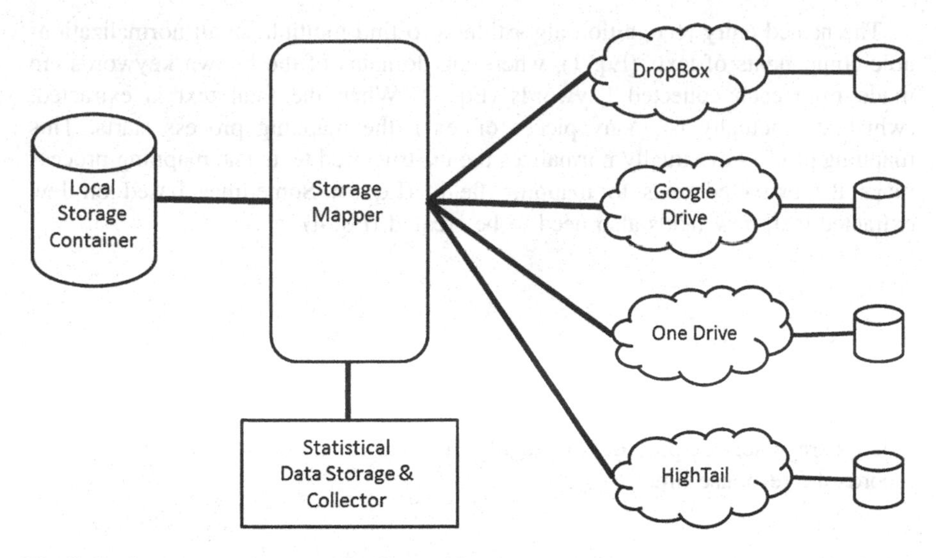

Fig. 5 Performance evaluator application architecture

and graphs, collaboration & text sharing documents and finally the instant messengers. The Local Storage is also equipped with a contexter component. The contexter component is basically a normalization component in this application. This component performs a few specific tasks in a specific order like Language detection, Named entity recognition, Anaphoric normalization and Text segmentation. It was assumed that the common text will appear in English and the rest of the process will start with this consideration. The first step is to compute each component block to extract the text by applying the formulas

$$Para_{(X)} = Para/D(Text_x) \tag{1}$$

$$D(Text_x) = \sum_{t=w_1}^{w_n} D_t \tag{2}$$

$$Para_{(X)_{i,j}} = Text_{(x)_{j,i}} \tag{3}$$

$$Para_{(X)_{i,j}} = Text_{(x)_{i+1,i-1}} \tag{4}$$

where, $Para_{(X)}$ is extracted text component

$Para$ is the total text block
$D(Text_x)$ is the domain of recognizable keywords
$Para_{(X)_{i,j}}$ is the extracted text component before mapping
$Text_{(x)_{j,i}}$ is the extracted text component after mapping

The named entry recognition algorithm is to find multiple small normalization-able components of texts (Eq. 1), where the domains of the known keywords are made from each collected keywords (Eq. 2). When the final text is extracted, (which can actually be many pieces of text), the mapping process starts. This mapping process eventually normalizes the unstructured text. The mapping process maps the extracted texts to mapping fields (Eq. 3). Sometimes based on few extracted text, new fields also need to be created (Eq. 4).

4.2 Storage Mapper

The simple storage mapper component of this application selects multiple different cloud storage service providers in regular interval for different types of loads and records the response time.

4.3 Statistical Data Storage and Collector

The statistical data storage and collector module collects the response time from different sources and generates a report for the all the services providers. The parameters considered in the resultant dataset are amount of data in the load, network speed, and type of action on the service provider data and the time for response.

4.4 Cloud Services and Containers

In this research we have considered multiple cloud service storage containers for the experiment. The configuration is demonstrated (Table 11).

Table 11 Cloud service provider instance configuration details

Instance type	Number of units	Type of architecture (bit)	Disk space (GB)	RAM (GB)
Small	2	32	160	1.7
Medium	2	32	350	1.7
Large	4	64	850	7.5
Extra large	5	64	1690	15
Extra large—high speed CPU	5	64	1690	15

5 Response Time Comparison

The response times recorded from multiple transactions on various data sources are recorded. The data load is tested on mentioned cloud service providers across multiple parameters as amount of data in the load, network speed, and type of action [20, 21]. We document the finds here (Table 12).

We closely observe there is no deviation in the response speed.

Table 12 Cloud service provider instance configuration details

Service provider	Data load (GB)	Network speed (MBPS)	Action type	Response time (min)
Dropbox	1	10	Write/read	1.4
Google Drive	1	10	Write/read	1.4
Hightail	1	10	Write/read	1.4
OneDrive	1	10	Write/read	1.4
SugerSync	1	10	Write/read	1.4
Dropbox	1	20	Write/read	0.7
Google Drive	1	20	Write/read	0.7
Hightail	1	20	Write/read	0.7
OneDrive	1	20	Write/read	0.7
SugerSync	1	20	Write/read	0.7
Dropbox	5	10	Write/read	7
Google Drive	5	10	Write/read	7
Hightail	5	10	Write/read	7
OneDrive	5	10	Write/read	7
SugerSync	5	10	Write/read	7
Dropbox	5	20	Write/read	3.5
Google Drive	5	20	Write/read	3.5
Hightail	5	20	Write/read	3.5
OneDrive	5	20	Write/read	3.5
SugerSync	5	20	Write/read	3.5
Dropbox	10	1000	Write/read	1.2
Google Drive	10	1000	Write/read	1.2
Hightail	10	1000	Write/read	1.2
OneDrive	10	1000	Write/read	1.2
SugerSync	10	1000	Write/read	1.2

6 Conclusions and Future Scope

The performance of all cloud service providers are analysed on a textual dataset, which is large in volume and the effect of the number of queries on the same dataset is studied. It is proven that there is a large difference for highly efficient latency time depending on the database used. We have also noticed that the latency time for the queries are heavily dependent on network speed or the network bandwidth, through which the services are accessed. But in both the cases we found that the effect of contexture is significant. Hence we conclude that the performance of multiple cloud service providers will be generating nearly same performance.

The further research needs to be carried out on multiple clustered storage containers arrangements where the virtualization factor to be considered. The redundancy control also to be considered to demonstrate the actual response time in the future works.

References

1. Kubiatowicz, J., Bindel, D., Chen, Y., Eaton, P., Geels, D., Gummadi, R., Rhea, S., Weatherspoon, H., Weimer, W., Wells, C., Zhao, B.: Oceanstore: an architecture for global-scale persistent storage. In: Proceedings of Ninth International Conference Architectural Support for Programming Languages and Operating Systems (ASPLOS), pp. 190–201 (2000)
2. Druschel, P., Rowstron, A.: PAST: a large-scale, persistent peer-to-peer storage utility. In: Proceedings of the Eighth Workshop Hot Topics in Operating System (HotOS VIII), pp. 75–80 (2001)
3. Adya, A., Bolosky, W.J., Castro, M., Cermak, G., Chaiken, R., Douceur, J.R., Howell, J., Lorch, J.R., Theimer, M., Wattenhofer, R.: Farsite: federated, available, and reliable storage for an incompletely trusted environment. In: Proceedings of the Fifth Symposium on Operating System Design and Implementation (OSDI), pp. 1–14 (2002)
4. Haeberlen, A., Mislove, A., Druschel, P., Glacier: highly durable, decentralized storage despite massive correlated failures. In: Proceedings of Second Symposium on Networked Systems Design and Implementation (NSDI), pp. 143–158 (2005)
5. Wilcox-O'Hearn, Z., Warner, B.: Tahoe: the least-authority filesystem. In: Proceedings of Fourth ACM Int'l Workshop Storage Security and Survivability (StorageSS), pp. 21–26 (2008)
6. Lin, H.-Y., Tzeng, W.-G.: A secure decentralized erasure code for distributed network storage. IEEE Trans. Parallel Distrib. Syst. **21**(11), 1586–1594 (2010)
7. Brownbridge, D.R., Marshall, L.F., Randell, B.: The newcastle connection or unixes of the world unite! Softw. Pract. Exp. **12**(12), 1147–1162 (1982)
8. Sandberg, R., Goldberg, D., Kleiman, S., Walsh, D., Lyon, B.: Design and implementation of the sun network filesystem. In: Proceedings of USENIX Association Conference (1985)
9. Kallahalla, M., Riedel, E., Swaminathan, R., Wang, Q., Fu, K.: Plutus: scalable secure file sharing on untrusted storage. In: Proceedings of Second USENIX Conference File and Storage Technologies (FAST), pp. 29–42 (2003)
10. Rhea, S.C., Eaton, P.R., Geels, D., Weatherspoon, H., Zhao, B.Y., Kubiatowicz, J.: Pond: the Oceanstore Prototype. In: Proceedings of Second USENIX Conference File and Storage Technologies (FAST), pp. 1–14 (2003)

11. Bhagwan, R., Tati, K., Cheng, Y.-C., Savage, S., Voelker, G.M.: Total recall: system support for automated availability management. In: Proceedings of First Symposium. Networked Systems Design and Implementation (NSDI), pp. 337–350 (2004)
12. Dimakis, A.G., Prabhakaran, V., Ramchandran, K.: Ubiquitous access to distributed data in large-scale sensor networks through decentralized erasure codes. In: Proceedings of Fourth in Symposium Information Processing in Sensor Networks (IPSN), pp. 111–117 (2005)
13. Dimakis, A.G., Prabhakaran, V., Ramchandran, K.: Decentralized erasure codes for distributed networked storage. IEEE Trans. Inf. Theory **52**(6), 2809–2816 (2006)
14. Mambo, M., Okamoto, E., Proxy cryptosystems: delegation of the power to decrypt ciphertexts. In: IEICE Transactions on Fundamentals of Electronics, Communication and Computer Sciences, vol. E80-A(1), 54–63 (1997)
15. Blaze, M., Bleumer, G., Strauss, M.: Divertible protocols and atomic proxy cryptography. In: Proceedings of International Conference on the Theory and Application of Cryptographic Techniques (EUROCRYPT), pp. 127–144 (1998)
16. Ateniese, G., Fu, K., Green, M., Hohenberger, S.: Improved proxy re-encryption schemes with applications to secure distributed storage. ACM Trans. Inf. Syst. Secur. **9**(1), 1–30 (2006)
17. Tang, Q.: Type-based proxy re-encryption and its construction. In: Proceedings of Ninth International Conference on Cryptology in India: Progress in Cryptology (INDOCRYPT), pp. 130–144 (2008)
18. Ateniese, G., Benson, K., Hohenberger, S.: Key-private proxy re-encryption. In: Proceedings of Topics in Cryptology (CT-RSA), pp. 279–294 (2009)
19. Shao, J., Cao, Z.: CCA-secure proxy re-encryption without pairings. In: Proceedings of 12th International Conference on Practice and Theory in Public Key Cryptography (PKC), pp. 357–376 (2009)
20. Ateniese, G., Burns, R., Curtmola, R., Herring, J., Kissner, L., Peterson, Z., Song, D.: Provable data possession at untrusted stores. In: Proceedings of 14th ACM Conference on Computer and Communication Security (CCS), pp. 598–609 (2007)
21. Ateniese, G., Pietro, R.D., Mancini, L.V., Tsudik, G.: Scalable and efficient provable data possession. In: Proceedings of Fourth International Conferences Security and Privacy in Communication Netowrks (SecureComm), pp. 1–10 (2008)

Attenuation of Broadband Problems Using Data Mining Techniques

Archana Singh, Soham Benerjee, Sumit Shukla and Shubham Singhal

Abstract The aim of this paper, is to provide appropriate solution of customer complaints on time and up to the mark has been considered a "defensive marketing" strategy or a "zero- defections" strategy, which reduces client disappointment. Overseeing client disappointment goes with web client grumbling administration. It is the discriminating issue for online client administration arrangements and e-CRM. The problems addressed in this paper are, (1) Scrutinize the sources and causes of online complaints; (2) Evaluating efficient ways of handling customer complaints in different categories and (3) Use of data mining techniques to provide solution for broadband issues. The paper proposed that e-businesses should provide excellent online customer services and response time of customers' requests/complaints must be much reduced in order to retain customers.

Keywords e-CRM · MTU · TRAI · Customer complaints

1 Introduction

The extant research and scenario explored that customers prefer to use internet at a very high speed. To attain high speed it incur heavy cost. However, for offices and lease lines broadband services are required at a reasonable cost. The effective and efficient e-CRM obliges adherence to a steady and reliable method that spotlights

A. Singh (✉) · S. Benerjee · S. Shukla · S. Singhal
Amity University, Noida, Uttar Pradesh, India
e-mail: archana.elina@gmail.com

S. Benerjee
e-mail: sohamhappy2help@gmail.com

S. Shukla
e-mail: sumit_shukla43@hotmail.com

S. Singhal
e-mail: ssinghal992@gmail.com

S.C. Satapathy and S. Das (eds.), *Proceedings of First International Conference on Information and Communication Technology for Intelligent Systems: Volume 2*,
Smart Innovation, Systems and Technologies 51, DOI 10.1007/978-3-319-30927-9_31

on the objectives of keeping up client steadfastness and of utilizing grievance addressing issues raised by clients. The main element in e-CRM [1] incorporates boost of consumer loyalty, expanding client dependability; expanding item/benefit quality; and determining client protestations. The grievance administration handle altogether enhances organization's quality execution by giving the capacity to react rapidly client protests. This thusly expands consumer loyalty and brand notoriety. Any proficient procedure will add to an abnormal state of Customer fulfillment and trust along these lines upgrading brand picture. A viable Complaints Management framework is indispensable to giving quality client administration. On the off chance that an organization does not handle its client objection legitimately and speedily, further negative client reactions will bring about real harm to the business. Frequently, clients are the first to distinguish holes between administration recommendations and administration conveyance when things are not living up to expectations appropriately. The client maintenance, consistence with industry regulations and evasion of potential item obligation claims are critical goals for all organizations. There are different wellsprings of online grumblings.

Mostly, organizations are giving online input structures to clients to exhibit their perspectives about the administrations they have gotten from any touch purpose of the organization. These structures are examined to accumulate data from clients about their item and client administration and help to enhance their administration in not so distant future. TRAI—Telecom Regulatory Authority of India. TRAI is an administration association who concentrates on telecom related dissensions confronted by Indian clients (Fig. 1).

The paper is divided into 5 sections. The Sect. 2, describes major problems faced by broadband customers. With research methodology In Sect. 3, Data mining techniques are discussed. In Sect. 4, the experiment, analysis and results of customer complaints is discussed regarding broadband. The paper concludes with the conclusion (Fig. 2).

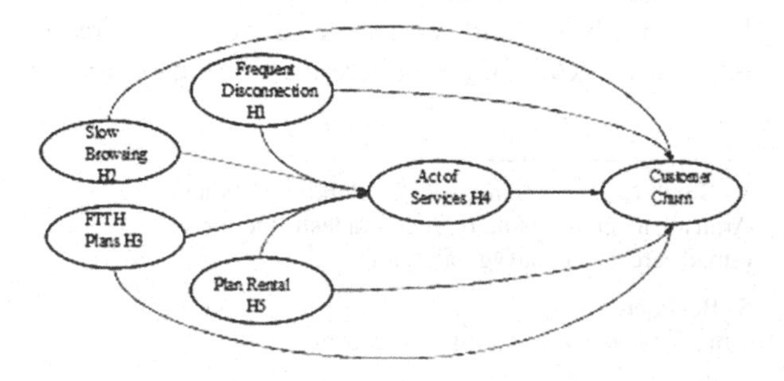

Fig. 1 Research model of customer churns with mediation effects

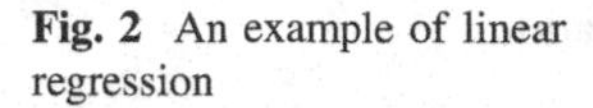

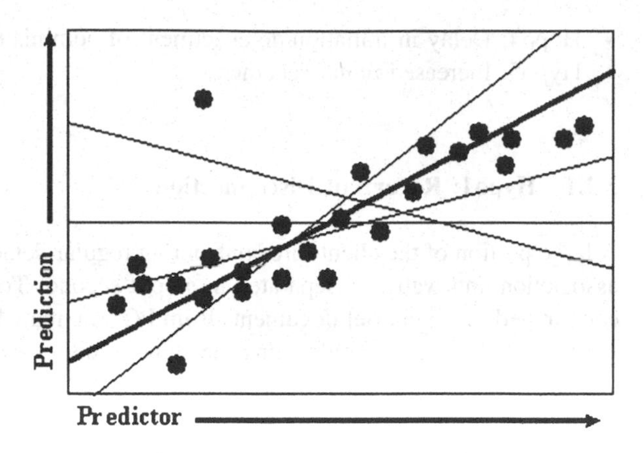

Fig. 2 An example of linear regression

2 Major Problems Faced by Broadband Customers

Broadband related issues found were related with frequent connection loss, the wifi connections were quite slow as compared with local area network connections. There was delay in action against complaints. Throttle of internet speed. The rental scheme was increased. Delay in introduction of Fibernet and VDSL plans [2]

2.1 About Internet Service Providers

Amid the presentation of simple cell administration and broadband administrations in 1984 and computerized administration in 1995, the Indian versatile information transfers and broadband administration business sector expanded altogether. As a sample, on 23rd December 2013, Airtel has gotten to be one of the top most administration suppliers in broadband web access suppliers. Be that as it may, the agitate rate of clients is expanding because of the most unmistakable issues of broadband administrations which we are centering to determine those issues. These real issues are: Slow scanning and incessant separation.

2.2 Customer Churn Determinants

The central point for client beat rate. Every component has its own effect on client stir rate and they are being arranged into diverse speculation highlights:

- Hypo1: Recurrent disconnection
- Hypo2: Slow browsing
- Hypo3: Fibernet and VDSL schemes

- Hypo4: Delay in initiation/de-enactment of administrations
- Hypo5: Increased rental schemes.

2.2.1 Hypo1: Recurrent Disconnection

A large portion of the clients are confronting regular detachment issue in broadband association and web gets separated in a split second. To determine this issue, it is encouraged to expel brief documents from PC and treats from program. Be that as it may, clients are as yet confronting the same issue, so we have isolated clients into two classifications (i) clients utilizing MTU esteem as 0 bytes and (ii) client utilizing MTU esteem as 1492 bytes.

In the wake of sparing MTU esteem as 1492 bytes in modem page, the vast majority of clients confronting incessant disengagement issue get corrected. In this way, it is standard estimation of MTU which should be set in all modem arrangement page to determine the successive disengagement issue in broadband.

2.2.2 Hypo 2: Slow Browsing

The next issue confronted by broadband clients is slow or moderate scanning resulting in a large portion of the clients are not ready to download information records from any URL and they additionally confront issue while perusing web. The download velocity is dependably 1/eighth of searching rate. When client tries to download any motion picture or melody from any site, so they generally concentrate on download velocity ought to be same as skimming pace which they are getting according to their broadband arrangement. On the other hand, while downloading any documents, it takes more than suitable time on the grounds that download sped is 1/eighth of perusing rate. Along these lines, clients ought to be mindful of this central point of moderate browsing.

2.2.3 Hypo3: Fibernet and VDSL Plans

Now a days, most of the customers are focusing on high internet speed either on mobile or broadband. Hence, Fibernet and VDSL have been introduced as an emerging technology to fulfill those requirements of customers. It is an advanced technology to resolve major factors which impacts broadband services i.e., frequent disconnection and slow browsing because special cabling is required to implement Fibernet and VDSL plans.

Fibernet is standardized as FTTH (Fiber to the Home). It requires special cabling from RSU (Remote switching unit) to customer's DP (Distribution Point) box through which broadband is supplied to customers premises.

2.2.4 Hypo4: Delay in Activation/De-activation of Services

When customer calls for activation or de-activation of any VAS (Value added service) on his account, so it is important to act/de-act the VAS within time frame or TAT (Turn Around time). If it takes longer than expected, so it is impacting customer churn rate and it is also impacting service provider representation.

2.2.5 Hypo5: Increased Plan Rental

If customers are using services of any service provider they usually focus on two major factors are: Services and Cost. So, it is really important to focus on cost factor along with service providing factor. It is mandatory to read market strategy and make plans accordingly which can fulfill requirements of all types of customers i.e., youth, children and old-age or personal and corporate users.

2.3 *Effects of Broadband Issues on Customer Status*

The effects of these broadband issues on customer churn rate. If any customer is using broadband service and he is facing any broadband related issues, so either he would complaint or he would change the service provider. Change of service provider would lead to customer churn rate and complaint gives a signal to company to focus on customer problems more thoroughly else it would impact customer base of that company.

Following are the factors affecting customer churn rate:

- Impact of successive separation in broadband administrations won't permit clients to peruse web with no intrusion and it may prompt client beat rate.
- Impact of moderate searching will set aside much time to finish the web related employments and unfavorably comes about into client stir rate.
- Impact of moderate execution of PC from which client is not mindful would likewise prompt moderate searching issues, so it ought to be unmistakably educated to client with respect to execution of PC.
- The presentation of new innovation i.e., FTTH and VDSL would have the capacity to determine broadband related issues, in any case it would require much investment to actualize this innovation. H3a0: the presentation of Fibernet arrangements would diminish client stir rate, on the other hand it need enormous measure of time to cover whole client base area.
- The presentation of VDSL arrangements would diminish client beat rate, be that as it may it need colossal measure of time to cover whole client base locale.
- The delay in act/de-demonstration of administrations generally affects client administrations in light of the fact that the administrations demonstration/de-act ought to be performed inside of SLA (Service Level Agreement) else it results into client dis-fulfillment.

- The expanded arrangement rental as contrast with other administration suppliers would likewise may prompt client stir rate on the grounds that each clients searches for best administrations as sensible costs and if some other administration suppliers is giving administrations on more prudent rates, so it may come about into lower client base. Along these lines, these are most imperative components which we have concentrated on to lessen client beat rate of broadband administrator.

3 Data Mining Techniques

An exchange database contains day by day executed things. In a value-based database, to accomplish crucial guidelines from center measurement diminishment of measurements are required. Selecting noteworthy number of important qualities, bringing out data from unexplored information can be accomplished by information mining procedures. Information using so as to mine investigates shrouded designs its different calculations for grouping, bunching and affiliation principle mining. Information mining can be ordered into six classes.

- Detection of anomalies: It includes the wrong information or loner information or veered off information or record of surprising information.
- Rule mining: This can be described as the technique of nontrivial extraction of certain, heretofore darken and possibly significant information.
- Clustering: It is the gathering of information of comparable sorts.
- Classification: It is the strategy for applying general structure to information.
- Regression: It is the technique helps in finding correlation between dependent and independent variables.

We have utilized direct relapse strategy to take care of this issue, Linear Regression [3] is a standout amongst the most vital information mining strategies which is utilized to determine the broadband related issues. It comprises of two parameters p and q which has one relapse variable X and it is utilized to minimize one parameter and impacts different parameters, toward the end the outcome considers

$Y = p + qX$, where p and q are variables.

4 An Analysis of Customers' Complaints

A rundown of 100 clients was analyzed confronting diverse broadband related issues [4]. They were confronting recurrent disengagement issue, many clients were confronting moderate skimming/browsing issue who needs to download information from two unique conventions are-SMTP (Simple mail exchange convention) and POP3 (Post office convention). Clients utilizing SMTP convention to download information records on their messages and POP3 convention are utilized to download information from other basic sites, for example, online networking and tunes.

4.1 Resolution of Regular Detachment Issue in Broadband Taking Indicator as MTU Worth

The MTU is most extreme transmission unit. It is utilized to alter estimations of send/get parcels for web network. Normally modems have MTU worth settled as 0 or 1500 bytes because of which after some time from modem establishment web begins dropping in light of the fact that this quality is not standard. The standard qualities to adjust send/get bundle status ought to be 1492 bytes and it is changed in modem design. When the worth is altered as 1492 bytes web association, begins working fine with no separation issue on little interims.

We have utilized a direct relapse [5] procedure, where, $Y = a + bX$

Where X = MTU as 1492 bytes, a clients utilizing MTU as 0 bytes, b: clients utilizing MTU as 1492 bytes

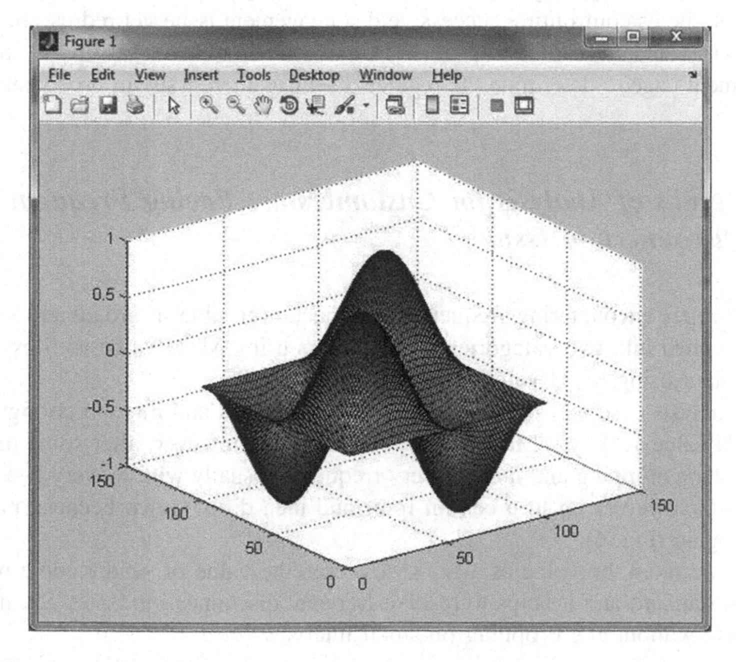

Fig. 3 Customers using MTU value as 1492 bytes

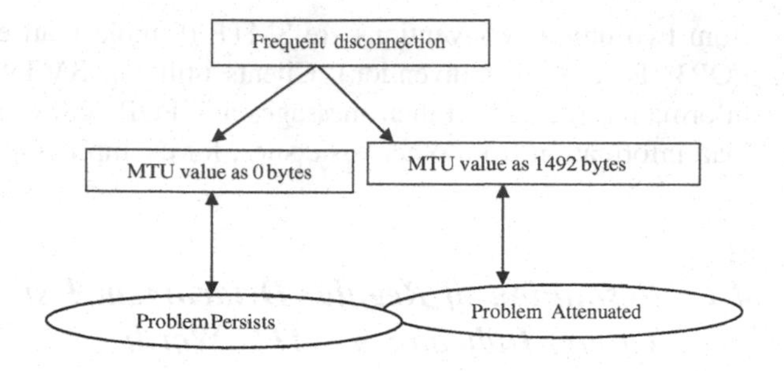

Fig. 4 Division of frequent disconnection issue on basis of MTU value

The above Fig. 3 speaks to the examination of clients who are utilizing MTU esteem as 1492 bytes. It demonstrates the investigation of straight relapse on continuous disengagement of web utilizing MTU as 1492 bytes. The greater part of the clients are confronting incessant disengagement issue in broadband association [6] and web gets separated immediately. To determine this issue, it is encouraged to expel makeshift documents from PC and treats from program. Nonetheless, clients are as yet confronting the same issue, so we have separated clients into two classes as appeared in Fig. 4.

(i) The clients utilizing MTU esteem as 0 bytes and
(ii) The client utilizing MTU esteem as 1492 bytes.

In the wake of sparing MTU esteem as 1492 bytes in modem page, a large portion of clients confronting successive disengagement issue get redressed. In this way, it is standard estimation of MTU which should be set in all modem arrangement page to determine the regular disengagement issue in broadband web.

4.2 *Results of Analysis for Customers are Facing Frequent Disconnection Issue*

The customers encountering frequent disconnection trouble in broadband services were classified into two categories (i) customers using MTU value as 0 bytes and (ii) customer using MTU value as 1492 bytes (Fig. 5).

The analysis customer data using broadband services and modem configuration has MTU value as 0 bytes. Initially it works fine and however, after some time the internet starts dropping and it drops very frequently usually within every 5–10 min. So, the graph moves up to a certain limit and then drops down because problem persists again (Fig. 6).

Once we fixed the value as 1492 bytes, then the value of send/receive packets becomes standard and it helps to resolve frequent disconnection issue. So, internet works fine without any dropping on small intervals.

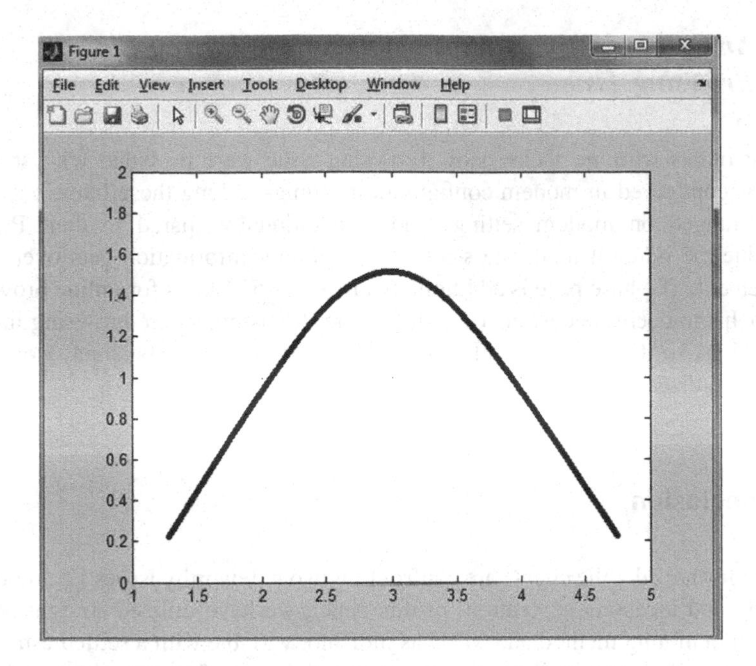

Fig. 5 MTU as 0 bytes persist frequent disconnection problem

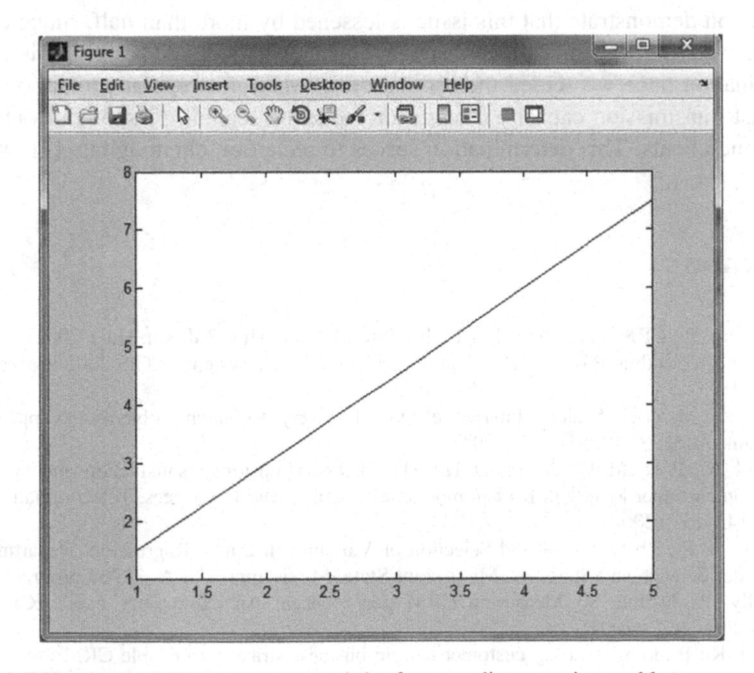

Fig. 6 MTU value as 1492 bytes attenuated the frequent disconnection problem

4.3 Analysis of Results for Customers are Facing Slow Browsing Issue

The customers who are facing slow browsing issue were provided with static IP which is configured in modem configuration settings. Along these lines, one static IP is arranged on modem settings and is additionally spared in their PCs, to determine the issue of moderate searching even after information gets over in the arrangement. The base pace is additionally altered as 512 kbps for online browsing. So, it helps to decrease customer churn [6] rate. If customers are browsing internet at 512 kbps, so it is standard value to perform any official or personal work. It is also standardized by TRAI (Telecom regulatory authority of India).

5 Conclusion

Without further ado, the clients are confronting two noteworthy issues i.e., moderate scanning and incessant separation. In this paper, we have utilized straight relapse information mining method and MTU as indicator variable with a settled estimation of 1492 bytes to determine the continuous separation issue. The static IP is utilized to determine moderate scanning issue. Our trials and results with 100 client grumbling information demonstrate that this issue is lessened by more than half. Since clients who were confronting successive disengagement and moderate searching issue got determination once we settled the estimation of MTU for regular separation issue and least transmission capacity esteem for moderate scanning issue confronted by broadband clients. This determination serves to reduction client agitate [7] rate.

References

1. Greenberg, P.: CRM at the Speed of Light. Tata McGraw-Hill, 3rd edn. (July 2004)
2. Freeman, P., Seddon, P.B.: Benefits from CRM Based Work Systems. ECIS 2005 Proceedings, 14 (2005)
3 Bowman, C.M., et al.: Scalable Internet resource discovery: Research problems and approaches. Commun. ACM. 37(8), 98–107 (1994)
4 Kim, M.-K., Park, M.-C., Jeong, D.-H.: The effects of customer satisfaction and switching barrier on customer loyalty in Korean mobile telecommunication services. Telecommun. Policy 28(2), 145–159 (2004)
5. Hocking, R.R.: The Analysis and Selection of Variables in Linear Regression. Department of Computer Science and Statistics, Mississippi State, Mississippi, U.S.A. 39766 program
6. O'Reilly, P., Dunne, S.: Measuring CRM performance: An exploratory case. ECIS 2004 Proceedings, 122 (2004)
7. Ballard, K.: Building a strong customer centric business strategy to enable CRM and develop customer loyalty. Intel Corporation, University of Oregon, Applied Information management Program

Chameleon Salting: The New Concept of Authentication Management

Binoj Koshy, Nilay Mistry and Khyati Jain

Abstract Entity authentication by means of keying in Username and Password has been adopted as the de-facto standard in many Internet based and enterprise based applications the world over. The service providers have always been bogged down by Information Security breaches and Cyber Attacks in the recent past, and the industry has time and again stood against odds and has been able to imbibe the best of security practices to beat back security breaches especially in the domain concerning Entity Authentication. Authentication process needs to be made robust and hardened; Salting of password is practiced by the application service provider to ensure security to the end-customer. Technology today offers protection and also provides the tools to the unscrupulous elements by means of cracking Username and Password. No amount of technology and processes is said to be adequate to keep away the perpetrator in this domain. Chameleon Salting is an innovative initiative to enhance security to the end user, where the end user is provided a secure environment for entity authentication even without the user having to implement or change the way the application is being used. Its implementation will lower impact of a loss, thereby providing better protection to the customers.

Keywords Authentication · User name · Password · Random key · String · Hashing · Password hash · Salting · Chameleon · Algorithm · Chameleon hashing · Chameleon salting · Pseudo-Random · Short salt · Salt reuse · Salt collisions · Double hashing · Wacky hash functions · Hash collisions · Trap-door · No-key exposure · Collision resistant · Semantically secure

B. Koshy (✉) · N. Mistry · K. Jain
Institute of Forensic Science, Gujarat Forensic Sciences University, Gandhinagar 382007, Gujarat, India
e-mail: binojkoshy1@gmail.com

N. Mistry
e-mail: nilaymistry30@gmail.com

K. Jain
e-mail: khyati792@gmail.com

S.C. Satapathy and S. Das (eds.), *Proceedings of First International Conference on Information and Communication Technology for Intelligent Systems: Volume 2*, Smart Innovation, Systems and Technologies 51, DOI 10.1007/978-3-319-30927-9_32

1 Introduction

Authentication has and will be an integral part of any system where the user is mandated to identify him/her self for access to the authorized component of an IT (Information Technology) eco-system. The use of 'Username' and 'Password' has been the de-facto practice in all web enabled and client-server based application. The security of information and its systems need to be safeguarded and protected; inturn ensuring security at all stages such as planning, implementation and production of the IT lifecycle. The misuse and abuse of these systems can be curbed by adopting adequate measures at the authentication level. The information security implementation in any organization or IT infrastructure mandates the use of robust and foolproof authentication system, which may be based on 'multi-factor' authentication. Use of 'Username' and 'Password' for initial access in many IT applications ensures CIA (Confidentiality, Integrity and Availability) to a user [1]. Notwithstanding the security overlay that is abinitio built into any integrated application, the presence of trespassers has been noticed even in the most secure IT system. The attempts made to obtain unauthorized access are again difficult to assimilate, as the effort both by the elite and the unskilled will carve a dent in the confidence level on the application or the 'Login' module. From the service providers' perspective or the owners' perspective; an attempt of sabotage or vandalism will tantamount to erosion of customer confidence.

Non-implementation of application driven 'Enforced Password Policy' mechanism, and also user awareness based practices in password policy leads to breaches. Such acts of information leak and unauthorized entry by elements will cause disruption and loss in terms of money and reputation. Any amount of security measure to ensure secure use of password in an application seems inadequate, as the technology/techniques that is used to prepare application security robustness can also be used by the perpetrators to accomplish Machiavellian tasks. For this reasons, applications have to be designed with inherent security features and also with high degree of 'Vulnerability Tolerance'. More so when the ownership of error-making has been absolved from the end-user and is made more 'corrective oriented' at the application level. Large investments in security overlay are not a feasible initiative, and critical industries resort to innovations to ensure security.

2 Access Control

The login or authentication security is an amalgamation of technology, procedure, practices and habits. It is seen that this ecosystem has been victim of such predators who are in sync with the technology and also using the same technology and strategies to circumvent the security policy of 'Login'.

E-mail is deemed to be the largest application on the Internet today. E-mail authentication, off late have been victim of password losses and authentication bypassing and allied attacks. The E-mail service providers, though provide robust authentication schemes like: HTTPs, 'Chameleon hash function framework' compliance [2], two-factor authentication, and "captcha" based login, etc; are yet been attacked by perpetrators resulting in loss of repudiation property.

3 Password Management

Password management has been a key concern area to all the sectors, especially in the critical applications like the banking sector. The banks are required to handle large volumes of customers and users, the concern is more aggrandized as the customer base of the bank starts widening and also when the e-banking application is rolled over for users to access multiple banking utilities across different financial institutions. Processes involving ATM PIN, Internet Banking Password, Transaction Password, Mobile Banking Password, Tele PIN, etc. have always been of concern to banks. The mismanagement of these will not only lead to loss of 'Customer Confidence' but also monetary losses. To manage these credentials and also to ensure 'Confidentiality', 'Integrity' and 'Availability'; banks have been taking due diligence in the IT applications that are being rolled out over the years. The cause of concern is not only to the end-user; but also to the Web Developer, the Database Designer, the System Administrator, the Data Base Administrator (DBA) and the IT Head [3]. The evolution of password management is elaborated below.

3.1 Password as Plain Text

In the legacy days of IT implementation, passwords were stored as 'Plain Text' along with the username as part of the database. Anyone who had access to the database was able to hack the complete column containing the password in 'Plain Text'. It was also obvious that any small part of the application that was vulnerable to 'SQL injection' would make the task of retrieval of password from the database easy.

3.2 Password Stored as Encrypted Entity with Standard Key

Encryption is the nascent process of encoding the content or text or raw ASCII (American Standard Code for Information Interchange) value, in such a manner that the resultant structure or value if obtained is not able to be deciphered or interpreted; unless the expert is able to decipher or obtain the algorithm that has gone into the encryption. The encryption entails two components: Firstly, the encryption algorithm or the cryptographic logic; and Secondly, the 'Input Key' that was used to undertake the encryption. If the attacker or perpetrator is privy to the encryption algorithm and the secret key (or even the pattern of generation of static key), it is assumed that the password can be decrypted; given the fact that breaking of password is much simpler today. Further, if the encrypted text is available, dictionary and brute-force method can be used to reproduce the password; which are also well automated these days.

3.3 Password Stored as Encrypted Entity with Random Key

The process of password management with encryption enabled storage (with random key), is the next step in the maturity model of password management. Again here, there is no way in which dictionary and brute force attack methods can be circumvented. The only complexity here is to generate the random key, which otherwise can be very easily deciphered with the automated and enhanced processing capability available today. This additional processing power and memory that is essential for password decryption, is otherwise available to the perpetrator due to advancement of technology and its exponential growth.

3.4 Storage of Hashed Password

'Hashing' is the complex process of generation of a random number or a set of ASCII values or hexadecimal values with a large fixed length from a simple plain text input. The 'Hash' of the password is unique and there was no means to generate the original password from the obtained Hash. Further, 'Hashing Password Management' was made more robust especially in the banking application by managing the Hash [4]. The 'Hash' was only stored in the database and no part or portion of password was stored anywhere in the application. If a user was to log-in at a later point of time, the application will generate the 'Hash' and then the generated 'Hash' is compared for similarity and uniqueness to the value already stored in the table containing the 'Password Hashes'. Further access was only granted when the condition returned a value 'True', when the 'Generated Hash' and 'Stored Hash' is found to be similar [4].

The 'Hash Algorithms' mostly are two-way functions and are otherwise difficult to reverse engineer. 'Hash functions' churn out fixed length formats' that is non-reversible [5], and also characterized by, an output, which is drastically changed even if the input is slightly dissimilar. Each time the user changes the password, the 'Hash' will differ. For 'Password Hashing', the access control modules have been using the regular framework of 'Cryptographic Hash Functions' [6] that have incorporated SHA256, SHA512, RipeMD and WHIRLPOOL. The 'Hashing' was further extended using the 'Cryptographically Secure Pseudo-Random Framework'.

The most elementary means to crack any Hash is to try and guess the password and then hashing each guess and then subsequently insertion of the hash to check its admissibility. Now, dictionary attack and brute-force attack are still admissible and hence the system is still vulnerable. There is no means to prevent these attacks and hence even the best 'Hashing Framework' is not secure. Apart from the two password cracking techniques that we have discussed earlier, there are other means; namely, 'Lookup Tables', 'Reverse Lookup Tables' and 'Rainbow Tables'.

3.5 Salting of Password Before It Is Hashed

The 'Cryptographically Secure Pseudo-Random Framework' provides provisions for secure authentication in many applications across the globe [6]. In this the calculated hash is not only stored separately but is also used for verification of the password acceptability and also incorporates complex hash computing through the technique of 'Salting'. The salting technique is a complex hash function, which is randomized based on the instance when the new password is created or when the password is changed. The random value, which is generated through a salting algorithm, is called a 'Salt'. This 'Salt' is appending to the password before it is hashed. In other words, Salting is a technique in which a random string or value is added to the user entered password and then resultant is hashed to produce a unique 'String' or 'Hash'. Here we can see that even if two users provide the same password, the resultant hash will be different as the salt that is received by the two users will be different. The Salt provided is based on instance and the Randomization Algorithm of the Salt. The Salt, that was provided to a particular user while generation of the password hash, will also be stored in a separate area of the database.

The service providers have taken efforts to ensure that no two users get the same 'Salt' while changing password, and also that the process of generation of 'Salt' is absolutely random. Notwithstanding the same, we see that the threat still exists in cases where the user does not prompt a change in password as a routine practice.

The user of such secure application is required to provide the correct password to establish a successful log-in. The Password is then taken from the user and the appropriate 'Salt', which has been stored abinitio, is fetched from the database. Further, the hash is computed using the hash function by applying the obtained 'Salt' and the provided 'Password', this value is computed post application of salt to the password. This Hash is then compared to the one stored in the database; if the hash is equal, then the user is allowed access. The 'Salting Algorithm' the 'Hashing Algorithm' and the 'Application Code' governing these functionalities is well concealed with adequate level of security.

4 Chameleon Salting

By means of this paper it is proposed to introduce the innovative, original, no cost solution to provide a more robust salting technique to ensure high level of security to the end user; in applications that require logging-in by means of 'Username' and 'Password' as part of entity authentication.

The first assumption here is that the System Architecture being followed has ensured implementation of the Salting processes and its techniques in its true idealism. Further, it is also clarified that the complete 'Nail-down' has been incorporated without any mythical assumptions while implementing Salting as part of the security overlay. It is also assumed that the wrongful approach, such as 'Short Salt', 'Salt Reuse', 'Salt Collisions', 'Double Hashing', 'Wacky Hash Functions', 'Hash Collisions' and the like, have been evaded; and idealism cinched while undertaking implementing of 'Salting' and 'Hashing' for a secure entity authentication process [7].

It does not matter if the Salt comes before or after the password; as the call is purely with the security implementer. However it is more important to pick one and stick on to the selected one. It is also important to use the correct reversal process, when the user is logging-in. The selection of this particular sequence will be vital, especially when there is a requirement of adhering to 'Interoperability Compliance'. Industry practice and adequate precedence, is seen in the practice of the Salt coming before the Password.

It is also observed that the client/customer of any application does tend to express "user fatigue", if the application enforces very frequent changing of password. As also the end-user will attempt multiple log-in effort to carry routine activity, but will not carry out 'password change' very frequently; especially while using the application where access is governed by entity authentication each time the user is intending to user the application [8].

4.1 The Concept of Chameleon Salting

The presently practiced 'Salting of Password' technique, in applications, is otherwise a strong form of authentication mechanism. It offers adequate security to those users who frequently change their 'Password'. However we cannot rule out the intelligent hacker still targeting the systems for gain. The loss of Salt to an attacker will make the perpetrator to compute the hash; again, this may be at a cost. It is also known that no level of security is adequate in the present scenario, where resources and intellectual capability is amply available to carry out such attacks.

Salting as part of Chameleon Salting: Salting of password, has been a proactive means to circumvent users who resort to passwords which are noncomplex and weak. The Salted password is a password generated after the user completes the process of entity authentication; in which the user input 'Password' is appended with a fixed length hash value called the 'Salt'. The Salt can be a string directly generated from a salt algorithm or a hash generated from the Salt string. Cracking of a salted password is now made difficult due to the fact that the cracker has to compute the hash from the scratch for each 'Guess Password' and also encounter a huge volume of 'Fixed Length String' that is concatenated with the password (E.g. if the hash is a 32 bit random number, the combination to be multiplied will be 2^{32}, i.e. 4294967296 combinations). In the 'Salting of Password', the plain text password is never stored anywhere in the system. However, the 'Algorithm' is stored in the web server or database server. The "Salt" is 'per user' and is unique. In case the user is using best practices where he/she is adhering to frequent changing of password; the 'Salt' is then found to be changing as and when the user makes a different password.

Chameleon Hash Function part of Chameleon Salting: Chameleon Hash Function is a 'Collision-Resistant' Hash Function. The algorithm produces Non-transferable Signatures called 'Chameleon Signatures'. It is a cryptographic hash function, which includes characteristics like, Non-repetitive pre-image, Collision-Resistant and Pseudo-randomized generation. The chameleon hash conceals the 'Public Key' by attaching the Function with a corresponding unique trapdoor. The name "chameleon" refers to the ability of the owner of the trapdoor information to change the input to the function, to any value of his/her choice without changing the resultant output. The Chameleon Hash Function is associated with a pair of public and private key with the following properties [9, 10]: (1) anyone who knows the public key can compute the associated hash functions. (2) for those who do not have the knowledge of the trapdoor (the secret key included), the hash function is collision-resistant, it is infeasible to find two inputs which are mapped to the same output. (3) The trapdoor information holder can easily find collisions for every given input.

In this new 'Chameleon Salting' process, a new salt will be generated each time the user logs-in. This new salt will be appended to the earlier stored salt, and the same will be called from the database during the users' next log-in. The salt is not only generated when the user changes the password, but also when the user attempts a log-in and is about to use the application through the user login interface. It is also clarified that when the user logs into the profile, the supplied password will now fetch the stored Salt and also initiate a trigger to generate a new Salt that will be used for the next log-in. The old Salt is also stored to facilitate handling of any subsequent error that may occur during the logging-in process. If in case a verification drill or process is to be initiated for cases involving non-generation of the Salt; the previous Salt will be fetched as part of the verification process that will ensue. However if the previous log-in was successful, the user will be provided with the previously refreshed Salt during its subsequent log-in attempts [11].

4.2 The Mechanics of Chameleon Salting

This process will happen each time the 'User' initiates a session for 'Log-in', and this renewed 'Salt' will be used for the next login attempt. The user is not aware of the back-end activity, but at the same time the user prompts the application to refresh the Salt, thereby generating a new Salt that will be used during the users' next log-in. This new salt is only generated if the log-in attempt has been successful, in which case the user had used his password and the system had computed the hash with the previously stored salt and the accesses was granted to the users account/profile. A successful hack is only achieved, if the same users has not logged-in after the instance of loss, even at least once.

The new Salt in the 'Chameleon Salting' technique is then generated based on the session that is created while a fresh successful log-in has been performed by the user. The initiation can be either based on the 'Session ID' or on the 'Time-Stamp' of that user, when attempting a successful login. Further, the logic is so built that the salt is unique and appropriate to ensure security of the password. Aspects like 'Salt Non-reusability', 'Length of the Salt Value (or string)', the 'Computation of Salt', the 'Randomization of Salt' and 'Salt Collisions' are addressed as part of the complete password management system.

4.3 The Nuances of Chameleon Salting and Its Implementation

The implementation of 'Chameleon Salting' is a three-step process and involves basic minimal cost effective measures. The System Administrator and the DBA, adhere to a well planned and derived process will be able to incorporate the changes, and its implementation can be carried out in less than 10 (ten) min. The same can also be implemented when there is a window of opportunity, especially during the routine maintenance outage.

Implementation of 'Chameleon Salting' will in any case be compliant to inter-operability standard; as the implementation does not in any way interfere with the interoperability modules. Further, 'Cross Platform' and 'Across Application' interfaces are only activated after a successful 'Log-in' by the user is undertaken. The effort to implement the 'Chameleon Salting' system is only a 'one-time' effort, and there is no maintenance attached to the complete ambit, subsequently.

4.4 *The Unique Feature of Chameleon in the Salting Process*

The 'Chameleon Hash' has evolved over time and the new "Session ID based Chameleon Hash Function" ensures 'No Key exposure', 'Collision Resistant' and also takes care of Multiple Recipient settings there by making it 'Semantically Secure'.

4.4.1 Reference Explanation 1

Chameleon Hashing schema is referred to that algorithm that returns a semantically secure hash if for all identity string ID and all paired number (value) (n, n!); the probability distribution of the random variables 'Hash (ID, n, r)' and Hash '(ID, n!, r!)' are computationally indistinguishable [12]. (Where r, r! are a random variable.)

4.4.2 Reference Explanation 2

Semantic Security is referred to that concept related to the hash value wherein it is infeasible to determine which number (value) is likely to have resulted in a new value by an application of the hash algorithm [13, 14].

4.4.3 Reference Explanation 3

Non-key exposure is the property of a Chameleon Hashing schema that tends to conceal the public key and the secret key; and also computational deny its exposure at any stage of the hashing process [9].

4.4.4 Reference Explanation 4

Collision Resistance refers to that conditional phenomena in which; there exists no efficient algorithm that accepts a number (value) n, a random integer r and another number (value) n! (n!≠n) as inputs, and output a random integer r! (r!≠r) that satisfy Hash (n!, r!) = Hash (n, r) with non- negligible probability, unless with the knowledge of trapdoor information 'SID' associated with ID [15].

4.4.5 Reference Explanation 5

Trapdoor commitment can be otherwise referred to chameleon hashing which is basically a non-interactive commitment schema in which the reversal is impossible and also the feature of the owner of the chameleon hash information

(Trapdoor Information) to change the input to the function to any value of the owner's choice without changing the resulting output [16].

As in the case of the Chameleon, this phenomenon of Chameleon Salting will ensure that the salt that is generated during each 'Log-in' is unique and is dynamic.

5 Security Analysis of Chameleon Salting Process

The Chameleon Salting process is a digital combination of integration, authentication and non-repudiation for a particular Log-in session. The uniqueness of the process is well verifiable, both mathematically and algorithmically. The security analysis of this process is verifiable based on six (6) 'Inferences' and subsequently its 'Proof'.

5.1 *Inference 1*

The Instance, "δ" and Session ID, "SID" in the Chameleon Salting process is unique.

Proof The Instance of log-in of any user being "δ" is unique; so is also the case of the Session ID, which is a relative value, generated from a function.

$$S_{ID} = {}_{X=1}^{N} f\mathbf{x}(\varepsilon)$$

[where ε is the epoch time elapsed in minutes • seconds]

Hence, a combination (or otherwise of) the Instance, "δ" and Session ID, "SID" will be unique.

5.2 *Inference 2*

In the Chameleon Salting process the 'Username' and 'Password' entered by the user at an instance 'δ' is resistant to Collision Forgery, as the discrete log-in is Non-repetitive due to the fact that 'Username' and 'Password' environment; i.e. "U·P" and "δ" cannot be similar to another occurrence of the combination of "(U·P)!" and "δ!".

Proof If a collision were to take place, then Hash (SID, δ, B, C) = Hash (SID!, δ!, B!, C!)

Where, $S_{ID} = {}_{X=1}^{N} f\mathbf{x}(\varepsilon)$ [where ε is the epoch time elapsed in minutes • seconds]

B = Hash (U·P) [where U is the username string and P is the password string]

C = Hash(σ) = Hash ($fy(\varepsilon)$) [where σ is the salt string generated by the Pseudo-Random function]

(Similar is the case of SID!, δ!, B! and C!)

Further; In the case of e-mail services B $\neq$ B! unique user-ID is provided to each client.

However; even if $\delta = \delta!$, B = B! and C = C! in a particular scenario,

$$\text{SID} = \sum_{x=1}^{N} fx(\varepsilon) \text{ will not be same as}$$

$$\text{SID}! = \sum_{x=1}^{N} fx(\varepsilon) \text{ [as 'x' will differ here]}$$

$$\text{Hence SID} \neq \text{SID}!$$

5.3 *Inference 3*

The Chameleon Salting process by means of the Pseudorandom Salt Generator, which is based on instance "δ" (initiated while a user is performing a 'log-in') as part of entity authentication without Key Exposure is Semantically Secure.

Proof The Chameleon Salting Scheme without Key Exposure is said to be semantically secure if, for all SID (Session ID String) and all pairs of Session/User ID and Password, i.e. SID, SID!, (U·P), (U·P)!; the probability distribution of the random variable of Hash (SID, δ, B, C) and Hash (SID!, δ!, B!, C!) are computationally indistinguishable, i.e. for a given (U·P), (U·P)!, Z=Hash (SID, δ, B, C) and Z!=Hash (SID!, δ!, B!, C!); an adversary cannot distinguish in polynomial time between (Z, Z!) of any pair of User ID and Password i.e. (U·P), (U·P)!.

Further, the proposal satisfies the 'Conditional Entropy' as at no 'Log-in the value of 'Z' is equal to Z!. Because, there is only one random element of $\text{SID} = \sum_{x=1}^{N} x(\varepsilon)$ that will generate different string even if the instance 'δ' is the same and also that the pair of (B, C) will not be equal to (B!, C!).

Hence, Hash (SID, δ, B, C) $\neq$ Hash (SID!, δ!, B!, C!) (Because of the non-degeneracy of bilinear pairing)

Consequently, It is proved that the inference that, Chameleon Salting follows Semantic Security is true.

5.4 *Inference 4*

The Chameleon Salting process is secure against decryption of the 'trap-door information' as there is no Collision Forgery occurrence which has been otherwise been ensured by the 'Pseudo-random Salt Generation' with a unique combination of instance "δ".

Proof It is understood that the only possibility of deciphering of the 'Trap-door Information' is when a Collision Forgery occurs.

Further, Trap-door commitments have the property that the knowledge of the trapdoor by itself is not sufficient to enable the computation of alternate de-commitments. (Assuming that Collision Forgery will result in the attacker recovering the information)

$$\hat{e}(\mathrm{P}, \delta, {}_{\mathrm{x}=1}^{\mathrm{n}} f\mathbf{x}(\varepsilon))$$

[Where P is a string value whose order are a prime q, which is used to generate the cyclic additive encryption set]

Again,

If Collision were to occur then Hash (SID, δ, B, C) = Hash (SID!, δ!, B!, C!)

Subsequently, this is not possible as 'P' is unique to a transaction whose order is decided by the prime 'q' that is generated for that instance 'δ' (δ ≠ δ!).

Hence proved that, Hash (SID, δ, B, C) ≠ Hash (SID!, δ!, B!, C!).

The attacker will not be able to recover the sum of the set of Secret Key $\sum_{i=1}^{x}$ SID or for that matter any set of the trap-door information. Also, it is proved that decryption of 'Trap-door information' is not possible in any attack scenario.

5.5 Inference 5

The Chameleon Salting process is secure against 'Hash Collision" as there is no occurrence of 'Collision Forgery' due to re-use of Hash or generation of a similar Hash from another user 'log-in' even if "δ" (instance) is the same.

Proof As per 'Inference 4', if there is Collision Forgery then the Trap-door Information is lost to the attacker. However this has been proved otherwise. Hence it is understood that Hash Collision will only happen if a Collision Forgery occurs. This is not a possibility.

Hence, it is proved from 'Inference 4' and based on the proof elaborated; that in Chameleon Salting; Hash Collision will not occur for any scenario.

5.6 Inference 6

The Chameleon Salting process provides provisions of 'Denying Unauthorized Authentication' and hence satisfies the properties of 'Non-transferability', 'Non-repudiation', 'Deniability', 'Key Exposure Freeness' and 'Non-interactive'.

Proof **Non-transferability**: It is pertinent to mention that the Semantic Security offered by the Chameleon Salting technique implies that the Security offers Non-transferability of the concerned Chameleon Signature Scheme; Hash (SID, δ, B, C). Therefore, in the Chameleon Salting Scheme; the attacker or the user cannot transfer/steal a signature of the complete Chameleon Salting setup.

Non-repudiation: Given a valid signature generated by the Chameleon Salting algorithm, the user or attacker deliberately or inadvertently cannot generate a valid Hash Collision which satisfies the Chameleon Hash Function as at any point $m \neq m!$. Hence, the Chameleon Salting technique is Non-repudiation.

Deniability: The protocols that govern Salting and Chameleon Hash, which is inherent to the Chameleon Salting Scheme are ‘Loss-less’ and ‘Error-less’ compliant; hence this inherently ensures compliance to Denial protocol.

Key Exposure Freeness: Given the likelihood of Collision (δ, B, C) and (δ!, B!, C!); the information $\hat{e}(P, \delta, {}^{n}_{x=1} f\mathbf{x}(\varepsilon))$ can be recovered. However, it is impossible for anyone to compute ‘SID’ from $\hat{e}(P, \delta, {}^{n}_{x=1} f\mathbf{x}(\varepsilon))$. Therefore, Collision Forgery cannot result in attacker or the user recovering the trap-door information. Further, due to the presence of strong Non-transferability property in the Chameleon Salting Scheme, Key Exposure is not possible.

Non-interactive: The Chameleon Salting technique is based on Session ID ‘SID’, and the instance of log-in ‘δ’. Further, the Chameleon Salting is based on trap-door. Hence, the reversal of process is completely denied. The password once input by the user, will only facilitate computation of the hashes at the client end or at an isolated server (as per configuration), and the application server will only be provide with the ‘Hashed Salt’ and ‘Hashed Password’ value. This inturn will only facilitate the penultimate verification of a mere logical function returning ‘True’ or ‘False’. It is concluded hence that the complete system is Non-interactive.

6 Advantages and Limitations of Chameleon Salting

6.1 *The Advantages of Chameleon Salting*

The solution proposed can be implemented with least effort and with no additional cost; hence the project implementer will be able to convince the management hierarchy to adopt the ‘Chameleon Salting’ solution. Further, the solution will only improve security and also ensure better protection to the end user. The end user, inspite of not adhering to best-practices, by means of changing password regularly, is still able to get the same protection as in the case where the users is presumed to be using ‘Change Password’ options. The mandate is that the user need to use the application, as the change of ‘Salt’ is only prompted when the user makes a successful ‘Log-in’.

6.2 The Limitations of Chameleon Salting

The need of the hour is to ensure user education and awareness. The proposed solution of 'Chameleon Salting' is only a small way forward. However the exploitation of brute-force attack is still possible on salted passwords. Notwithstanding the fact; that there exists a limited protection threshold against dictionary attacks. 'Chameleon Salting' is only activated when the user 'Logs-in' into the application. The inactivity of a user profile will still have the same vulnerability and hence the threat to 'password theft' is persistent. Again, loss of salt database to a hack is detrimental, if the user has not attempted a log-in during the window between the loss and the final breaching activity by the attacker.

7 Conclusion

Authentication based on 'Username-Password' is the most common and user-friendly means to the end user; however we see that there are many such web-based authentication methods to provide stronger security. It is also to be remembered that any complex authentication procedure will be at the cost of 'Availability'. Also, complex and protracted procedure will result in user fatigue and wearisome mood swings. The 'Chameleon Salting' technique is another step in this direction. This initiative still can be claimed as puny, as the perspective attackers are wittier than what technology can offer today. Accession of password through Social Engineering and Phishing means are still a threat and will remain in the environment despite any application security measures. The answer to a holistic security measure considering 'Human Resource' as the epic-centre is the right step forward. This again culminates at 'User Awareness'. The web service and other service providers also need to address this aspect of 'User Awareness' in its security campaigns to ensure a holistic security mechanism while considering entity authentication and also for a more secure IT eco-system.

References

1. Matt, B., Introduction to Computer Security. Addison-Wesley Professional (2004)
2. Sadeghi, A., Steiner, M.: Assumptions related to discrete logarithms: why subtleties make a real difference. In: Advances in Cryptology Eurocrypt 2001, LNCS 2045. Springer, Berlin, pp. 243–260 (2001)
3. Drevin, L., Kruger, H., Steyn, T.: Determinants of password security: some educational aspects, information assurance and security education and training. IFIP Adv. Inf. Commun. Technol. **406**, 241–248 (2013)
4. Provos, N., David, M.: A future-adaptable password scheme. In: USENIX Annual Technical Conference, FREENIX Track, USENIX, pp. 81–91 (1999)

5. Dodis, Y., Katz, J., Xu, S., Yung, M.: Key-insulted public-key cryptosystems. In: Advances in Cryptology-Eurpcrypt 2002, LNCS 2332, pp. 65–82, Springer, Berlin (2002)
6. Krawczyk, H., Rabin, T.: Chameleon hashing and signatures. In: Network and Distributed System Security Symposium. The Internet Society, pp. 143–154 (2000)
7. Theoharoulis, K., Papaefstathiou, I.: Implementing rainbow tables in high end FPGAs for superfast password Cracking. In: International Conference on field programmable logic and applications (FPL). ISBN: 978-1-4244- 7842-2, Aug 2010
8. Madero, A.: Password secured systems and negative authentication. In: System Design and Management Program, Engineering Systems Division, Massachusetts Institute of Technology, pp. 50–52 (2013)
9. Chen, X., Zhang, F., Kim, K.: Chameleon Hashing without key exposure, information security. Lect. Notes Comput. Sci. **3225**, 87–98 (2004)
10. Chaum, D., van Antwerpen, H.: Undeniable Signatures. In: Advances in Cryptology- Crypto 1989, LNCS 435, Springer, Berlin, pp. 212–216 (1989)
11. Bellare, M., Ristenpart, T., Tessaro, S.: Multi-instance security and its application to password-based cryptography. In: Advances in Cryptology—CRYPTO 2012. Lecture Notes in Computer Science, vol. 7417, 2012, pp. 312–329, May 2013
12. Gao, W., Li, F., Wang, X.: Chameleon hash without key exposure based on Schnorr signature. Comput. Stand. Interfaces **31**(2009), 282–285 (2009)
13. Pointcheval, D., Stern, I.: Security proof for signature scheme. In: Eurocrypt'96. Lecture Notes computer Science vol. 1070. pp. 387–398 (1996)
14. Paillier, P.: Public key cryptosystems based on composite degree residuosity classes. In: Advances in Cryptology EUROCRYPT99. LNCS 1592, Springer, Berlin, pp. 223–238 (2000)
15. Zhang, J., Chen, H., Geng, Q.: A secure Chameleon hash function without key exposure from pairings. In: Proceedings of the 2009 International Symposium on Web Information Systems and Applications (WISA'09) Nanchang, P. R. China, May 22–s24, pp. 015–018 (2009)
16. Jakobsson, M., Sako, K., Impagliazzo, R.: Designated verifier proofs and their applications. Advances in Cryptol

Identification of the Choice of Drugs in Epilepsy by the Method of Classification and Regression Tree

Vivek Kshirsagar, Anil Karwankar, Meghana Nagori and Kailas Elekar

Abstract Data Mining helps its users deduce important information from huge databases. In medical stream, practitioners make use of huge patient data. Any effective medical treatmentis achieved after complete survey of ample amount of patient data. But practitioners usually faced with the obstacle of deducing pertinent information and finding certain trend or pattern that may further help them in the analysis or treatment of any disease. Data Mining is such a tool which sifts through that voluminous data and presents the data of essential nature. In this paper, we have designed a five-step data mining model that will help medical practitioners on determining the appropriate drug to be used in ministration for epilepsy. Most of the epileptic seizures are managed through drug remedy, particularly anti-convulsant drugs. The choice is most often related to other aspects particular to every patient. The trick to building a successful predictive model is to include parts of data in your database that describes what has happened in the past. There are a wide range of older as well as recent anticonvulsants present in market. Our paper will take into consideration both the older and the recent anticonvulsants and other factors to justify the use of a drug suitable for treatment in epilepsy. To determine the drug choice for treatment in different epilepsy, we have selected the classification method. Decision trees are a sort of data mining technology that has been around for almost 20 years now. They are now increasingly being used for prediction.

Keywords Data mining · Decision tree classification · Drug choice · Epilepsy

V. Kshirsagar (✉) · A. Karwankar · M. Nagori
Government Engineering College, Aurangabad, India
e-mail: vkshirsagar@gmail.com

A. Karwankar
e-mail: akarwankar@gmail.com

M. Nagori
e-mail: kshirsagarmeghana@gmail.com

K. Elekar
National Informatics Center, Pune, India
e-mail: ekailas@gmail.com

S.C. Satapathy and S. Das (eds.), *Proceedings of First International Conference on Information and Communication Technology for Intelligent Systems: Volume 2*, Smart Innovation, Systems and Technologies 51, DOI 10.1007/978-3-319-30927-9_33

1 Introduction

Data mining tools like classification are designed to learn from the past successes and failures and then predict the outcome. The trick to building a successful predictive model is to have some data in your database that describes what has happened in the past. Data mining tools like classification are designed to learn from the past successes and failures and then predict the outcome. The medical data is usually a voluminous data and in order to effectively treat a patient of disease it is often important to consider the patient's entire medical history. Here the data under consideration is that of epilepsy and the factors on which the choice of drug depends. Epilepsy is a chronic disorder distinguished by repeated unprovoked convulsion. About 40 varieties of seizure are identified. Epilepsy is a physical state arisen by unanticipated, quick transforms in how the brain works. Epilepsy is a genetic disorder that affects the sections of the brain alike to a computer. These sections communicate electronically. When this can't take place a seizure occurs. Most of the epileptic seizures are monitored through drug treatment, particularly anticonvulsant drugs. Root research has yielded some of the often prescribed anticonvulsant drugs to the market [1–5]. Our model will determine the most effective anticonvulsant. Epilepsy varies with its sources, indications and therapy. Usually, for a stated type of epilepsy there are slight differences among appropriate drugs. The choice is most often based on other aspects particularly for each patient. It is often misunderstood that low-aged suffer epilepsy. But it is equally prevalent in 65 above age group as in minors below 13. Idiopathic epilepsy is that group of repetitive seizures that cannot be characterized. Symptomatic epilepsy can be characterized by physical cause, like brain cancer, head injury or cerebral hypoxia. The symptoms which classify the Idiopathic mean genetic causes, true epilepsy is generally obtained from biological parents, or are mostly derived from heredity. In our paper we have considered the factors such as age, inherited disorder and category of epilepsy. These factors are critical to choosing the best treatment. Usually, medications can control captures in about 70 % of patients. In general, low portions of suitable anticonvulsants are required to successfully treat any type of epilepsy. The drugs used for treatment should be selected carefully as some drugs are known to increase seizure frequency. In most cases an appropriate choice showed an 80 % effect in making the patient seizure free. Through this model and supported by the training set data we will show that the data mining process can be effective for not only predicting the drug choice in epilepsy, but the model can be implemented in various other medical fields to determine the choice of therapy in cancer patient or choice of antibiotics in individual infections. The algorithm that is implemented in our paper is CART. The techniques applied for building the predictive model in this paper are attributes pertinence, rule generation and CART classification. The general idea behind decision tree technique is that they are built from historical data. The first step in the process is that of growing the tree. The algorithm seeks to create a tree that works on all the data that is available. The main aim in the process of growing the tree is to frame appropriate questions at every branch point of the tree. The goal is to have the leaves of the tree as homogenous as possible with respect to the prediction value.

1.1 Preparation of Data Marts

A data mart consist a part of the medical-records that may prove beneficial to some researchers. Only specifically intended objects are taken into consideration. The process of preparing a data mart is bifurcated into three stages. The first stage is to extract and inspect the properties of the epileptic patient's data and respective treatment. The second stage is to gather rows corresponding to the properties of the epileptic patient data and respective treatment. In the third stage, all the rows within each property element are abstracted.

1.2 Pre-Processing of Data

Data pre-processing methods help in acquiring qualitative data for subsequent quality data mining. This is an important part of the pattern extraction, as effective decisions must be based on standard data. Detecting abnormality in data, rectifying them, and aggregating data to be examined can lead to efficient and effective decisions. Data cleaning refers to removal of redundant, unnecessary and recurring data from those beneficial for analysis.

1.3 Measures for Selecting the Best Split

In our paper we have used the CART algorithm. The impurity measure used in selecting the variable in CART is Gini Index. For categorical target variables, we can choose GINI. The Gini index lies between 0 and 1, where perfect similarity goes on decreasing from 0 to 1. The Gini index at node t, GINI (t) is defined as:

$$GINI(t) = \sum j \neq ip\left(\frac{j}{t}\right)p\left(\frac{i}{t}\right) \tag{1}$$

Where i and j are categories of the target variable. This can also be written as:

$$GINI(t) = 1 - \sum jp^2(j/t) \tag{2}$$

Where p (j/t) refers to proportion of target category j present in node t. The minimum value is zero and it occurs when all the data at a node belongs to one target category.

1.4 CART Classification Rules

The diagrammatic representation of decisions can be achieved. Each node indicates an attribute. Each branch deals with possible values of that attribute. Every leaf node represents the decision. The main objective of Decision tree is to find optimal solution. At first complete training set is at root. Then the best attribute is detected using some measures as explained in the next section. Similarly while traversing from root to leaf nodes are split into branches. The stages of CART Algorithm are as follows:

(1) Beginning with t = 1 root node, determine a split from all possible data that improves the purity. The split 1st node into 2nd and 3rd nodes, using split.
(2) Repeat the split searching process for 2nd, 3rd, and henceforth. Continue with the same until a rule is derived.

The diagram of obtained tree can then be converted into conditional rules. Each path from root to leaf node defines a rule. The if part consists of the attribute and related values in the path and the Then part includes the decision represented by leaf node.

1.5 Application of Model for Determining Choice of Drugs

In this part, the complete data is divided into two parts namely training data and test data. Both of the data sets are selected in random to avoid over fitting of data. The decision rules are produced using training data. Once the decision rules are derived, it is important to determine the efficiency of the model before its application. The ratio of samples from the testing set that are correctly classified by the model to complete test set determines the accuracy rate. Greater accuracy rate symbolizes acceptable model. Precisely, the properties relevant to a disease in training set will be input and the prediction of drug will be the output.

2 Description of the Application

This section uses the epileptic data as an specimen to show how the prescribed model can be applied for deciding the drug in the successful treatment of epilepsy. The epileptic data is described as follows. There are different categories under which a particular type of epilepsy falls. In our sample illustration we have three different categories of epilepsy which are denoted as {PS: Partial Seizures},{AS: Absence seizures}, and {GS: Generalized seizures}.Generally epilepsy is predominant in children than adults, but some epileptic types are found to be present in adults too. So accordingly we have created five age groups of different range types: {A1: 0-9}, {A2: 10-19}, {A3: 20-29}, {A4: 30-39}, {A5 >40}. The cause of the epilepsy can be mainly due to two different reasons which are idiopathic meaning Genetic/Hereditary disorder, or due to an injury to brain which is symptomatic. The cause attribute is thus represented by the values {GE: Genetic disorder/Hereditary}, {SM: Symptomatic}. For the successful treatment of epilepsy there are many antiepileptic or anticonvulsants available in market.

2.1 *Building The Epileptic Data Mart*

The available information on epilepsy can be characterized by the following eight properties: Epilelpsy name, Category, Symptoms, Inherited disorder, Age, Anticonvulsant 1, Anticonvulsant 2, Drug choice. Sample data mart is shown in Table 1. The epilepsy name describes the type of epilepsy. Furthermore, whether the disease is hereditary or otherwise is indicated by the inherited disorder attribute. The anticonvulsants used against each disease type are mentioned by the attribute anticonvulsant 1 and anticonvulsant 2. The drug choice is our output variable which will describe the most commonly used anticonvulsant among the two for different epilepsies.

2.2 *Data Pre-Processing*

Of the available information on epilepsy as shown in Table 1 the relevant information needed for our analysis is extracted. The original data is cleaned by

Table 1 The available information on epilepsy

Epilepsy name	Juvenile Myoclonic Seizure	Gelastic seizures
Category Symptoms	Generalized, sporadic, jerking movements	Focal Forced eye Movements Chewing or grinding of teeth
Cause	Genetic	Genetic (migraines)
Age	6 yrs–36 yrs	–
Anticonvulsant 1	Keppra	Dextroamphetamine
Anticonvulsant 2	Carbamazepine	Phenobarbital
Drug choice		

removing the attribute Symptoms. The succeeding task of pre-processing is to Bucket into groups the other attributes. The data that is randomly selected for illustration purposes in this paper is as shown in Table 2.

2.3 Property Selection Analysis

For attribute relatedness inspection the data in Table 2 is used. The Drug choice is the output variable. The data in Table 3 gives the details of the calculation of the variables selected for splitting at the root node. The calculations are as per the formula (2) mentioned *above*. The minimum gain is obtained for the split based on category. So, at the root node, we split the data set based on the variable category.

Table 2 The training data set

Sr. No.	Epilepsy Name	Category	Age	Cause	Drug choice
1	Simple	PS	A2	GE	Phenytoin
2	Complex	PS	A5	GE	Phenytoin
3	Clonic	GS	A2	GE	Valproate
4	Tonic	GS	A2	GE	Valproate
5	Atonic	GS	A3	GE	Valproate
6	Neonatal	GS	A1	GE	Phenytoin
7	Grand Mal	GS		SM	Phenytoin
8	Status Epilepticus	GS	A1	SM	Phenytoin
9	Lennox Gastauat	GS	A1	SM	Valproate
10	Rolandic Epilepsy	GS	A1	GE	Phenytoin
11	Juvenile Myoclonic Epilepsy	AS	A2	GE	Valproate
12	Petit-mal Absense Epilepsy	AS	A1	GE	Valproate

Table 3 The gain in splitting the data at root node for various variables

Drug Choice	Category			Age				Cause	
	GS	AS	PS	A1	A2	A3	A5	GE	SM
Valproate	4	0	0	2	3	1	0	5	1
Phenytoin	4	2	2	4	1	0	1	4	2
Gini(i)	0.5	0	0	0.4457	0.375	0	0	0.495	0.446
Gain	0.25			0.4103				0.4705	

2.4 Decision Tree Classification Rule

The resulting tree from which to derive classification rules in shown in Fig. 1. On the tree, N refers to number of data points in that node and V refers to the variable based on which data on the node is split to its child nodes. The rules so formed after the classification process are as follows. These rules will guide us in selecting drugs in the following step.

(1) If Category = "PS" then drug choice = "Phenytoin".
(2) If Category = "GS" and age = "A3" then drug choice = "Phenytoin"
(3) If category = "GS" and age = "A2" and cause = "GE" then drug choice = "Valproate"
(4) If Category = "GS" and age = "A1" and cause = "SM" then drug choice = "Phenytoin"
(5) If Category = "AS" and age = "A4" then drug choice = "Valproate"

2.5 Applying the Model for Determining the Choice of Drugs

As observed from the training data set and rules extracted from classification tree it is observed that valproate may be the most potential antiepileptic drug in the ministration of the Epilepsy. This anticonvulsant is potential for wide scope of seizure types. The next best choice is that of phenytoin which is one of oldest drugs known for treatment of seizures. These anticonvulsants are used both in adults as well as in children, only the difference is in the treatment dose for adults and for children. Using the *above* mentioned calculations, we can generate the CART trees

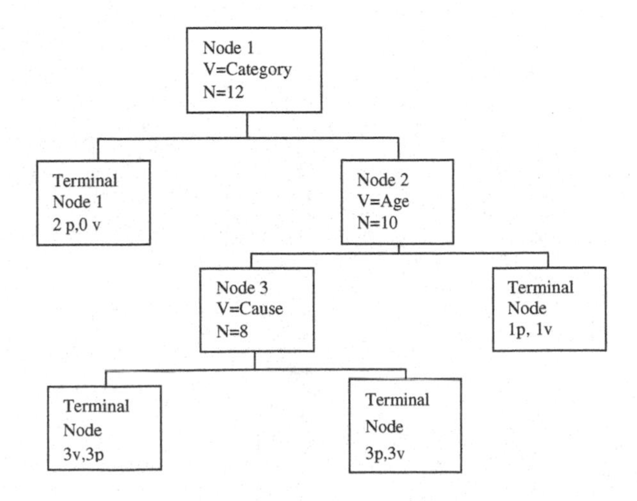

Fig. 1 The epilepsy drug choice CART tree

and relevant rules for the remaining properties, like "contraindications" and "prior illnesses". Using the similar mechanisms it is thus possible to identify different drug treatments that can be used in the successful treatment of Epilepsy.

3 Concluding Remarks

In this article, we represent a way to determine the choice of drugs that can be expected to be used for effective treatment in epilepsy using some data mining techniques. The techniques are subdivided into five tasks. The Epileptic drug prediction symbolizes that this data mining technique overcomes the drawbacks of conventional prediction techniques and help medical practitioners a deep and accurate peek into the variety of drugs and the choice among them.

References

1. Hand, D.J., Mannila, H., Smyth, P.: Principles of data mining, Prentice-Hall India (2004)
2. Marakas, G.M.: Modern data warehousing, mining, and visualization, Pearson Education (2003)
3. Han, J., Kamber, M.: Data mining:concepts and techniques, Morgan Kaufmann publishers Inc (2001)
4. Snyman, J.R.:Monthly index of medical Specialities, 27(4): (2007)
5. Shorvon, S.D., Perucca, V., Fish, V., Dodson, W.E.: Thetreatement of epilepsy, Blackwell publishing (2006)

Part III
Intelligent Managerial Applications for ICT

Encryption of Motion Vector Based Compressed Video Data

Manoj K. Mishra, Susanta Mukhopadhyay and G.P. Biswas

Abstract Enormous size of video data for natural scene and objects is a burden, threat for practical applications and thus there is a strong requirement of compression and encryption of video data. The proposed encryption technique considers motion vector components of the compressed video data and conceals them for their protection. Since the motion vectors exhibit redundancies, further reduction of these redundancies are removed through run-length coding prior to the application of encryption operation. For this, the motion vectors are represented in terms of ordered pair (val, run) corresponding to the motion components along the row and column dimensions, where 'val' represents value of the motion vector while 'run' represents the length of repetition of 'val'. However, an adjustment for having maximal run is made by merging the smaller run value. Eventually we encrypted the 'val' components using knapsack algorithm before sending them to the receiver. The method has been formulated, implemented and executed on real video data. The proposed method has also been evaluated on the basis of some performance measures namely PSNR, MSE, SSIM and the results are found to be satisfactory.

Keywords Motion vector · Motion estimation · Motion compensation · Accordion matrix · Run-length coding · CR · MSE · PSNR

M.K. Mishra (✉) · S. Mukhopadhyay · G.P. Biswas
Department of Computer Science, Christ University, Bangalore 560029, India
e-mail: manojkumar.mishra@christuniversity.in

S. Mukhopadhyay
e-mail: msushanta2001@gmail.com

G.P. Biswas
e-mail: gpbiswas@gmail.com

M.K. Mishra · S. Mukhopadhyay · G.P. Biswas
Department of CSE, Indian School of Mines, Dhanbad, Jharkhand 826004, India

S.C. Satapathy and S. Das (eds.), *Proceedings of First International Conference on Information and Communication Technology for Intelligent Systems: Volume 2*, Smart Innovation, Systems and Technologies 51, DOI 10.1007/978-3-319-30927-9_34

1 Introduction

Video data has become an indispensable and integrated part of modern communication system as it can convey any message or information with wider, deeper and quicker impact. Almost all the popular fields like news, infotainment, entertainment, education, sports, e-classroom, medical instrumentation, video conferencing etc. involve significant amount of video data processing and handling. For high quality video data the volume of the data is enormous and is a big burden for the channel transmitting the video signal for on-line/offline cast. In most cases the data is to be made secure from the group of unauthorized users, in some cases the video data could be intended to be confidential. All these requirements are full filled by applying compression and encryption techniques on the video data at the transmitting end and decompression and decryption at the receiving end.

Security of the video data is always a big challenge for both aspects of efficiency and soundness. To overcome such problem different video encryption algorithms [1–4] are proposed for different purposes utilizing different properties of the multimedia data. The general assumption is that, image pixel value are more important than motion vectors (MV) which is nothing but the difference between the two video frame, some algorithms [5–7] are proposed to perform encryption on pixel data only and motion vector data are still kept unencrypted. But motion vector data are equally important than that of pixel data because motion vectors are sufficient enough to reveal the movement of object and the shape of the object in video communication, so it is always important to encrypted it with the pixel data for secure visual communication.

In this work we have devised a method for compression and encryption of the motion vector of the video data using runlength encoding and knapsack encryption algorithm subsequently. The rest of this paper is organized as follows. Next to this introductory section we present a preliminaries in which we have done brief review of the motion vector, runlength encoding along with accordion representation and knapsack encryption in Sect. 2. Section 3 presents the theoretical formulation of the proposed work. The experimental results, performance analysis along with security are furnished in Sect. 4. Finally, concluding remarks are presented in Sect. 5.

2 Preliminaries

2.1 Motion Vectors

The motion vector [5, 7] is basically an ordered pairs integrating the pixels movement in x and y dimension of the video frames in course of the time. Since video frame of the natural scene are exuberant in spatial redundancy. The motion vector of same blocks in different frames are likely to present the same object in video sequence because it are spatially related with respect to both intraframe and inter-frames. The component of the motion vectors along the row and column manifests such redundancies which can be exploited to achieved compression. To enhance the degree of compression the temporal redundancy of the motion vectors is combined with their spatial redundancy by representing the components of motion vector in the form of the separate accordion matrices constructed from group of frames of thc input video data.

2.2 Accordion Representation

Video data essentially consist of temporal sequence of spatial frames [8, 9]. The three dimensional video data (two spatial and one temporal dimension) can be given an alternative two dimensional representation, often termed as accordion. The set of motion vector frames which also consist spatial and temporal redundancies are extracted from standard coder like MPEG, are the input to the encoder. A single pixel model is denoted as P(x, y, t) where x, y are pixel spatial coordinates; t is video instance at time and p is pixel value. So, by using this method, we can put pixels together so that we can achieve temporal redundancies in spatial frame.

$$P(x,y,t) - P(x,y,t+1) < P(x,y,t) - P(x+1,y,t) \quad (1)$$

So, it is formed by collecting the motion vector cube pixels where the columns of the accordion matrix come from the column of the frames in periodic fashion.

2.3 Knapsack Encryption

The Merkle-Hellman knapsack crypto-system [10, 11], which was invented in 1978, is based on the super-increasing subset problem sum. Suppose there is two k-tuples, $a = [a_1, a_2, \ldots, a_k]$ and $x = [x_1, x_2, \ldots, x_k]$ where the first tuple is the pre-defined set; the second tuple, in which x_i is only 0 or 1, defines which elements of a are to be dropped in the knapsack. The sum of element in the knapsack is,

$$S = knapsackSum(a, x) = a_1x_1 + a_2x_2 + \cdots + a_kx_k \quad (2)$$

S forms the encrypted message/data and the original super-increasing vector is used as the private key which is used to decipher the message/data. The knapsack encryption contains two steps: one is key generation and other one is encryption of the data [10].

3 Motion Vector Knapsack Based Encryption (KNAPMVE)

We have devised a novel algorithm (KNAPMVE) for compression and encryption of motion vector extracted from the MPEG video. The proposed two pass algorithm first construct the accordion of the motion vectors extracted from the successive video frames and employs runlength coding on them in a prescribe manner as shown in Fig. 1. The result of run-length encoding is further adjusted by neglecting minor entries and the adjusted output is subjected to another pass of run-length coding. In the last step the encoded result is encrypted using knapsack encryption method. Let $F(x, y, t)$ be the given video data and $(F_x^{mv}(x,y,t), F_x^{mv}(x,y,t))$ are components of the motion vector along the row and column dimension at the spatial location (x, y) and temporal instance t. Let $ACC - F_x^{mv}$ and $ACC - F_x^{mv}$ be the two sets of the accordion matrices constructed with the row (x) and column (y) components of the motion vectors in a group of k successive frames as depicted in Fig. 1.

As a result the motion vectors of the entire video data is converted into sets of array of accordion matrices for the row and column dimension. The size of each accordion matrix is $m_x \times (m_y k)$ and the number of such accordion matrices for row (column) dimension is $m_x \times (m_y k) \cdot (\frac{P}{k})$. Now each accordion matrix representing the row component of the motion vector is subjected to a scan as illustrated below with the directional straight line. At the end of this scan the two dimensional accordion matrix of dimension $m_x.m_y.k$ is converted to a one dimensional array of length $m_x.m_y.k$. This one dimensional array which have values in the range limited by $-b_x$ to $-b_y$ i.e. $(2b_x + 1)$ and same values are repeated within this array. We employ run-length coding on this one dimensional array to represent this array in terms of sequence of values and the corresponding run-length (val(D), run(C)). If the run-length of an entry is below a prescribed threshold, then we absorb the value and its run-length with its preceding entry and this process is repeated as long as all

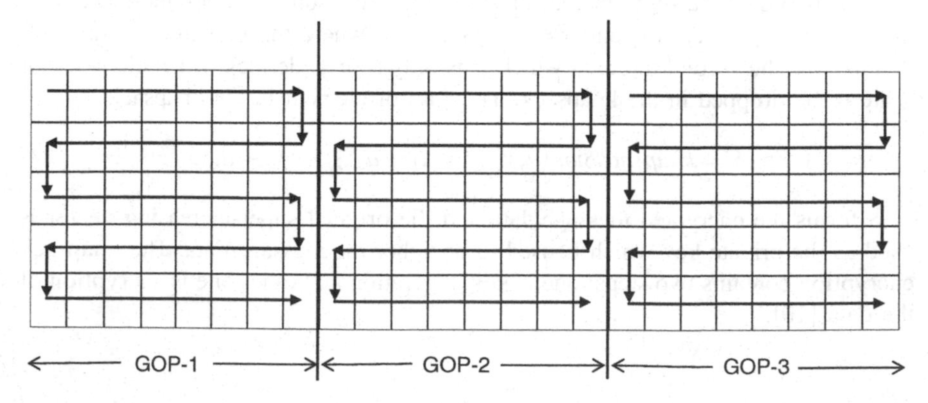

Fig. 1 Effective scanning order in case of accordion of motion vector


```
Algorithm: Knapsack based Motion Vector Encryption (KNAPMVE)
Input: 1) D array (values) and 2) C array (run).
Output: Compressed and modified, encrypted D Array (ND) and C Array.
----------------------------------------------------------------------
Step 1: i =1;
    while (C(i) ≠ NULL)
        If (C(i) = = 1)
            If ((diff( D(i), D(i-1)) > diff(D(i), D(i+1)))then
                C(i+1) = C(i+1) + 1;
                Delete (D(i));
                Delete( C(i));
            else
                C(i-1) = C(i-1) + 1;
                Delete (D(i));
                Delete (C(i));
            end if
        end if

        If (D(i) == D(i+1)) then
            C(i) = C(i) + C(i + 1)
            Delete (C(i+1))
            Delete (D(i+1))
        end if
            i = i+1;
    end while
Step 2: Apply ND = Knap_Encryp(D).
Step 3: Stop.
```

Fig. 2 Proposed algorithm (KNAPMVE)

the run-lengths do not go above threshold. The modified 'val' component is then subjected to knapsack encryption (as illustrated in Sect. 2.3). The encrypted data with modified 'val' (ND) component and unaltered 'run' component is then transmitted over the channel. At the receiving end the 'val' components are first decrypted using knapsack algorithm. After receiving the 'val' and 'run' finally we reconstruct the row and column components of the motion vectors in the form of accordion matrices. The final video is reconstructed by compensating the reconstructed motion vectors with the I-frame. The proposed algorithm is shown in Fig. 2.

4 Algorithm Assessment

4.1 Security

Our method consists of simple run-length encoding in hierarchal fashion followed by knapsack encryption operations using permutation based key which scramble mapping process. In our method the inter/intra redundancies of the motion vector data are highly concealed. The security of our proposed method depends on the size of the permutation key as well as on the private key used in knapsack algorithm. By choosing a n-length permutation key, which will highly randomized the D-array. So without knowing the correct secret key it will be impossible to decrypt the data correctly which as such guaranteed the system security. In our method, although attackers can get the result but they still cannot attack it because they do not know the correct sequence order and the corresponding private key used in algorithm. Here, only brute-force attack could be a only method to decrypt this but it would be practically impossible.

4.2 Performance Analysis

In the video data, the frames are highly related; It has very close relationship in terms of pixel value dependencies on each other, and these dependencies carry useful information about the outline of the objects in the video data. The experiment results in Table 1 has proved our assumption. The visual quality of the reconstructed frames gives a subjective evaluation of the proposed algorithm and for objective evaluation we furnish a few quantitative measures [12] namely: *Mean Square Error* (*MSE*), *Peak Signal to Noise Ratio* (*PSNR*) and *Structural Similarity Index Measure* (*SSIM*).

4.3 Simulation Results

Figure 3 shows one of the tested video frames, which we have used for our proposed KNAPMVE algorithm. It is a clip from a Table Tennis sequence, which contains some objects with fast movement. We have used the motion vectors which are estimated by the MPEG compression method for the validity of our proposed method. A motion vector is use to measure the difference between the two block of two frames and when it playback, P frame would be built upon I frame using the motion vectors.

Table 1 Performance of proposed algorithm on the test sequence

Video sequence	Frame number	MSE	PSNR	SSIM
Miss America	17	7.0390	39.7857	0.9981
	18	7.1527	39.6541	0.9976
	19	7.3412	39.5218	0.9984
	20	7.3512	39.3562	0.9974
	21	7.3152	39.4038	0.9981
Table Tennis	41	9.8520	36.4630	0.9631
	42	9.8986	36.2487	0.9613
	43	9.8383	36.5214	0.9617
	44	9.8783	36.2758	0.9608
	45	9.7753	36.5386	0.9658

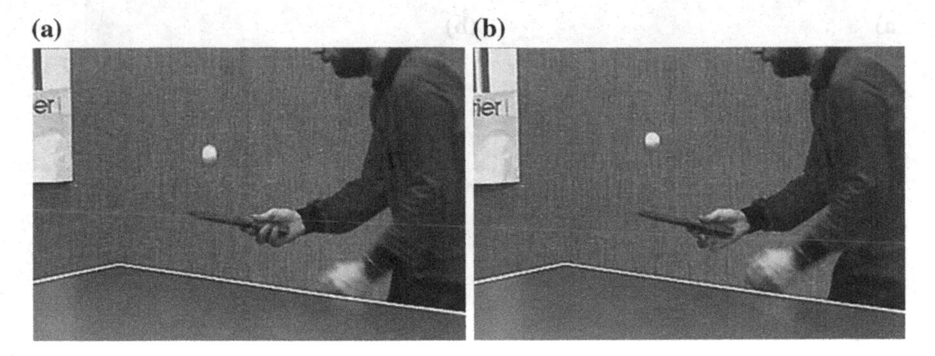

Fig. 3 *Source* **a** I-frame, **b** P-frame

In this work, basically we are interested only for the compression and encryption of the motion vectors data, which consume a good amount bandwidth in the communication of MPEG stream. A real frame and test results are shown in Figs. 3 and 4 corresponding to Table Tennis sequence. In Fig. 4, picture on left is the real frame (I-frame), picture on right is a reconstructed frame (P-frame) by the proposed algorithm. After applying KNAPMVE encryption method, motion vectors in the matrix became a highly random, as shown in Fig. 5b corresponding to original motion vector. So by this result, we can conclude that it is next to impossible to predict the correct P-frame without knowing the correct motion vectors. The histogram of distribution of the motion vector matrix is shown in Fig. 6 which shows that encrypted motion vectors has an random distribution. Hence it will be very difficult to construct the video frame exactly without knowing the correct values of the motion vector.

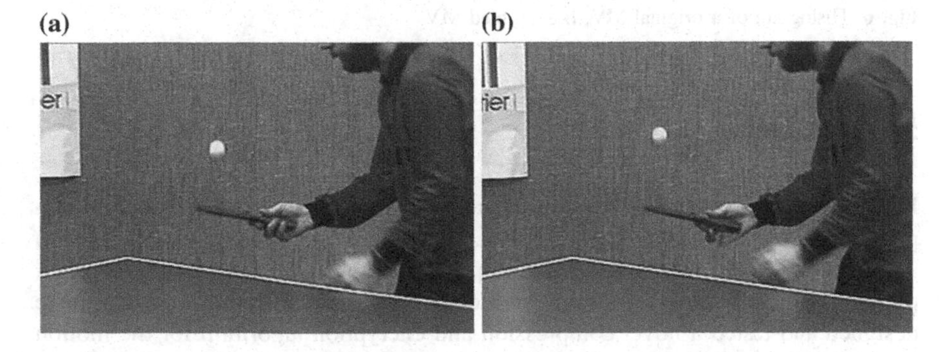

Fig. 4 *Source* **a** I-frame, **b** Decrypted P-frame from the proposed method

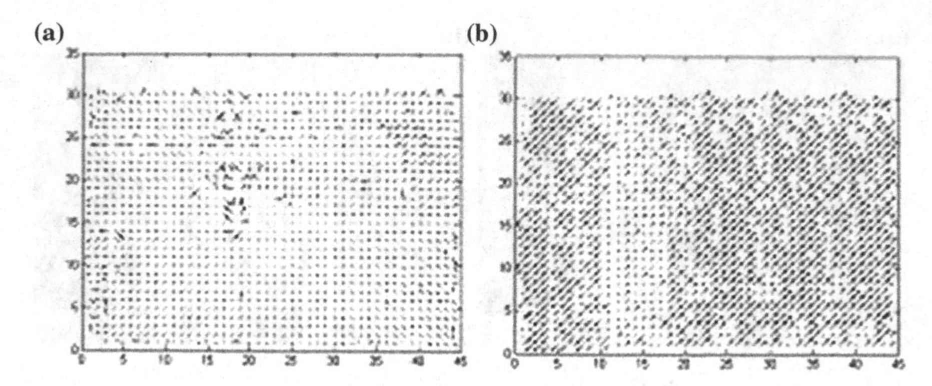

Fig. 5 Resultant motion vector **a** Original motion vector **b** Encrypted motion vector

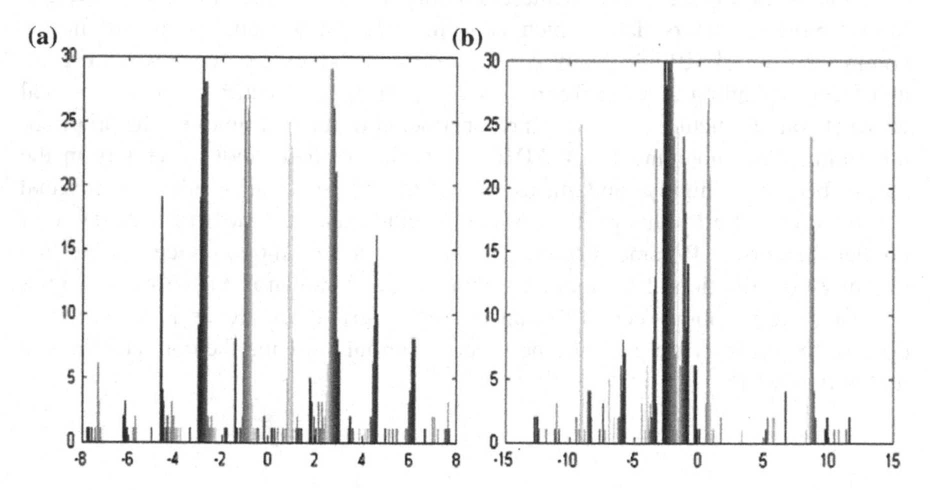

Fig. 6 Histogram of **a** original MV, **b** encrypted MV

5 Conclusion

In video sequence communication, the motion vector data should also be compressed and secured as the pixel data. But, almost all of the video data compression and encryption algorithms ignored this part because of the inefficiency of compression and encryption algorithms on motion vectors data. In this work, we have designed and tested a novel compression and encryption algorithm for the motion vector data to provide a security for the network communication. Our experiments and evaluations show that KNAPMVE can satisfy performance and security requirements for video communication and/or storage.

References

1. Mishra, M.K., Mukhopadhyay, S., Biswas, G.P.: Architecture and secure implementation for video conferencing technique. In: Third International Conference on Computer, Communication, Control and Information Technology (C3IT), pp. 1–6. IEEE (2015)
2. Hong, G.M., Yuan, C., Wang, Y., Zhong, Y.Z.: A quality-controllable encryption for H. 264/AVC video coding. In: Advances in Multimedia Information Processing-PCM, pp. 510–517. Springer, Berlin (2006)
3. Qian, Z., Jin-mu, W., Hai-xia, Z.: Efficiency video encryption scheme based on H. 264 coding standard and permutation code algorithm. In: 2009 WRI World Congress on Computer Science and Information Engineering, vol. 1, pp. 674–678. IEEE, Mar 2009
4. Shahid, Z., Chaumont, M., Puech, W.: Fast protection of H. 264/AVC by selective encryption of CAVLC and CABAC for I and P frames. IEEE Trans. Circuits Syst. Video Technol. **21**(5), 565–576 (2011)
5. Agi, I., Gong, L.: An empirical study of secure MPEG video transmissions. In: Proceedings of the Symposium on Network and Distributed System Security, pp. 137–144. IEEE, Feb 1996
6. Qiao, L., Nahrstedt, K.: Comparison of MPEG encryption algorithms. Comput. Graph. **22**(4), 437–448 (1998)
7. Shi, C., Wang, S.Y., Bhargava, B.: MPEG video encryption in real-time using secret key cryptography. In: Proceedings of International Conference on Parallel and Distributed Processing Techniques and Applications (1999)
8. Mishra, M.K., Mukhopadhyay, S.: Scheme for compressing video data employing wavelets and 2D-PCA. In: Information Systems Design and Intelligent Applications, pp. 409–417. Springer, India (2015)
9. Ouni, T., Ayedi, W., Abid, M.: New low complexity DCT based video compression method. In: International Conference on Telecommunications, ICT'09, pp. 202–207. IEEE May 2009
10. Merkle, R.C., Hellman, M.E.: Hiding information and signatures in trapdoor knapsacks. IEEE Trans. Inf. Theory **24**(5), 525–530 (1978)
11. Menezes, A.J., Van Oorschot, P.C., Vanstone, S.A.: Handbook of Applied Cryptography. CRC Press, Boca Raton (1996)
12. Wang, Z., Bovik, A.C., Sheikh, H.R., Simoncelli, E.P.: Image quality assessment: from error visibility to structural similarity. IEEE Trans. Image Process. **13**(4), 600–612 (2004)

A Hybrid Technique for Hiding Sensitive Association Rules and Maintaining Database Quality

Nikunj H. Domadiya and Udai Pratap Rao

Abstract In this digital world, data mining is a decisive for innovation and better services for user, but it raises the issues about individual privacy. Privacy can be achieved by hiding sensitive or private information in database before publishing it for innovation. This paper presents a hybrid technique for hiding sensitive association rules, which combines two heuristic based techniques viz. Decrease Support and Decrease Confidence of sensitive rule for selection and modification of items from the transactions. The proposed hybrid technique combines advantages of both algorithms to maintain the quality of the database and preserve the privacy of database. From the experimental results it is observed that proposed algorithm is competent to maintain privacy and database quality.

Keywords Data mining · Association rule · Sensitive pattern · Privacy preserving data mining and association rule hiding

1 Introduction

Association rule mining is very useful technique to find the hidden relationship among data in large database. It can be used for many applications as healthcare, business policy and marketing, e-commerce etc. As a part of business improvement or decision, many organizations share their database for mutual benefits. In many case, due to privacy law or some policy, organization don't want to disclose the private information stored in database. It raises the problem of securing this private data from the adversary to maintain the privacy. The privacy is major constrain during the mining of database which contains private information. Here privacy is

N.H. Domadiya (✉) · U.P. Rao
Sardar Vallabhbhai National Institute of Technology, Surat 395007, Gujarat, India
e-mail: domadiyanikunj002@gmail.com

U.P. Rao
e-mail: udaiprataprao@gmail.com

S.C. Satapathy and S. Das (eds.), *Proceedings of First International Conference on Information and Communication Technology for Intelligent Systems: Volume 2*, Smart Innovation, Systems and Technologies 51, DOI 10.1007/978-3-319-30927-9_35

to hide private information which organization don't want to disclose with other, to solve this issue PPDM technique is very useful to enhance the security of database. In a centralize database, PPDM techniques are used to transform the database such that private information or sensitive pattern cannot be mined from the data mining techniques and maintain the quality of final result. In 1996, Clifton et al. [1, 2] had discussed the privacy in data mining. They also discussed the hiding of association rule to maintain the privacy. In 1999 Atallah et al. [3] proposed heuristic approach to prevent disclosure of sensitive patterns.

1.1 Problem Description

The problem of association rule hiding can be describe as: Transform the original database such that data mining techniques will results only non sensitive rules and all sensitive rules must not mined from transformed database. This transformed database is known as sanitized database.

In general the problem can be defined as:

Given transnational database D, Minimum support threshold, Minimum confidence threshold. Association rules R can be generated as result of data mining technique from D. Sensitive rules (SR, SR $\subset$ R) selected from given set of rules R by database owner. The problem is to transform database D into D′ in such way that, data mining results from D′ will hide all sensitive rules.

The objective of proposed algorithm is to achieve the following conditions

1. Transformed database D′ must hides all sensitive rules.
2. Mining of transformed database D′ must results in all non sensitive rules.
3. Transformed database must not introduce any artificial rules, which are not present in D.

The problem of finding an optimized sanitized database, which satisfies all these conditions has been proved as NP-hard in [3].

The structure of remaining paper is as follows: Sect. 2 presents theoretical background and related work. In Sect. 3, proposed hybrid technique for hiding sensitive association rule is discussed. In Sect. 4, we analyze the algorithm with some evaluation parameter of database quality. Finally, last Sect. 5 concludes our work and gives the future direction.

2 Theoretical Background and Related Work

2.1 Association Rule Mining

Association rule mining [4] with given minimum support threshold (MST) and minimum confidence threshold (MCT) is defined as follows: let I = $\{i_1, \ldots i_N\}$ be distinct literals called items. Given a database D = $\{T_1 \ldots T_m\}$ is a set of transaction where each transaction T is a set of items as $T_i \subset I$ (1 = i = m).

A rule X → Y is mined from database if *support(X → Y)* ≥ MST and *confidence (X → Y)* ≥ MCT.

Researchers have proposed different approaches for hiding the sensitive association rule to preserve the privacy of sensitive information and these approaches can be classified in Heuristic based approach, Reconstruction based approach and Cryptography based approach. Here we propose a hybrid algorithm for sensitive rule hiding, which is based on heuristic approach.

2.2 Heuristic Based Approach for Sensitive Rule Hiding

The heuristic based approach is used to hide association rules as many as possible by modifying the transactions, while minimizing side effects which can be generated by hiding process. Following side effects can be generated by hiding process.

1. Some sensitive rules which can be mined from sanitized database, called *hiding failure* (HF) effect.
2. Some non-sensitive rules are hidden accidently in sanitized database, called lost rule or missing cost (MC) effect.
3. Some rules are newly created, called ghost rule or artificial rule (AP) effect.

In this type of approach, no algorithm can satisfy all condition of association rule hiding. It produces some side effect as described above. In [3], author proved that finding an optimal solution of association rule hiding algorithm is NP-hard. Algorithms included in this approach, which hide sensitive knowledge by sanitizing selected transactions from the database to decrease the support or confidence of sensitive rule and tries to minimize the side effect. These approaches use either *Data Perturbation* which permanently remove some items from selected transaction of database or *Data Blocking* which replace some items by unknown value (ex. ?).

Next, we see some existing efforts to solve association rule hiding based on data perturbation approach.

Oliveira et al. [5] proposed a graph base sanitization approach to hide sensitive rules. The author also presents the forward inference and backward inference attack on sanitize database. They maintain 2-itemset paring set which is all possible subsets of sensitive frequent itemset. Then algorithm deletes one by one 2-itemset to hide sensitive frequent items set. This algorithm soles both above attack and also minimizes the side effect factors.

Wu and Wang [6] proposed algorithm that compare all three techniques for hiding sensitive rule in terms of number of transaction modification. The algorithm selects the one with the lowest number of modifications of transaction in database to maintain database quality.

Oliveira et al. [7] proposed different flexible algorithm which considers the disclosure threshold for each sensitive rules. Secondly they focus on discovery of maximum non-sensitive rules after database sanitization. This algorithm is not on memory based so it is more suitable for large database size.

Domadiya and Rao [8] modifies the algorithm of Modi et al. [9] to reduce the side effect on sanitized database. They select the items from the R.H.S parts of the sensitive rules based on their frequency. This algorithm modify the original database in such way that total number of modification are reduced and hide all the sensitive rules to maintain the privacy.

In context of heuristic based approach, there is another area called *data blocking approach*, where researchers have worked to address NP-hardness of PPDM.

All the above techniques comes under wither D_SUPP or D_CONF1. we can calculate total number of modification require in database to hide any sensitive rule using following equation.

$$N_{supp} = suppcount(Rule) - Minsuppcount + 1$$
$$N_{conf} = suppcount(Rule) - \lceil Min\_Conf * Supp\_count(L.H.S.\ of\ Rule) \rceil + 1$$

From the analysis, we had observed that, Some sensitive association rules had $N_{supp} < N_{conf}$ and other had $N_{supp} > N_{conf}$. We can find the better technique for hiding any sensitive rule based on the relation of N_{supp} *and* N_{conf}. For any sensitive rule if

- $N_{supp} < N_{conf}$ then D_SUPP technique hide sensitive rules with less number of modification.
- $N_{supp} > N_{conf}$ then D_CONF1 technique hides sensitive association rule with less number of modification.

In all existing research, only single D_SUPP or D_CONF1 technique is used for hiding all sensitive rules.

In this paper, we propose a hybrid technique to combine the advantage of both D_CONF1 and D_SUPP techniques. We do not include D_CONF2 technique to reduce the artificial pattern (AP) as it insert some new items in database. So we have not used this technique. In next section we describe the proposed hybrid technique for hiding sensitive rule.

3 Proposed Hybrid Technique for Hiding Sensitive Rules

The framework of proposed hybrid technique is shown in Fig. 1. It starts with applying the apriori algorithm [4] to mine the association rules from original database D. Association rules which disclose some private information are

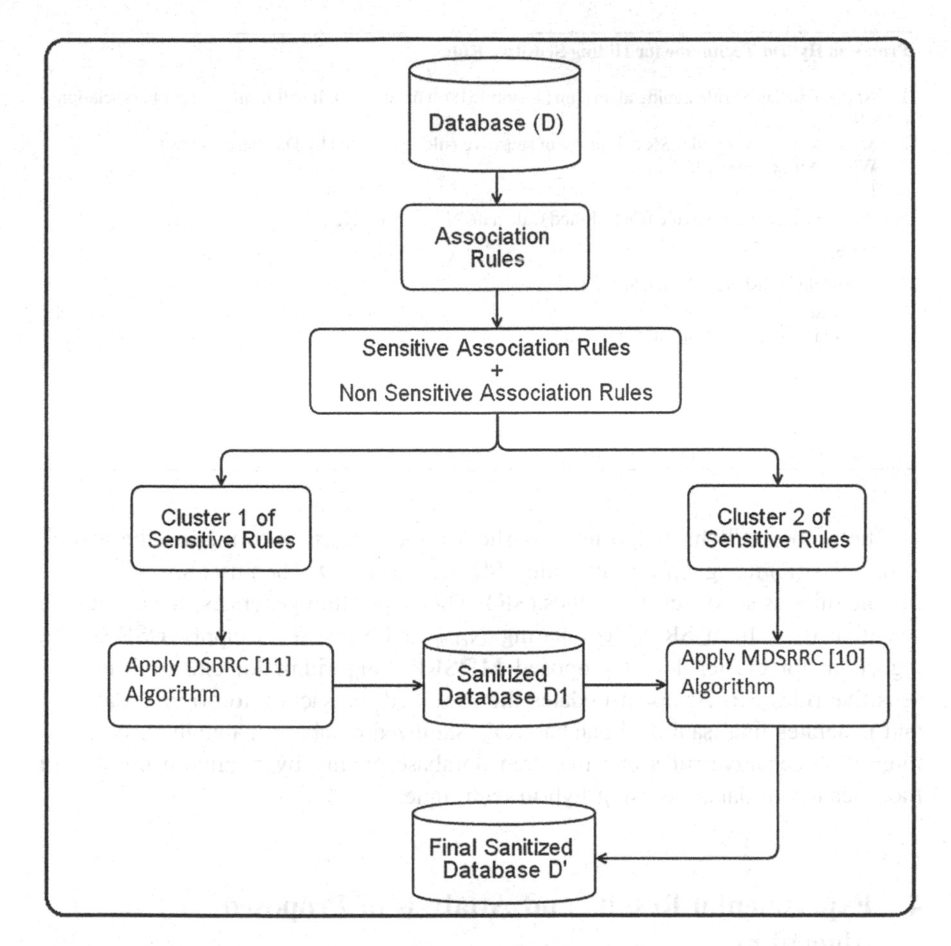

Fig. 1 A framework of proposed hybrid technique for sensitive rule

consider as sensitive rules. Then algorithm divide the sensitive rules based on N_{supp} = number of modification require to reduce the support below min_supp and N_{conf} = reduce the confidence below min_conf. Now we have two clusters of sensitive rules. Cluster 1 with all sensitive rule having $N_{supp} \leq N_{conf}$ and cluster 2 having $N_{supp} > N_{conf}$. Now for hiding sensitive rules in cluster 1, apply DSRRC [9] algorithm based on decreasing support below min_supp. For cluster 2, apply our MDSRRC [8] algorithm based on decreasing confidence below min_conf. It generates the final sanitized database. Sanitized database maintain privacy by hiding all sensitive rules [10].

Proposed Hybrid Technique for Hiding Sensitive Rules

1. Apply association rule mining algorithm (apriori [5]) on database D. It will results in set of association rules R.
2. Select set of private rules SR $\subset$ R as set of sensitive rules. (Selected by Database Owner)
3. While (SR not empty)
4. {
5. Remove one sensitive rule from SR and Calculate N_{supp} and N_{conf}
6. If ($N_{supp} \leq N_{conf}$)
   Add this sensitive rule in Cluster 1.
   Else
   Add this sensitive rule in Cluster 2.
7. }
8. For cluster 1 apply DSRRC [6] algorithm on original database D and generate sanitized database D1.
9. For cluster 2 apply our MDSRRC [4] algorithm on sanitized database D1 and generate final sanitized database D'.
10. Update all modified transaction in original database and generate final sanitized database D'

The proposed hybrid algorithm as shown above starts with mining the association rules (R) using Apriori algorithm [4] in database D. Then user specifies some private rules as set of sensitive rules (SR). Then algorithm generates two clusters of sensitive rules from SR by comparing N_{conf} and N_{supp}. Then apply DSRRC [9] algorithm for cluster 1 and proposed MDSRRC algorithm for cluster 2 to hide sensitive rules SR. At last it updates the modified transaction to original database and generates final sanitized database D′. Sanitized database maintains privacy by hiding the sensitive rules and maintain database quality by minimum number of modification in database using hybrid technique.

4 Experimental Results and Analysis of Proposed Algorithm

For performance analysis of proposed algorithm, we have compared proposed algorithm with MDSRRC [8] algorithm using a retail database with total 88,162 transactions given in [11]. We have applied apriori algorithm [4] with MST = 5 % or MST (in count) = 4408 and MCT = 10 % to generate all possible association rules. In our experiment, we choose three rules ($\{32, 39\} \rightarrow \{48\}$, $\{32, 48\} \rightarrow \{39\}$, $\{39\} \rightarrow \{38, 48\}$) as sensitive rules from all possible rules. All sensitive rules with its MST, MCT and N_{supp} and N_{conf} values are shown in Table 1. As shown in Table 1, first two sensitive rules have $N_{supp} < N_{conf}$ and for last one $N_{supp} > N_{conf}$. After applying hybrid algorithm on retail database given in [11], we have evaluated our proposed Hybrid algorithm by considering the following evaluation parameter discussed in [12]. (a) HF (hiding failure), (b) MC (misses cost), (c) AP (artificial patterns), (d) DISS (dissimilarity) and (e) SEF (side effect factor). Experimental results show that proposed hybrid algorithm works better compared to existing MDSRRC algorithm [8] based on any single basic techniques (D_SUPP, D_CONF1,

Table 1 Sensitive rules with MST, MCT, N_{supp}, N_{conf}

Rule no.	Sensitive rules	MST (in count)	MCT (%)	N_{supp}	N_{conf}
1	{32, 39} → {48}	5402	63.89	995	4557
2	{32, 48} → {39}	5402	67.24	995	4599
3	{39} → {38, 48}	6102	12.04	700	140

Table 2 Performance results of MDSRRC [8] and proposed algorithm

Evaluation parameter	MDSRRC [8] (%)	Proposed hybrid algorithm (%)
HF (hiding failure)	0	0
MC (missing cost)	26.66	13.33
AP (artificial pattern)	0	0
DISS (D, D′)	2.77	0.78
SEF (side effect factor)	26.66	13.33

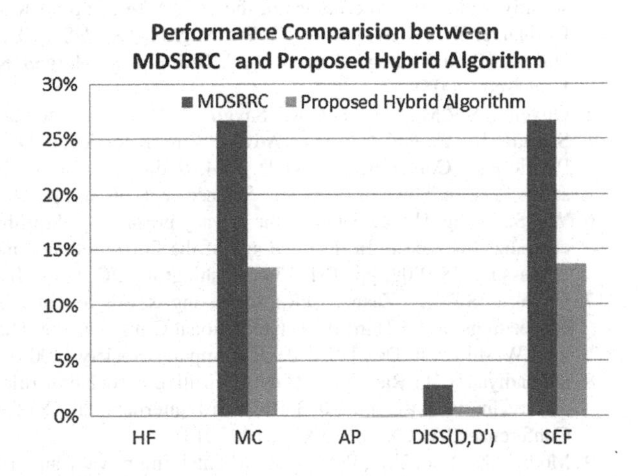

Fig. 2 Performance results of MDSRRC [8] and proposed algorithm

D_CONF2). It hides all sensitive rules (HF = 0 %) without generating any artificial rules (AP = 0 %) and maintain database quality by minimizing the modification on database. Performance results in terms of evaluation parameters are shown in Table 2 (Fig. 2).

5 Conclusion and Future Scope

This paper presented a hybrid technique to hide the sensitive rule by combining the advantage of both D_SUPP and D_CONF1 approaches. The proposed algorithm tries to remove the limitation of these both algorithms by comparing the number of modifications N_{supp} and N_{conf} to minimize the modification on the database.

We have demonstrated our proposed algorithm using a sample database. The proposed hybrid technique modifies a minimum number of transactions to maintain database quality. An artificial pattern (AP) gives the wrong direction to analysts, so the proposed algorithm does not insert any new items in transactions and maintains an artificial pattern (AP) 0 %. In future, the proposed hybrid technique can be improved in terms of missing cost (MC) and side effect factor (SEF).

References

1. Clifton, C., Kantarcioglu, M., Vaidya, J.: Defining privacy for data mining. In: National Science Foundation Workshop on Next Generation Data Mining, vol. 1, p. 1. Baltimore, MD (2002)
2. Clifton, C., Marks, D.: Security and privacy implications of data mining. In: ACM SIGMOD Workshop on Research Issues on Data Mining and Knowledge Discovery, pp. 15–19 (1996)
3. Atallah, M., Elmagarmid, A., Ibrahim, M., Bertino, E., Verykios, V.: Disclosure limitation of sensitive rules. In: Proceedings of the 1999 Workshop on Knowledge and Data Engineering Exchange, KDEX'99, pp. 45–52, Washington, DC, USA. IEEE Computer Society (1999)
4. Han, J.: Data Mining: Concepts and Techniques. Morgan Kaufmann Publishers Inc., San Francisco (2005)
5. Oliveira, S.R.M., Zaane, O.R., Saygin, Y.: Secure association rule sharing. In: Dai, H., Srikant, R., Zhang, C. (eds.) Advances in Knowledge Discovery and Data Mining, 8th Pacific-Asia Conference, PAKDD 2004, Sydney, Australia, 26–28 May 2004. Proceedings, series Lecture Notes in Computer Science, vol. 3056, pp. 74–85. Springer, Berlin (2004)
6. Wu, S., Wang, H.: Research on the privacy preserving algorithm of association rule mining in centralized database. In: Proceedings of the 2008 International Symposiums on Information Processing, ISIP'08, pp. 131–134, Washington, DC, USA. IEEE Computer Society (2008)
7. Oliveira, S.R.M., Zaane, O.R.: Protecting sensitive knowledge by data sanitization. In: Proceedings of the Third IEEE International Conference on Data Mining, ICDM'03, pp. 613–618, Washington, DC, USA. IEEE Computer Society (2003)
8. Domadiya, N.H., Rao, U.P.: Hiding sensitive association rules to maintain privacy and data quality in database. In: 2013 IEEE 3rd International Conference on Advance Computing Conference (IACC), pp. 1306–1310 (2013)
9. Modi, C.N., Rao, U.P., Patel, D.R.: Maintaining privacy and data quality in privacy preserving association rule mining. In: 2010 International Conference on Computing Communication and Networking Technologies (ICCCNT), pp. 1–6, July 2010
10. Wu, Y.H., Chiang, C.M., Chen, A.L.P.. Hiding sensitive association rules with limited side effects. IEEE Trans. Knowl. Data Eng. **19**(1), 29–42 (2007)
11. Retail database. http://fimi.ua.ac.be/data/retail.dat
12. Verykios, V.S., Gkoulalas-Divanis, A.: A survey of association rule hiding methods for privacy, vol. 34. In: Advances in Database Systems, pp. 267–289. Springer, Berlin (2008)
13. Verykios, V., Elmagarmid, A., Bertino, E., Saygin, Y., Dasseni, E.: Association rule hiding. IEEE Trans. Knowl. Data Eng. **16**(4), 434–447 (2004)
14. Saygin, Y., Verykios, V.S., Clifton, C.: Using unknowns to prevent discovery of association rules. SIGMOD Rec. **30**(4), 45–54 (2001)
15. Saygin, Y., Verykios, V., Elmagarmid, A.: Privacy preserving association rule mining. In: Twelfth International Workshop on Research Issues in Data Engineering: Engineering E-Commerce/E-Business Systems. RIDE-2EC 2002. Proceedings, pp. 151–158 (2002)

16. Wang, S.L., Jafari, A.: Using unknowns for hiding sensitive predictive association rules. In: IEEE International Conference on Information Reuse and Integration, IRI, pp. 223–228, Aug 2005
17. Modi, C., Rao, U.P., Patel, D.R.: An efficient approach for preventing disclosure of sensitive association rules in databases. In: Arabnia, H.R., Hashemi, R.R., Vert, G., Chennamaneni, A., Solo, A.M.G. (eds.) Proceedings of the 2010 International Conference on Information and Knowledge Engineering, IKE 2010, 12–15 July 2010, Las Vegas Nevada, USA, CSREA Press, 2010, pp. 303–309

A Review on Load Balancing of Virtual Machine Resources in Cloud Computing

Pradeep Kumar Tiwari and Sandeep Joshi

Abstract An effective load balance (LB) management achieves high performance computing (HPC) and green computing. Users can run their jobs on virtual machines (VMs). Virtual machine (VM) has own resources (CPU and memory). VM migrates from host to another host during fail of VM, hot spot and high resource demand. Effective LB management is based on scheduling policy and management Strategies. In this paper it is discussed the available scheduling mechanisms, goals and strategies of load balancing techniques. The aim of this work to elaborate the key analysis of research works on LB.

Keywords Virtual machine · Physical machine · Load balancing

1 Introduction

Cluster, grid and cloud computing using the fundamental concept of distributed system to achieve HPC (High Performance Computing). Distributed paradigms depend on distributed application. Applications and operating system can run separately on VMs. The core concept of resource pool and management is virtualization. Hardware virtualization (CPU partitioning and memory) can be achieved by Hypervisor. It is divided into Type 1 (hosted hypervisor) and Type 2 (bare-metal hypervisor). Type 1 as explore in Fig. 1, hypervisor is directly installed on the ×86 based hardware and provide direct access to the hardware resources. Type 2 explored in Fig. 2, hypervisor is installed and run as an application on top of an operating system and it is run on an operating system. Type 1 hypervisor more efficient rather than type 2. VM has own operating system and applications and they do not interfere with each others. Resource

P.K. Tiwari (✉) · S. Joshi
Department of Computer Science and Engineering, Manipal University Jaipur, Jaipur, India
e-mail: pradeeptiwari.mca@gmail.com

S. Joshi
e-mail: sandeep.joshi@jaipur.manipal.edu

S.C. Satapathy and S. Das (eds.), *Proceedings of First International Conference on Information and Communication Technology for Intelligent Systems: Volume 2*, Smart Innovation, Systems and Technologies 51, DOI 10.1007/978-3-319-30927-9_36

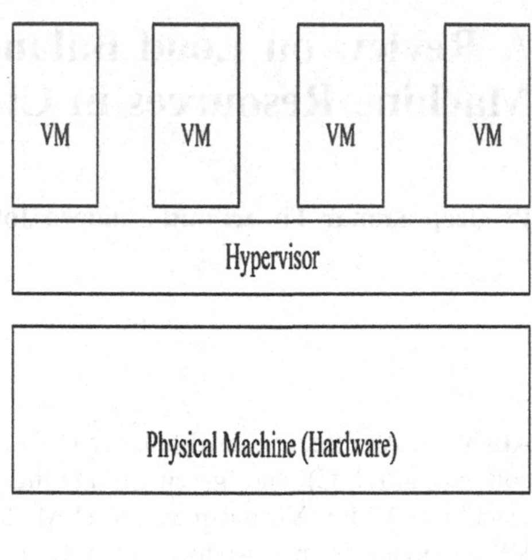

Fig. 1 Type 1 hypervisor

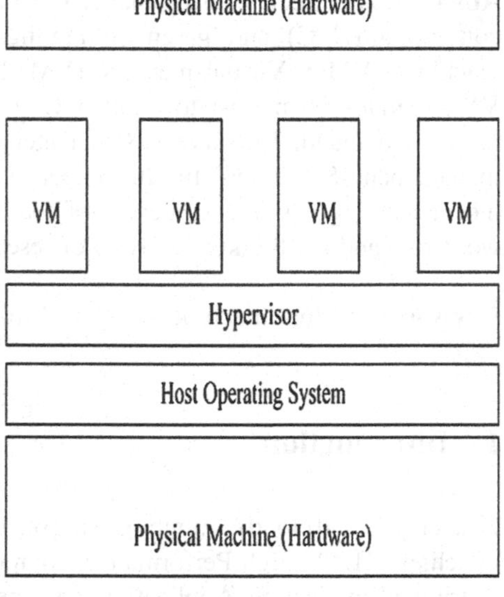

Fig. 2 Type 2 hypervisor

distribution among VMs not affected by VM failure and VM will be migrated without downtime. The LB policy should manage quality of service (QoS) and service level agreement (SLA). Week management of resources, load imbalance, hot spot and scattered information among heterogeneous servers are the main cause of SLA violation. The load imbalance problem occurs during frequently changing demand in heterogeneous environments. Load balance between low load to high load machine can be manages load imbalance. LB is not easy to manage on frequently changing high resource demand. Transfer the load, selection of VM, Location to migrate VM and information policies are responsible to manage load balance [1].

Researchers proposed numerous mechanisms to manage LB but still have many recommendations. Network, Compute and memory management in multitenant environments with scattered servers not easy to manage. This review is based on process (CPU) and memory management policies and its effect on load management. Available approaches are static, dynamic and dynamic consolidation. The modern load balancing mechanism can be automated the load. Recent mechanism provides high performance, energy saving, dynamic load requirement and minimize the cost. Effective dynamic load management during frequently changing environment and high resource demand does the vital role for resource management. VMs can have only available resources on PM. Resource demands from VMs grater then available resource on PM is called a hot spot. The opposite situation, when resources are underutilized of PM is termed as cold spot. Hotspot and cold spot can be managed by moving VMs. Hop spot mitigation selected VM and moved to less loaded PM [2].

2 Live Migration of Virtual Machines

Management of live migration technique must follow these two policies (a) Energy efficient migration technique and (b) Load balanced with fault tolerance technique.

2.1 Energy Efficient Migration Technique

Servers usually consume 70 % of total access energy. Live migration technique must follow the energy efficient mechanism. If VMs loads in an ideal situation and no need to use of others VMs in this situation remain VMs should go into sleep (energy saving) mode [1].

2.2 Load Balance with Fault Tolerance Technique

VMs loads should be scalable and monitored during load transfer and when resources are being utilized. VMs must be able to take loads of crashed VMs or not working VMs. Fault tolerance should gives assurance to clients, if any machine or system will be crashed then it must manage the clients' jobs according to SLA. Fault tolerance improves efficiency of VMs and makes an effective, robust mechanism during crashes of VMs [1, 2].

3 Resource Management Goals and Management Strategies

Resource management is a semantic relationship between resource availability and resource distribution. Network, compute and storage are a main resource component. Resource manager manages the resources according the availability of resources and provide to end user as per of SLA agreement. Resource management problems include allocation, provisioning, requirement mapping, adaptation,

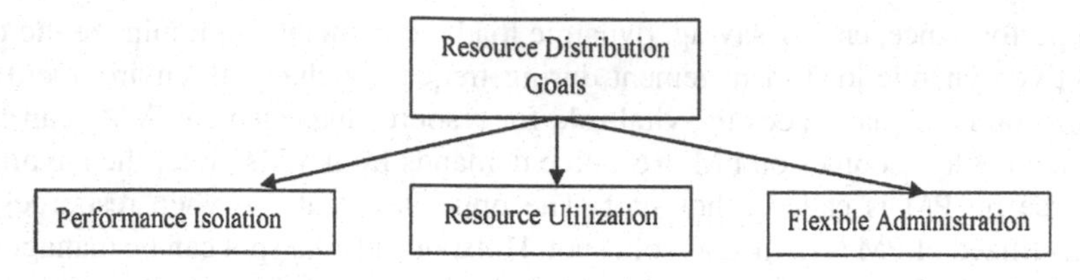

Fig. 3 Goals of resource management

discovery, brokering, estimation, and modeling. The main goals as shown in the Fig. 3 of resources distribution are (a) Performance Isolation, (b) Resource Utilization and (c) Flexible Administration.

Performance Isolation: VMs are isolated from each other and do not affect to another VM capacity. Failure of VM does not affect the performance and load. The load will migrate to another VM. Hyper-V provide the facility of quick migration and VMware provide the facility of live migration. The Load should migrate VM to VM or VM migrate to another physical server due to occurrences of high resource demand or failure of VM [3, 4].

Resource Utilization: Dynamic resource allocation management does the effect on resource utilization and provides resources from PM. Resource utilization is based on a minimum consumption of available resources without down time. Effective resource management managers maximize the resource availability and minimize the energy consumption. Resource manager must give the attention to SLA based on demand resource requirements. Load Manager mapped and measured load of individual VM. VMs resource utilization can be map with PM resources and manager can utilize unused resources of PM [5].

Flexible Administration: Resource availability administrator must able to manage high load resource demands and utilize in synchronized manner. VMware uses distributed resource scheduler (DRS) to manage VMs capacity (resource reservations, priorities and limit) and VM migration. VMware, distributed power management (DPM) system manages power on or off management of used and not used VM to achieve energy efficient resource management [6]. Microsoft Hyper-V uses system center virtual machine manager (SCVMM) support system to manage virtual machine manager 2008 R2. SCVMM increases the flexibility in storage management and migration management of VMs among hosts.

4 Load Balancing Approaches

4.1 Static Consolidation

Defines pre-reserved dedicated resource allocation to VM according the need of end users. VMs resource allocation based on total capacity of PM and migration does not perform till all demands are changing. Optimal and sub optimal scheduling belongs to static scheduling. Optimal scheduling has the information about the job and resources. Job scheduling and resource allocation decision can be taken on feasible time. If any problem occurs during feasible job scheduling then suboptimal manage [7].

4.2 Dynamic Consolidation

This is periodically, current demand based VM migration approach. If required VMs resource demands are higher from physical available capacity than VM migrate to another PM. Dynamic scheduling is based on distributed and centralized policy. Distributed approach handle the scheduling and rescheduling of a job. The dynamic centralized approach takes the decision by centralized resource management manager [7].

4.3 Dynamic Consolidation with Migration Control

This approach gives the stability during high resource demand, hotspot and frequently change resource demands. This approach reduces the required number of physical servers and save the consumed energy. This approach based on heuristic and round robin mechanism.

5 Analysis of Load Balancing Mechanism

Researchers proposed a different mechanism to manage load balance. Many researchers approach ants, agent, genetic, heuristic honey bee, round robin, token routing, sender initiative, receiver initiative, max min, min min and many more static and dynamic approaches. Ants deal with to achieve a diversity of complicated tasks with consistency and reliability. In spite of the fact that this is generally self association as opposed to learning, ants need to adapt to a phenomenon that looks all that much like over preparing in learning methods [8].

5.1 Agent Based

Aarti et al. [9], proposed an autonomous agent based load balancing algorithm (A2LB) which provides dynamic load balancing for cloud environments. The proposed mechanism has been implemented and found to provide acceptable outcome.

Omar et al. [10], proposed the most ideal approaches to consequently adjust the load among machines in large scale circumstances. The proposed mechanism is based on the performance of two dissimilar applications with two diverse conveyance approaches, and experimental results demonstrates that few applications can consequently adjust the load among the machines and get alone a superior performance in large scale simulations with one distribution approach than the other.

5.2 Genetic Algorithm

Joseph et al. [11], proposed genetic algorithm based technique to allocate VMs using the Family Gene approach. Experimental results show that the proposed mechanism minimize energy consumption and the migrations. A user specifies the requirement of resources and service to a provider and makes a contract with the service provider. This user requirement is called the SLA. A cloud service provider ought to be to manage a superior level of SLA. The proposed mechanism is able to minimize energy consumption and the VM migrations. A proposed mechanism increases the SLA level, while keeping the number of active hosts at a minimal level.

Hu et al. [12] also proposed genetic algorithm based scheduling mechanism for load balancing among VMs. This mechanism selects the least loaded virtual machine for load transfer and optimizes the high migration cost. However, due to a large number of virtual machines and frequent service requests in the data center, there is chance of inefficient service scheduling.

5.3 Heuristic Approach

Ferreto et al. [13] proposed LP-Heuristic linear programming (LP) based heuristic worst fit decreasing (WFD), best fit decreasing (BFD), first fit decreasing (FFD) and AWFD almost worst fit decreasing (AWFD). The proposed mechanism is two ways resource management approach. First approach identifies VMs and maps the capacity from existing physical machine capacity and another approach short physical machine increasingly according to their capacities. LP goal is minimization of required physical machine and map VMs resource availability form hosted physical machine.

Beloglazov et al. [14] proposed energy aware data center location approach modified best fit decreasing (MBFD) and minimization of migration (MM) approach. MBFD optimize the current VM allocation and choose the most energy efficient nearest physical servers to migrate VM and MM approach minimize the VM migration needs.

Table 1 Load balancing mechanism analysis

Author(s)	Technique	Strength	Scheme	Recommendations
Andreolini et al. [19]	Dynamic load management of virtual machines	Robust and selective reallocations	Performance based	Heterogeneous infrastructures and platforms
Beloglazov et al. [14]	Dynamic consolidation of virtual machines	SLA based load management	Energy efficient	Researchers can focus on multi-core CPU architectures
Ferreto et al. [13]	LP formulation and heuristic	Server consolidation with migration control	Energy efficient	Migration control without downtime
Forsman et al. [20]	Push and pull strategy	Rebalance the load when VMs added and removed	Load management	Downtime can be less
Jin et al.	Pre-copy model	Maximize the CPU utilization, Reduce the downtime from previous approaches	Live migration	Add network bandwidth controlling and memory writing pattern to optimize current pre-copy algorithm
Joseph et al. [11]	Genetic approach	SLA based resource management, maximize the hardware resource utilization with performance	Resource management	Modify the algorithm to decrease the calculation time in terms of prediction process to improve the genetic algorithm convergence speed
Lau et al. [15]	Guarantee reservation (GR) protocol and task batch composition (TBC) scheme	Maximize the processing speed of sender and receiver	Demand based load balanced	Performance on real time system
Li et al. [18]	Dynamic VM placement	Minimizing the total completion time	Performance based	Hybrid scheme of integrating off-line placement into online scenario
Zhang et al. [17]	Dynamically CP and heuristic allocation of cloud data center's resources	Maximize the resource utilization from) first-fit and best-fit	Performance based	(i) Refining the model to account for the energy consumption for providing data center services (ii) Dynamically determine the reservation ratio and the duration threshold for long jobs

5.4 Dynamic Approach

Lau et al. [15] proposed adaptive load distribution algorithm. This research shows previous result analysis of sender and receiver initiative algorithm and manages the work in heterogeneous load. The proposed mechanism is able to handle adaptive load distribution algorithms for heterogeneous distributed systems.

Beloglazov et al. [16] proposed a novel technique for dynamic consolidation of VMs based on adaptive utilization thresholds. The proposed technique validates the high efficiency different kinds of workloads.

Zhang et al. [17] proposed an optimization based approach that manages long jobs to dynamically allocate a cloud data center's resources. This mechanism can achieve considerably better utilization by increasing the number of jobs. The authors use a constraint programming (CP) solution to schedule the long jobs, and use simple heuristics mechanism schedule the short jobs. The proposed mechanism is able to increase the number of jobs accommodated using dynamic scheduling by 18 %. It also compares the performance of CPU and memory.

Li et al. [18] proposed an off-line VM placement method through an emulated VM migration process, while the on-line VM placement is solved by a real VM migration process. The migration algorithm is a heuristic approach. This approach uses place the VM placement to its best PM directly, if this PM has enough capacity. Otherwise, it migrate another VM from this PM to accommodate the new VM. Outcomes results validate the high efficiency of algorithms.

Andreolini et al. [19] Authors proposed reallocations of VMs management algorithms in a large number of hosts. The novel algorithms identify the real critical instances and take decisions without recurring to typical thresholds. Experimental results show that proposed algorithms are truly selective and robust even in variable contexts, thus reducing system instability and limit migrations when really necessary. Table 1, show the analysis of researchers work area and his/her future

Table 2 Research Support analysis

Organization	Support system	Scheme	Support model
Eucalyptus	Linux based frame work	Open source	Execution control of VM in heterogeneous environment
GENI (Global Environment for Network Innovations)	Apache HTTP server	Free of charge for research and classroom use	Experimental heterogeneous network structure, distributed system and security
Google App Engine	Execution of web application, Java and Python	Freeware platform	Experimental heterogeneous network structure, distributed system and security
Grids Lab Aneka	.NET-based framework	On-demand	Multiple application models, persistence, and security solutions
Open Stack	Web-based dashboard	Open source	Large network of VM, storage system, resource management
Sun Network.com (Sun Grid)	C, C++ and FORTRAN based application	Open source	Job management

recommendation. Table 2, is based on a research support system and research model for load management.

6 Conclusion

In this paper, have discussed the Strategies, goals, and polices of load balance. Effective load management policies can work in heterogeneous workload with maximum utilization of the CPU. The resources are determined by, which VM will utilize and where the target host to migrate the VM. Energy efficient migration technique and load balance with fault tolerance technique make green computing. Performance isolation, resource utilization and flexible administration are mail goal to manage load balance. We have discussed many policies to manage resources and VM migration. Researchers can do the work on the CPU utilization and VM migration in minimum downtime. The recommendations for researchers are, they can do work on optimal resource management to improve performance, scalability of resources with the future prediction of resources need, and minimize the migration of VMs.

References

1. Zhang, Q., Cheng, L., Boutaba, R.: Cloud computing: state-of-the-art and research challenges. J. Internet Serv. Appl. **1**(1), 7–18 (2010)
2. Vinothina, V., Sridaran, R., Ganapathi, P.: A survey on resource allocation strategies in cloud computing. Int. J. Adv. Comput. Sci. Appl. **3**(6) (2012)
3. Gupta, D., Cherkasova, L., Gardner, R., Vahdat, A.: Enforcing performance isolation across virtual machines in Xen. In: Middleware, pp. 342–362. Springer, Berlin (2006)
4. Nathan, S., Kulkarni, P., Bellur, U.: Resource availability based performance benchmarking of virtual machine migrations. In: Proceedings of the 4th ACM/SPEC International Conference on Performance Engineering, pp. 387–398. ACM (2013)
5. Isci, C., Liu, J., Abali, B., Kephart, J.O., Kouloheris, J.: Improving server utilization using fast virtual machine migration. IBM J. Res. Dev. **55**(6), 1–4 (2011)
6. Resource management policy. Retrieved from https://pubs.vmware.com/vsphere-50/index.jsp#com.vmware.vsphere.vm_admin.doc_50/GUID-E19DA34B-B227-44EEB1AB-46B826459442.html, July 2015
7. Rathore, N., Chana, I.S.: Load balancing and job migration techniques in grid: a survey of recent trends. Wireless Pers. Commun. **79**(3), 2089–2125 (2014)
8. Mishra, R., Jaiswal, A.: Ant colony optimization: a solution of load balancing in cloud. Int. J. Web Semant. Technol. (IJWesT) **3**(2), 33–50 (2012)
9. Singh, A., Juneja, D., Malhotra, M.: Autonomous agent based load balancing algorithm in cloud computing. Proc. Comput. Sci. **45**, 832–841 (2015)
10. Rihawi, O., Secq, Y., Mathieu, P.: Load-balancing for large scale situated agent-based simulations. Proc. Comput. Sci. **51**, 90–99 (2015)
11. Joseph, C.T., Chandrasekaran, K., Cyriac, R.: A novel family genetic approach for virtual machine allocation. Proc. Comput. Sci. **46**, 558–565 (2015)

12. Hu, J., Gu, J., Sun, G., Zhao, T.: A scheduling strategy on load balancing of virtual machine resources in cloud computing environment. In: Proceedings. PAAP, pp. 89–96 (2010)
13. Ferreto, T.C., Netto, M.A.S., Calheiros, R.N., De Rose, C.A.: Server consolidation with migration control for virtualized data centers. Future Gener. Comput. Syst. **27**(8),1027–1034 (2011)
14. Beloglazov, A., Abawajy, J., Buyya, R.: Energy-aware resource allocation heuristics for efficient management of data centers for cloud computing. Future Gener. Comput. Syst. **28**(5), 755–768 (2012)
15. Lau, S.M., Lu, Q., Leung, K.S.: Adaptive load distribution algorithms for heterogeneous distributed systems with multiple task classes. J. Parallel Distrib. Comput. **66**(2),163–180 (2006)
16. Beloglazov, A., Buyya, R.: Adaptive threshold-based approach for energy-efficient consolidation of virtual machines in cloud data centers. In: Proceedings of the 8th International Workshop on Middleware for Grids, Clouds and e-Science, vol. 4. ACM (2010)
17. Zhang, Y., Fu, X., Ramakrishnan, K.K.: Fine-grained multi-resource scheduling in cloud datacenters. In: 2014 IEEE 20th International Workshop on Local and Metropolitan Area Networks (LANMAN), pp. 1–6. IEEE (2014)
18. Li, K., Zheng, H., Wu, J., Du, X.: Virtual machine placement in cloud systems through migration process. Int. J. Parallel Emergent Distrib. Syst. 1–18 (ahead-of-print, 2014)
19. Andreolini, M., Casolari, S., Colajanni, M., Messori, M.: Dynamic load management of virtual machines in cloud architectures. In: Cloud Computing, pp. 201–214. Springer, Berlin (2010)
20. Forsman, M., Glad, A., Lundberg, L., Ilie, D.: Algorithms for automated live migration of virtual machines. J. Syst. Softw. **101**, 110–126 (2015)

Design of Square Shaped Polarization Sensitive Metamaterial Absorber

Vandana Jain, Sanjeev Yadav, Bhavana Peswani, Manish Jain, H.S. Mewara and M.M. Sharma

Abstract In this paper, the polarization sensitive metamaterial absorber is proposed. The metamaterial absorber structure consist of a plus shaped structure surrounded by square loop has been proposed. The proposed metamaterial absorber unit cell provides approximately unity (99.99 %) absorption at 16.25 GHz. The design of an ultrathin polarization sensitive absorber based on Ku-band applications. Absorber design has been simulated for different incident and polarization angles to tune the single band and dual band response for different frequency slots of Ku-band. To obtain best response, FR-4 substrate material having substrate height of 1 mm is used. This metamaterial absorber is used in stealth technology for battlefield, solar cell and airborne radar.

Keywords Metamaterial · Dual band response · Absorber · Airborne radar · Stealth technology

V. Jain (✉) · S. Yadav · B. Peswani
Department of ECE, Government Women Engineering College Ajmer (Raj), Ajmer, India
e-mail: j.vandana26@gmail.com

S. Yadav
e-mail: sanjeev.mnit@gmail.com

B. Peswani
e-mail: bhavnapesswani21@gmail.com

M. Jain
Rockwell Collins India Pvt. Ltd, Gachibowli, Hyderabad, India
e-mail: manish.jain@rockwellcollins.com

H.S. Mewara
Department of EICE, Government Engineering College Ajmer, Ajmer, India
e-mail: hsmewara@gmail.com

M.M. Sharma
Malaviya National Institute of Technology, Jaipur, India
e-mail: mmsjpr@gmail.com

S.C. Satapathy and S. Das (eds.), *Proceedings of First International Conference on Information and Communication Technology for Intelligent Systems: Volume 2*, Smart Innovation, Systems and Technologies 51, DOI 10.1007/978-3-319-30927-9_37

1 Introduction

In recent years, electromagnetic metamaterial are the exotic artificially designed structure that have large number of attractive properties such as group velocity of all the resulting waves, phase opposition. The other properties of the metamaterial such as permittivity, permeability and refraction index have negative values [1, 2]. Different applications of metamaterial include collimation of lenses which create highly directive beams [3], rejecting the light incoming from the space region [4], provide sub wavelength imaging technique in a form of lenses [5] etc.

To reduce the interference in microwave component the standard form of microwave absorbers has been use recently. But conventional microwave absorber have one major disadvantage of larger thickness($\sim\lambda/4$). To overcome this disadvantage metamaterial (MTM) structures has been used, which provides maximum absorption in case of dielectric and ohmic loss by using the ultra-wavelength based thin structure [6]. To fulfill the goal of maximum absorption the necessary condition is to balance the properties of magnetic and electric resonances in the specified frequency band which simultaneously minimize reflection as well as transmission due to which maximum absorption peak for perfect absorber design is obtained.

In this paper a polarization sensitive perfect absorber have been presented over a FR-4 substrate material using ANSOFT HFSS and various variation for parameters have been simulated. Metamaterial-based absorbers can be used as an alternative of conventional absorbers because of their near unity absorption properties, thickness in the range of ultra-wavelength, highly effective and simple manufacturing process [2, 7]. These metamaterial based absorber designed structures are mainly composed of metallic patches on top surface and metal plane on ground surface separated from dielectric substrate.

At resonance frequency of operation, both dielectric and top layer of the FSS excited. The dielectric layer is excited magnetically and top layer by the electric field. The result of the both excitation produces surface current which circulate through the FSS structure. At certain frequency effective permeability and effective permittivity becomes almost the same due to the manipulating the material property of the structure by the electromagnetic fields [5]. This results in matching the free space impedance of air with the input impedance of that structure closely and reduction in reflection from metamaterial designed absorber.

In most applications, the imaginary parts of the dielectric contribute in absorption loss so it will degrades the performance of metamaterial so that the smallest possible value is to be defined for the imaginary parts. To design a perfect absorber material absorption loss becomes useful.

Firstly perfect designed metamaterial absorber, consist of cut wire and a metallic split ring, having 88 % absorptivity has been reported [2]. After this design, metamaterial conceptual based absorbers have been studied in different application including microwave [8, 9], terahertz [13], rays of infrared frequency [10] and optical range frequencies [11, 12]. These all application based absorber have high absorptivity, but single band absorption of absorber limit their use in other application. To enhance the limited use of absorber dual band [13] and triple band [14] absorptions have been made.

These absorbers are basically thin structure compared with the operating wavelength thereby used in RCS reduction of microwave structure, thermal imaging, bio-chemical detection, FSS bolometers and FSS solar cells. In different field of research metamaterial are used such as electromagnetic cloaking and super lenses started from fabrication of left-handed materials. Such devices have examined for different slots of frequency range started from radio frequency, infrared and optical frequencies.

2 Design of Metamaterial Absorber

Figure 1a shows the top surface of the designed unit cell structure. This structure consist of three layer in which dielectric layer is used in between the top layer of copper metallic patch and bottom copper ground layer. Patch consists of two square scaled versions of unit cell, one embedded inside another, where each unit cell comprises one complete square loop and one split loop. A plus shaped unit cell is used inside the two square loop. FR-4 substrate has been selected as dielectric having thickness of 1 mm (loss tangent tanδ = 0.025 and relative permittivity $\varepsilon_r = 4$). The copper layer have conductivity of ($\sigma = 5.8 \times 10^7$ S/m) and layers thickness of 0.035 mm (Table 1).

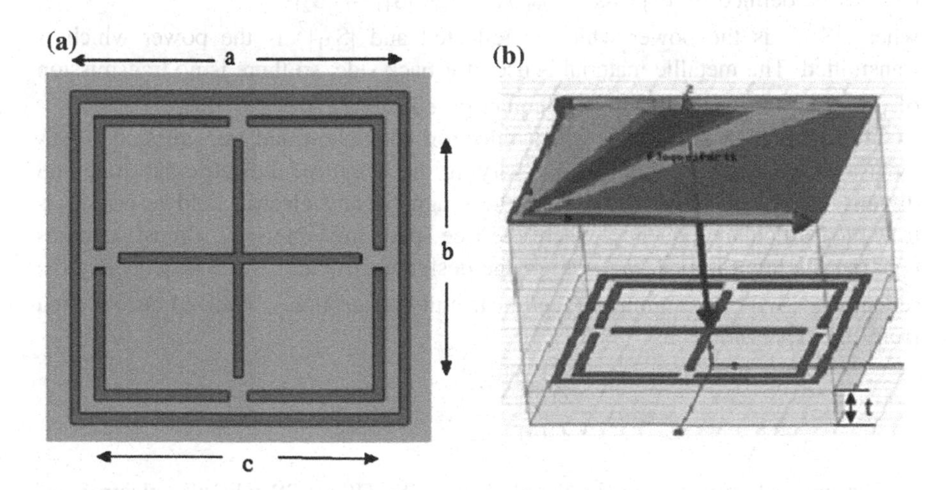

Fig. 1 **a** Front view of the unit cell structure. **b** Floquet port direction (a, b) in the structure

Table 1 Parameter values of the proposed design

Parameters	Values (mm)
a	7
b	5
c	6
t	1

3 Absorption Phenomena of MTM Structure

The absorption phenomena of a MTM structure are defined by two factors:

1. By decreasing the value of reflectance ($|S11|^2$): The bottom surface of the structure is made up of copper, so the transmitted power $|S21|^2 = 0$. By adjusting the dimensions of that structure, the impedance $Z(\omega)$ is approximated close to the free space impedance $\eta(\omega)$, resulting in reduction of absorptivity $A(\omega)$.
2. By exceeding the imaginary value of refractive index: As $|S_{21}| = 0$ and imaginary part should have high value [6].

4 Simulated Results

When the structure is excited by plane electromagnetic wave then, the absorptivity (A) can be defined by expression as: $A = 1 - |S_{11}|^2 - |S_{21}|^2$
where $|S_{11}|^2$ is the power which is reflected and $|S_{21}|^2$ is the power which is transmitted. The metallic material is used for back side, so there is no transmission of power ($|S_{21}|^2$) and absorptivity equation reduces as $A = 1 - |S_{11}|^2$

Therefore, high absorption when reflection coefficient and transmission coefficient are minimized is obtained. By varying the height of dielectric substrate and dimension of top copper layer, both the magnetic and electric field responses is tuned to particular frequency, where the free space impedance Z_0 almost becomes equal to the input impedance $Z(\omega)$ of the designed structure. At this frequency the reflectivity $|S_{11}|^2$ gets minimized and near unity absorption is realized, as obtained from the expression.

$$S_{11}(\omega) = \frac{Z(\omega) - Z_0}{Z(\omega) - Z_0} (\text{since } |S_{11}| = 0)$$

The unit cell structure is simulated with ANSOFT HFSS which is electromagnetic solver based on finite element. It is observed that one absorption peak is visible at 16.25 GHz with absorptivity of almost 99.99 % and reflectivity attain its minimum value, which make it a perfect absorber respectively as shown in Fig. 2.

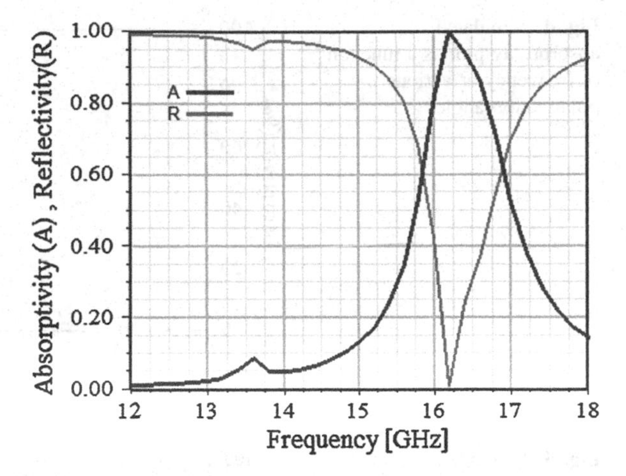

Fig. 2 Simulated absorptivity and reflectivity plot for perfect absorber

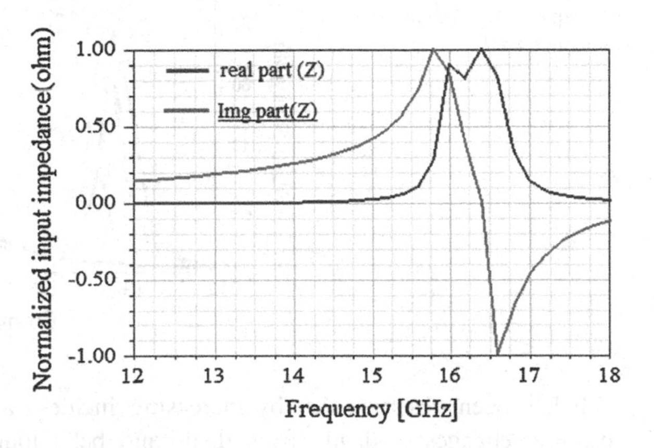

Fig. 3 Normalized input impedance for single absorption peak with imaginary and real part

At absorption peaks, the impedance Z(ω) of the metasurface matched closely to the impedance Zo as shown in Fig. 3. It has been observed from the plot that the real parts of the impedance Z(ω) becomes unity and the imaginary parts of the impedance Z(ω) are close to zero for the peak value of absorption at frequency 16.25 GHz. In order to investigate the polarization nature of the absorber, the structure has been simulated and observed under different polarization angles and different incident angle. The structure has been simulated for different polarization angles from 0° to 45° in the steps of 15° as shown in Fig. 4. The simulated absorptivity plot has dual band behaviour response and one peak have maximum absorptivity up to 93 %.

The different incident angle response varied from 0° to 60° in steps of 20° as shown in Fig. 5. The simulated incident angle plot shows that all the response for different incidence angle have two absorption peak, but one extra absorptivity peak gradually increases with higher incident angles, reaches maximum at 60° as shown in Fig. 5.

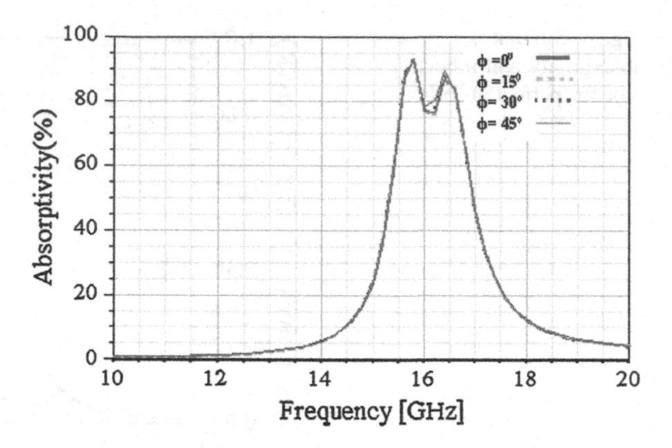

Fig. 4 Simulated absorptivity plot as a function of frequency at various polarization angles

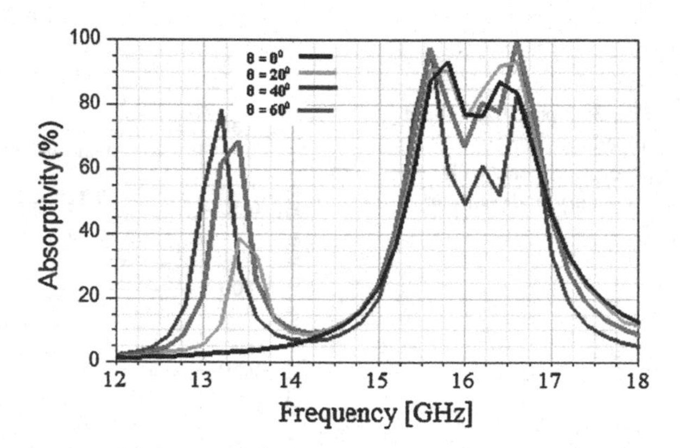

Fig. 5 Simulated absorptivity plot as a function of frequency at various incidence angles

It has been observed that by increasing incident angle, the absorption result response changes gradually from dual-band behaviour to four peak and attains 99.99 % absorption.

5 Conclusion

In this paper, we have presented a polarization sensitive ultra-thin metamaterial based square shaped absorber, using FR-4 as substrate material and having total thickness of 1.07 mm. For the narrow band at 16.25 GHz, the structure shows highly absorption. The absorber structures have been simulated for different combination of incident angle and polarization angle. We observed that by increasing value of incident angle, the absorption results changes gradually from dual-band behaviour to four peak and attain 99.99 % absorption peak and polarization angle is used to change the single narrow band response to dual band response,both close to 16.25 GHz. The structure can be modified into broadband absorber by using the concept of array.

References

1. Lai, A., Caloz, C., Itoh, T.: Composite right/left- handed transmission line metamaterials. IEEE Microwave Mag. **5**(3), 34–50 (2004)
2. Landy, N.I., Sajuigbe, S., Mock, J.J., Smith, D.R., Padilla, W.J.: Perfect metamaterial absorber. Phys. Rev. Lett. **100**, 207402 (2008)
3. Enoch, S., Tayeb, G., Sabouroux, P., Guerin, N., Vincent, P.: A metamaterial for directive emission. Phys. Rev. Lett. **89**, 213902 (2002)
4. Pendry, J.B., Schurig, D., Smith, D.R.: Controlling electromagnetic fields. Science **312**, 1780–1782 (2006)
5. Pendry, J.B.: Negative refraction makes a perfect lens. Phys. Rev. Lett. **85**, 3966–3969 (2000)
6. Gu, S., Barrett, J.P., Hand, T.H., Popa, B.I., Cummer, S.A.: A broadband low-reflection metamaterial absorber. J. Appl. Phys. **108**, 064913 (2010)
7. Saville, P.: Review of Radar Absorbing Materials. Defense R & D, Canada-Atlantic, 2005, pp. 5–15
8. Luo, H., Cheng, Y.Z., Gong, R.Z.: Numerical study of metamaterial absorber and extending absorbance bandwidth based on multi-square patches. Eur. Phys. J. B **81**(4), 387–392 (2011)
9. Liu, Y.H., Gu, S., Luo, C.R., Zhao, X.P.: Ultra-thin broadband metamaterial absorber. Appl. Phys. A **108**(1), 19–24 (2012)
10. Mason, J.A., Allen, G., Podolskiy, V.A., Wasserman, D.: Strong coupling of molecular and mid-infrared perfect absorber resonances. IEEE Photon. Technol. Lett., **24**(1):31–33 (2012)
11. Hu, C., Zhao, Z., Chen, X., Luo, X.: Realizing near-perfect absorption at visible frequencies. Opt. Express **17**(13), 11039–11044 (2009)
12. Ye, Y., Jin, Y., He, S.: Omni-directional, broadband and polarization insensitive thin absorber in the terahertz regime. J. Opt. Soc. Amer. B **27**(3), 498–503 (2010)
13. Ma, Y., Chen, Q., Grant, J., Saha, S.C., Khalid, A., Cumming, D.R.S.: A terahertz polarization insensitive dual band metamaterial absorber. Opt. Lett. **36**(6), 945–947 (2011)
14. Shen, X., Cui, T.J., Zhao, J., Ma, H.F., Jiang, W.X., Li, H.: Polarization-independent wide-angle triple-band metamaterial absorber. Opt. Express **19**(10), 9401–9407 (2011)

RSM and MLR Model for Equivalent Stress Prediction of Eicher 11.10 Chassis Frame: A Comparative Study

Tushar M. Patel and Nilesh M. Bhatt

Abstract The main objective of the study is to compare the prediction accuracy of Response Surface Methodology (RSM) and Multiple Linear Regressions (MLR) model for the Equivalent stress of the chassis frame. The chassis frame is made of two sidebars connected with a series of crossbar. The web thickness, upper flange thickness and lower flange thickness of sidebar becomes the design variables for the optimization. Since the number of parameters and levels are more, so the probable models are too many. The variants of the frame are achieved by topology modification using the orthogonal array. Then Finite Element Analysis (FEA) is performed on those models. RSM model and MLR model are prepared using the results of FEA to predict equivalent stress on the chassis frame. The results indicate that predictions of RSM model are more accurate than predictions of MLR model.

Keywords Chassis frame · FE analysis · RSM · MLR · Equivalent stress

1 Introduction

Structural optimization using computational methods has become a major field of research in recent years. The methods used in structural analysis and optimization may require considerable computational time and cost, depending on the problem complexity. To overcome this problem FEA is used along with the mathematical modelling tools viz. RSM, Artificial Neural Network (ANN), MLR etc.

Box et al. [1] determined the optimum conditions in chemical investigation. Brey et al. (1996) compared the prediction accuracy of biomass production by MLR model and ANN model. The result indicates that the accuracy of MLR and ANN

T.M. Patel (✉)
Mewar University, Gangrar, Chittorgarh, Rajasthan, India
e-mail: tushar.modasa@gmail.com

N.M. Bhatt
Gandhinagar Institute of Technology, Moti Bhoyan, Gandhinagar, Gujarat, India
e-mail: nmbhatt19@gmail.com

S.C. Satapathy and S. Das (eds.), *Proceedings of First International Conference on Information and Communication Technology for Intelligent Systems: Volume 2*, Smart Innovation, Systems and Technologies 51, DOI 10.1007/978-3-319-30927-9_38

may be used to estimate the production of larger population assemblages [2]. Anwar et al. [3] compared the performance accuracy of ANN, MLR and GARCH model for Islamic banking. Besalatpour et al. (2012) compared predictive capabilities of ANN and adaptive Neuro-fuzzy inference system in estimating soil shear. The results showed that the prediction of ANN and ANFIS techniques were more accurate in comparison with the conventional MLR technique [4]. Nelofer et al. (2012) used ANN and RSM models to optimize the lipase production for four independent variables, viz. sodium chloride, glucose, temperature and induction time. The prediction capabilities of ANN and RSM were then compared [5]. Mouhibi et al. (2013) prepared ANN and MLR models to predict for CCR5 binding affinity of amides and ureas with four descriptors. Comparison of the MLR and ANN model shows that the effect of some of the descriptors to activity may be non-linear [6]. Fatiha et al. (2013) have been investigated optimal conditions for immobilization, through RSM and ANN. They have concluded that both models provided good quality predictions [7]. Vasundara et al. (2014) carried out research to predict the maximum elastic deformation of the workpiece during machining for the fixture layout. The FEM has been used to evaluate the workpiece elastic deformation. ANN and RSM both are used to predict the position of the fixturing elements. In this study, a numerical example has been considered from the literature to compare the performance of ANN and RSM [8]. Maran et al. (2014) evaluated the prediction accuracy of RSM and ANN models on fatty acid methyl esters yield achieved from muskmelon oil. The influence of process variables was investigated by central composite rotatable design of RSM and multi-layer perceptron neural network. They have concluded that both the models showed better predictions in this study [9].

Number of gaps have been observed during the comprehensive study of the literature. Literature review reveals that the researchers have carried out most of the work on FEA and design of various automobile parts but very limited work has been reported on optimization of the design variables of automobile. The comparison of MLR and RSM model's prediction capability has not been fully explored using FEA results as a data. The study presented here shows the validity and effectiveness of hybrid modeling for prediction using FEA based RSM and MLR methods, which have effectively decreased the efforts and time, required for evaluating the variables. The RSM model and MLR model are constructed using datasets of FEA result in Minitab. The main objective of the study is to compare the RSM model and the MLR model to predict Equivalent Stress for Eicher 11.10 chassis frame.

2 Analysis Data Sets for Experiments

The selected process variables are varied up to five levels and Central Composite Design (CCD) was used to design the experiments. Response Surface Method is used to develop a second order regression equation relating response characteristics and process variables [10]. The three parameters considered for this study are web thickness (mm), upper flange thickness (mm) and lower flange thickness (mm). The parameters are set at five levels each. The summary of the parameters is shown in Table 1.

The value of equivalent stress for all variants is measured using ANSYS for finding out the optimum thickness of web, upper flange and lower flange. Experiments were designed according to the test conditions specified by the second order CCD [10, 11]. The FEA were conducted for all data sets, to study the effect of parameters over the output parameters. Altogether 20 analysis were conducted using Ansys to prepare data set for response surface model. FEA analysis results for equivalent Stress are given in Table 2.

Table 1 Parameters and their levels

Process parameter	Level				
	(−2)	(−1)	(0)	(1)	(2)
Web thickness (X_1)	3	4	5	6	7
Upper Flange thickness (X_2)	3	4	5	6	7
Lower Flange thickness (X_3)	3	4	5	6	7

Table 2 Coded values of the variables and the response

Run	X_1	X_2	X_3	Equivalent stress (MPa)
1	1	1	−1	97.6
2	−1	−1	1	107.53
3	−1	1	−1	118.83
4	1	−1	−1	98.8
5	0	0	0	106.16
6	0	0	−2	124.86
7	0	0	2	93.031
8	1	1	1	91.002
9	−1	−1	−1	121.83
10	0	0	0	106.16
11	−1	1	1	108.39
12	0	0	0	106.16
13	0	0	0	106.16
14	0	0	0	106.16
15	2	0	0	88.478
16	−2	0	0	118.16
17	0	0	0	106.16
18	0	−2	0	108.04
19	0	2	0	106.45
20	1	−1	1	90.7

X_1, X_2 and X_3 represent coded values of various factors

3 RSM Model for Equivalent Stress Prediction

Response surface methodology (RSM) is a statistical technique for experimental model building. By design of experiments, the objective is to optimize a response (dependent variable) which is influenced by several input variables (independent variables) [10, 11].

Response Surface Method is used to examine the relationship between a set of quantitative variables or factors and a response. All the coefficients are to be estimated using experimental data as shown in Table 3.

The Model F-value of 26.75 implies the model is significant. The "Lack of Fit P-value" is not significant relative to the pure error. Non significant lack of fit is good. The coefficient of determination (R^2) and adjusted coefficient of determination (R^2 adj) were 96.01 % and 92.42 %, respectively, which indicated that the estimated model fits the experimental data satisfactorily. Values of "P" less than 0.0500 indicates model terms are significant. In this case web thickness A and lower flange thickness C are significant model terms.

The general equation for the proposed second order regression model to predict the response can be written as Eq. (1).

$$Y = b_0 + b_1X_1 + b_2X_2 + b_3X_3 + b_{11}X_1^2 + b_{22}X_2^2 + b_{33}X_3^2 + b_{12}X_1X_2 + b_{13}X_1X_3 + b_{23}X_2X_3 \quad (1)$$

The second-order polynomial models used to express the Equivalent Stress as a function of independent variables Eq. (2) is shown below in terms of coded level:

$$\text{VMS (CODED)} = 105.781 - 8.61X_1 - 0.38X_2 - 6.44X_3 - 0.89X_1^2 + 0.08X_2^2 + 0.5X_3^2 + 0.15X_1X_2 + 1.25X_1X_3 + 0.67X_2X_3 \quad (2)$$

Table 3 Regression coefficients for equivalent stress

Term	Coef	SE coef	P
Constant	105.781	1.1219	0.000
A	−8.615	0.7032	0.000
B	−0.389	0.7032	0.593
C	−6.443	0.7032	0.000
A*A	−0.900	0.5609	0.140
B*B	0.082	0.5609	0.887
C*C	0.507	0.5609	0.387
A*B	0.155	0.9944	0.879
A*C	1.255	0.9944	0.235
B*C	0.670	0.9944	0.516
R-Sq = 96.01 % R-Sq(adj) = 92.42 %			
Model F-value of = 26.75 Lack-of-fit P = 0.6444			

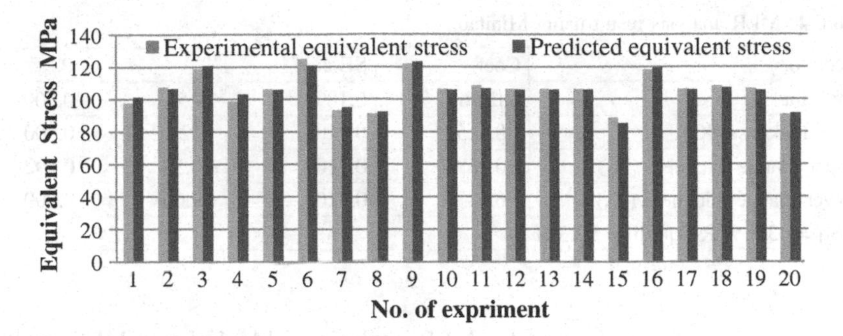

Fig. 1 Comparison of experimental vesus predicted VMS results

This equation is used to predict the equivalent stress for set of independent variable. The comparison of experimental equivalent stress and predicted equivalent stress is given in terms of standard error and R squared error.

Figure 1 shows the comparison of experimental versus predicted VMS obtained from Eq. (2). A linear distribution is observed, which is indicative of a well-fitting model. The values predicted were close to the FEA values of equivalent stress.

4 MLR Model for Equivalent Stress Prediction

Linear regression is an approach to modeling the relationship between a scalar dependent variable $\hat{Y}$ and one or more independent variables denoted X. The case of one independent variable is called simple linear regression. For more than one independent variable, it is called multiple linear regression [10, 12]. A multiple linear regression equation expresses a linear relationship between a response variable ($\hat{Y}$) and three-predictor variables (X_1, X_2 and X_3). The multiple regression is given in Eq. (3) [11, 12].

$$\hat{Y} = B_0 + B_1X_1 + B_2X_2 + B_3X_3 \tag{3}$$

where,

$\hat{Y}$	Predicted value of Equivalent Stress
X_1	Web Thickness
X_2	Upper Flange Thickness
X_3	Lower Flange Thickness
B_0	Estimate value of y-intercept
B_1, B_2 and B_3	Estimate value of the independent variable coefficient.

Table 4 MLR analysis result using Minitab

Predictor	Coef	SE coef	T	P
Constant	182.769	6.189	29.53	0.000
Web thickness (X_1)	−8.6151	0.7109	−12.12	0.000
Upper Flange thickness (X_2)	−0.3886	0.7109	−0.55	0.592
Lower Flange thickness (X_3)	−6.4435	0.7109	−9.06	0.000
R-Sq = 93.5 % R-Sq(adj) = 92.3 %				

Minitab 16 software is used for MLR analysis. Table 4 shows MLR analysis result using Minitab. Relation of equivalent stress with web thickness, upper flange thickness and lower flange thickness is shown in Eq. (4).

$$\text{Equivalent Stress } \hat{Y} = 183 - 8.62X_1 - 0.389X_2 - 6.44X_3 \tag{4}$$

Based on 95 % confidence level, web thickness and lower flange thickness has a statically significant impact on equivalent stress, since its p-value is smaller than 5 %. Table 4 shows the highest R Square (93.5) and Adjusted R Square (92.3) values. Hence, as it is found that prediction capacity of MLR formula for the equivalent stress is good.

5 Comparison of RSM and MLR Predicted Results

Table 5 shows MLR and RSM prediction model error and Fig. 2 shows a regression plot of RSM and MLR. The coefficient of determination (R^2) between the FEA and predicted responses were respectively equal to 0.96 and 0.93 for the RSM model and MLR model. Both models provided predictions with accuracy, yet the RSM model is more efficient than the MLR model for predicting the equivalent strength. This might be due to the large amount of data required for developing a sustainable regression model, while the RSM could recognize the relationships with less data. A second reason is the effect of the predictors on the response variable, which may not be linear in nature. In other words, the RSM model could probably predict equivalent stress with a better accuracy owing to their greater capability as well as greater flexibility to model nonlinearity. Therefore, RSM may be preferred in the case of less numbers of datasets.

Table 5 Comparison of RSM and MLR Prediction model error

Run	X_1 (mm)	X_2 (mm)	X_3 (mm)	FEA equivalent stress a_j (MPa)	RSM predicted equivalent stress p_{rj} (MPa)	MLR predicted equivalent stress p_{mj} (MPa)	RSM error $e_{aj} = a_j - p_{rj}$ (MPa)	MLR error $e_{mj} = a_j - p_{mj}$ (MPa)
1	6	6	4	97.6	101.14	102.941	−3.54	−5.341
2	4	4	6	107.53	106.26	108.072	1.27	−0.542
3	4	6	4	118.83	120.57	120.181	−1.74	−1.351
4	6	4	4	98.8	102.95	103.719	−4.15	−4.919
5	5	5	5	106.16	105.78	105.5065	0.38	0.6535
6	5	5	3	124.86	120.7	118.3935	4.16	6.4665
7	5	5	7	93.031	94.92	92.6195	−1.889	0.4115
8	6	6	6	91.002	92.1	90.054	−1.098	0.948
9	4	4	4	121.83	123	120.959	−1.17	0.871
10	5	5	5	106.16	105.78	105.5065	0.38	0.6535
11	4	6	6	108.39	106.51	107.294	1.88	1.096
12	5	5	5	106.16	105.78	105.5065	0.38	0.6535
13	5	5	5	106.16	105.78	105.5065	0.38	0.6535
14	5	5	5	106.16	105.78	105.5065	0.38	0.6535
15	7	5	5	88.478	84.95	88.2665	3.528	0.2115
16	3	5	5	118.16	119.41	122.7465	−1.25	−4.5865
17	5	5	5	106.16	105.78	105.5065	0.38	0.6535
18	5	3	5	108.04	106.89	106.2845	1.15	1.7555
19	5	7	5	106.45	105.33	104.7285	1.12	1.7215
20	6	4	6	90.7	91.23	90.832	−0.53	−0.132

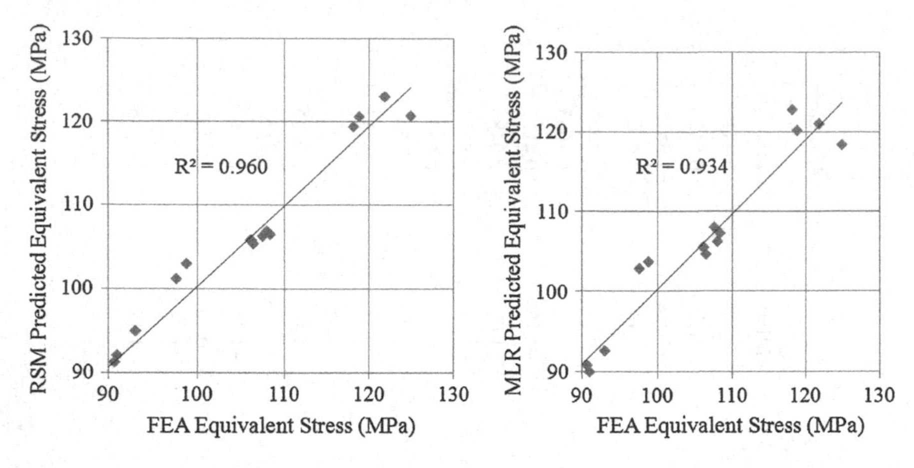

Fig. 2 Regression plot of RSM and MLR

6 Conclusions

The present study aimed at the comparative analysis of RSM and MLR model for Equivalent stress prediction. This optimization is carried out by developing equivalent stress models based on 20 data sets of CCD. An RSM model and MLR model are developed to predict equivalent stress of Eicher 11.10 chassis frame. The conclusions drawn by the comparative study of MLR model and the RSM model for equivalent stress prediction are given below.

- The RSM model is a more promising than the MLR model for accurately estimating Equivalent Stress.
- The RSM has greater capability and flexibility to model nonlinearity.
- The RSM model has better prediction capability even with less number of observations data sets.

Thus, the RSM model is preferred over the MLR method.

References

1. Box, G.E., Wilson, K.B.: On the experimental attainment of optimum conditions. J. Roy. Stat. Soc.: Ser. B (Methodol.) **13**(1), 1–45 (1951)
2. Brey, T., Jarre-Teichmann, A., Borlich, O.: Artificial neural network versus multiple linear regression prediciting P/B ratios from empirical data. Mar. Ecol.-Prog. Ser. **140**, 251–256 (1996)
3. Anwar, S., Mikami, Y.: Comparing accuracy performance of ANN, MLR, and GARCH model in predicting time deposit return of Islamic bank. Int. J. Trade Econ. Finance **2**(1), 44–51 (2011)

4. Besalatpour, A., Hajabbasi, M.A., Ayoubi, S., Afyuni, M., Jalalian, A., Schulin, R.: Soil shear strength prediction using intelligent systems: artificial neural networks and an adaptive neuro-fuzzy inference system. Soil Sci. Plant Nutr. **58**(2), 149–160 (2012)
5. Nelofer, R., Ramanan, R.N., Rahman, R.N.Z.R.A., Basri, M., Ariff, A.B.: Comparison of the estimation capabilities of response surface methodology and artificial neural network for the optimization of recombinant lipase production by *E. coli* BL21. J. Ind. Microbiol. Biotechnol. **39**(2), 243–254 (2012)
6. Mouhibi, R., Zahouily, M., El Akri, K., Hanafi, N.: Using multiple linear regression and artificial neural network techniques for predicting CCR5 binding affinity of substituted 1-(3, 3-Diphenylpropyl)-Piperidinyl amides and ureas. Open J. Medicinal Chem., **3**(01), 7 (2013)
7. Fatiha, B., Sameh, B., Youcef, S., Zeineddine, D., Nacer, R.: Comparison of artificial neural network (ANN) and response surface methodology (RSM) in optimization of the immobilization conditions for lipase from Candida rugosa on AMBERJET® 4200-Cl. Prep. Biochem. Biotechnol. **43**(1), 33–47 (2013)
8. Vasundara, M., Padmanaban, K.P., Sabareeswaran, M., RajGanesh, M.: Machining fixture layout design for milling operation using FEA, ANN and RSM. Procedia Eng. **38**, 1693–1703 (2014)
9. Maran, J.P., Priya, B.: Comparison of response surface methodology and artificial neural network approach towards efficient ultrasound-assisted biodiesel production from muskmelon oil. Ultrason. Sonochem. **23**, 192–200 (2015)
10. Swift, R.J.: A course in mathematical modeling. Cambridge University Press (1999)
11. Eriksson, L. (Ed.). Design of experiments: principles and applications. MKS Umetrics AB. (2008)
12. Patel, T.M., Bhatt, N. M.: ANN and MLR model for shear stress prediction of Eicher 11.10 chassis frame: a comparative study. Int. J. Mech. Eng. Technol. (IJMET). **4**(5), 216–223 (2013)

Gender Classification by Facial Feature Extraction Using Topographic Independent Component Analysis

Shivi Garg and Munesh C. Trivedi

Abstract Recognition of gender from face image has attracted a huge attention now a days. Many identification systems are being developed to identify a person, as most of the technique for gender classification stand on facial features. In this paper, we presented a gender classification framework consist of a series of phases for determining the gender as the final output. Initially we start by detecting the face from an image using Viola Jones and then extract the facial feature using the Topographic Independent Component Analysis. The features extracted here are used to train the SVM classifier for the classification step. Our experimental result gives the best accuracy in determining the images as of male or female and gives average performance of 96 % correct gender identification on images.

Keywords Facial feature extraction · TICA · Support vector machine

1 Introduction

Recent years have been tremendous amount of research being carried out in the field of biometrics. The idea of feature extraction by physical attributes like face, fingerprint, voiceprint or any of other characters to prove human identification has huge importance. Any property of human beings that is peculiar can serve as a specification measure for verification, recognition and for categorization.

Gender classification is arguable one of the most important tasks for human beings- many interactions directly depend on the correct gender identification of the persons involved. Many of the researchers have focused on gender recognition in several application fields such as biometric authentication, smart human identification, safety monitoring systems, computer vision approaches for monitoring

S. Garg (✉) · M.C. Trivedi
ABES Engineering College, Ghaziabad, India
e-mail: Shivigarg16@gmail.com

M.C. Trivedi
e-mail: Munesh.trivedi@abes.ac.in

S.C. Satapathy and S. Das (eds.), *Proceedings of First International Conference on Information and Communication Technology for Intelligent Systems: Volume 2*, Smart Innovation, Systems and Technologies 51, DOI 10.1007/978-3-319-30927-9_39

people, human computer interface etc. For example, for person identification gender recognition acts as a pre-processing step. A gender recognition system extracts person's gender information and then plays suitable advertisements. Gender identification methods can be broadly classified into four categories as Voice Based, Face Based [1], Hand Based and gait Based.

We have tremendous amount of information in human faces which is very helpful in social interaction between people. In today's time for security purpose biometric is an essential and very important feature and human face is one of the most interesting and important biometric feature, so for many researchers it's a challenging research problem.

Gender Classification via face area improves speed and accuracy as it only needs to match the face in database to nearly half. Gender classification involves extracting features from faces and classifying those features using labelled data. Approaches to gender classification can be categorized based on:

(I) Feature Extraction Methods [2].
(II) Classification Methods.

Feature extraction can be broadly broken down into:

(i) Appearance-based methods
(ii) Geometry-based methods

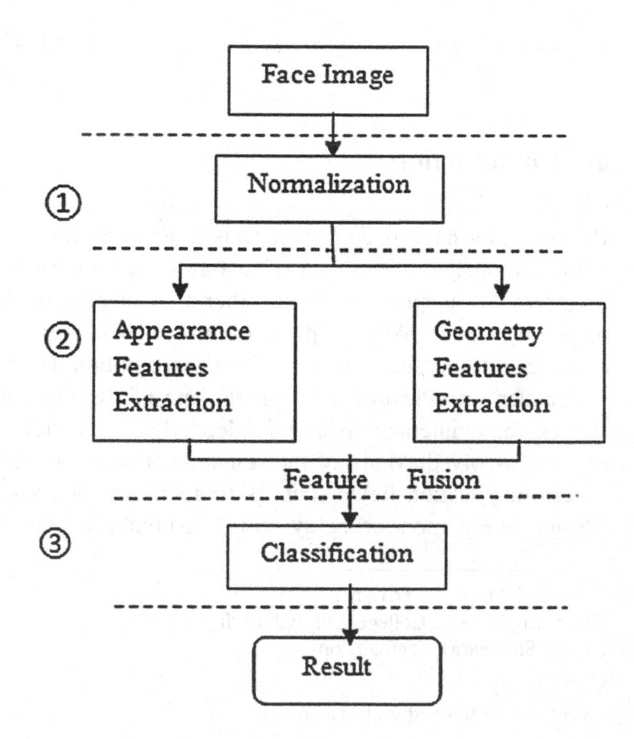

Fig. 1 Architecture of gender classifier [2]

Geometry-based features represent high-level face description such as distances between nose, eyes and mouth, face width, face length, eyebrow thickness and so on whereas Appearance-based features use low-level information about face images areas based on pixel values. For the case of classification, most of the works use nearest neighbours, neural network and support vector machine. The framework of gender classifier can be depicted using the following Fig. 1 in which it is shown that any face image is taken and then it is normalized. After that the appearance and geometry based features are extracted and they are fused together, then any classifier is applied to obtain the results.

2 Literature Review

Several different approaches have been reported for solving the problem of gender classification. Many people have made significant improvements in this area as:

Lin et al. [3] has proposed a gender recognition method based on color information which consists of four parts: face detection, then eye detection through this second phase feature extraction take place and finally gender classification. They have used three color features and SVM classifier for gender recognition. They have worked on different size of face images which are captured by using low cost webcam. Their result shows 80 % accuracy in classification of gender recognition.

Xu et al. [4] has proposed a hybrid approach for gender classification which consists of three phases as firstly face image get normalized then from that they extract the features from the feature vector and then used it as input and reaches output. They have used SVM and RBF kernel as classifier. Xu states that the success rate is 92.38 % which is already high but if optimization can be made then localization time get reduced.

Wang et al. [5] has proposed a technique for gender classification namely Local Circular pattern (LCP) which is an improvement of LBP (Local Binary pattern). They have used clustering based quantization which is more robust to noise. They have used SVM with a linear kernel for the classification.

Rahman et al. [6] gave a face detection and gender identification technique in non- uniform background. Is has been done by detecting the human face area in an image and by detecting facial features based on measurements in pixels. The facial reason was preprocessed by convolving with Gabor Filter. This technique can handle huge variations in static color of images.

Jain et al. [7] introduced Independent Component Analysis (ICA) as to represent an image in low dimension space. It was used to symbolize face images as linear speculation of different basis function and for dimensionality reduction. Many classifier has been used as cosine classifier, LDA and SVM. On combining SVM and ICA they have achieved an accuracy of 96 % in gender classification.

3 Proposed Approach

The proposed gender classification framework consists of a series of different phases that helps to produce the final output as:

In the first phase, face is detected from the whole image so as to make sure that our work is mainly focused on the face not on the background information. In the second phase, facial features are extracted automatically from the facial images then a classifier is trained to identify gender based on the extracted features. This whole process of proposed system can be depicted as: (Fig. 2).

3.1 Face Detection

Face detection can be defined as detection of face from an image or from a scene. Many face detection techniques have been proposed in last few years [8]. In this field the most significant advances in the last decade are because of the work of Viola and Jones who proposed a face detector based on rectangular Haar Classifier [9]. This detector is the fastest face detector reported so far. The Viola Jones face detector is presently the gold standard against which other face detection techniques are benchmarked.

The main idea of Viola Jones algorithm is detecting faces by scanning the sub-window of an input image. The first step is to convert the input image into an integral image. By integral image we can sum up the pixel values under the rectangular region (e.g., a region defined by point A, B, C, D), which becomes very efficient if we need to sum up the pixels within many regions of interest within an image [9]. Sum of rectangle = D − (B + C) + A (Fig. 3).

Now figure below shows a frontal facial image on which Viola Jones is applied to detect the face only and detected face is depicted in the figure using a rectangle (Figs. 4 and 5).

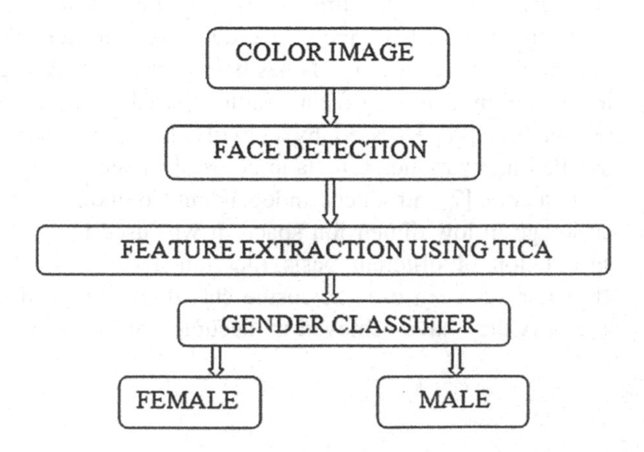

Fig. 2 Block diagram of proposed gender recognition scheme

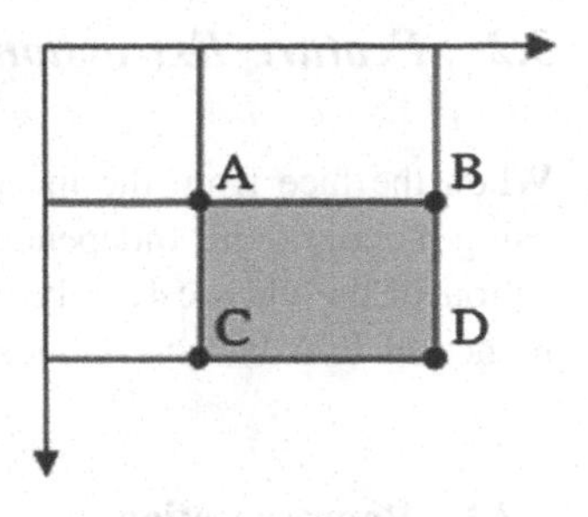

Fig. 3 Sum calculations [9]

Fig. 4 Frontal face image

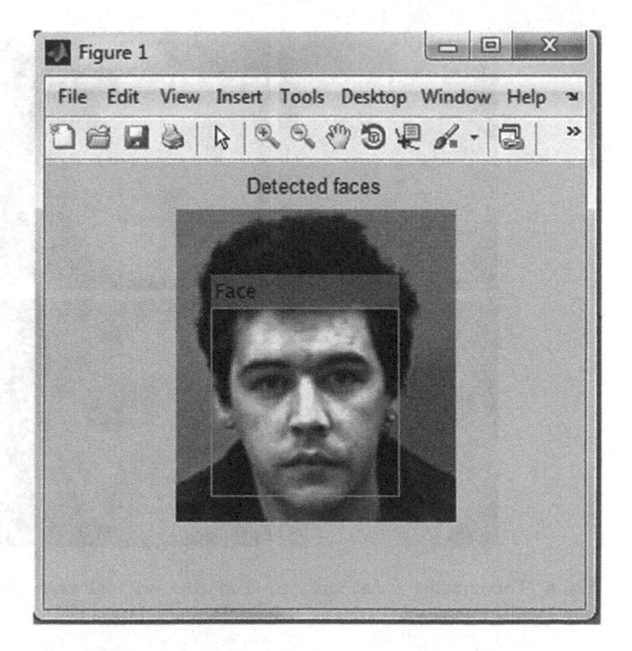

Fig. 5 Detected face

3.2 *Feature Extraction*

When the face from the image is detected, the features of the faces are extracted using Topographic Independent Component Analysis (TICA). TICA is a modification of the classic ICA based feature extractions methods enhancing the performance of ICA-based methods [10].

3.2.1 Representation

For Face Detection we have two types of labels as Convex Hill and Convex saddle hill. These labels are generated by analysis of facial surface and form a Topographic Mask (TM). Convex hill compromises the area like eyes, eyebrows, nose, mouth etc., we refer these textures as expressive textures (ET) and convex saddle hill covers the rest of the part of the face [11]. Since our main aim is to classify the gender and mainly convex saddle hill label is used for emotion classification, we choose convex hill area and in that we also extract the eyes, mouth and nose of a face. For a particular subject, we have TM and ET here TM is used for facial expression pattern. For a specific purpose, on taking (TM, ET) together as of someone's neutral face with its expressive face we can easily identify expressions (Fig. 6).

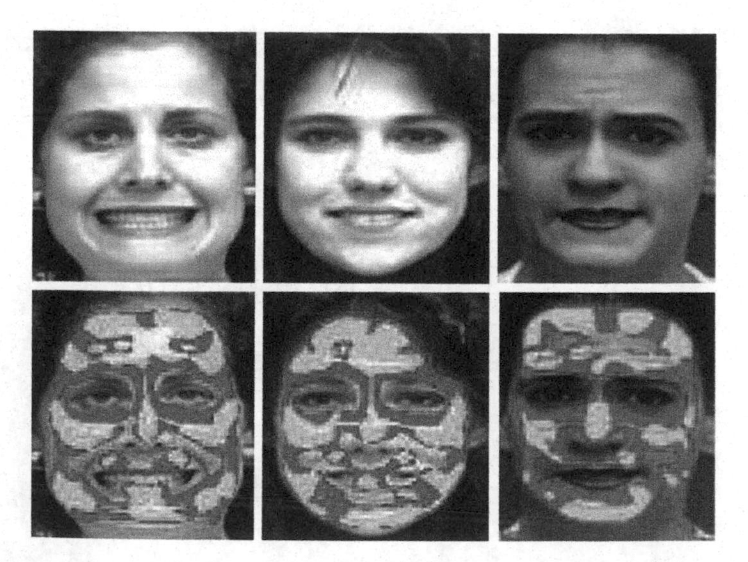

Fig. 6 Topographic facial analysis (*Top row* original faces, *bottom row* labelled faces with convex hill (*in red*) and convex saddle hill (*in pink*)) [11]

3.2.2 Classification

For each and every individual expression of a person, we have a unique TM. Here TM denotes labelled and non-labelled region in binary form as '1' for labelled region (red and pink region) and '0' as non-labelled region. The disparity of two faces can be expressed as:

$$D_{TM} = \frac{1}{N}\sum TM_n \, xor \, TM_e \tag{1}$$

Here,

N	total number of pixels
TM_n	Neutral Mask
TM_e	Expressive Mask

Now correlation C_{ET} defined as:

$$C_{ET} = E(t_n \, t_e) - m_{tn} \, m_{te} / \sigma(t_n)\sigma(t_e) \tag{2}$$

Here

t_n and t_e	are neutral textures and expressive textures
E()	is mean operation
m_{tn}, $\sigma(t_n)$, $m(t_e)$ and $\sigma(t_e)$	are mean and variances of two textures

Similarity is defined as Similarity Score S_{exp} as:

$$S_{exp} = (1 - D_{TM}) + C_{ET} \tag{3}$$

On calculating the similarity score we get a range within [0,2] in a descending order. This similarity curve is for universally defined seven facial expressions namely neutral, sad, fear, angry, disgust, surprise and happy. We also have a monotonicity similarity curve as: (Fig. 7).

Now, In the feature extraction phase, firstly the input image given is resized so as to increase the speed of the extraction thus giving us a mixed image and then a restored image and this is depicted in figure. The features finally extracted i.e. eyes, nose and mouth are depicted in figure. The features are represented in terms of matrices (Figs. 8 and 9).

3.3 Gender Classification

The features obtained in the form of matrices are used to train a classifier, which classifies any test image to be of male or female. In present work, Support Vector Machine has been used for this purpose. Support vector Machine is a learning

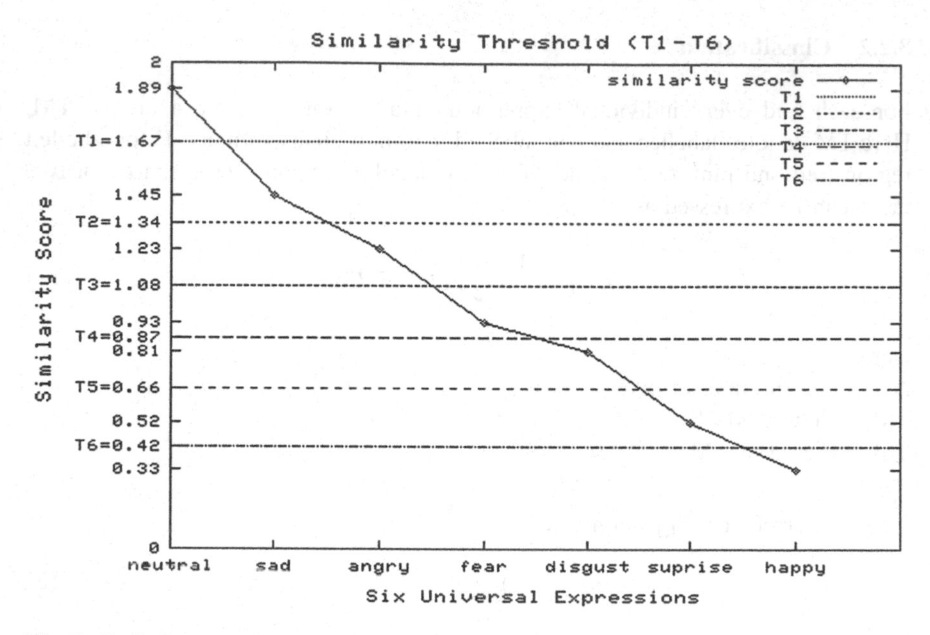

Fig. 7 Similarity curve [11]

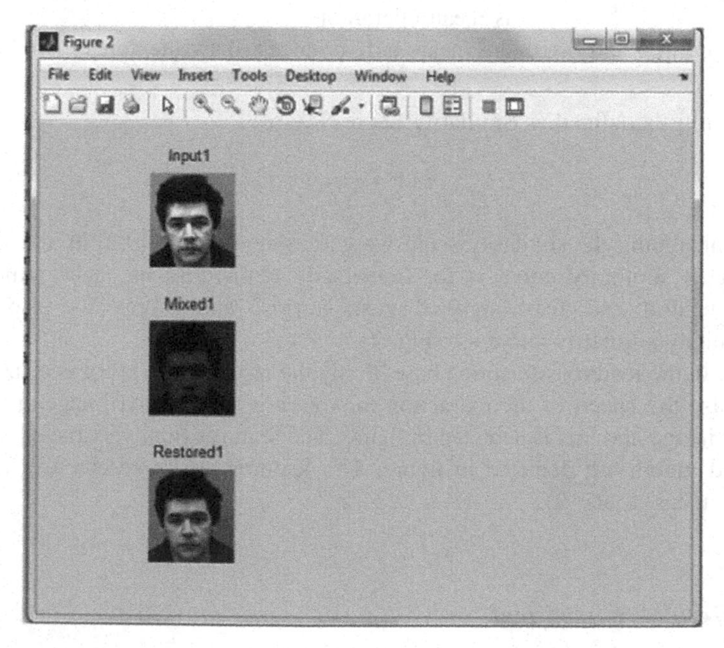

Fig. 8 Resize image

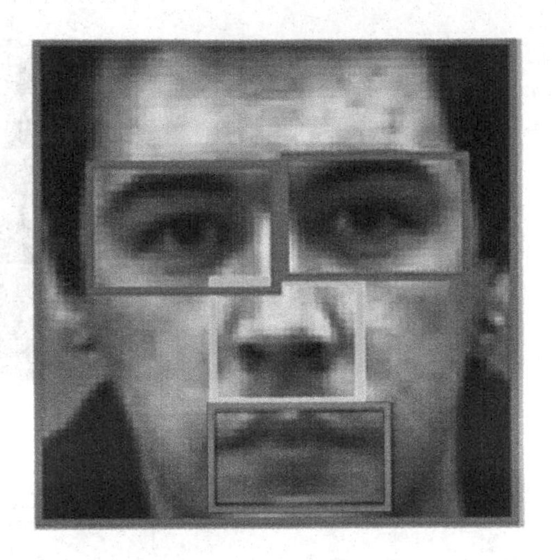

Fig. 9 Extracted features

algorithm for regression and pattern classification widely used in neural network. Given a set of training examples, each marked as belonging to one of two categories, an SVM training algorithm builds a model that assigns new examples into one category or the other, making it a non-probabilistic binary linear classifier [12]. For minimizing classification error in unseen samples SVM generates an Optimal Hyper Plane VLFeat library [13] is used for SVM classification. The classifier is trained to output 0 for female and 1 for male. A test image is taken and then from this, face is detected and then features are extracted. The extracted features are in the form of matrices. Support Vector Machine classifier combines these features to distinguish male persons from female ones. These matrices are compared with the training database matrices and it is seen that if the test image matrices match with which image matrix in the training dataset. If it matches with the male image, then the image is classified as male or if with a female, the test image is classified as female. If the test image extracted features matrices value match with the training image matrices value less than 0.5, then the image is classified as 0 and that means female otherwise it gives output gender = 1 and i.e., male. In the end what we are expecting is depicted in the Fig. 10.

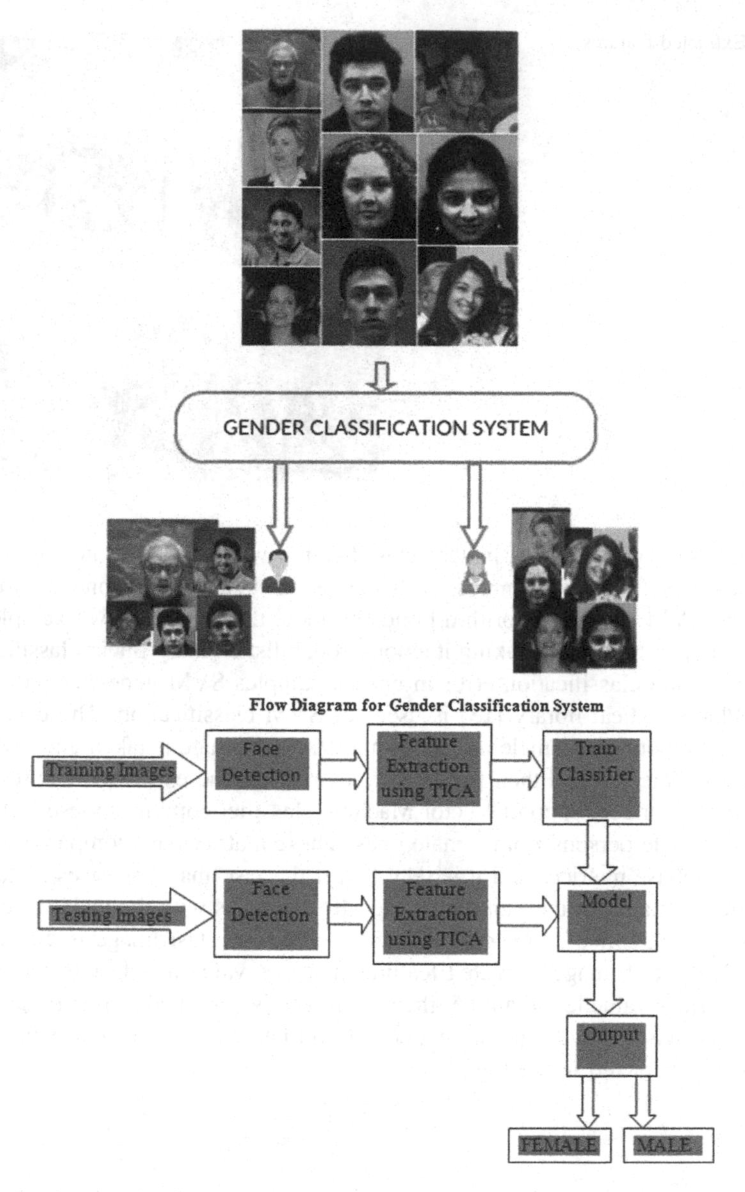

Fig. 10 Gender classification system

4 Experimental Results

In this work, facial images are collected from the database Essex Face 94 Database and some images are collected randomly by Google search. All these images are gathered in different illuminations, different lighting conditions and different facial

Table 1 The classification rate of the proposed gender classification scheme is as

Total no of images used in testing			
Gender	Total images used in testing	Correctly classified images	Accuracy ratio in %
Male	100	98	98
Female	100	96	96

Table 2 5 different expression images used/head

5 Different expression images used/head		
Gender	Avg. recognition time (in sec)	Recognition time diff. expressions
Male	5	1.5
Female	4	2

poses. All images are resized to 200 × 250. The database contains 600 facial images in which 300 are of male and 300 are of female. Out of 600 images, training set has 400 images, 200 each in number for male and female both respectively while remaining 200 images have used for testing. Testing dataset has 100 male images and 100 female images. The result of the gender classification system can be depicted in the form of table shown: (Tables 1, 2) (Figs. 11 and 12).

Out of the 100 male images, the proposed gender classification system correctly identifies 98 male images and incorrectly classifies 2 male images to be of female. And out of the 100 female images, the system correctly classifies 96 female images and incorrectly classifies 4 female images to be of male. Combining the results of the classification, the gender identification system accuracy comes out to be 96 %. This has been validated through SVM classifier (Fig. 13).

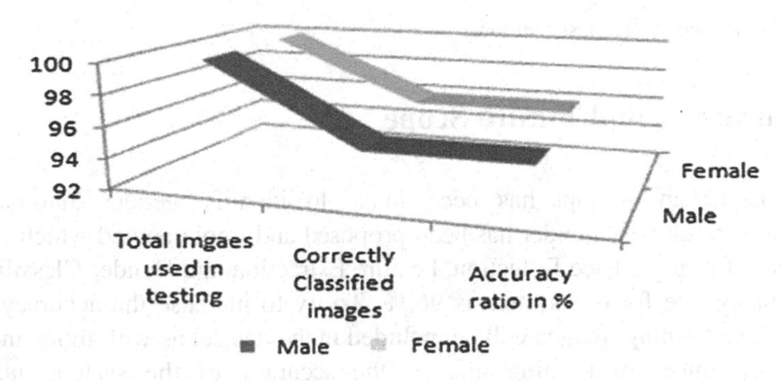

Fig. 11 Testing dataset has 100 male images and 100 female images

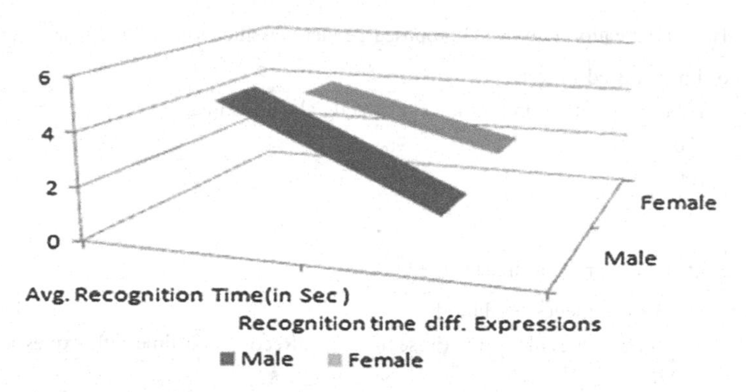

Fig. 12 Out of the 100 male images, the proposed gender classification system correctly identifies 98 male images and incorrectly classifies 2 male images to be of female

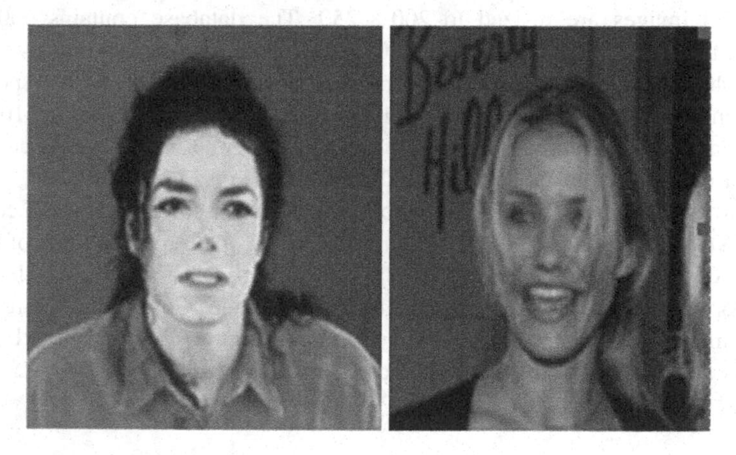

Fig. 13 Some incorrectly classified faces

5 Conclusion and Future Scope

In this paper an attempt has been made to identify gender automatically. A technique to classify gender has been proposed and implemented which consists of a series of steps as Face Detection, Feature Extraction and Gender Classification. The accuracy rate for our system is 96 %. To try to increase the accuracy of the system, more training images will be included in the dataset as with more and more increasing number of training images, the accuracy of the system increases. Classification can be done using some other classifier also, so as to check the difference in accuracy of the system.

References

1. Khryashchev, V., Priorov, A.., Shmaglit, L., Golubev, M.: Gender recognition via face area analysis. In: World Congress on Engineering and Computer Science (2012)
2. Saber, E., Tekalp, A.M.: Frontal-view face detection and facial feature using color, shape and symmetry based cost functions. Pattern Recogn. Lett. **19**(8), 669–680 (1998)
3. Lin, G.-S.., Zhao, Y.-J.: A feature- based gender recognition method based on color information. In: First International Conference on Robot, Vision and Signal Processing (RVSP), pp. 40–43 (2011)
4. Xu, Z., Lu, L., Shi, P.: A hybrid approach to gender classification from face images. In: 19th International Conference on Pattern Recognition, ICPR, pp. 1–4 (2008)
5. Wang, C., Huang, D., Wang, Y., Zhang, G.: Facial image-based gender classification using local circular patterns. In: 21st International Conference on Pattern Recognition (ICPR), pp. 2432–2435 (2012)
6. Rahman, M.H., Bashar, M.A., Rafi, F.H.M., Rahman, T., Mitul, A.F.: An automatic face detection and gender identification from color images using logistic regression In: International Conference on Informatics, Electronics and Vision (ICIEV), pp. 1–6 (2013)
7. Jain, A., Huang, J.: International independent components and support vector machines. In: Proceedings of the 17th International Conference on Pattern Recognition (ICPR), pp. 558–561 (2004)
8. Yang, M.H, Kriegman, D.J., Ahuja, N.: Detecting faces in images: a survey. IEEE Trans. Pattern Anal. Mach. Intell. **24**(1), pp 34–59 (2002)
9. Viola, P.A., Jones, M.J.: Robust real-time face detection. Int. J. Comput. Vis. **57**(2), 137–154 (2004)
10. Hyvarinen, A., Hoyer, P.O., Inki, M.: Topographic independent component analysis. Neural Comput. **13**(7), 1527–1558 (2001)
11. Wei, X., Loi, J., Yin, L.: Classifying Facial Expression based on Topo-Feature Representation, Affective Computing. In Tech, China (2008)
12. (Online). Available: http://en.wikipedia.org/wiki/support Vector machine/
13. (Online). Available: http://www.vlfeat.org/

Provision of XML Security in E-Commerce Applications with XML Digital Signatures Using Virtual Smart Card

Joannah Ravi and Balamurugan Balusamy

Abstract The paper aims at enhancing XML security by generating an XML digital signature capable of providing the major security features such as authentication, integrity, non-repudiation and confidentiality [14]. It also extends the concept of Information hiding which overcomes the hidden problem of traditional XML digital signature generation called "MID-WAY READING". The security of the document is ensured by a process called 'information hiding'. The document to be sent is digitally signed as well as encrypted and thereby ensuring excellent security level during the business transactions in an e-commerce environment and in addition to that, the private key used for signing the document is stored in a virtual smart card that provides enhanced security.

Keywords XML · Digital signatures · Virtual smart cards · Security · Cryptography · Information hiding

1 Introduction

Cryptography is a prominent way to encrypt information. Digital signature is a cryptographic element in electronic security to ensure integrity, authenticity, and non-repudiation of data [2]. XML stores information in its structure and data nodes [3]. XML technology is used in securing web oriented business transactions and applications [4]. Use of XML based security provides confidentiality. XML Digital signature is widely used to authenticate E-commerce Documents [5]. The Digital signature protocol identifies forgery and alteration [6]. But in the existing digital signature process, the message digest is encrypted but not the message, so it has the

J. Ravi · B. Balusamy (✉)
School of Information Technology and Engineering,
Vellore Institute of Technology, Vellore, Tamil Nadu, India
e-mail: balamuruganb@vit.ac.in

J. Ravi
e-mail: joannah.ravi2013@vit.ac.in

S.C. Satapathy and S. Das (eds.), *Proceedings of First International Conference on Information and Communication Technology for Intelligent Systems: Volume 2*,
Smart Innovation, Systems and Technologies 51, DOI 10.1007/978-3-319-30927-9_40

danger of "MID-WAY READING" [7] allowing the third party to read the message [8]. In order to avoid this, Information Hiding technology based on XML encryption is used [9]. The intention of the paper is to overcomes the problem of MIDWAY-READING in the traditional way of XML digital signature generation. The Electronic signatures are used to sign an entire document, but with the help of XML Digital Signature, portions of a document can be signed. XML digital signatures are specially designed for XML document transactions, finally adding up to the level of security, a virtual smart card is used to share the Private key. It is a type of embedded device that is easily accessible [10]. The key that is used to sign the XML document is generated and stored in the virtual smart card. This gives the user full control over the business transaction.

A. ***Problems in the Traditional Method***

The literature review on the existing scenario and current developments lead to the finding of three problems.

1. The message digest is only encrypted and not the message [11].
2. The Traditional Electronic signatures are used to sign an entire document, but with the help of the proposed XML Digital Signature, portions of a document can be signed. This encrypts the vital information of the document selectively by using document splitting techniques [3], reducing the time of encryption and decryption of the document [12].
3. Encryption using asymmetric key algorithms is comparatively slow, according to the metrics and is not optimal when the data size is high. Hence, not advisable for bulk encryption [13] and symmetric algorithms are employed, that use the same key for encryption as well as decryption, instead. These problems are and not limited to, vulnerabilities to different kinds of attacks, frauds [14].

B. ***Contributions of the paper***

The paper resolves certain critical issues associated with the E-commerce environment and proposes new methods for increasing the existing security.

1. To create a secure e-commerce environment that enables secure B-B (Business to Business transaction) by generating an XML based digital signature for an XML input data using ECDSA and SHA-1.
2. To overcome the problem of MIDWAY-READING. Facilitating the business transactions between the buyer and the seller to occur securely over the network.
3. To eliminate the cost of implementing EDI in VANS [15]. Security throughout the environment mainly focusing on key management level of security at server side and end to end encryption at client side.

2 Related Work

If the document is signed and transmitted, it can be subjected to MIDWAY-READING. Hence, the document must be encrypted before signing. This solution is called Information Hiding. Works to solve similar problems are discussed. In [16] the implementation of Digital Signature Algorithms on 8-bit Smartcards was explained. It describes the implementation of 1152-bit RSA and ESIGN digital signature algorithms on 8-bit smartcards. The result shows that it takes less than 0.5 s to generate 1152-bit ESIGN signature on H8/3113 Smartcards without a coprocessor while 1152-bit RSA signature takes more than 150 s. In [17] the solution to the DPA attacks is explained. The solution describes the use of Stenography and smart cards. Stenography provides the method of hiding the private key inside the smart card. The user has to provide his password before retrieving the key from the smart card. In [6, 1] the Differential Power Analysis (DPA) attacks [18] are described [19]. The techniques to overcome the DPA attacks have been explained [20]. Paper [1] introduced a novel high-speed methodology for security evaluations of integrated processor systems, using power emulation [21].

3 Comparative Analysis

The concept of homomorphic encryption was proposed in [1]. Digital Signature uses public key cryptography or Asymmetric key cryptography. It uses two keys; one to create digital signature and other to verify it [22]. For verification of the XML signature, we calculate the message digest using hash algorithm and compare this with the message digest decrypted. Java supports a wide spectrum of cryptographic capabilities. One such capability is out-of-the-box JDK support for digital signatures [23]. The Core Java 2 platform has an extensible infrastructure when compared to other languages and it has the ability to generate digital signatures and also includes the key management system for managing the keys [24]. SHA-1 hashing is used as it has the interoperability and backward compatibility features [25]. It also has anti-exhaustive capability and anti-cryptanalysis capability. The cryptographic algorithm used here for signing and verifying the message digest is elliptic curve cryptography. ECDSA has the upper hand in enhancing efficiency of the algorithm. In devices with less processing capability, storage and bandwidth, the efficiency of the device is important. ECDSA uses less resources when compared to other algorithms like RSA and so it is used in embedded devices. It also has a smaller key length which helps complete cryptographic operations fast due to fewer processor cycles [13, 26].

Upon analyzing, it was found that RSA can be used for both signature generation and encryption [27], whereas DSA supports only signature generation process. The usage of Elliptic curve cryptography [28, 29], an Enhancement of DSA can be used for both signature generation and validation. RSA and DSA uses 1024 bit and 512-bit key respectively, whereas ECDSA public key uses 160 bits providing more security than the previous algorithms [30]. In all three conditions key, signature and certificate files, the size increases significantly with key size for the RSA signature scheme whereas the ascent is relatively moderate for the ECDSA signature scheme. ECDSA uses private key to sign and public key to validate whereas DSA and RSA use a public key to verify the signature.

3.1 Structuring the Analysis

The architecture diagram in Fig. 1 describes the process at buyer and the process at the seller. A server is established at seller side and the buyers can buy the products

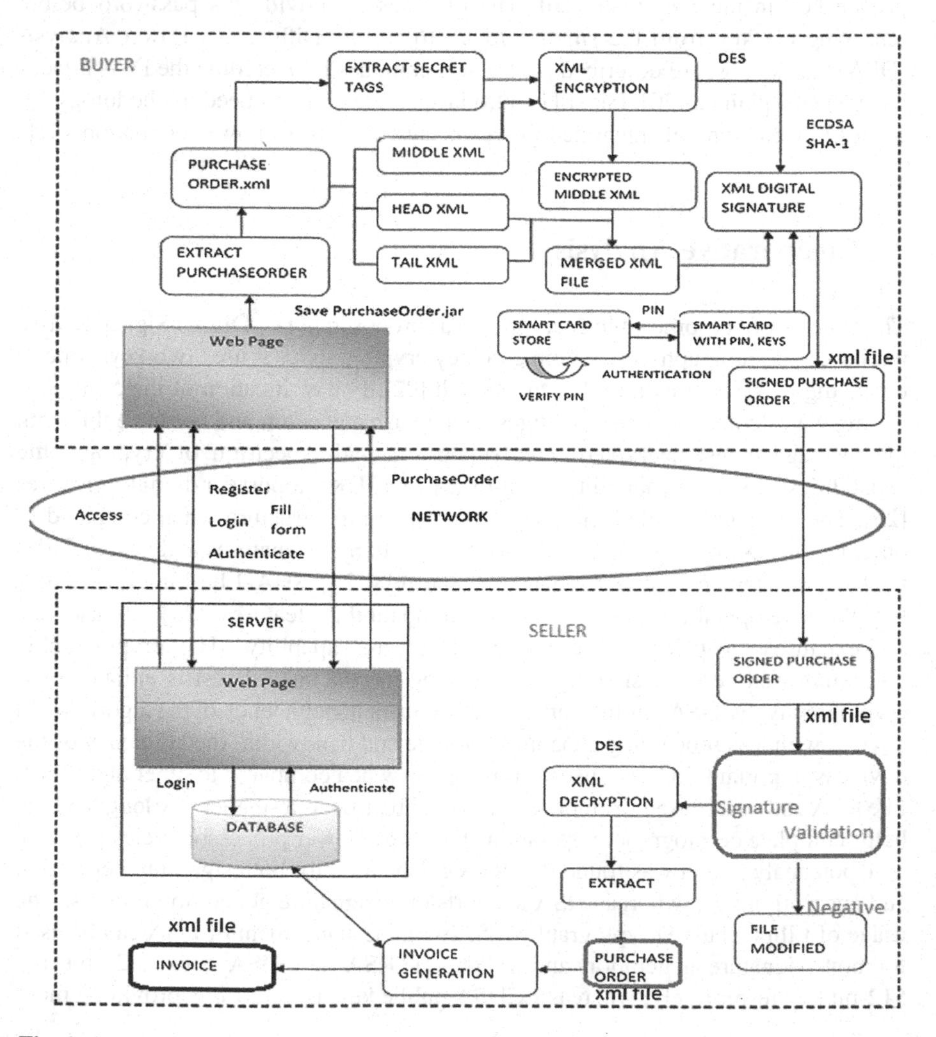

Fig. 1 System architecture

of their own choice by logging onto the seller website [31]. The process involves creation of the purchase order in XML format based on the details filled in by the buyer. This is XML encrypted and signed by the buyer, and is sent to the seller. The keys used for signing are stored in virtual smart card. The USB Flash drive is used as a virtual smartcard [10]. The sender verifies the signed purchase order and decrypts it. The Sender generates the invoice in which he specifies the price details of the product. The invoice in turn is sent to the buyer after encrypting and signing it. The Buyer verifies and decrypts the document. The final invoice in XML format is obtained. The main purpose of the project is to focus on protecting the purchase order and invoice document that contain sensitive data like card number, type and pin that need security from hackers and from MIDWAY-READING. The purchase order that is to be signed by the buyer is zipped with the secret code given by the buyer during purchase order filling. So, only the buyer could access the purchase order document. The process of network establishment is explained [32]. The buyer will have to log on to the seller website to view the product type, design and price. In case he wants to purchase, one has to make registration first; as access is denied for non-registrants. After registering using user name and password the user can log-in and fill the form with required details for buying the product. Based on the details filled the purchase order is created. The purchase order can be downloaded from the server in the zipped file. Once the user needs to open the file, he/she is prompted to give the secret code. Only if he/she enters the secret code, the purchase order can be extracted or can be viewed. So, the sensitive data entered like card number is secured from external web hacking attacks [14]. The network establishment is done using the WLAN.

1. **Working of Buyer**

Buyer after logging in and filling the details can download the purchase order file and also the jar file provided by the seller in order to sign and send the XML file. The jar file has the GUI that can be used for encryption and signing. In case of encryption, two methods are possible; either the buyer can encrypt only the secret tags of the XML file or can split the XML file into three parts in which middle contains the secret tags and encrypt only the middle part. Then three parts, the head, encrypted middle and tail merge. If the buyer wants to sign the document, he must insert his virtual smart card and get authenticated through his PIN. Now, the encrypted file is signed with the keys that are stored in the virtual smart card [33]. The virtual smart card used here is USB drive. The virtual smartcard is created, using software called safe house. The signing is done, using ECDSA [34, 11] and SHA1 [35] algorithms. This purchase "order.xml" signed using "purchaseorder.jar" is sent to the seller through the WLAN network.

2. **Working of Seller**

The seller after receiving the signed purchase order.xml verifies the signature and if it is valid, he decrypts the XML document. The purchase order is used for the generation of invoice by adding the price, details of the products that were purchased. The verification process involves calculating the hash for both the existing original document and received document. If both the hashes match, the document is authorized else is considered modified, and buyer is asked to resend the document. Traditional XML encryption/decryption are used. Also the digital signature is XML digital signature. The invoice is again signed using ECDSA XML digital signature and sent to the buyer. The buyer again verifies and decrypts the invoice and gets the final invoice.xml file.

Information Hiding:

If an XML DOC is signed directly and transmitted in the public network, the contents in the document can be viewed by the third party, if the document is hacked. Hence, the XML document is encrypted, using XML encryption (XMLENC) which is used to encrypt the data and produce the cipher text in the form of XML document. The XMLENC doc is signed using XMLDSIG SHA-1 and ECDSA algorithm and then transmitted. Thus confidentiality of the document is maintained and this provides.

Splitting an XML Document:

The XML document can contain secret information which is present in the form of tags. If the XML document contains set of continuous secret tags, then that particular set of secret tags are extracted in the form of XML document. The XML document thus extracted is entirely encrypted. This encrypted content is merged with the remaining part of non-encrypted document.

Extracting Secret Tags:

Search for the secret tags like Credit Card Number, Security Code, and Credit Card Type, extract them and encrypt only those particular tags using XML encryption. Thus, it saves time rather than encrypting the entire document.

XML Encryption:

The secret XML tags and the split XML document which is to be kept confidential are encrypted using AES XML Encryption [17]. Both XML Signature and XML Encryption use the Key Info element, which appears as the child of a Signed Info. Encrypted Data or Encrypted Key element provides information to a recipient about what key is to be used in validating a signature or decrypting encrypted data. The Key Info element is optional: it can be attached in the message, or be delivered through a secure channel.

Modular Description:

The design of the XML digital signature generation uses the following modules of implementation as seen in the process.

4 Process

Key Generation:

The buyer will provide his virtual smart card to the seller during direct interaction. The seller generates the key for ECDSA XML digital signature and stores the key in virtual smart card of buyer. The buyer can use the key from his virtual smart card, whenever he wants to sign the document. The key cannot be lost or obtained by any other users. The buyer can access the key only when he is authenticated. Information Hiding is achieved by hiding the data in the XML document. The process of information hiding is as follows:

Encryption of the content:

There are two ways to encrypt the content:

1. Splitting the XML document into three XML files, one with secure data and other two with non-secure data and small XML file which has important data is encrypted.
2. Extracting the secret tags from XML document and encrypt only the secret data. If the XML document is entirely secret then the first method can be chosen excluding splitting. If the document contains only certain secret tags like card number, security code and other details and key information, then the second method is chosen.

XML digital signature:

The process of XML digital signature is as follows:

1. The encrypted XML document is used to calculate the hash value or message digest using SHA-1.
2. The message digest thus calculated is then encrypted (signed), using ECDSA.
3. The signature is then attached with the XML document which is encrypted using information hiding.

Signature Verification:

The XML digital signature thus obtained is again parsed and validated. This shows whether the document is modified or not. If the document is valid and not modified, it is decrypted and the original XML file is obtained.

The overall description of the modules of the proposed XML digital signature is as follows.

Module 1: Key Generation:

The buyer will provide his virtual smart card to seller, during direct interaction. The seller generates the key for ECDSA XML digital signature and stores the key in virtual smart card of buyer. The buyer can use the key from his virtual smart card, whenever he wants to sign the document. The key cannot be lost or obtained by any other users. The buyer can access the key only when he is authenticated.

Module 2: Xml Encryption:

The input XML document is encrypted with the key generated by using DES cryptographic algorithm. If the document has a set of continuous secret tags, then the XML document is split and then encrypted. If there are a few secret tags, then the important tags are obtained and are merely encrypted. The Encryption method used here is XML encryption which gives the Encrypted file also in the form of XML. The algorithm used for encryption is Data encryption standard.

Submodules:

In case splitting is used, there are two sub modules: Splitting and Merging. Splitting is done using vtd-XML parser [36] based on secure element tags. The XML document is split into smaller fragments based on the child elements of the document and then the important fragments are only encrypted. All these fragments both encrypted and non encrypted ones are merged into a single XML document.

Module 3: Xml Signature Generation:

For the XML document, a hash value or message digest is calculated using a hash algorithm and the hash value is encrypted using a private key from virtual smart card based on asymmetric cryptographic algorithm. The algorithm used for calculating the hash value is Secure Hash Algorithm-I and the algorithm used for Encryption of the Hash value is Elliptic curve cryptography [37].

Module 4: Xml Signature Verification:

The XML signature verification process involves two steps. In the first step: For the signed XML document, a hash value or message digest is calculated using the hash algorithm SHA1 (Secure Hash Algorithm).

In the second step, the XML signature is decrypted with the public key generated by using EDSA algorithm (Elliptic curve cryptography) and a hash value is obtained. The hash values obtained in both steps are compared. The signature is said to be verified when both the hash values are same. Thus the document sent by the sender is secure and authenticated.

Module 5: Xml Decryption:

The encrypted document attached with the signature is received. In the received XML document, the important information is secure as it is encrypted. So the XML document is decrypted with the key generated by using AES (Advanced encryption standard) algorithm or with the des algorithm which uses the key info stored in encrypted document for decryption.

Module 6: Virtual Smartcard:

Virtual Smartcard is used to store the private key used for signing. This key is retrieved from the virtual smart card only if the pin that is entered by the user is correct.

Module 7: Xml Invoice Generation:

Based on the details given by the buyer while filling the form, purchase order in XML format is created using PHP and this purchase order.xml is zipped, password protected by the secret code and is uploaded at the server. The file is signed and sent to the seller by the buyer and seller verifies and converts the purchase order into invoice by quoting the total price of the products purchased by the buyer. The conversion is done using PHP code.

5 Performance Measure of the Signature Validation

To measure the performance of the signature validation, a typical shopping portal is created and the transaction XML details are signed and validated. The system is compared to other traditional XML digital signature creation system available and the results showed better performance of the proposed system in terms of processing time and space occupied [38].

6 Conclusion and Future Enhancements

The new system of XML digital signature provides integrity, confidentiality, authentication, non-repudiation, minimum use of resources, less power consumption and eaves-dropper free communication. It also overcomes the hidden problem in the traditional method of digital signature, called MIDWAY-READING through information hiding. The key used for digital signature generation is stored in virtual smart card [39]. An E-commerce application needs security at purchase level. If the purchase of the products is done through credit card, then the details in the purchase order are encrypted and signed, using XML security. Moreover, the project focuses on using XML as input. Data XML is used because any data can be converted to XML. It provides secure storage of document. It also provides granularity. Thus, the project provides security to XML documents by using XML Encryption and XML Digital Signature.

References

1. http://www.oasisopen.org/committees/download.php/20508/oasis-dss-1.0-interop-wd-07.doc
2. Dournaee, B., Dournee, B.: XML Security. Mcgraw-Hill, New York (2002)
3. Groz, B., et al.: Static analysis of XML security views and query rewriting. Inf. Comput. **238**, 2–29 (2014)
4. www.w3.org/TR/XMLdsig-core
5. Barhoom, T.S.M, Shen-Sheng, Z.; Trusted exam marks system at IUG using XML-signature. In: The Fourth International Conference on Computer and Information Technology, CIT'04. IEEE (2004)
6. Rao, W., Gan, Q.: The performance analysis of two digital signature schemes based on secure charging protocol. In: International Conference on Wireless Communications, Networking and Mobile Computing. Proceedings, vol. 2. IEEE (2005)
7. ESA-02: SOAP Interfaces vulnerable to XML signature element wrapping attacks. Retrieved Apr 2012, from http://www.eucalyptus.com/eucalyptus-cloud/security/esa-02
8. Tao, H., Qihai, Z., Le, Z., Zhongjun, L., Xun, L.: An improved scheme for e-signature techniques based on digital encryption and information hiding. In: 2008 International Symposiums on Information Processing (ISIP), pp. 593, 597, 23–25 May 2008
9. Jie, Y.: Algorithm of XML document information hiding based on equal element. In: 2010 3rd IEEE International Conference on Computer Science and Information Technology (ICCSIT), vol. 3. IEEE (2010)
10. How to Enable Smartcard Support. Retrieved Apr 2012. http://www.safehousesoftware.com/manual/SafeHouse.htm#user_s_guide/SMARTCARD_Virtual.htm
11. Bedi, H., Yang, L.: Fair electronic exchange based on fingerprint biometrics. Int. J. Inf. Secur. Privacy (IJISP) **3**(3), 76–106 (2009)
12. Gómez, J.M., Lichtenberg, J.: Intrusion detection management system for ecommerce security. J. Inf. Priv. Secur. **3**(4), 19–31 (2007)
13. Grabher, P., Großschädl, J., Page, D.: Light-weight instruction set extensions for bit-sliced cryptography. In: Cryptographic Hardware and Embedded Systems–CHES 2008, pp. 331–345. Springer, Berlin (2008)
14. Chan, G.Y., Lee, C.S., Heng, S.H.: Defending against XML-related attacks in e-commerce applications with predictive fuzzy associative rules. Appl. Soft Comput. **24**, 142–157 (2014)
15. Meadors, K.: Secure electronic data interchange over the Internet. IEEE Internet Comput. **9**(3), 82–89 (2005)
16. Wajih, E.H.Y., Mohsen, M., Rached, T.: A secure elliptic curve digital signature scheme for embedded devices. In: 2nd International Conference on Signals, Circuits and Systems, SCS 2008, pp. 1, 6, 7–9 Nov 2008
17. Masoumi, M., Mohammadi, S.: A new and efficient approach to protect AES against differential power analysis. In: 2011 World Congress on Internet Security (WorldCIS). IEEE (2011)
18. Hasan, M.A.: Power analysis attacks and algorithmic approaches to their countermeasures for Koblitz curve cryptosystems. IEEE Trans. Comput. **50**(10), 1071–1083 (2001)
19. Mahmoud, H., Alghathbar, K.: Novel algorithmic countermeasures for differential power analysis attacks on smart cards. In: 2010 Sixth International Conference on Information Assurance and Security (IAS). IEEE (2010)
20. Kocher, P., et al.: Introduction to differential power analysis. J. Crypt. Eng. **1**(1), 5–27 (2011)
21. Krieg, A., et al.: Accelerating early design phase differential power analysis using power emulation techniques. In: 2011 IEEE International Symposium on Hardware-Oriented Security and Trust (HOST). IEEE (2011)
22. Karras, D.A., Zorkadis, V.: Neural network based benchmarks in the quality assessment of message digest algorithms for digital signatures based secure Internet communications. In: Proceedings of the International Joint Conference on Neural Networks, vol. 2. IEEE (2003)

23. Lesson: Generating and verifying signatures. Retrieved Apr 2012. http://docs.oracle.com/javase/tutorial/security/apisign/index.html
24. Appendix A Key Management. Retrieved Apr 2012 from http://docs.oracle.com/cd/E19316-01/820-3748/gghyb/index.html
25. Michail, H.E., et al.: Optimizing SHA-1 hash function for high throughput with a partial unrolling study. In: Integrated Circuit and System Design. Power and Timing Modeling, Optimization and Simulation, pp. 591–600. Springer, Berlin (2005)
26. Großschädl, J., Page, D., Tillich, S.: Efficient java implementation of elliptic curve cryptography for J2ME-Enabled mobile devices. In: Information Security Theory and Practice. Security, Privacy and Trust in Computing Systems and Ambient Intelligent Ecosystems, pp. 189–207. Springer, Berlin (2012)
27. RSA Laboratories|Cryptography FAQ: http://www.rsasecurity.com/rsalabs/faq/index.html
28. Caelli, W.J., Dawson, E.P., Rea, S.A.: PKI, elliptic curve cryptography, and digital signatures. Comput. Secur. **18**(1), 47–66 (1999)
29. Brown, D.R.: Standards for efficient cryptography. SEC 1: Elliptic curve cryptography. Released Standard Version 1.0 and Working Draft v1.5, 2005. Available online http://www.secg.org. Last accessed 3 Apr 2012
30. Koblitz, N., Menezes, A., Vanstone, S.: The state of elliptic curve cryptography. In: Towards a Quarter-Century of Public Key Cryptography, pp. 103–123. Springer, US (2000)
31. Bensheng, Y., Qiaoyun, W., Fangming, Z.: Security architecture design of bidding MIS based on B/S. In: 2009 International Workshop on Information Security and Application (IWISA 2009) (2009)
32. Dhawan, P: Performance comparison: security design choices. Microsoft Developer Network, Oct 2002. Retrieved Apr 2012: http://msdn.microsoft.com/en-us/library/ms978415.aspx
33. Takase, T., Uramoto, N., Baba, K.: XML digital signature system independent of existing applications. In: 2002 Symposium on Applications and the Internet (SAINT) Workshops. Proceedings. IEEE (2002)
34. Poulakis, D.: Some lattice attacks on DSA and ECDSA. Appl. Algebra Eng. Commun. Comput. **22**(5–6), 347–358 (2011)
35. Teat, C., Peltsverger, S.: The security of cryptographic hashes. In: Proceedings of the 49th Annual Southeast Regional Conference. ACM (2011)
36. Lam, T.C.B., Ding, J.J., Liu, J.C.: XML document parsing: operational and performance characteristics. Computer **9**, 30–37 (2008)
37. Chang, M.H., Chen, I.T., Chen, M.T.: Design of proxy signature in ECDSA. In: Eighth International Conference on Intelligent Systems Design and Applications. ISDA'08, vol. 3. IEEE (2008)
38. Lu, W., et al.: A streaming validation model for SOAP digital signature. In: 14th IEEE International Symposium on High Performance Distributed Computing. HPDC-14. Proceedings. IEEE (2005)
39. Yang, C.H., Morita, H., Okamoto, T.: Fast implementation of digital signature algorithms on smartcards without coprocessor. J. Int. Technol. Inf. Manag. (JITIm) **2**, 82–90 (2002)

Fuzzy Weighted Metrics Routing in DSR in MANETs

Vivek Sharma, Bashir Alam and M.N. Doja

Abstract Communication path from source node to destination node in MANET is affected due to arbitrary movement, connection break down, power supply, bottleneck traffic, security etc. Hence, while designing the routing protocols these aspects are taken into account. Therefore, in this paper, we improve the existing DSR routing protocol by applying fuzzy inference system that offers a natural way of accepting multiple input constraints which are uncertain and imprecise in nature. The fuzzy weighted logic input metrics are hop count, stability factor and output metric fuzzy cost is dynamically calculated using MATLAB Fuzzy logic toolbox. Simulations are carried out using NS2.35 simulator and results show better performance of modified DSR protocol than DSR in terms of Packet delivery ratio and delay.

Keywords Fuzzy logic system · M-DSR · Soft computing technique · Multimetric

1 Introduction

Over the last few years, there has been a proliferation in the field of mobile communication because of the rise of economic, rapid increase in usage of wireless devices. All these devices, applications and protocols are entirely based on cellular or wireless local area networks (WLANs), without considering the great perspectives of mobile ad hoc networks. In various sectors like education, networking,

---

V. Sharma (✉) · B. Alam · M.N. Doja
Department of Computer Engineering, Jamia Millia Islamia, New Delhi 110025, India
e-mail: vivek2015@gmail.com

B. Alam
e-mail: babashiralam@gmail.com

M.N. Doja
e-mail: mndoja@gmail.com

S.C. Satapathy and S. Das (eds.), *Proceedings of First International Conference on Information and Communication Technology for Intelligent Systems: Volume 2*, Smart Innovation, Systems and Technologies 51, DOI 10.1007/978-3-319-30927-9_41

e-commerce, entertainment, military, civil etc., we can easily see the implementation of MANET's technology.

MANET is infrastructure less, provide freedom of movement and flexibility to network nodes to connect and disconnect at any time and there is no long term guaranteed path from one node to other node. So the routing protocols play a vital role in communication. Canon of routing protocols was developed for different applications, but none of the single routing is suited to all. Based upon the mechanism of updating the routing information [1, 2] routing protocols are categorized into three classes. First class is proactive, in which each node update its routing table based upon the present network topology information periodically. Second class is reactive routing protocol, it obtains necessary path information only when needed. Third class is Hybrid routing protocol that combines the best of above two.

In MANETs it is not easy to take dynamic shortest path from source to destination as the optimal route because parameters like link quality between the nodes, propagation path loss, multiuser interference, energy usage and topological updates become important issues to handle. The routing paths should be altered dynamically to handle above said issues. In the last few years, the routing protocols have attracted attention from both academic and industry researchers. In order to improve existing routing protocols, some of them proposed new routing protocol that emphasized fuzzy logic system as soft computing technique using one of the parameters like path selection, security, mobility and traffic congestion. Ad hoc environment is insecure and hence susceptible to attack. Earlier the routing protocol employ cryptographic algorithm in the network establishment and its operation. Security in MANET's routing is an important concern but few approaches or algorithms for securing the routing message have been proposed by the researchers. Authors in [3] implement security enabled routing protocol that embedded fuzzy logic system to it. The proposed protocol is compared with AODV routing protocol and results shows that proposed technique select more secure route than AODV under different circumstances and its performance has also been studied. Authors in [4] uses Fuzzy logic system decision algorithm for traffic management in ad hoc networks which measure the distinct links cost for the path that is being constructed to destination. Authors in [5] presents a routing strategy that include fuzzy logic and applied for DSR and SMR protocols. The proposed protocols picked the paths based on output of fuzzy logic system that considered network state and routine as input parameter. The experimental result shows that fuzzy logic based DSR and SMR routing protocol is more reliable and gives lower delay for important traffic. A number of routing protocols exists for mobile ad hoc networks but none of it consider all the variable which affect the routing decision. Fuzzy logic provides a feasible tool to solve the multi-metric QoS problem. Authors in [6] proposes an efficient protocol that consider various input parameter to calculate the cost of route and select the route that provide minimum cost.

This paper intends to provide a stable route routing protocol by embedding fuzzy logic system that considered input metric hop count and stability factor. The simulations show the effectiveness of the considered metrics on the system performance and improvement in main routing measurement parameter such as packet delivery ratio and delay. Following is the structured overview of the paper: Sect. 2 describes the fuzzy inference system. Section 3 gives the proposed description. The performance evaluation is depicted in Sect. 4 which is followed by conclusions in Sect. 5.

2 Fuzzy Inference System

Dr. Lofti Zadeh of university of California at Berkely in the 1965 proposed the concept of fuzzy logic [7–9]. It describes how human language which involves imprecision and ambiguity can be represented and reasoned naturally. Fuzzy logic has been applied in DSR routing protocol to improve performance. The working process of Fuzzy inference system [8, 10] are as follows:

Step 1. Input and Output variable selection for Fuzzy inferencc system

In order to take dynamically route decision, fuzzy logic system input variable are Hop count and stability factor that has fuzzy sets either Low (L), medium (M) or high (H) and range value of fuzzy set is in between 0 and 1. The output variable Fuzzy cost has fuzzy sets either very low (VL), low (L), medium (M), high (H), or very high (VH) within the range value between 0 and 1. Figure 2 shows that Triangle membership's function is used for Hop count, stability factor.

Step 2. Applying Fuzzification processes

In this step, Mamdani minimum inference method is used where 'and' operation is used to set the minimum.

Step 3. Design fuzzy IF-Then rule base

Based on [7–10] Study shows that if number of hop count increase in the path formation of traditional AODV routing then cost of routing is increased. Here stability factor is also considered along with hop count and it affects the routing cost i.e. fuzzy cost. The constructed set of IF THEN rules are as follows:

Rule1: If hope count is L and Stability factor is L then fuzzy cost must be VH.
Rule2: If hope count is L and Stability factor is M then fuzzy cost must be H.
Rule3: If hope count is L and Stability factor is H then fuzzy cost must be M.
Rule4: If hope count is M and Stability factor is L then fuzzy cost must be H.
Rule5: If hope count is M and Stability factor is M then fuzzy cost must be M.
Rule6: If hope count is M and Stability factor is H then fuzzy cost must be L.
Rule7: If hope count is H and Stability factor is L then fuzzy cost must be M.
Rule8: If hope count is H and Stability factor is M then fuzzy cost must be L.
Rule9: If hope count is H and Stability factor is H then fuzzy cost must be VL.

Step 4. Aggregation of rule output

The fuzzified input values are combined by selecting the proper rule to generate the output.

Step 5. Applying Defuzzication processes

Defuzzification process is carried out using centroid defuzzifier. The conclusion given by the inference engine is expressed in terms of fuzzy set, a single real number to define the action of fuzzy system is converted by the defuzzifier.

3 Proposed DSR

It is proposed that the use of Fuzzy logic system that receive two input metrics namely hop count h_i, and stability factor sf_i as fuzzy input parameters and one output metric namely fuzzy cost fc_i as shown in Fig. 1. The membership functions are shown in Fig. 2 and the rule base are defined above, respectively. In next step, these input parameters are fuzzified using linguistic variables. Finally, it's performed the defuzzifiction process using the centroid method. This defuzzifiction process produces a single value output name fuzzy cost fc_i that is used as a decision metric.

In DSR, when a node wants to transmit data packet to destination node it broadcast a RREQ packet to the network and when an intermediate node receive the RREQ packet, it calculate input parameter h_i and sf_i currently in path. Then fuzzy logic system evaluates these parameters and determines route is available or not if the route is available, then RREQ is again broadcasted and the node extracts route record. This process is to be done for each intermediate node until it reaches to the destination. Route reply message generated from destination is sent to source node via the path stored in route record.

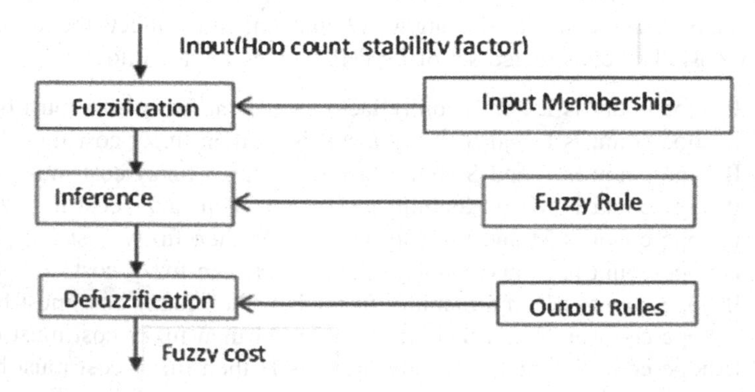

Fig. 1 Fuzzy inference system

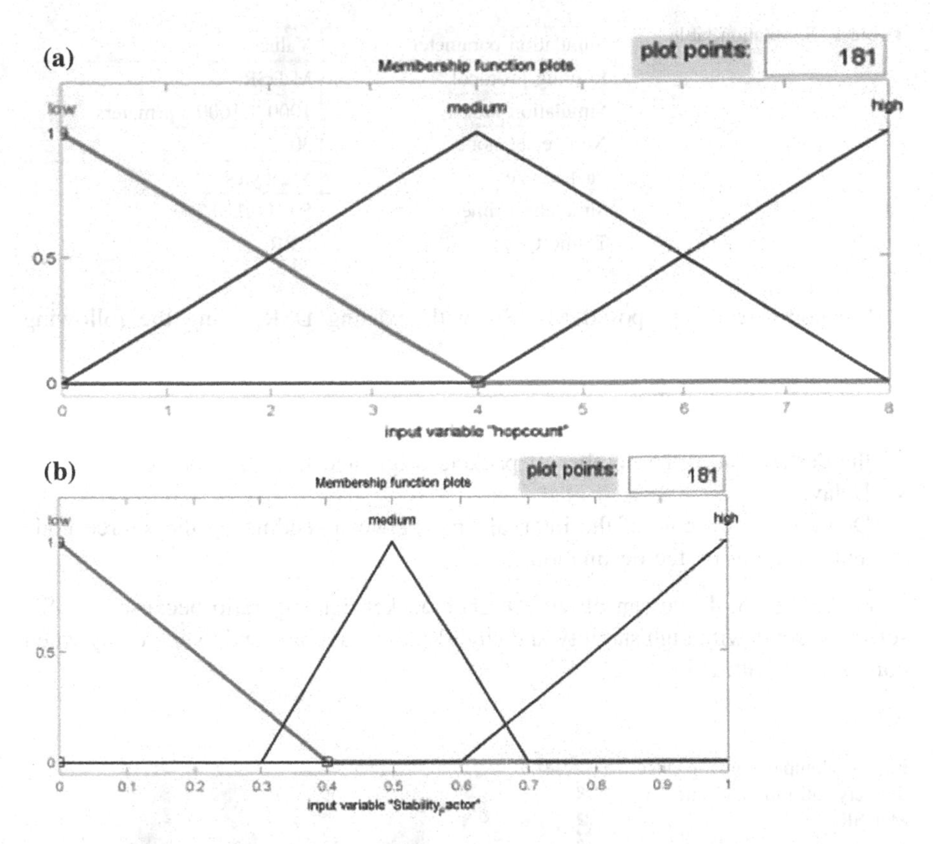

Fig. 2 **a** Membership functions of M-DSR (hop count), **b** Membership functions of M-DSR (stability factor)

4 Performance Evaluation

4.1 Performance Metrics

In this simulation, networks of 30 nodes are considered that are placed within a 1000 m × 1000 m area. Each node in the network has transmission range of 250 m. For mobility model, random waypoint model is used. The speed of each node for the movement ranges from 0 mps to 5 mps. The traffic type is CBR. The size of UDP packet is 512 bytes (Table 1). For implementation of fuzzy inference system MTLAB toolbox fuzzy logic is used. It receive input from the simulation environment of NS and produced output in the form of numeric result is given back to NS environment (Fig. 1).

Table 1 Simulation table

Simulation parameter	Values
Routing protocol	M-DSR
Simulation area	1000 * 1000 sq. meters
Number of nodes	30
Packet size	512 bytes
Simulation time	50,100,150,200 s
Traffic type	CBR

We compare the proposed M-DSR with existing DSR, using the following metrics:

- Packet Delivery Ratio:
  Packet delivery ratio is obtained by dividing the number of data packets reached the destination to the number of packets originated from the source.
- Delay:
  Delay is the average of the interval time between sending by the source node and receiving by the destination node.

From Figs. 3, 4, we can observe higher packet delivery ratio because M-DSR select the route with high stability and slightly lower average end to end delay when compared it with DSR.

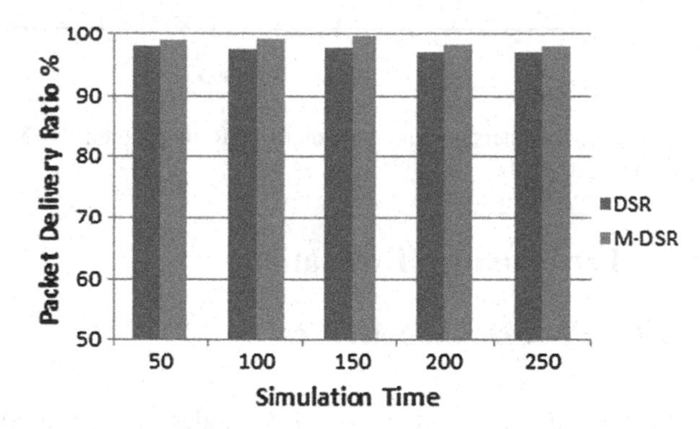

Fig. 3 Comparison of packet delivery ratio in DSR and M-DSR

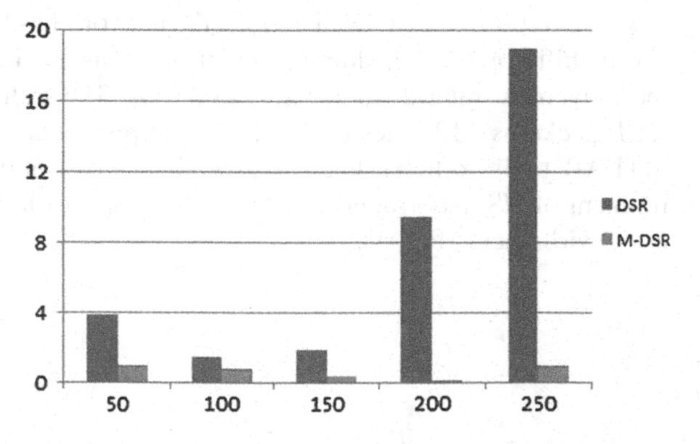

Fig. 4 Comparison of delay in DSR and M-DSR

This result shows that by utilizing fuzzy logic in the routing decision, better performance obtained.

5 Conclusion

In this paper an improvement has been made in DSR routing protocol that dynamically consider the stability factor. The modified DSR protocol based upon fuzzy logic system take routing decision. It is tested for 50 nodes, simulation time and the performance analysis is done and evaluated in terms of packet delivery ratio and average delay. The input for fuzzy logic toolbox is hop count and stability factor and the output is fuzzy cost. This output is provided as input to DSR routing protocol that runs on NS 2.35. The performance analysis shows that fuzzy logic base proposed scheme has better Packet delivery ratio and delay than DSR.

References

1. Murthy, C.S.R., Manoj, B.S.: Ad Hoc Wireless Networks: Architecture and Protocols. Pearson Ltd (2004)
2. Sharma, V., Alam, B.: Unicaste routing protocols in mobile ad hoc networks: a survey. Int. J. Comput. Appl. pp. 148–153, USA (2012)
3. Jin, L., Zhang, Z., Lai, D., Zhou, H.: Implementing and evaluating secure routing protocol for mobile ad hoc network. In: IEEE Wireless Telecommunication symposium, pp. 1–10 (2006)
4. Rea, S., Pesch, D.: Multi-metric routing decisions for ad hoc networks using fuzzy logic. In: 1st IEEE Symposium on Wireless Communication Systems, pp. 403–407 (2004)
5. Alandjani, G., Jhonson, E.E.: Fuzzy routing in ad hoc networks. In: IEEE International Conference on Performance computing and Communication, pp. 525–530 (2003)
6. Santhi, G., Nachiappan, A.: Fuzzy-cost based multiconstrained QoS routing with mobility prediction in MANETs. Egypt. Inform. J. pp. 19–25 (2012)
7. Wang, C., chen, S., yang, X., Gao, Y.: Fuzzy logic-based dynamic routing management policies for mobile ad hoc networks. In: Workshop on High Performance Switching and Routing, pp. 341–345 (2005)
8. Ross, T.J.: Fuzzy Logic with Engineering Application. Macgraw hills, New York (1995)
9. Doja, M.N., Alam, B., Sharma, V.: Analysis of reactive routing protocol using fuzzy inference system. AASRI Procedia pp. 164–169 (2013)
10. Torshiz, M.N., Amintoosi, H., Movaghar, A.: A fuzzy energy-based extension to AODV routing. In: International Symposium on Telecommunications, pp. 371–375 (2008)

Semantic Model for Web-Based Big Data Using Ontology and Fuzzy Rule Mining

Sufal Das and Hemanta Kumar Kalita

Abstract A huge amount of data is being generated everyday through different transactions in industries, medicals, social networking, communication systems etc. This data is mainly of unstructured format in nature. Transformation of the large heterogeneous datasets into useful information is very much required for society. This huge unstructured information should be easily presented and made available in a significant and effective way to obtain semantic knowledge so that machine can interpret them. In this paper, we have introduced a novel approach for semantic analysis with web based big data using rule based ontology mapping. To handle social data with natural language terms, we have proposed the fuzzy rule based resource representation. After that, a refined semantic relation reasoning mining is applied to obtain overall knowledge representation. Finally semantic equivalent of these unstructured data is stored in structured database using Web Ontology Language (OWL) based ontology system.

Keywords Big data · Semantic analysis · Ontology · OWL · Fuzzy rule · Association rule mining

1 Introduction

Amount of information on the Web rapidly increases everyday through different transactions, communications etc. in social network, medical, research activities, industries etc. Thus end users like researchers, retailers, customers etc. are interested to deal with this information. Therefore, information resources are collected by diverse strategies with specific algorithms. Due to diversity in e-World, information sources are heterogeneous with respect to data content, structural concept etc.

S. Das (✉) · H.K. Kalita
Department of Information Technology, North-Eastern Hill University, Shillong 79022, India
e-mail: sufal.das@gmail.com

H.K. Kalita
e-mail: kalita.hemanta@gmail.com

S.C. Satapathy and S. Das (eds.), *Proceedings of First International Conference on Information and Communication Technology for Intelligent Systems: Volume 2*,
Smart Innovation, Systems and Technologies 51, DOI 10.1007/978-3-319-30927-9_42

The idea of Big Data is mainly the being of a very large, heterogeneous and dynamic volume of data from unstructured data sources. Traditional relational databases systems are not effective and suitable to handle Big Data. The processing of Big Data typically occurs only through human intervention. So, it is very challenging task to make machines only to analysis Big Data.

Different heterogeneous sources with information can be linked through semantic annotations. This has not been widely adopted. Therefore, end users are still struggling with searching for relevant documents as well as filtering through large volumes of content for actual information relevant to their interests. Semantic predications therefore recommend a mechanism for transforming from unstructured dataset to structured dataset.

In this paper, we describe how different datasets from heterogeneous data sources can be converted into single structured dataset with equivalent semantic meaning.

2 Background Study

2.1 Big Data Concept

Data is being collected every second in our day to day life. Mainly different sectors like communication, corporate, medical, social network etc. are generation tremendous amount of data regularly. Due to that, all types of researcher have to consider this large volume datasets. Since, data is being generated continuously from different sources; researchers have to also handle dynamic and heterogeneous characteristics of data. Thus data sets become so large and complex that traditional database system and its applications are inadequate.

The Big data [1–3] can be described in 4 V's: Volume, Velocity, Veracity and Variety.

Volume: It relates to the quantity of data that is generated everyday in large scale and its size is increasing continuously. It refers the large size of input data that can't be handled by traditional database system.

Velocity: Since data is being generated continuously, end users have to consider for online data sets. Velocity refers the characteristic of data with speed of generation of data as well as processing of that data to meet the demands and challenges which is related to the path of growth and development.

Veracity: It is very important to consider relevant information from very large datasets. Veracity refers to the quality of the data being captured can vary significantly. Accuracy of analysis depends on the veracity of the input data.

Variety: There are different types of data and data sources available and all these refer to this term. The availability of information to analyze is of different types, such as mainly coming from social media and communication devices. The term 'variety' includes structured data like tabular data (databases), transactions etc. and unstructured and semi-structured data like hierarchical data, documents, e-mail, video, images, audio etc.

2.2 Ontology and Semantic Web

Ontology is a precise explanation of terms and reasoning in subject areas. In a broad sense, Ontology can be referred as a finite set of rules for interpreting a system of symbols. It defines a set of inference rules with which a machine illustrates inference over symbols. It is mostly used as a tuple-base knowledge representation. As semantic reasoning is a system of symbols, machines would need ontology for drawing inferences. Ontology can be defined as an explicit specification of a concept, which is a simplified view of a knowledge domain.

Semantic web involves more involved questions, relationships and trust. It is used to show related items with new relationship instead of word matching in web search. Semantic web is used to create machine readable data from human understandable uncertainties facts. Thus why, semantic web technologies are acting a vital role in enhancing traditional web search. For semantic web, ontology can be described as exact description of web information and relationship between them. The semantic web is an extension of the current Web. In this web, resources are represented using ontology-based concept representation languages for automated machine processing within heterogeneous sources. A framework is needed to enable humans and machines to make and understand statements. There are several languages for formally representing ontology mainly OWL and Resource Description Framework (RDF) [4, 5].

2.3 Fuzzy Theory

Fuzzy sets were introduced by Zadeh in 1965 to represent or manipulate data and information for possessing non-statistical uncertainties. For example, in “the man is tall” is difficult to represent “tall” by using some IF-ELSE rules or discrete representation. Fuzzy rule concept is used to represent quantifiers which can’t be defined by discrete representation. Fuzzy logic is used to getting computers to make decisions like human. This logic uses fuzzy rules and fuzzy sets to model the world and make decisions. Fuzzy logic allows computers to make decisions like the human brain. Knowledge is represented using fuzzy sets combined using rules. When everything is considered, a decision can be made. This theory proposed membership function whose value ranges with real numbers from 0 to 1 to define the uncertainties of information. Highest membership value implies strongest and lowest value represents weakest behaviors of the input information [6, 7].

2.4 Association Rule Mining

Association rule mining [8–10] is a method for finding out hidden data dependencies. It is one of the popular data mining techniques which is applied in transactions database. The basic concept is to discover different item-sets with high probability and whether the occurrence of specific items implies also the occurrence of other items with dependency. It is expected to recognize strong rules discovered in databases using different measures of interestingness. There are several applications of this rule mining like market analysis, decision system, stock prediction etc. [11–13].

An association rule is an expression of the form IF X THEN Y (or X→Y), $X \cap Y = \varphi$, where X and Y are non-empty sets of tractions. The meaning of this expression is that transactions of the dataset, which contain X, tend to contain Y. This rule or relation shows the dependencies or connections between X and Y. Let consider for super market example. Buyers have common interest to buy milk while they are pursing bread and butter. Shop manager can take decision like to make similar stock in future for bread, butter and milk. Here association rule can be presented as IF {bread, butter} THEN {milk}, i.e. {bread, butter}→{milk}.

Two quality measurements for an association rule are considered for assessment the rule. They are support and confidence. To define efficiency of the rule, the number of occurrences of X and Y should be considered as well as chances of Y when X is occurring.

The support of a rule X→Y is the support of XUY, where XUY means both X and Y is occurring in the same transactions. An itemset X in a transaction database D has a support, denoted as Supp(X) that is the probability of occurrences of transactions in D which contains X. So, Supp(X) = |X(t)| / |D|, where X(t) = {t in D / t contains X}. An itemset X in a transaction database D is called a frequent itemset if its probability of occurrence (Support) is equal to, or greater than, a threshold of value, which is predefined by users or experts.

Similarly, confidence or accuracy of the rule X→Y is conf (X→Y) is represented as the ratio: |(XUY)(t)| / |X(t)| or Supp(XUY) / Supp(X). This measurement represents the strength of the association rule where the occurrence of Y is considered as dependent of X.

Association rule mining technique is divided into the following two sub-problems [14].

Phase I: To generate all frequent itemsets from the given dataset whose support is greater than, or equal to, the user specified threshold support.

Phase II: To generate all the association rules by considering all frequent itemsets which are generated in first phase. These rules should have minimum threshold confidence and considered as strong rules.

3 Related Works

Several approaches for automatic generation of ontology have been described in the literature for few years. Researchers have tried to establish common link between different heterogeneous data sources.

Correa at el. [15] have proposed a framework in which heterogeneous human-readable documents and information have been considered as data sources.

In this work, meaningful and machine-readable data are obtained from different sources using automatic generation of ontology from unstructured information.

Ma at el. [16] have presented a novel approach for semantic search which combines Multi-Categorization Semantic Analysis (MCSA) with personalization technology. The MCSA approach can classify documents into multiple categories, which is distinct from the existing approaches of classifying documents into a single category. Then, the search history and personal information for users are significantly considered in analyzing and matching the original search result by Term Vector Database (TVDB). A series of personalization algorithms are proposed to match users' personal information and search history.

Zou at el. [17] have introduced semantic extension search with ontology-based framework. In this work, Semantic extension logic-based algorithm is presented with constructed domain ontology.

Our approach proposed in this paper is distinct from existing methods. We introduce a complete framework to convert a single structured dataset from different unstructured date sources.

4 Proposed Semantic Model

In this paper, we propose a novel model for semantic analysis of Big Data based on compact ontology. Figure 1 shows a conceptual overview proposed framework with four main components: Automated Web Scraper, Rule based Information Extraction, OWL Module and Structured Data Store. Details of these components are mentioned as follows:

Automated Web Scraper: The purpose of this component is to capture all information available in different sources. Basically it is a mechanism used as data collection. After preprocessing those to form intermediate documents are stored as temporally.

Rule Based Information Extraction: It is very much important to collect the related domain information and store it according to domain ontology. As the improvement of the responding time of retrieval is concern, we need to classify the information by clustering formula. This component is used to combine all information into large semantic dataset. Firstly, different data from different sources are represented in fuzzy rule based resource representation. The purpose of using fuzzy rule is to handle uncertainty as well as proper domain representation. After that, a refined semantic relation reasoning mining is applied to obtain overall knowledge representation in semantic form.

OWL Module: OWL is a part of semantic web vision. The purpose of this component is to make the input dataset understandable for the computer machine with semantic reasoning. OWL assists better computer interpretability of Web content. This web content is supported by XML, RDF, and RDF Schema by providing additional vocabulary along with a formal semantics. OWL is designed to be interpreted by computers by processing information on web. Using RDF, ontology

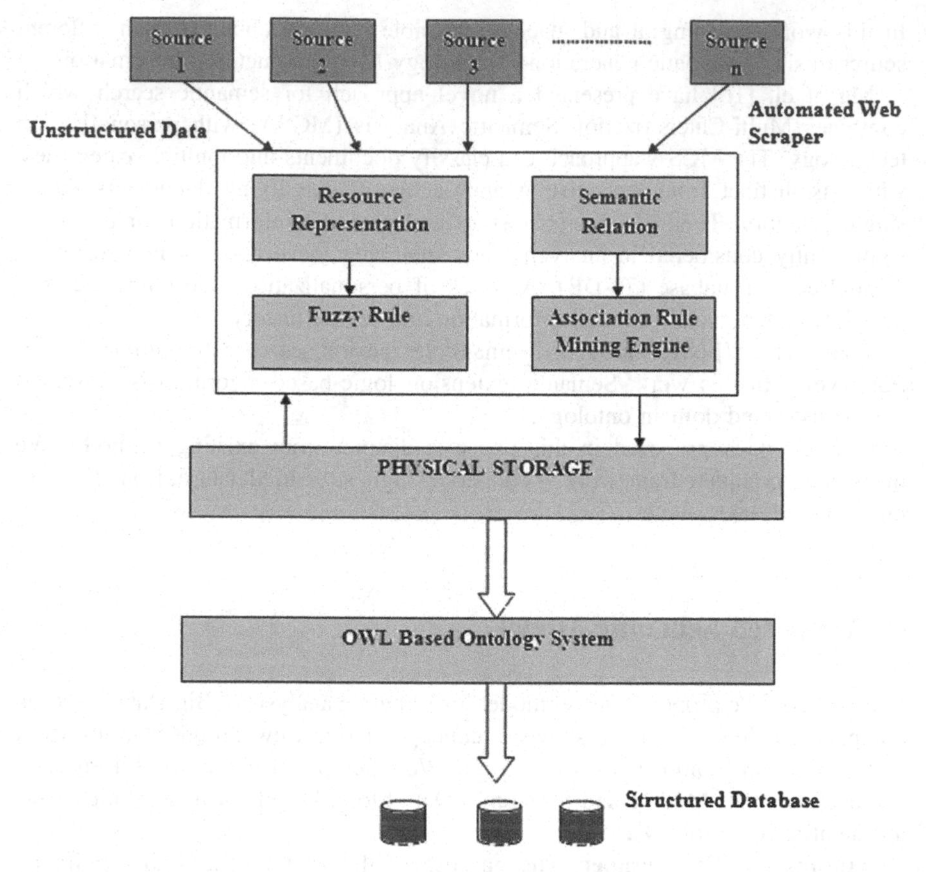

Fig. 1 Proposed semantic model for web-based resources

and OWL, this dataset can be made understandable to the machine. Ontology provides some knowledge implicit in the annotations of the web resources to take into account in the query processing. It defines associations between domain concepts and between domain relations as well as other semantic relations between domain concepts, domain axioms or rules.

Structured Data Store: This component is used to store all semantic knowledge which can be processed by machine. Since, different knowledge or concepts are represented with predefined structure, any data accessing techniques can be applied based on application domain.

5 Conclusion

In this paper, we have proposed a novel methodology for design a semantic model for Big Data. Big Data is a hot topic in current research. Due to generation of huge information from different sources, researchers are facing difficulties to handle this huge data. The main challenge occurs due to different characteristics of this data like large volume, velocity and heterogeneity. To overcome these difficulties, researchers need to have a system to process these data. The proposed method integrates all heterogeneous information available in different sources based on semantic mapping. This model can be useful to handle text data as fuzzy rule mining is applied for the model. In future work, we will implement the proposed model with web-based large data.

References

1. Gopalkrishnan, V., Steier, D., Lewis, H., Guszcza, J.: Big data, big business: bridging the gap. In: Proceedings of the 1st International Workshop on Big Data, Streams and Heterogeneous Source Mining: Algorithms, Systems, programming Models and Applications, Big-Mine'12, pp. 7–11, New York, NY, USA (2012)
2. Zikopoulos, P., Eaton, C., De Roos, D., Deutsch, T., Lapis, G.: Understanding Big Data: Analytics for Enterprise Class Hadoop and Streaming Data. McGraw-Hill Companies, Incorporated (2011)
3. Liang, Z., Li, W., Li, Y.: A parallel probabilistic latent semantic analysis method on mapreduce platform. In: Proceeding of the IEEE International Conference on Information and Automation, Yinchuan, China (2013)
4. Arakawa, N.: Semantic analysis based on ontologies with semantic web standards. In: International Conference on Computer-aided Acquisition of Semantic Knowledge (CASK-2008), Sorbonne (2008)
5. Papadopoulos, G.T., Mezaris, V., Kompatsiaris, I., Strintzis, M.G.: Ontology-Driven Semantic Video Analysis Using Visual Information Objects in Semantic Multimedia, pp. 56–69. Springer, Berlin Heidelberg (2007)
6. Kuok, C.M., Fu, A., Wong, M.H.: Mining fuzzy association rules in databases. ACM Sigmod Rec. **27**(1), 41–46 (1998)
7. Ishibuchi, H., Yamamoto, T.: Rule weight specification in fuzzy rule-based classification systems. Fuzzy Syst. IEEE Trans. on **13**(4), 428–435 (2005)
8. Imielinski, T., Agrawal, R., Swami, A.: Mining association rules between sets of items in large databases. In: Proceedings ACM SIGMOD Conference Management of Data, pp. 207–216 (2006)
9. Das, S., Saha, B.: Data quality mining using genetic algorithm. Int. J. Comput. Sci. Secur. **3**(2), 105–112 (2009)
10. Agrawal, R., Srikant, R.: Fast algorithms for mining association rules. In: Proceeding of the 20th International Conference on Very Large Databases, Chile (1994)
11. Pei, M., Goodman, E.D., Punch, F.: Feature Extraction using genetic algorithm. Case Center for Computer-Aided Engineering and Manufacturing W, Department of Computer Science (2000)
12. Wu, X., Zhang, S.: Synthesizing high-frequency rules from different data sources. IEEE Trans. Knowl. Data Eng. **15**(2), 353–367 (2003)
13. Su, K., Huang, H., Wu, X., Zhang, S.: Logical framework for identifying quality knowledge from different data sources. Decis. Support Syst. **42**(3), 1673–1683 (2006)
14. Das, S., Nath, B.: Dimensionality reduction using association rule mining. In: IEEE Region 10 Colloquium and Third International Conference on Industrial and Information Systems, IIT Kharagpur, India (2008)

15. Corrêa, A.S., Borba, C., da Silva, D.L., Corrêa, P.A.: Fuzzy ontology-driven approach to semantic interoperability in e-government big data. Int. J. Soc. Sci. Humanity **5**(2), 178–181 (2014)
16. Ma, Y., Shi, M.: Using multi-categorization semantic analysis and personalization for semantic search. arXiv preprint arXiv:1406.7093 (2014)
17. Zou, G., Zhang, B., Gan, Y., Zhang, J.: An ontology-based methodology for semantic expansion search. In: IEEE Fifth International Conference on Fuzzy Systems and Knowledge Discovery, pp. 453–457 (2008)

Face Verification Across Ages Using Discriminative Methods and See 5.0 Classifier

K. Kishore Kumar and P. Trinatha Rao

Abstract Identifying a person through face biometric and analysis of Facial image has been drawing interest of researchers in the field of Machine learning and Pattern recognition. Face Recognition Across Ages (FRA) is a challenging task due to aging effects like changes in facial shape and texture. In this paper, an attempt is made to describe a schematic using two different discriminative approaches for feature extraction and a see5.0 classifier for classification purposes. One of the feature finding approaches is based on Gradient Orientation Pyramids (GOP) that includes finding of gradient orientations in Gaussian pyramids, the later one is based on Local Binary Patterns (LBP) calculated at each stage of Gaussian pyramid decomposition. These we have used FG-NET database and accuracies of both the approaches are compared.

Keywords Face verification · Age progression · FRA · GOP · LBP · See5.0 classifier

1 Introduction

Recent advancements in pattern analysis and computer vision are throwing challenges in face recognition. Age invariant face recognition is a complex and challenging process as it involves changes in texture and shape of the face with age. These variations are unpredictable as it depends on three main factors like food habits, race, family circumstances etc which makes the face recognition against

K. Kishore Kumar (✉)
Department of ECE, Faculty of Science and Technology, IFHE University, Hyderabad, India
e-mail: kkishore@ifheindia.org

P. Trinatha Rao
Department of ECE, GITAM School of Technology, GITAM University, Hyderabad, India
e-mail: trinath@gitam.in

S.C. Satapathy and S. Das (eds.), *Proceedings of First International Conference on Information and Communication Technology for Intelligent Systems: Volume 2*, Smart Innovation, Systems and Technologies 51, DOI 10.1007/978-3-319-30927-9_43

aging as a very difficult task. The first one is biometric change over years of growth (shape and structure), facial hair changes with age, presence of glasses, scars and wrinkles over the age progression. The second one is source of illumination, image quality change caused by cameras of different times etc. The last one is the all the images are not the taken from the camera, rather scanned ones from hard copies or new papers etc. Though the aging based face recognition is a complex problem, it has various important applications like Passport verification, Identifying missing children and avoiding multiple enrollment issues etc.

1.1 Literature Review

Gibson's ecological approach [1] and Thompson's [2] works on study of cranio-facial growth against aging has been used in FRA with some morphological operations on human faces. Shaw et al. [3] proposed the method of remodeling human skull using cardioidal strain transformation. Pittenger and Shaw [4] studied the aging phenomenon of the faces as a series of viscal elastic events. Todd et al. [5] made some changes to cardioidal strain transformation proposed by Shaw considering the key landmark features of the facial image for computing the ratios of distances. Kwon et al. [6] developed a method which classifies a face image into three categories based on age as (i) infants (ii) young adults (iii) senior adults. Lanitis et al. [7] proposed an aging function based on a parametric model [8] of face images. Further they performed a quantitative evaluation [9] of different classifiers for age estimation problem. Burt and Perett [10] generated facial prototypes [8] for different age groups by averaging the texture and shape of faces. Tidderman et al. [11] extended the work done by Burt and Perett by using the wavelet based methods for compensating the loss of texture in facial prototypes. Wu et al. [12] represented skin deformations as a plastic-visco-elastic process. O'Toole et al. [13] proposed a standard facial caricaturing algorithm to model the 3D views of faces. Gandhi [14] developed age prediction based on support vector machine. Givens et al. [15] analyzed the performance of different facial recognition algorithms due to different co-variants like age, gender, facial hair etc and observed that younger faces are difficult to recognize than the older faces.

1.2 FG-NET Database

FG-NET and MORPH are two public domain datasets contains only age information unlike various datasets. In this paper, our experiments are done on FG-NET database which contains 82 subjects and 1002 images. These images have variations in pose, illumination, expressions and distractions like spectacles facial hair etc. Given two images if they belong to the same subject called as intra-pairs otherwise extra-pairs.

1.3 See5.0 Classifier

See5.0 Classifier [16] is developed by Quinlan which is the latest version of the induction systems. This algorithm uses the entropy criterion; i.e. at each step the classification tree grows based upon the variable which has the highest amount of information or entropy and it is calculated by

$$Entropy = -\sum_{j=1}^{k} \frac{n_j}{N} \times Log_2\left(\frac{n_j}{N}\right) \quad (1)$$

where k is the number of classes in the total number of observations N and n_j is the number of observations belonging to same class. SEE5 classifier [17] uses a more advanced functions such as the possibility of converting the classification tree into a set of simpler classification rules [16]. The algorithm by default selects a classification to assign the cases which do not satisfy the conditions of any classification rules and this class by default will be calculated so that classification mistakes are lowest.

2 Gradient Orientation Pyramids

Firstly we propose Gradient orientation pyramid for the mission of face verification across ages. We prove that, when GOP combined with see5.0 classifier gives better performance for different age gaps. The pyramid technique [18] is used to capture hierarchal information which improves the image information further. For a given a face image pair, we use the cosines between gradient orientations at all scales to build the feature vectors.

We proposed to use Gradient Orientation (GO) [19, 20] for face verification across age progression [21] because of its robustness to illumination and face color changes over age progression. As per the skin anatomy [22], the change in face color across age progression is due to two mechanisms: melanin and hemoglobin. Gradient Orientation of each color channel of human faces is robust under age progression. To retain most visual information, we have collected gradient orientation in a hierarchical way (Fig. 1).

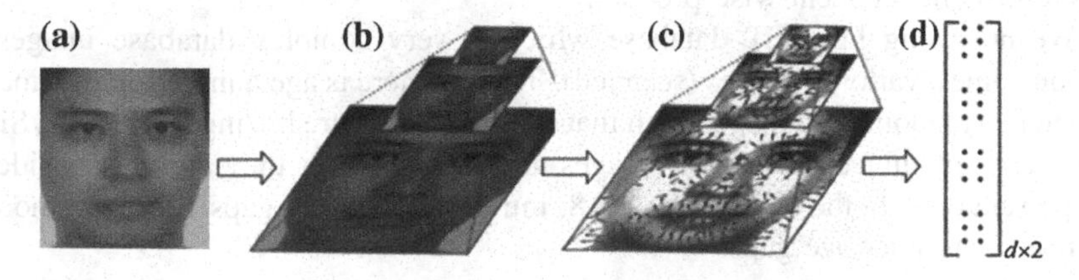

Fig. 1 Input image I and computation of a P(I), GOP, G(I). **a** Input Image I. **b** P(I). **c** GOP. **d** G(I)

In the proposed work, gradient magnitude information is discarded and only orientations are used, which demonstrates significant enhancement. Hierarchical representations are done by combining the gradient directions at different scales.

For the given image $I(q)$, where q = (u; v) denotes pixel locations, pyramid [21, 23] of image I is defined as

$$P(I) = \{I(q;\sigma)\}\ S_\sigma = 0$$

with

$$I(q;0) = I(q);$$

$$I(q;\sigma) = [I(q;\sigma - 1) * \Phi(q)]\downarrow_2 \quad \sigma = 1\ldots s \tag{2}$$

where $\Phi(q)$—Gaussian kernel; 0.5—standard deviation value; *—convolution operator; $\downarrow_2$—half size down sampling and s is the number of pyramid layers. The notation I represents both the original image and the images at different scales for convenience. The gradient orientation [18, 21] at each scale σ is defined by its normalized gradient vectors at each pixel.

$$g(I(q;\sigma)) = \frac{\Delta(I(P,\sigma))}{|\Delta(I(P,\sigma))|} \quad \text{if } |\Delta(I(P,\sigma)| < \tau$$

Otherwise

$$= (0, 0)^T \tag{3}$$

where τ is a threshold of flat pixels. The gradient orientation pyramid (GOP) of I, is defined by the mapping function that maps I to a $dX2$ representation as G(I) = stack $(\{g(I(q;\sigma))\}_{\sigma=0}^{s}) \in R^{dX2}$), where stack(.) is used for stacking gradient orientations of all pixels across all scales and d is the total number of pixels [20].

For a given image pair (I_1; I_2) and resultant GOPs ($G_1 = G(I_1)$; $G_2 = G(I_2)$), the feature vector $X = F(I_1; I_2)$ is computed as the cosines of the difference between gradient orientations [23] at all pixels over scales.

$$X = F(I_1; I_2) = (G_1 \Theta G_2) \tag{4}$$

where Θ is the element-wise product.

We are using FG-NET database which is very complex database images of various times, various sources (scanned). Though there is age annotation the images are picked randomly with ages such that no two images are having same ages. Since the studies of subjects less than 18 ages are difficult hence they are not considered for present case. In the case of above 18, intra pairs with age gaps 1 to 5 are more in number than other age gaps.

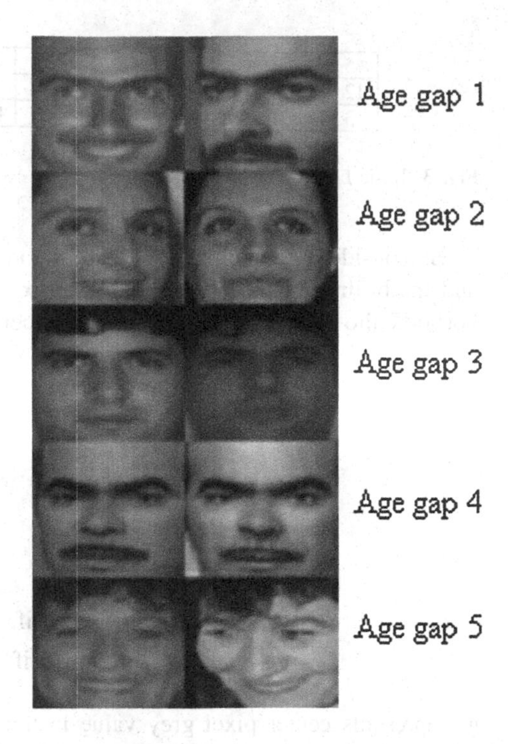

Fig. 2 Sample age gap wise FG-NET dataset

In our current experiment we are dealing intra pairs with age gaps 1–5 separately (Fig. 2).

In our experiments, 362 intra pairs, 492 extra pairs are used for training and 309 intra pairs, 395 extra pairs are used for testing. These intra pairs and extra pairs are with age gaps 1–5. With this dataset 83.48 % accuracy is observed.

3 Local Binary Pattern (LBP)

The second discriminative approach for feature extraction is Local Binary Pattern [24] which is an efficient operator for texture description. First Ojala et al. [25] proposed Local Binary Pattern operator which is used to extract the features of a given set of images by applying hierarchically and map to the Local Binary Pattern feature space to construct See5.0 classifier.

LBP labels the pixels of an image by binary number which is achieved by thresholding the neighborhood of each pixel with the center pixel value [24]. LBP Labels are calculated in a single iteration of the given image and histogram of the LBP labels can be used as a descriptor for obtaining the texture information. Grey-scale invariance and scaling invariance are achieved by considering a local neighborhood of each pixel and sign differences in the pixel values respectively. The LBP [24] operator can also be used with different neighborhood sizes (Fig. 3).

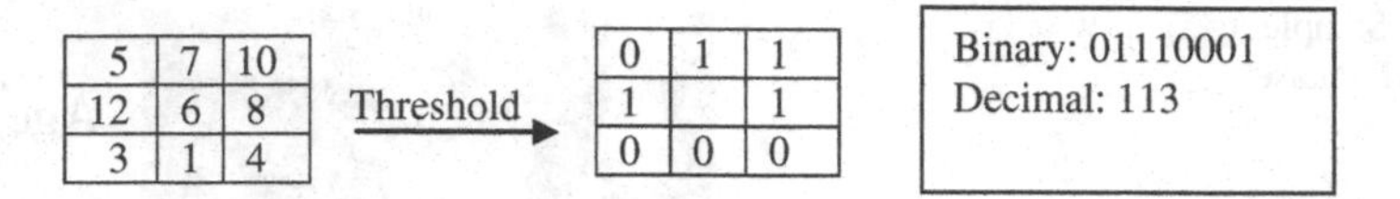

Fig. 3 Basic LBP operator

By considering the circular neighborhoods, the labels of each pixel are obtained and the bi-linear interpolation of these pixel values obtained from circular neighborhood allows the treatment of any number of pixels in the neighborhood of any radius.

For a given N sampling points on a circle of radius R, LBP operator is given by,

$$LBP_{N,R} = \sum_{n=0}^{N-1} s(g_n - g_c)2^n \tag{5}$$

where

$$\begin{aligned} S(x) &= 1 \quad \text{if } x > 0 \\ &= 0 \quad \text{if } x < 0 \end{aligned}$$

g_c represents center pixel grey value in the local neighborhood pixels with grey values of g_n, where n = 0, …, N − 1 (Fig. 4).

Ojala et al. [25] introduced uniform LBP, an LBP is said to be uniform if its exhibits at most two bitwise transitions from 1 to 0 or vice versa. Uniform patterns can be used in reducing the dimensions significantly since they represent at local level micro-patterns like edges, spots etc of the image. This would be beneficial for the applications like face verification. In the same reference LBP operator can be extended for building the rotational invariance features. The idea proposed is to obtain the least binary value for the operator by rotating the grey values of the neighboring pixels of an image.

In the proposed work, uniform Local Binary Pattern operator $LBP^{u,2}_{n,r}$ of window size of 5 × 5 centered around each pixel of the image was used to extract the hierarchically the LBP features, so that the most visual information is retained.

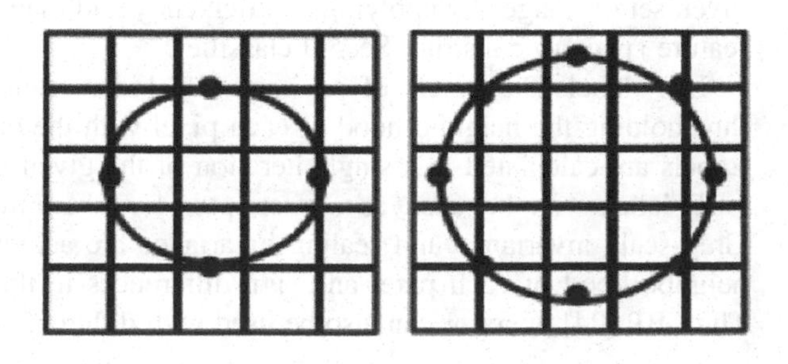

Fig. 4 Circular LBP operators of (4, 1) and (8, 2)

For a given image I(u; v), where (u; v) denotes pixel locations, the pyramid [24] of Image I is defined as

$$G_k(I) = I(u, v, k); \quad k = 0\ldots s$$

with

$$G_0(I) = I(u, v, 0)$$
$$G_k(I) = [I(u, v, k-1) \otimes \phi(u, v)] \tag{6}$$

where Φ(p, q)—Gaussian kernel; s—number of pyramid levels. By convolving the Gaussian kernel Φ(p, q) with the image at previous level k − 1, gives the next level pyramid image at k. At each level of the pyramid image it is broken into blocks with sizes of eight. A LBP histogram from each block size of 8 is obtained by applying the LBP operator on to each block. The respective pyramid level image will generate a net histogram obtained by cascading individual histograms.

LBP pyramid for the given input image I is given as

$$L(I_0) = [LBP(G_0(I)); \quad LBP(G_1(I)); \quad :::; \quad LBP(Gs(I))] \tag{7}$$

and $L(I) \in R^{dxs}$ is the mapping function which maps the image I into a dxs representation, where d is the size of the collective LBP histogram (Fig. 5).

For a given image pair (I_i, I_j), their equivalent LBP pyramids are $L(I_i)$ and $L(I_j)$, then feature vector x is given by

$$x = S(I_i; I_j) \tag{8}$$

$$= S(I_i; I_j) = (L(I_i) * L(I_j)) \begin{bmatrix} 1 \\ . \\ . \\ 1 \end{bmatrix}_{sx1} \tag{9}$$

where * is the element-wise product.

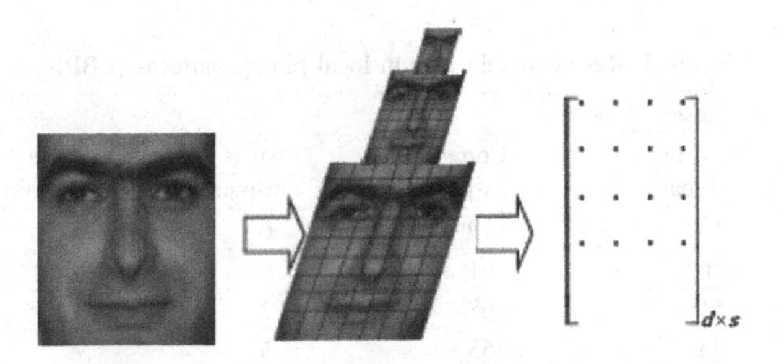

Fig. 5 Computation of LBP pyramid

4 Results

In our experiments, 671 intra-pairs, 887 extra-pairs of images are used for classification using see5.0 classifiers which are having age gaps from 1 to 5. Age gaps above 5 years are not considered for the proposed work due to the lack of sufficient number of intra-pairs in the FGNET database for constructing the classifier. We have sorted database images age gap wise and performed classification using the both discriminative approaches i.e using GOP and LBP. The results are tabulated and the performance is evaluated using Correct In-Pairs (In-pairs that are classified correctly), Correct Ex-Pairs (Ex-pairs that are classified correctly), False In-Pairs (Ex-pairs that are classified as In-pairs), and False Ex-Pairs (In-pairs that are classified as Ex-pairs) (Tables 1, 2 and 3).

Table 1 Dataset taken for the training of see5.0 classifier

Age gap	Training		Testing	
	In-pairs	Ex-pairs	In-pairs	Ex-pairs
1	128	140	105	125
2	93	107	80	100
3	60	100	54	70
4	56	95	50	60
5	25	50	20	40

Table 2 Results obtained from gradient orientation pyramids (GOP)

Results				Accuracy (%)	
Correct-in-pairs	Correct ex-pairs	False in-pairs	False ex-pairs	In-pairs	Ex-pairs
98	116	9	7	93.33	92.80
72	90	10	8	90.00	90.00
48	63	7	6	88.89	90.00
41	52	8	9	82.00	86.67
14	32	8	6	70.00	80.00

Table 3 Results obtained from local binary patterns (LBP)

Results				Accuracy (%)	
Correct in-pairs	Correct ex-pairs	False in-pairs	False ex-pairs	In-pairs	Ex-pairs
100	119	6	5	95.24	95.20
76	91	9	4	95.00	91.00
51	67	3	3	94.44	95.71
43	55	5	7	86.00	91.67
16	30	10	4	80.00	75.00

5 Conclusion

In this proposed work, two feature based methods are presented to handle the face recognition against ages. First we have discussed robust face descriptors GOP and LBP. We have evaluated performance of both on FG-NET Database. Both the operators are applied hierarchically on texture of face image to extract features. In GOP, Gradients describes intensity differences along X and Y directions only. Whereas LBP operator describes the intensity variations of a pixel radially with respect to surrounding pixels. Hence, LBP performs better than GOP. Future scope of this work in face verification task is to further examine the effects of disguise facial hair, scars etc.

References

1. Gibson, E.J.: Principles of Perceptual Learning and Development. Appleton-Century-Crofts, New York (1969)
2. Thompson, D.W.: On Growth and Form. Dover, New York (1992)
3. Shaw, R.E., McIntyre, M., Mace, W.: The role of symmetry in event perception. In: Perception: Essays in Honor of James J. Gibson, pp. 276–310 (1974)
4. Pittenger, J.B., Shaw, R.E.: Aging faces as viscal-elastic events: implications for a theory of nonrigid shape perception. J. Exp. Psych. Hum. Percept. Perform. **1**(4), 374–382 (1975)
5. Todd, J.T., Mark, L.S., Shaw, R.E., Pittenger, J.B.: The perception of human growth. Sci. Amer. **242**(2), 132–144 (1980)
6. Kwon, Y.H., da Vitoria Lobo, N.: Age classification from facial images. Comput. Vis. Image Underst. **74**, 1–21 (1999)
7. Lanitis, A., Taylor, C.J., Cootes, T.F.: Toward automatic simulation of aging effects on face images. IEEE Trans. Pattern Anal. Mach. Intell. **24**(4), 442–455 (2002)
8. Ramanathan, N.: Face verification across age progression. IEEE Trans. Image Process. (2006)
9. Lanitis, A., Draganova, C., Christodoulou, C.: Comparing different classifiers for automatic age estimation. IEEE Trans. Syst. Man Cybern. B Cybern. **34**(1), 621–628 (2004)
10. Burt, M., Perrett, D.I.: Perception of age in adult Caucasian male faces: computer graphic manipulation of shape and colour information. J. Roy. Soc. **259**, 137–143 (1995)
11. Tiddeman, B., Burt, D.M., Perret, D.: Prototyping and transforming facial texture for perception research. IEEE Comput. Graph. Appl. **21**(5), 42–50 (2001)
12. Wu, Y., Thalmann, N., Thalmann, D.: A dynamic wrinkle model in facial animation and skin aging. J. Vis. Comput. Anim. **6**, 195–205 (1995)
13. OÕToole, A.J., Vetter, T., Volz, H., Salter, M.: Three-dimensional caricatures of human heads: distinctiveness and the perception of facial age. Perception **26**, 719–732 (1997)
14. Gandhi, M.: A method for automatic synthesis of aged human facial images. M.S. thesis, McGill University, Montreal, QC, Canada (2004)
15. Givens, G.H., Beveridge, J.R., Draper, B.A., Grother, P., Phillips, P.J.: How features of the human face affect recognition: a statistical comparison of three face recognition algorithms. In: Proceedings of International Conference on Pattern Recognition, vol. 2, pp. 381–388 (2004)
16. www.uhu.es
17. De Andrés, J.: Statistical techniques vs. SEE5 algorithm. An application to a small business environment. Int. J. Dig. Account. Res. **1**(2), 153–179. ISSN: 1577-8517
18. Ling, H., Soatto, S., Ramanathan, N., Jacobs, D.: A study of face recognition as people age. In: Proceedings of IEEE International Conference on Computer Vision, pp. 1–8 (2007)

19. Chen, H., Belhumeur, P., Jacobs, D.: In search of illumination invariants. In: IEEE Conference on Computer Vision and Pattern Recognition (CVPR), vol. 1, pp. 254–261 (2000)
20. Ling, H., Soatto, S., Ramanathan, N., Jacobs, D.: Face verification across age progression using discriminative methods. IEEE Trans. Inf. Forensics Secur. **5**(1), 82–91 (2010)
21. Ling, H., Soatto, S., Ramanathan, N., Jacobs, D.W.: Face verification across age progression using discriminative methods. In: IEEE Trans. Inf. Forensics Secur. (2010)
22. Tsumura, N., Haneishi, H., Miyake, Y.: Independent component analysis of skin color image. J. Opt. Soc. Am. **16**, 2169–2176 (1999)
23. Pandey, D.: Hybrid algorithm using fuzzy C-means and local binary patterns for image indexing and retrieval. Stud. Comput. Intell. (2012)
24. Mahalingam, G., Kambhamettu, C.: Face verification with aging using AdaBoost and local binary patterns. In: Proceedings of the Seventh Indian Conference on Computer Vision Graphics and Image Processing—ICVGIP 10 (2010)
25. Ojala, T., Pietikainen, M., Maenpaa, T.: A generalized local binary pattern operator for multiresolution gray scale and rotation invariant texture classification. In: Second International Conference on Advances in Pattern recognition, pp. 397–406 (2001)

Autonomous Robot Navigation Using Fuzzy Logic

Sagar Nandu, Nikhil Nagori and Alpa Reshamwala

Abstract The paper starts with explaining in brief about fuzzy logic and then goes on to explain a Fuzzy Logic Controller (FLC). It is the basic unit or a block of any application of fuzzy logic. We mainly focus on how to design a Fuzzy Logic Controller (FLC), its main functions and all the parameters or different factors involved. We then move on to our main objective i.e. implementing fuzzy logic in robots for making it autonomous in terms of navigation. After understanding basics of FLC, we focus on steering and obstacle avoidance of robot by assuming certain conditions regarding the environment, in other words putting restrictions on the behaviors that the robot can display. Some of these restrictions are then lifted and improvements are made to our previous model. We then focus on controlling the speed of the robot as well, considering the environment complexity. So the main objective of this paper is to get clear understanding on how to make a robot navigate autonomously using fuzzy logic controller. The paper then concentrates on why fuzzy logic will be a good approach to do so giving some examples.

Keywords Fuzzy logic controller · Autonomous robots · Robot navigation

1 Introduction

Fuzzy Logic is a mathematical method that when implemented helps computers make human like decisions. It uses fuzzy sets and fuzzy rules to model the world and make decisions about it. It relies on fuzzy sets. Fuzzy sets allows to deal with situations that are not precise. Fuzzy set theory can be referred as generalization on classical set theory. But there exists many properties that are unique to fuzzy set theory. The paper starts by giving a generic explanation on building any Fuzzy

S. Nandu (✉) · N. Nagori · A. Reshamwala
Computer Engineering Department, Mukesh Patel School of Technology
Management and Engineering, NMIMS University, Mumbai, India
e-mail: sagarnandu.nmims@gmail.com

S.C. Satapathy and S. Das (eds.), *Proceedings of First International Conference on Information and Communication Technology for Intelligent Systems: Volume 2*,
Smart Innovation, Systems and Technologies 51, DOI 10.1007/978-3-319-30927-9_44

Logic Controller. It explains why a fuzzy controller is necessary. It gives basic working of any fuzzy controller. Then it goes in detail and explains how a fuzzy controller is built. It explains each and every step in detail i.e. from designing a fuzzifier, explaining how knowledge base is created, how a rule base is created and what all should it contain, explaining importance of decision making logic to designing a defuzzifier. After explaining all the basic requirements for developing any application that incorporates fuzzy logic, it moves on to the most important part of the paper i.e. designing a self-navigating robot using fuzzy logic. Paper show two such implementations. The first one concentrates on steering and obstacle avoidance capability of the robot. It uses vision as well as SONAR for doing so. Second implementation concentrates on speed, steering and obstacle avoidance capabilities of the robot. It also uses vision and SONAR for doing so. But both implementations have different algorithms for achieving its goals. It also shows some simulation details done for testing the accurate working of FLC in MATLAB. These results are purely taken from the research paper [1]. The paper will be concluded by explaining the flaws of the designs explained in the paper and implemented, and consequences that might occur if restrictions are lifted. Observations thus obtained will be considered as further work for this project. The paper also explains why the author has chosen fuzzy logic for creating an autonomous navigating robot. So basically paper tries to answer following questions: Fuzzy mathematics, properties of fuzzy Logic, designing fuzzy logic controller, how to implement fuzzy logic, how to implement steering of any robot autonomously, how to control speed of robot autonomously, how to avoid obstacles autonomously.

2 Brief Description

Fuzzy Logic based on mathematical theory of fuzzy sets is generalization of classical set theory. By assigning a degree to participation of each element within a set, thus enabling the state of elements other than true or false, fuzzy logic gives good level of flexibility in handling inaccuracies and uncertainties. So this helps fuzzy logic in formalizing human like reasoning.

2.1 Fuzzy Logic Controller [2]

Basic component involved in any application of fuzzy logic is Fuzzy Logic Controller (FLC). Figure 1 gives block diagram of a generic FLC which can be optimized or modified according to need of an application. The efficient, accurate, sensitive and precise design of an FLC will ensure proper and smooth functioning of the controller. FLC consists of fuzzifier, defuzzifier, a knowledge base and decision making logic. Block diagram is given in Fig. 1.

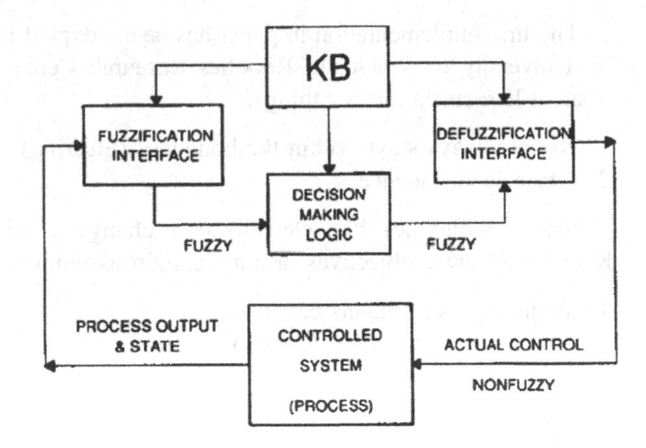

Fig. 1 Fuzzy logic controller (FLC) [2]

Main purpose of fuzzifier is to get the input data and fuzzify it which means that data is converted to linguist form. Input data helps in forming Universe of discourse. Knowledge Base consists of all the databases which helps in understanding the data and a rule base which characterizes control goal and control policies i.e. these rules are responsible in making decisions. Decision making logic is the heart of any FLC. It analyses input from the source by using rule base and helps in making human like decisions. Defuzzifier is responsible for converting fuzzy output to crisp output.

The changing behavior of any environment is being handled by a fuzzy system because of a strong rule base which consists of set of linguistic description rules based on expert knowledge. These rules are of the form

$$\begin{gathered} \textit{If}\,(\text{set of conditions}) \\ \textit{Then}\,(\text{consequences when the conditions are true}) \end{gathered}$$

And these rules are termed as fuzzy rules [2]. For more detailed understanding related to what fuzzification strategies are used or how a knowledge base is formed or how to design a decision making logic accurately and how to defuzzify the data then refer to [2].

3 Implementation of Fuzzy Logic for Building an Autonomous Navigating Robot

For a robot to navigate by itself it must be able to sense its environment, then it should be capable of planning to make decisions and then act on those decisions. Such an environmental model requires a human-like decision making. This can be accomplished if real time sensing along with some stored information together combined may give us the desired results.

The first implementation in paper has been adopted from a project implemented by University of Cincinnati Robotics Research Center. Robot developed named Bearcat II focused on two things:

1. Robot always stays within the boundary (Steering).
2. Obstacle Avoidance.

Path complexities like elevations or change in directions were taken care. Keeping all these objectives in mind certain assumptions were made:

1. Width of path remains constant.
2. Speed of robot remains constant.

So basically, this robot will be able to reach to its destination avoiding all the obstacles on the path and will not cross the boundaries laid down. This can be achieved in three ways:

3.1 Model Based Approach [3]

In this method, an accurate description of the environment will be needed by the robot to generate an obstacle free path. A path can be generated from initial position to its destination using environmental model. But it is highly unlikely to get an accurate environment description.

3.2 Sensor Based Approach [3]

In this method, operations are performed by evaluating data obtained from the sensors. Behavioral architecture, a strategy of sensor based approach, can be used. It consists of multiple behaviors which respond to different types of sensor inputs. This approach is feasible since a robot can travel on a path in changing environment. But robot does not have any destination.

3.3 Hybrid Approach [3]

In this paper we will be focusing on the hybrid approach which incorporates the advantages of both approaches and eliminates their limitations. First step will be generate path using model based approach and then using sensor based approach to navigate itself along the path. This will ensure that robot will reach its destination avoiding all the obstacles on the path. Figure 2 shows how the working environment of robot will look like.

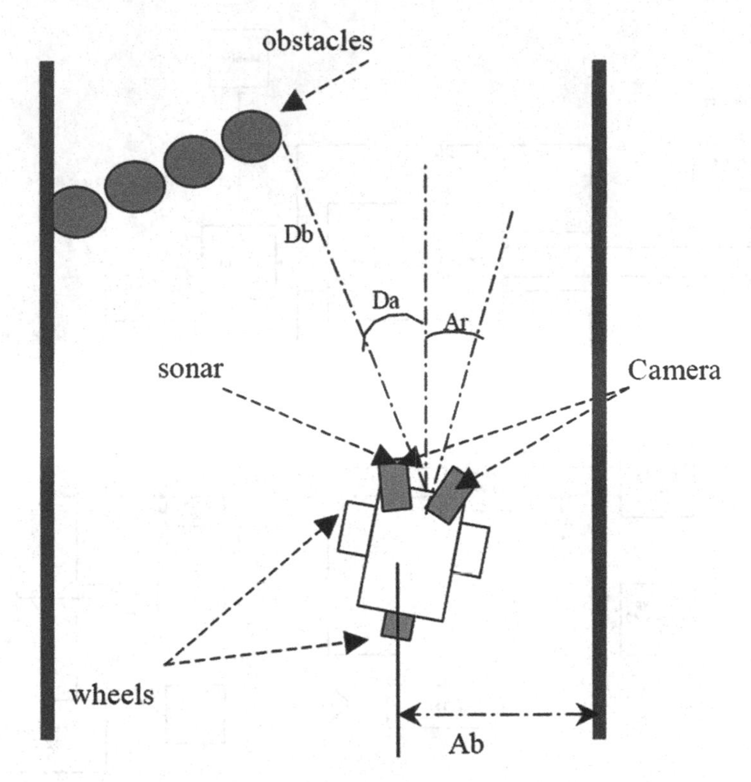

Fig. 2 Bearcat II in its working environment [3]

3.4 System Design Description

The physical structure of Bearcat II is divided as Digital Control, Vision Guidance, Ultrasonic Distance sensor and Emergency stop. The main components of the robot include: the central CPU, Iscan and two CCD camera as vision sensor, Plariod ultrasonic sonar, two 12 V batteries as motor power source, GALIL digital controller and two DC motors. Relationships among component are illustrated in Fig. 3 [3]. Description of their functions can be read in detail in reference [3].

3.5 Fuzzy Controller Design for Bearcat II

This will be the most important block of the system as it is going to help analyze the data and make decisions. For designing any controller we need to know what the system has to perform to reach its destination.

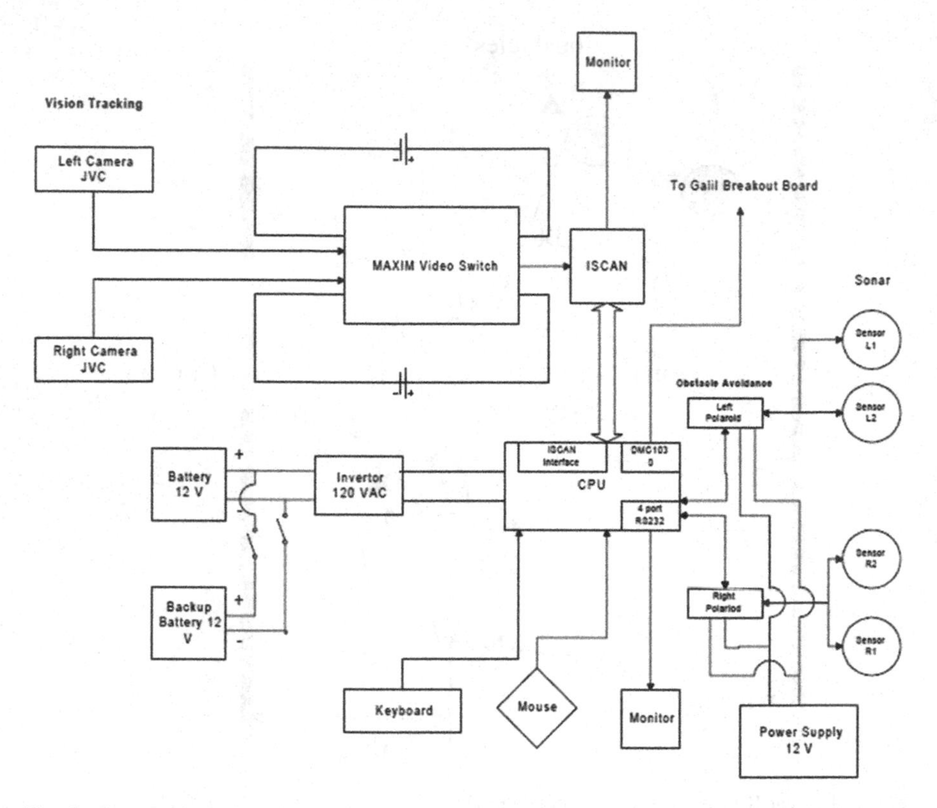

Fig. 3 System block diagram (Bearcat II) [3]

1. At any point of time system must be aware of its position i.e. distance to the border of the road and angle of its body with respect to orientation of the path.
2. If system encounters an obstacle then it should know about the distance from the obstacle and its relative angle to its body.

Figure 4 shows the way in which fuzzy controller is designed for this implementation. So in this system CCD cameras will track the position of the robot with respect to border line in terms of errors Ae (angular error i.e. difference of body angle and orientation of the path) and De (distance error from the border line). If both of these errors exists which will be identified by the controller then track following fuzzy rules will generate a result giving feasible angle range which on defuzzification will give the exact angle to align the robot to its natural state. In case if there is an obstacle place on the path then controller will also receive input as distance of the object and its angle with respect to its body. Now obstacle avoidance rules will be applied and this will limit our range of directions and as some direction available when obstacle was absent will be blocked. Now important thing to understand is what if both the cases occur at the same time (command fusion) then

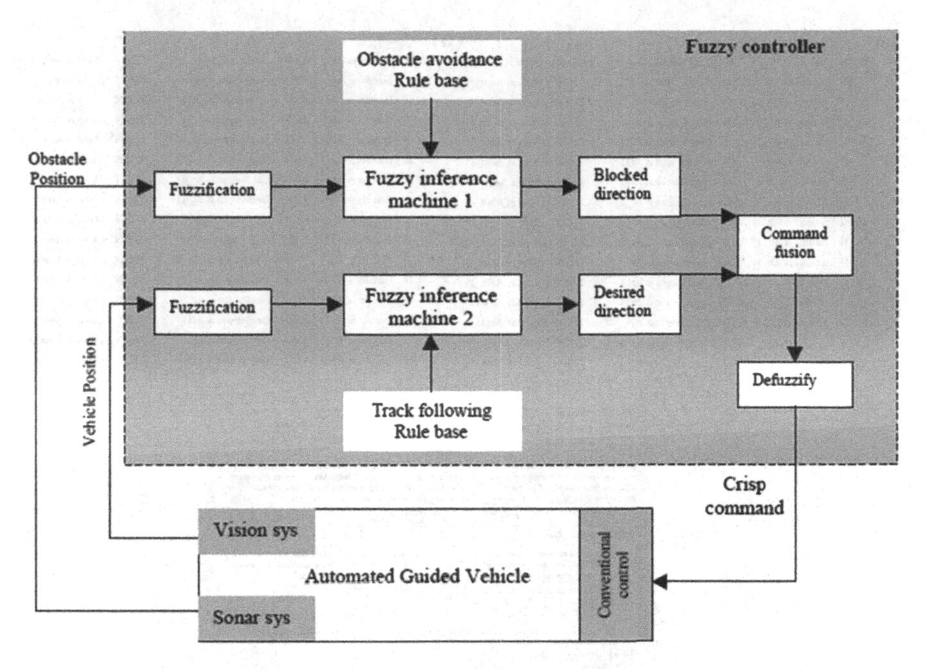

Fig. 4 Fuzzy controller block diagram [3]

we will have to take an intersection of directions that will ensure both the constraints are fulfilled. Figure 5 shows how fuzzy sets are defined in reference [3] for this particular implementation

3.6 Fuzzy Rules

Two behaviors that has to be implemented:

Track following Rules: These rules are concerned with values of *Ar* (angle of orientation) and *Ab* (distance from border). The controller gets values of these variables as input and output will be de-fuzzified range of directions having membership function $\mu A(x)$.

Obstacle Avoidance Rules: These rules are concerned with values of *Dr* (angle with respect to obstacle) and *Db* (Distance from the obstacle). The controller gets values of these variables as input and output will be de-fuzzified range of directions having membership function $\mu_B(x)$.

Command Fusion: For this we get all four variables as inputs and a new membership function is defined as

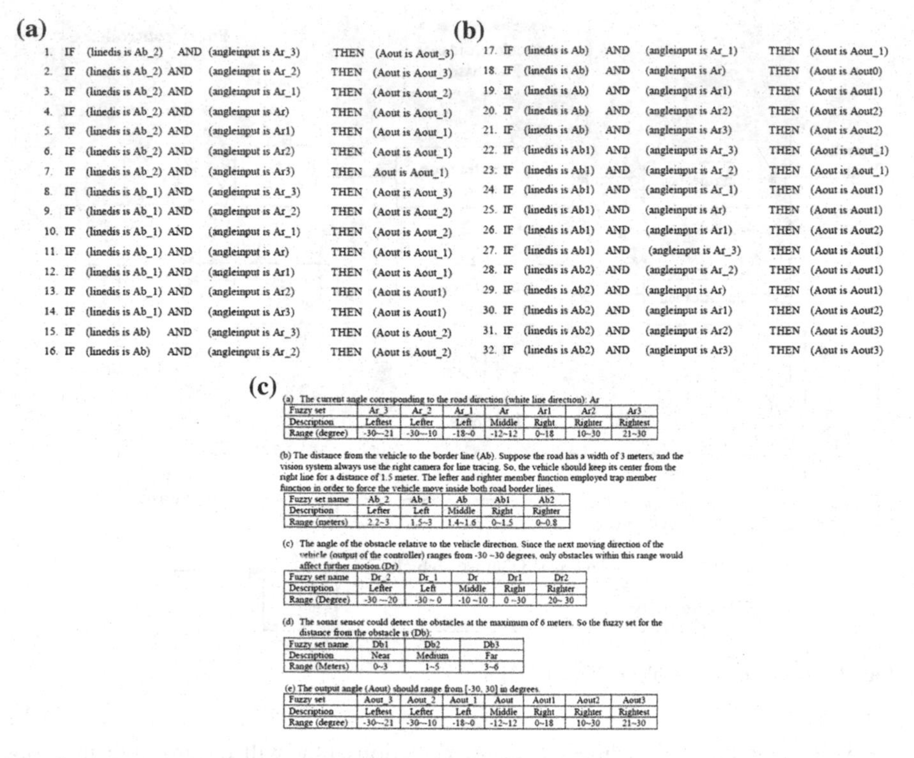

(a)

1. IF (linedis is Ab_2) AND (angleinput is Ar_3) THEN (Aout is Aout_3)
2. IF (linedis is Ab_2) AND (angleinput is Ar_2) THEN (Aout is Aout_3)
3. IF (linedis is Ab_2) AND (angleinput is Ar_1) THEN (Aout is Aout_2)
4. IF (linedis is Ab_2) AND (angleinput is Ar) THEN (Aout is Aout_1)
5. IF (linedis is Ab_2) AND (angleinput is Ar1) THEN (Aout is Aout_1)
6. IF (linedis is Ab_2) AND (angleinput is Ar2) THEN (Aout is Aout_1)
7. IF (linedis is Ab_2) AND (angleinput is Ar3) THEN Aout is Aout_1)
8. IF (linedis is Ab_1) AND (angleinput is Ar_3) THEN (Aout is Aout_3)
9. IF (linedis is Ab_1) AND (angleinput is Ar_2) THEN (Aout is Aout_2)
10. IF (linedis is Ab_1) AND (angleinput is Ar_1) THEN (Aout is Aout_2)
11. IF (linedis is Ab_1) AND (angleinput is Ar) THEN (Aout is Aout_1)
12. IF (linedis is Ab_1) AND (angleinput is Ar1) THEN (Aout is Aout_1)
13. IF (linedis is Ab_1) AND (angleinput is Ar2) THEN (Aout is Aout1)
14. IF (linedis is Ab_1) AND (angleinput is Ar3) THEN (Aout is Aout1)
15. IF (linedis is Ab) AND (angleinput is Ar_3) THEN (Aout is Aout_2)
16. IF (linedis is Ab) AND (angleinput is Ar_2) THEN (Aout is Aout_2)

(b)

17. IF (linedis is Ab) AND (angleinput is Ar_1) THEN (Aout is Aout_1)
18. IF (linedis is Ab) AND (angleinput is Ar) THEN (Aout is Aout0)
19. IF (linedis is Ab) AND (angleinput is Ar1) THEN (Aout is Aout1)
20. IF (linedis is Ab) AND (angleinput is Ar2) THEN (Aout is Aout2)
21. IF (linedis is Ab) AND (angleinput is Ar3) THEN (Aout is Aout2)
22. IF (linedis is Ab1) AND (angleinput is Ar_3) THEN (Aout is Aout_1)
23. IF (linedis is Ab1) AND (angleinput is Ar_2) THEN (Aout is Aout_1)
24. IF (linedis is Ab1) AND (angleinput is Ar_1) THEN (Aout is Aout1)
25. IF (linedis is Ab1) AND (angleinput is Ar) THEN (Aout is Aout1)
26. IF (linedis is Ab1) AND (angleinput is Ar1) THEN (Aout is Aout2)
27. IF (linedis is Ab1) AND (angleinput is Ar_3) THEN (Aout is Aout1)
28. IF (linedis is Ab2) AND (angleinput is Ar_2) THEN (Aout is Aout1)
29. IF (linedis is Ab2) AND (angleinput is Ar) THEN (Aout is Aout1)
30. IF (linedis is Ab2) AND (angleinput is Ar1) THEN (Aout is Aout2)
31. IF (linedis is Ab2) AND (angleinput is Ar2) THEN (Aout is Aout3)
32. IF (linedis is Ab2) AND (angleinput is Ar3) THEN (Aout is Aout3)

(c)

(a) The current angle corresponding to the road direction (white line direction): Ar

Fuzzy set	Ar_3	Ar_2	Ar_1	Ar	Ar1	Ar2	Ar3
Description	Leftest	Lefter	Left	Middle	Right	Righter	Rightest
Range (degree)	-30--21	-30--10	-18~0	-12~12	0~18	10~30	21~30

(b) The distance from the vehicle to the border line (Ab). Suppose the road has a width of 3 meters, and the vision system always use the right camera for line tracing. So, the vehicle should keep its center from the right line for a distance of 1.5 meter. The lefter and righter member function employed trap member function in order to force the vehicle move inside both road border lines.

Fuzzy set name	Ab_2	Ab_1	Ab	Ab1	Ab2
Description	Lefter	Left	Middle	Right	Righter
Range (meters)	2.2~3	1.5~3	1.4~1.6	0~1.5	0~0.8

(c) The angle of the obstacle relative to the vehicle direction. Since the next moving direction of the vehicle (output of the controller) ranges from -30 ~30 degrees, only obstacles within this range would affect further motion.(Dr)

Fuzzy set name	Dr_2	Dr_1	Dr	Dr1	Dr2
Description	Lefter	Left	Middle	Right	Righter
Range (Degree)	-30 --20	-30 ~ 0	-10 ~10	0 ~30	20~ 30

(d) The sonar sensor could detect the obstacles at the maximum of 6 meters. So the fuzzy set for the distance from the obstacle is (Db):

Fuzzy set name	Db1	Db2	Db3
Description	Near	Medium	Far
Range (Meters)	0~3	1~5	3~6

(e) The output angle (Aout) should range from [-30, 30] in degrees.

Fuzzy set	Aout_3	Aout_2	Aout_1	Aout	Aout1	Aout2	Aout3
Description	Leftest	Lefter	Left	Middle	Right	Righter	Rightest
Range (degree)	-30--21	-30--10	-18~0	-12~12	0~18	10~30	21~30

Fig. 5 Fuzzy set definition and rule base of Bearcat II [3] **a** Fuzzy rules **b** Fuzzy rules **c** Fuzzy sets

$$\mu_{new} = Min\{\mu_A(x), NOT\mu_D(x)\} \quad (1)$$

Fuzzy rules are mentioned in reference [3]. By applying these rules we get unique fuzzy output which on defuzzification gives us a crisp result.

3.7 *Defuzzification*

The output of an FLC must be an exact value i.e. crisp as robot requires just an angle value for navigating. In Bearcat II, centroid method has been used. There are other methods as well which can be found in [3].

Formula is given by [4]

$$Z_0 = \frac{\int z\mu_e(z)dz}{\int z\mu_e(z)dz} \quad (2)$$

Z_0 is final crisp output.

Now our second implementation has some of the constraints lifted i.e. assumptions decreases. In this we also concentrate on:

1. Varying speed of the robot.
2. Steering along with obstacle avoidance.

Assumption that width of path remains constant has been lifted but it always remains wider than the robot. Now in this method, obstacle avoidance is same but for path tracking a different algorithm is used.

For design of any autonomous guided robot, one has to design a controller that will be responsible of making decisions. This controller must be aware of its inputs, its state variables and its output. All these will be dependent on the environment where the robot is going to work. Now the main objective of this paper is to understand how this controller is designed and to do so all the inputs are analyzed in detail i.e. how each input will be obtained and how it is going to be transformed so that the controller can accept it. The main goal of this research is to model a modular Fuzzy Logic Control for an automated guided vehicle and test the performance of the vehicle Simulation in a MATLAB Simulation. The research is focused on the design of the Fuzzy Controller for vision and sonar navigation of the automated guided vehicle. Design of this controller is executed in three stages: Flowchart in Fig. 6

In the first stage, the universe of discourse is identified and fuzzy sets are defined. The rule base (Fuzzy Control Rules) for the control is then defined through a human decision making process. The membership functions and their intervals are defined. Aggregation and de-fuzzification methods are selected. In the second stage, the Fuzzy Controller is implemented on the autonomous guided vehicle. In the third and final stage, performance of the controller is tested through a series of simulations and real time running of the vehicle.

Now for path tracking vision guidance is used. In this, an image processing algorithm has been used to convert 3-D world co-ordinates to 2-D image co- ordinates. These received 2-D image co-ordinates are used to evaluate current status of robot. Evaluation results obtained are then converted to 3-D world co-ordinates and accordingly robot position is adjusted. In this way, steering of the

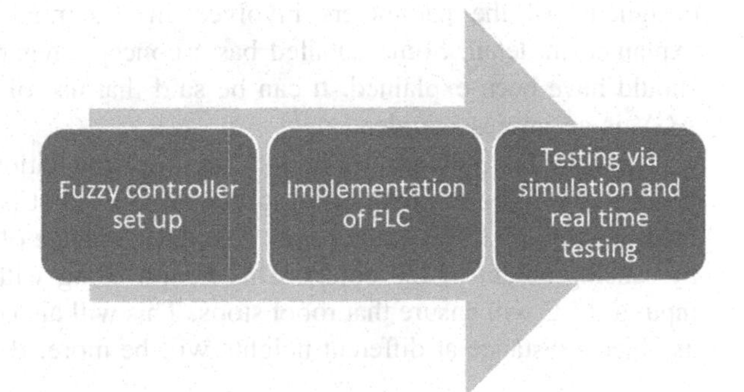

Fig. 6 Stages of second implementation [1]

robot is managed and robot stays on the path. Obstacle avoidance technique is similar to the one in the first implementation.

This method has been implemented in MATLAB. This implementation gives various results for every case in consideration. These results can be seen in reference [1]. These results given doesn't have corresponding inputs or rule base mentioned.

4 Advantages/Disadvantages

The main reason of using fuzzy logic is that it has the power to deal with imprecise data. It can do so because of its well defined rule base and measurements it receives from sensors. This ability to handle imprecision in changing environment can also be applied by other methods but fuzzy logic provides a lot more flexibility whereas in other methods the complexity keeps on increasing and re usability gets compromised. It is one of the most convenient choice when we cannot find a linear model of the system. Fuzzy logic helps in solving uncertainties that cannot be solved by classical probability theory. It is a MIMO (Multiple input and Multiple Output). Now using vision guidance and ultrasonic sensors for navigation will provide a better accuracy as inputs from both will help the robot make much better decisions as rule base will be more distinguished. Using just one of the method will not provide that much flexibility.

Disadvantage is that fuzzy logic is very vast. Design of any controller is complex and making rule base may depend on experience. Precision on values for making rule base i.e. selecting the range of data might be difficult. Complex mathematical calculations are involved. For designing any robot safety measurements must be included that may complicate the designing of controller.

5 Inference

The basics of fuzzy logic and has then explained how fuzzy logic controller is designed. All the parameters involved in designing the controller have been explained in detail. Some detailed basic concepts regarding theory of fuzzy logic should have been explained. It can be said that use of fuzzy logic for making an AGV is an impressive idea.

There are many loopholes in both the implementations. In first implementation, if there is wall on the path, as an obstacle then robot is going to crash because of modification of SONAR algorithm. The disadvantage of this system can be tackled by sensing height of the object. This height along with the distance of object as input to FLC will ensure that robot stops. This will also ensure an upward elevation as objects distance at different heights will be more. But yet there might be some flaws to this solutions which can be considered as further work. Proper algorithms can be devised to handle such problems.

These implementations have so many assumptions i.e. the environment complexity increases, which always will just like in real world, nothing is constant. Secondly, what are the safety mechanisms or different techniques if there is no specified boundaries or robot malfunctions? This must be mentioned in further work. But one way to do that would be using vision as well as ultrasonic sensors to determine obstacle on both of its side. But since all the calculations are with respect to boundaries then some mathematical mechanism must be introduced so that existence of boundaries can be eliminated.

6 Conclusion

In the first implementation, paper explains the improvised strategies for making an autonomous robot. The paper describes component functions of the robot and developed a fuzzy logic based on two behaviors. Two fuzzy logic approaches in mobile robot control have been discussed and one is used. The developed fuzzy logic let the mobile robot have the ability to follow the track and avoid obstacles [3]. The design and implementation of a modular fuzzy logic based controller [4, 5] for an autonomous mobile robot for line following along with position control with respect to an obstacle course has been presented [1].

References

1. Shukla, S., Tiwari, M.: Fuzzy Logic of Speed and Steering Control System for Three Dimensional Line Following of an Autonomous Vehicle. ArXiv preprint arXiv:1004.1675 (2010)
2. Lee, C.C.: Fuzzy logic in control systems: fuzzy logic controller. IEEE Trans. Syst., Man Cybern. **20**(2), 419–435 (1990)
3. Cao, M., Hall, E.L.: Fuzzy logic control for an automated guided vehicle. In: Photonics East (ISAM, VVDC, IEMB), International Society for Optics and Photonics, pp. 303–312 (1998)
4. Tanaka, K.: Design of model-based fuzzy controller using lyapunovs stability approach and its application to trajectory stabilization of a model car. In: Theoretical Aspects of Fuzzy Control, San Francisco, CA, 2nd IEEE Conference on fuzzy system, pp. 31–50 (1995)
5. Von Altrock, C., Krause, B., Zimmermann, H.J.: Advanced fuzzy logic control technologies in automotive applications. In: IEEE International Conference on Fuzzy Systems, IEEE, pp. 835–842 (1992)

Improved Non-Linear Polynomial Filters for Contrast Enhancement of Breast Tumors

Vikrant Bhateja, Mukul Misra, Shabana Urooj and Aimé Lay-Ekuakille

Abstract Non-Linear Polynomial Filters (NPF) consists of a framework of weighted coefficients of low-pass and high pass filters. This paper explores the applicability of NPF for the contrast enhancement of breast tumors in mammograms. NPF algorithm in the present work has been improved to provide controlled background suppression during the mammogram enhancement. This is because, in the process to control overshoots and visualization of tumor margins; the uncontrolled background suppression may lead to loss of finer details in the vicinity of the lesion region. Simulation results have shown that the response of the proposed NPF has been reasonably good on mammograms containing tumors embedded in different types of background tissues.

Keywords Background tissues · Contrast · Mammograms and NPF

V. Bhateja (✉)
Department of Electronics and Communication Engineering,
Shri Ramswaroop Memorial Group of Professional Colleges (SRMGPC),
Lucknow 227105, Uttar Pradesh, India
e-mail: bhateja.vikrant@gmail.com

V. Bhateja · M. Misra
Faculty of Electronics and Communication Engineering,
Shri Ramswaroop Memorial University, Deva Road, Lucknow
Uttar Pradesh, India
e-mail: mukul.katyayan@gmail.com

S. Urooj
Department of Electrical Engineering, School of Engineering,
Gautam Buddha University, Greater-Noida, Uttar Pradesh, India
e-mail: shabanabilal@gmail.com

A. Lay-Ekuakille
Department of Innovation Engineering, University of Salento, Lecce, Italy

S.C. Satapathy and S. Das (eds.), *Proceedings of First International Conference on Information and Communication Technology for Intelligent Systems: Volume 2*,
Smart Innovation, Systems and Technologies 51, DOI 10.1007/978-3-319-30927-9_45

1 Introduction

The 2013 statistics from American Cancer Society (ACS) has reported more than 1000 (approx.) breast cancer related deaths in females less than 40 years of age. Breast cancer is the primary cause of mortality (excluding cancers of skin and lungs) in US women claiming 29 % of new diagnosed cases [1]. Breast cancer is characterized as an uncontrolled growth of malignant cells inside lobules (milk glands), ducts (which connect lobules to the nipple), fatty as well as lymphatic tissues. Breast cancers are primarily detected during the initial Mammography screening; which comprises of a low-dose X-ray of the breast region. Breast masses or lumps often referred to as tumor detected during mammographic screening are reported benign upon examination by the radiologists. This cancer generally has no visible symptoms and is often painless when the tumor is small [2]. ACS therefore recommends that annual mammographic screening is necessary for women above the age of 40 [1]. The severity and stage of the cancer could be depicted from the careful examination of the structural and morphological details of these masses. Apart from this, an increased breast tissue density on mammograms is also a strong predictor of growing cancer. Among, the invasive or malignant masses, cells penetrate through the walls of glandular tissues and spread into the neighboring breast tissues. Once, the mammographic mass is being reported malignant; it calls for surgical microscopic analysis of the breast tissues via biopsy. It is therefore necessary that women above 40 are screened on routine basis in order to have early detection of breast cancer prior to the detection of symptoms [3, 4]. Earlier detection leads to a range treatment possibilities involving less-extensive surgery. However, there exist some potential drawbacks as well for this popular screening modality. Mammographic images suffer from poor contrast and noises leading of false positive detection outcomes. As a result, unnecessary biopsies are conducted leading to ample of patient discomfort. In-spite of various limitations like false alarms, over diagnosis etc., mammography is still the potential method for early detection; as it can identify the tumors years before the development of physical symptoms [5, 6]. Computer algorithms for pre-processing mammograms are therefore on growing call for the purpose of enhancement of contrast and suppress noise in the background in order to improve the lesion region in mammographic images [7–9]. Mammogram enhancement for the purpose of Computer-aided detection (CAD) has been achieved with a variety of approaches: Wavelet based Multi resolution approaches [10, 11], Morphological filtering [12, 13], Unsharp Masking (UM) [14, 15] and hybrid combinations of the aforesaid enhancement approaches [16, 17]. This paper explores the applicability of Non-Linear Polynomial Filtering (NPF) algorithm for the contrast enhancement of the suspicious regions in mammograms. The approach has been distinguished from its earlier versions [18, 19] in the sense to have controlled background suppression during the enhancement. This is because, in the process to control overshoots and visualization of lesion margins; the uncontrolled background suppression may lead to loss of finer details in the vicinity of the lesion region. These background tissues also play a vital role in the diagnosis by the radiologists. Further, it is also necessary that the enhancement process should not introduce any noise and artifacts. The rest of the paper is organized as follows. The proposed NPF methodology is detailed in Sect. 2. Further, the simulation results and discussions are presented in Sect. 3 whereas the conclusions are drawn in Sect. 4.

2 Proposed NPF Framework

2.1 Background

Bilinear realizations of Volterra series expansions are found suitable for filtering systems with Poisson noise inputs [20]. Quadratic filters are sub-class of bilinear Volterra filters whose output is linearly dependent upon filter coefficients. Alpha Weighted Quadratic Filter (AWQF) is the modified version of Volterra filter for enhancement of mammograms used by only implementing the quadratic component (of the Volterra Model) [21]. The Polynomial filter can be fundamentally modeled using second order truncation of the classical Volterra series and can be stated in general form in Eq. (1) [22, 23].

$$y(n) = h_{Poly}[x(n)] = h_0 + h_\theta[x(n)] + h_\phi[x(n)] \tag{1}$$

The Non-linear Polynomial Filtering (NPF) [24, 25] can be considered as a combination of linear and quadratic filter components.

$$y(n) = \sum_i \theta(i)x^{2\gamma(i)}(n-i) + \sum_i \sum_j \phi(i,j)x^{\lambda(i)}(n-i)x^{\lambda(j)}(n-j) \tag{2}$$

The NPF model of Eq. (2) possesses combined high-pass and low-pass filtering; providing noise smoothening and contrast improvement respectively.

2.2 NPF for Mammogram Contrast Enhancement

The proposed NPF explores the combination of Type-0 and Type-I Polynomial filter as a versatile platform to perform enhancement of mammograms. The degree of enhancement can be controlled in an automated manner by minimal tuning of NPF parameters without any variations with the type of abnormality, nature of lesion or the background breast tissues. As discussed in [24, 25], NPF may be classified into Type-0, Type-I and Type-II. Type-0 NPF can be mathematically stated in Eq. (3) and (4) respectively.

$$y_{linear} = \theta_0 x_5^{2a} + \theta_1 (x_1^{2b} + x_3^{2b} + x_7^{2b} + x_9^{2b}) + \theta_2 (x_2^{2c} + x_4^{2c} + x_6^{2c} + x_8^{2c}) \quad (3)$$

$$y^0_{quadratic} = \phi_0 x_5^{2a} + \phi_1 (x_1^{2b} + x_3^{2b} + x_7^{2b} + x_9^{2b}) + \phi_2 (x_2^{2c} + x_4^{2c} + x_6^{2c} + x_8^{2c}) \quad (4)$$

The improvement in contrast and denoising is performed majorly by this component of the proposed NPF. On the other hand, sharpening of features can be performed via the Type-I component given in Eq. (5).

$$\begin{aligned} y^I_{quadratic} = {} & \phi_3 (x_1^b x_2^c + x_1^b x_4^c + x_2^c x_3^b + x_3^b x_6^c + x_4^c x_7^b + x_6^c x_9^b + x_7^b x_8^c + x_8^c x_9^b) \\ & + \phi_4 (x_1^b x_5^a + x_3^b x_5^a + x_5^a x_7^b + x_5^a x_9^b) \\ & + \phi_5 (x_2^c x_5^a + x_4^c x_5^a + x_5^a x_6^c + x_5^a x_8^c) + \phi_6 (x_2^c x_4^c + x_2^c x_6^c + x_4^c x_8^c + x_6^c x_8^c) \end{aligned} \quad (5)$$

The proposed NPF framework for contrast enhancement can be therefore stated as a superposition of Type-0 and Type-I NPF responses respectively as given by $y(n)$ in Eq. (6).

$$y(n) = y_{linear}(n) + y^0_{quadratic}(n) + y^I_{quadratic}(n) \quad (6)$$

3 Results and Discussions

3.1 Enhancement Metrics for IQA

In this paper, different performance metrics are deployed to estimate the fruitfulness of the enhancement performed using the proposed algorithm and also to compare the results from different state-of-art algorithms [26, 27]. These performance metrics not only serve as a tool for quality evaluation of obtained mammograms but also help in optimal tuning and selection of NPF parameters [18, 19]. It has been ascertained from literature in the past that any single quality measure cannot be just sufficient to quantify the performance of mammogram enhancement approaches in consistency with human visual perception [28]. Therefore, three distinct Image Quality Assessment (IQA) metrics to evaluate the effectiveness of the proposed NPF have been used in this work. Firstly, the evaluation of proposed NPF algorithm is carried out using Contrast Improvement Index (CII) as the figure of merit. Higher values of CII indicate improvement in contrast of tumor region with respect to soft tissues in the background. Secondly, to assess amount of denoising carried out and to evaluate further that the contrast enhancement is not leading to enhancement in background noise levels; Peak Signal to Noise Ratio (PSNR) is estimated. Higher values of PSNR are indicative of suppression of background noise levels during contrast enhancement [24, 25, 29].

3.2 Simulation Results

The mammograms used in the present work for simulations are obtained from DDSM [30] and MIAS [31] databases respectively. On the basis of their subtlety, these mammograms are rated from normal to suspicious in terms of the relevant stage of cancer. The test images used for experiments include three sets of mammograms from DDSM and another three from MIAS databases respectively. Figure 1 shows the enhancement results using proposed NPF on mammograms of DDSM database. The test data from DDSM contained mammograms with malignant tumor (Subtlety Grade-5), irregular shape and ill-defined margins. Similarly, Fig. 2 shows the enhancement response on test mammograms from MIAS database. These mammograms each consists of tumor surrounded with different categories of background tissues like fatty, glandular and dense respectively. Figure 1 and 2 clearly shows the enhanced tumor region (in the foreground) with a very reasonable suppression of the background. This is the highlighting aspect of the proposed algorithm that it provides not just the contrast improved version of the tumor but also a clear visualization of the neighboring background (tissues, arteries and other diagnostic details). The surrounding region of the lesion also has a significant contribution in the mammogram analysis; that has been realized by avoiding

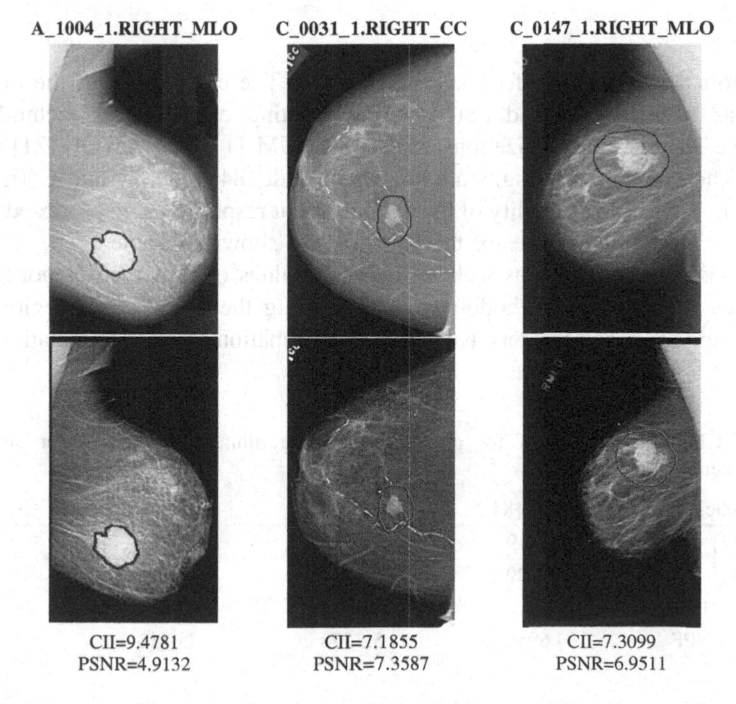

Fig. 1 Enhancement Response of Proposed NPF Algorithm on Mammograms from DDSM Database

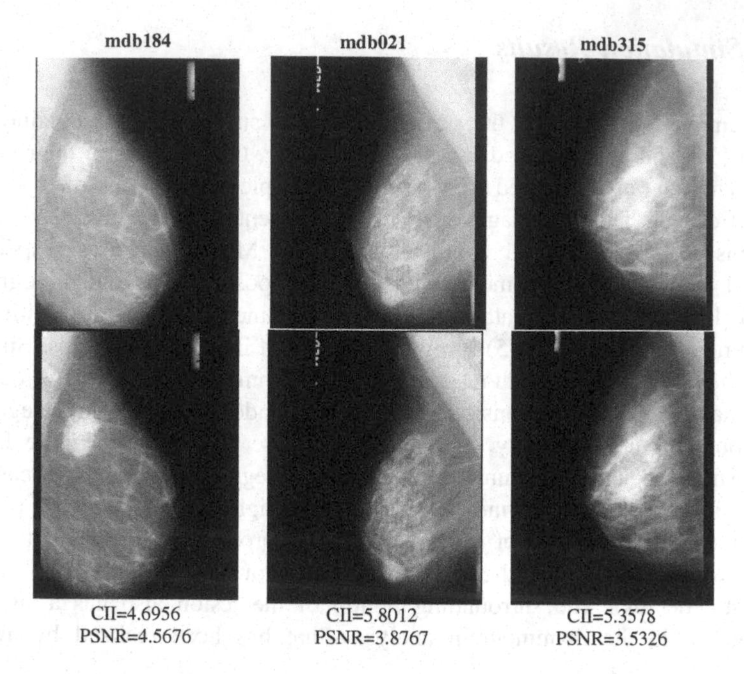

Fig. 2 Enhancement Response of Proposed NPF Algorithm on Mammograms from MIAS Database

uncontrolled suppression of background tissues. The comparison of the proposed NPF has been also carried out with the existing enhancement methods like: Adaptive Histogram Equalization (AHE) [32], UM [14] and AWQF [21] respectively. The comparative results are shown on mdb184, mdb315 and C_0147 test images in Table 1. The quality of the enhancement response using proposed NPF is evident from the high values of the IQA metrics shown in Table 1.

The visual comparison as well as the high values of IQA metrics portrays the efficiency of proposed methodology in enhancing the nodal tumor region along with preservation of necessary background information (diagnostic details).

Table 1 CII values computed for proposed NPF algorithm along with other state-of-art enhancement Approaches

Approaches	Mdb184	Mdb315	C_0147_1. right_CC
AHE [32]	0.9219	1.0044	1.5943
UM [14]	0.9999	1.4455	1.3970
AWQF [21]	17723	1.6978	2.8060
Proposed NPF	4.6956	5.3578	7.6511

4 Conclusion

Computer-aided detection (CAD) of mammograms serves as a potential alternative to accurately locate the features associated with breast tumors' morphology. It is also evident that not just the tumor region, but the background also contains the necessary diagnostic information for the radiologists. The proposed NPF algorithm has been improved to provide controlled background suppression during the enhancement using a superposition of Type-0 and Type-I NPF. Simulation results have been demonstrated on test mammograms from DDSM and MIAS databases. The enhancement responses are visually better in terms of reduced blurring of tumor region and also there is ample suppression of the background noise. The proposed methodology do not contains any hybridization of enhancement approaches; this further relieves of any severe computational loads.

References

1. Ghafoor, A., Samuels. A., Jemal, A.: American cancer society breast cancer facts & figures. American Cancer Society Inc, Atlanta (2013)
2. Rovere, G.Q., Warren, R., Benson, J.R. (eds.): Early breast cancer from screening to multidisciplinary management, 2nd edn. Florida, USA, Taylor & Francis Group (2006)
3. Bhateja, V., Devi, S.: An improved non-linear transformation function for enhancement of mammographic breast masses. In: 3rd IEEE International Conference on Electronics & Computer Technology, vol. 5, pp. 341–346 (2011)
4. Bhateja, V., Devi, S., An improved non-linear transformation function for enhancement of mammographic breast masses. In: IEEE 3rd International Conference on Electronics & Computer Technology (ICECT-2011), vol. 5, pp. 341–346. Kanyakumari, India April 2011
5. Jain, A., Singh, S., Bhateja, V.: A Robust approach for denoising and enhancement of mammographic breast masses. Int.J. on Convergence Comput. **1**(1), 38–49 (2013)
6. Bhateja, V., Vikrant, Devi., S.: Mammographic image enhancement using double sigmoid transformation function. In: International Conference on Computer Applications (ICCA-2010), pp. 259–264. Pondicherry, India (2010)
7. Feig, S.A., Yaffe, M.J.: Digital mammography, computer-aided diagnosis, and tele-mammography. Radiol. Clin. North Am. **33**(6), 1205–1230 (1995)
8. Bhateja, V., Urooj, S., Pandey, A., Misra, M., Lay-Ekuakille., A.: Improvement of masses detection in digital mammograms employing non-linear filtering. In: IEEE International. Multi-Conference on Automation, Computing, Control, Communication. and Compressed. Sensing (iMac4 s-2013), pp. 406–408. India March 2013
9. Gupta, R., Bhateja, V.: A new unsharp masking algorithm for mammography using non-linear enhancement function. In: International Conference on Information Systems Design and Intelligent Applications (INDIA 2012), pp. 779–786. India, January 2012
10. Diekmann, F., et al.: Evaluation of a wavelet-based computer-assisted detection system for identifying microcalcifications in digital full-field mammography. Acta Radiol. **45**(2), 136–141 (2004)
11. Shrivastava, A., Raj, A., Bhateja, V.: Combination of wavelet transform and morphological filtering for enhancement of magnetic resonance images. In: International Conference on Digital Information Processing and Communications Part-I. pp. 460–474 (2011)
12. Raj, A., Shrivastava, A., Bhateja, V.: Computer aided detection of brain tumor in MR images. Int. J. on Eng. and Technol. 3, 523–532 (2011)

13. Mustra, M., Grgic, M., Rangayyan, R.M.: Review of recent advances in segmentation of the breast boundary and the pectoral muscle in mammograms. Med. & Biolog, Eng. & Comput. 1–22 (2015)
14. Panetta, K., Yicong, Z., Agaian, S., Jia, H.: Nonlinear unsharp masking for mammogram enhancement. IEEE Trans. Inf Technol. Biomed. **15**(6), 918–928 (2011)
15. Gupta, R., Bhateja, V.: A log-ratio based unsharp masking (UM) approach for enhancement of digital mammograms. In: The International Information Technology Conference on CUBE, pp. 26–31. ACM (2012)
16. Anand, S., Kumari, R., Jeeva, S., Thivya, T.: Directionlet transform based sharpening and enhancement of mammographic X-ray images. Biomed. Signal Process. Control **8**(4), 391–399 (2013)
17. Srivastava, S., Sharma, N., Singh, S.K., Srivastava, R.: A combined approach for the enhancement and segmentation of mammograms using modified fuzzy C-means method in wavelet domain. J. of Med. Phys./Assoc. of Med. Phys. of India **39**(3), 169–183 (2014)
18. Pandey, A., Yadav, A., Bhateja, V.: Contrast improvement of mammographic masses using adaptive volterra filter. In: 4th International Conference on Signal and Image Processing. 2, 583–593 Springer (2012)
19. Pandey, A., Yadav, A., Bhateja, V.: Design of new volterra filter for mammogram enhancement. In: International Conference on Frontiers in Intelligent Computing Theory and Applications. 199, 143–151 Springer (2012)
20. Ramponi, G.: Quadratic filters for image enhancement. In: 4th European Signal Processing Conference (EUSIPCO-88), pp. 239–242. Grenoble, France, September 1988
21. Zhou, Y., Panetta, K., Agaian, S.: Mammogram enhancement using alpha weighted quadratic filter. In: Annual International Conference on IEEE Engineering in Medicine and Biology Society, 3681–3684. Minneapolis, Minnesota (2009)
22. Mathews, V.J.: Adaptive polynomial filters. IEEE Signal Proc. Mag. **8**(3), 10–26 (1991)
23. Bhateja, V., Urooj, S., Misra, M., Pandey, A., Lay-Ekuakille, A.: A polynomial filtering model for enhancement of mammogram lesions. In: IEEE International Symposium on Medical Measurements and Applications (MeMeA), pp. 97–100. IEEE (2013)
24. Bhateja, V., Misra, M., Urooj, S., Lay-Ekuakille, A.: A robust polynomial filtering framework for mammographic image enhancement from biomedical sensors. IEEE Sens. J. **13**(11), 4147–4156 (2013)
25. Bhateja, V., Urooj, S., Misra, M.: Technical advancements to mobile mammography using nonlinear polynomial filters and ieee 21451-1 ncap information model. Sens. J. IEEE **15**(5), 2559–2566 (2015)
26. Srivastava, H., Mishra, A., Bhateja, V.: Non-linear quality evaluation index for mammograms. In: 3rd Students Conference on Engineering and Systems. 269–273 (2013)
27. Gupta, P., Tripathi, N., Bhateja, V.: Multiple distortion pooling image quality assessment. Int. J. on Convergence Comput. **1**(1), 60–72 (2013)
28. Bhateja, V., Patel, H., Krishn, A., Sahu, A., Lay-Ekuakille, A.: Multimodal medical image sensor fusion framework using cascade of wavelet and contourlet transform domains. Sens. J. IEEE **15**(12), 6783–6790 (2015)
29. Trivedi, M., Jaiswal, A., Bhateja, V.: A novel HVS based image contrast measurement index. In: 4th International Conference on Signal and Image Processing (ICSIP 2012) pp. 545–555, Springer, India. January 2013
30. Heath, M., et al.: The digital database for screening mammography. In: Yaffe M.I., (eds.): 5th International Workshop on Digital Mammography, pp. 212–218, Medical Physics Publishing (2001)
31. Suckling, J., et al.: The mammographic image analysis society mammogram database. In: 2nd International Workshop Digital Mammography, pp. 375–378, U.K (1994)
32. Cheng, H.D., Shi, X.J., Min, R., Hu, L.M., Cai, X.P., Du, H.N.: Approaches for automated detection and classification of masses in mammograms. Pattern Recognition. 39(4): 646–668 (2006)

Sparse Representation Based Face Recognition Under Varying Illumination Conditions

Steven Lawrence Fernandes and G. Josemin Bala

Abstract In this paper we have developed novel technique to recognize faces across Illumination. Illumination is a condition where an image of same individual looks different due to varying lighting conditions. Recognizing faces across Illumination is proposed using Sparse Representation technique and tested using Extended Yale B database which consist of images across varying lighting conditions. Here we have considered images of 16 individuals at different lighting conditions. From our analysis we have found that proposed system to recognize faces across various lighting conditions using Sparse Representation Technique gives the best recognition rate of 95 % on Extended Yale B database.

Keywords Face recognition · Illumination · Sparse representation

1 Introduction

Face recognition alludes to the procedure of recognizing people in view of their facial features. It has as of late turned into a standout amongst the most prominent examination zones in the fields of computer vision, machine learning, and pattern recognition in light of the fact that it has different customer applications, for example: reconnaissance, security of computer frameworks, Visa check in the e-business, and criminal ID [1]. It's being used in many gadgets. These gadgets incorporate the home gadgets, for example, savvy TV, video and music players [2, 3]. Amid the most recent decades, numerous methodologies have been created and proposed for fruitful face recognition system.

S.L. Fernandes (✉) · G.J. Bala
Department of Electronics and Communication Engineering,
Karunya University, Coimbatore, Tamil Nadu, India
e-mail: steva_fernandes@yahoo.com

G.J. Bala
e-mail: josemin@karunya.edu

S.C. Satapathy and S. Das (eds.), *Proceedings of First International Conference on Information and Communication Technology for Intelligent Systems: Volume 2*,
Smart Innovation, Systems and Technologies 51, DOI 10.1007/978-3-319-30927-9_46

Specifically, change in lighting conditions that happens on face pictures debases the execution of face recognition system under pragmatic situations. For recognition of face it's been watched that the varieties among pictures of the same face within illumination changes are bigger than image variations in pictures because of variations in identities of faces [4]. Chen et al. [5] delineated this unpredictability by demonstrating that there is no discriminative illumination invariant for Lambertian questions on source of light which are placed apart. Which means that it is impractical to figure out if two pictures were made by the same protest under two different sources of light or by separate objects? Therefore issued one picture of an object, it is hard to anticipate anything positive about this item or how will it show up under varying lighting conditions.

2 Previous Work

To conquer the issue created by illumination variation, different methodologies have been presented, for example: pre-processing and illumination normalization procedures [6–9], feature extraction methods in fixed lighting conditions [4, 10, 11], and strategies for modeling 3D faces [12–15].

Despite the overwhelming way of this lighting issue numerous techniques were proposed in the writing to handle lighting varieties. Histogram adjustment is the a bit easiest and most regular method to decrease the impacts of lighting variations [16]. In the model of 3-D morphable system [17], every face model is spoken by a linear combination of set of 3-D model faces. To the input image morphable model was fitted and was utilized effectively as a part of face recognition [18] and face synthesis [17–19]. The system in [18] is sure to images which are obtained under different angular light sources and requires the learning of light course, which is not simple to gauge by and large. Zhang et al. [20] further coordinated a more normal light representation into a morphable model methodology. In their technique, the spherical harmonic bases takes care of the illumination variations of general lighting conditions. In spite of the fact that 3-D morphable model gives superb execution in both synthesis and recognition applications, it obliges an extensive arrangement of 3-D outputs of individual faces, which is not simple to gather; the investigation by blend methodology requests a ton of processing in light of the fact that both the surface and 3-D shape has to be evaluated, and additionally every single other parameter. These make it infeasible in a few circumstances.

Barrow and Tenenbaum contended that, even after not expressing all the physical reasons for image appearance, such a depiction to a midlevel can be to a great degree valuable for some visual inferences.

3 Proposed System

To recognize images across various lighting Conditions we enhance the image with different lighting conditions which is a test image and images in database for performing pre-processing using Median filter. And then we and we try to prove that our new methodology using sparse representation provides us solution to this problem. We propose a general classification model by using sparse representation calculated by L1—minimization.

Modules of proposed system:

- Preprocessing
- Normalization
- Sparse Representation via L1—minimization
- Sparse Representation based Classifier

3.1 Preprocessing

In our methodology we are using median filter to remove noise. This filter selects the matrix of pixels arranges them in ascending order and finds the median value among these pixel values. Finally we replace the center pixel value by this median value.

3.2 Normalization

Normalization is a process that changes the range of pixel intensity values. Illumination variation caused by changes in sources of light at different positions and various intensities causes large variations. To overcome this problem we obtained a new method of performing image normalization. This method removes shadows and specularities from images. All shadowed regions are given a uniform color and then it eliminates the soft shadows and specularities and thus creates an illumination invariant copy of the original image.

3.3 Sparse Representation via L1-Minimization

Recent developments in sparse representation say that if x0 solution is very sparse then the solution of l0-minimization problem is same as the solution of l1-minimization problem. Sparse Representation is used for feature extraction and the extracted features are assigned unique class id. The class id is unique for each person.

3.4 Sparse Representation Based Classifier

The extracted features after Sparse Representation are classified using Sparse Representation classifier. The shortest distance is calculated between the features of train and test. Based on the least minimum distance obtained between the test and train features the images are matched.

4 Results and Discussions

Recognizing faces across varying light conditions is tested using Extended Yale B database which consists of illuminated images. This system has five folders they are: test, train, cropped images, norm and smooth. Test folder is where test (input) images are stored, train folder is where train images are stored, cropped images folder is where the cropped test image is stored, smooth is where the smoothened test images are stored and norm is where normalized test image is stored. Here in our work we have made use of Extended Yale B database.

4.1 Implementation Steps of the Proposed System

1. A GUI pops up with 6 buttons. They are: Load Me (Training data), Test image, Test Image Enhancement, smoothening, Gradient, Normalization, Identified Image and Accuracy.
2. When Load Me is pressed another GUI pops up asking us to select the folder containing train images. After selecting that folder train images get loaded. Test image button is highlighted.

Table 1 Indicates that proposed system gives better recognition rate

Author	Method	Database	Recognition rate (%)
Punnappurath et al. [21]	TSF (Trading support facility) model	Extended Yale B	81.2
Chan et al. [22]	Multiphase local phase quantization (LPQ), kernel fusion of multiple descriptors	Combined extended Yale B and Yale B	78.6
Zhang et al. [23]	Riemannian framework	Yale B	87.1
Yucel et al. [24]	Gaussian process regression, neural networks, saliency schemes	Videos	80.2
	Sparse representation (proposed system)	Extended Yale B	95

3. When Test image button is pressed a GUI pops up asking us to select the test image. After the image is selected it gets loaded into the main GUI. Now Test Image Enhancement button is highlighted.
4. When Test Image Enhancement button is pressed the enhanced test image of required region is loaded. After this smoothening button is highlighted.
5. When smoothing button image is pressed the smoothening of enhanced image takes place and the smoothened image is loaded to the GUI. After this Gradient button is highlighted.
6. When Gradient button is pressed Smoothened image is loaded in gradient form. Now the Normalization button is highlighted.
7. When Normalization button is pressed the Gradient image is normalized and is loaded. Now Identified image butto is highlighted.
8. When Identified Image button is pressed the Recognized image from the training set is displayed. Now Accuracy button is displayed.
9. When accuracy button is pressed the accuracy level is displayed with a graph (Table 1).

4.2 Step by Step Implementation of the Proposed Approach

See Fig. 1.

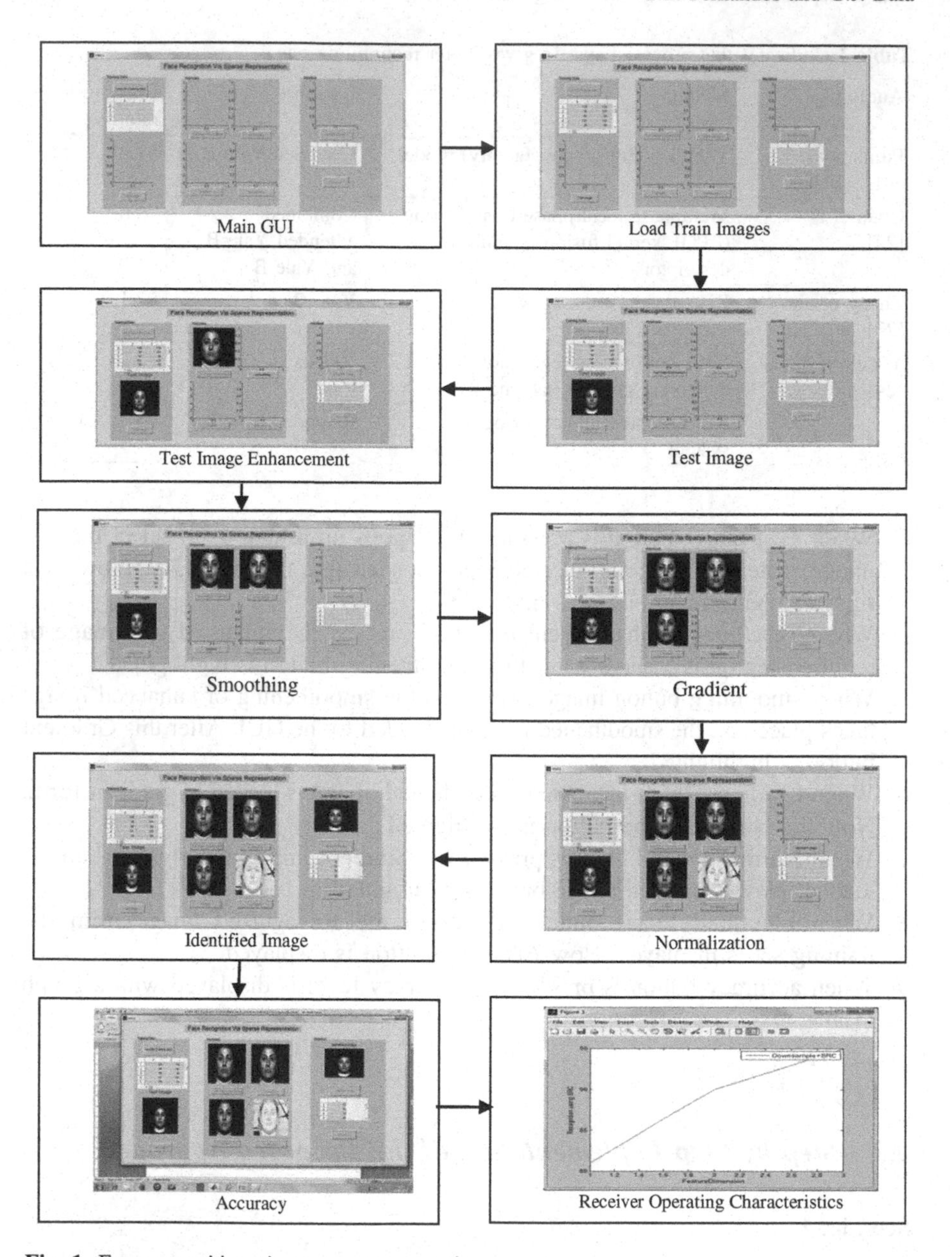

Fig. 1 Face recognition via sparse representation

5 Conclusion

In this paper we have developed novel technique to recognize faces across Illumination. Illumination is a condition where images of same individual look different due to varying lighting conditions. Recognizing faces across Illumination is proposed using Sparse Representation technique and tested using Extended Yale B database which consist of images across varying lighting conditions. Here we have considered images of 16 individuals at different lighting conditions. From our analysis we have found that proposed system to recognize faces across various lighting conditions using Sparse Representation Technique gives the best recognition rate of 95 % on Extended Yale B database.

Acknowledgments The proposed work was made possible because of the grant provided by Vision Group Science and Technology (VGST), Department of Information Technology, Biotechnology and Science and Technology, Government of Karnataka, Grant No. VGST/SMYSR/GRD-402/2014-15 and the support provided by Department of Electronics and Communication Engineering, Karunya University, Coimbatore, Tamil Nadu, India.

References

1. Kachare, N.B., Inamdar, V.S.: Survey of face recognition techniques. Int. J. Comput. Appl. **1**(1), 29–33 (2010). (Published By Foundation of Computer Science)
2. Zuo, F., de With, P.H.N.: Real-time embedded face recognition for smart home. IEEE Trans. Consum. Electron. **51**(1), 183–190 (2005)
3. An, K.H., Chung, M.J.: Cognitive face analysis system for future interactive TV. IEEE Trans. Consum. Electron. **55**(4), 2271–2279 (2009)
4. Adini, Y., Moses, Y., Ullman, S.: Face recognition: the problem of compensating for changes in illumination direction. IEEE Trans. Pattern Anal. Mach. Intell. **19**(7), 721–732 (1997)
5. Chen, H.F., Belhumeur, P.N., Jacobs, D.W.: In search of illumination invariants. In: Proceedings of IEEE Conference on Computer Vision and Pattern Recognition, vol. 1, pp. 254–261 (2000)
6. Chen, W., Er, M.J., Wu, S.: Illumination compensation and normalization for robust face recognition using discrete cosine transform in logarithm domain. IEEE Trans. Syst. Man Cybern. Part B: Cybern. **36**(2), 458–466 (2006)
7. Savvides, M., Vijaya Kumar, B.V.K.: Illumination normalization using logarithm transforms for face authentication. In: Proceedings of the 4th International Conference on Audio- and Video-Based Biometric Person Authentication (2003)
8. Xie, X., Lam, K.-M.: Face recognition under varying illumination based on a 2D face shape model. Pattern Recogn. **38**(2), 221–230 (2005)
9. Shan, S., Gao, W., Cao, B., Zhao, D.: Illumination normalization for robust face recognition against varying lighting conditions. In: IEEE International Workshop on Analysis and Modeling of Faces and Gestures, pp. 157–164, 17 Oct 2003
10. Sanderson, C., Paliwal, K.K.: Fast features for face authentication under illumination direction change. Pattern Recogn. Lett. **24**, 2409–2419 (2013)
11. Shashua, A., Riklin-Raviv, T.: The quotient image: class-based rendering and recognition with varying illuminations. IEEE Trans. Pattern Anal. Mach. Intell. **23**(2), 129–139 (2011)
12. Belhumeur, P.N., Kriegman, D.J.: What is the set of images of an object under all possible illumination conditions. Int. J. Comput. Vis. **28**(3), 1–16 (1998)

13. Georghiades, A.S., Belhumeur, P.N., Jacobs, D.W.: From few to many: illumination cone models for face recognition under variable lighting and pose. IEEE Trans. Pattern Anal. Mach. Intell. **23**(6), 630–660 (2001)
14. Basri, R., Jacobs, D.W.: Lambertian reflectance and linear subspaces. IEEE Trans. Pattern Anal. Mach. Intell. **25**(2), 218–233 (2003)
15. Zhang, L., Samaras, D.: Face recognition under variable lighting using harmonic image exemplars. In: Proceedings of the IEEE Computer Society Conference on Computer Vision and Pattern Recognition, vol. 1, pp. 19–25 (2003)
16. Phillips, P.J., Vardi, Y.: Efficient illumination normalization of facial images. Pattern Recognit. Lett. **17**(8), 921–927 (1996)
17. Blantz, V., Vetter, T.: Face recognition based on fitting a 3D morphable model. IEEE Trans. Pattern Anal. Mach. Intell. **25**(9), 1063–1074 (2003)
18. Blanz, V., Scherbaum, K., Vetter, T., Seidel, H.P.: Exchanging faces in images. In: Proceedings of EUROGRAPHICS, pp. 669–676 (2004)
19. Zhang, L., Wang, S., Samaras, D.: Face synthesis and recognition from a single image under arbitrary unknown lighting using a spherical harmonic basis morphable model. In: Proceedings of IEEE Computer Society Conference Computer Vision Pattern Recognition, vol. 2, pp. 209–216 (2005)
20. Barrow, H., Tenenbaum, J.: Recovering intrinsic Scene Characteristics from Images, in Computer Vision System. Academic, New York (1978)
21. Punnappurath, A., Rajagopalan, A.N., Taheri, S., Chellappa, R., Seetharaman, G.: Face recognition across non-uniform motion blur, illumination, and pose. IEEE Trans. Image Process., **24**(7) 2015
22. Chan, C.H., Tahir, M.A.., Kittler, J., Pietikainen, M.: Multiscale local phase quantization for robust component-based face recognition using kernel fusion of multiple descriptors. IEEE Trans. Pattern Anal. Mach. Intell., **35**(5) (2013)
23. Zhang, Z., Klassen, E., Srivastava, A.: Gaussian blurring-invariant comparison of signals and images. IEEE Trans. Image Process. **22**(8) (2013)
24. Yucel, Z., Salah, A.A., Mericli, C., Mericli, T., Valenti, R.: Theo Gevers. Joint attention by gaze interpolation and saliency. IEEE Trans. Cybern. **43**(3) (2013

Part IV
Content Based and Applications

Interference Minimization Protocol in Heterogeneous Wireless Sensor Networks for Military Applications

Kakelli Anil Kumar, Addepalli V.N. Krishna and K. Shahu Chatrapati

Abstract Wireless sensor Network (WSN) is an emerging technology has significant applications in several important fields like military, agriculture, healthcare, environmental, artificial intelligence and research. All these applications demands high quality data transmission from resource constraint WSNs. But interference is one of the severe problem in WSN which can degrade the quality data transmission. Various interference minimization techniques have been proposed but not results the expected degree of quality enhancement in WSNs. This research paper investigates various types of heterogeneous wireless sensor networks (HTWSN) deployment techniques, interference and its effects, existing interference minimization techniques with limitations. We propose interference minimization (IM) protocol for heterogeneous wireless sensor networks. IM protocol can efficiently minimize the interference and enhance the quality data transmission in WSN.

Keywords Heterogeneous wireless sensor networks · Military wireless sensor networks · Interference minimization protocol · Quality of service

K.A. Kumar (✉)
Indore Institute of Science and Technology, RGPV,
Indore, MP, India
e-mail: anilsekumar@gmail.com

A.V.N. Krishna
Navodaya Institute of Technology, VTU, Raichur, KA, India
e-mail: hari_avn@rediffmail.com

K. Shahu Chatrapati
COE, Manthani, Jawaharlal Nehru Technological University,
Hyderabad, TS, India
e-mail: shahujntu@gmail.com

S.C. Satapathy and S. Das (eds.), *Proceedings of First International Conference on Information and Communication Technology for Intelligent Systems: Volume 2*,
Smart Innovation, Systems and Technologies 51, DOI 10.1007/978-3-319-30927-9_47

1 Introduction

Military wireless sensor network (M-WSN) [1] consists of tiny autonomous sensor nodes having sensing, processing and transmission capabilities. These nodes can be useful for data transmission from source sensor node (SN) to destination node (DN) through flexible network. M-WSN is to set up with the sensor nodes which are low cost, low power and multifunctional. WSN is having vast number of applications in various important fields like military, agriculture and industry [2]. But M-WSN is having many open challenges like sensor nodes are low battery power, limited memory storage, less computational capability, short distance communication range and high network traffic [3]. Due to these limitations, M-WSN is unable to extend its quality of services (QOSs) for military and other applications. Apart from these, interference is another biggest challenge in M-WSN which severely affects the QOS. In M-WSN, sensor nodes are deploying randomly in to the area of operation (AO). After the deployment, each sensor node in the network would be autonomous, self-controlled, self-protected and self-resistance. So M-WSN requires strong and efficient routing protocols to overcome these challenges for an effective communication and QOS [4]. M-WSNs are classified as homogeneous WSN and heterogeneous WSN. Homogeneous wireless sensor networks (HOWSN) consists of large number of similar low configured sensor nodes. Due to that, HOWSN may not handles the complex tasks, large data processing due to low memory and low processing power of sensor nodes. HOWSN can require more number of nodes which consumes high cost. Heterogeneous wireless sensor networks (HTWSN) is having few special sensor with large number of low configure sensor nodes. HTWSN can classify in to three types, computational heterogeneous wireless sensor networks (C-HTWSN), link heterogeneous wireless sensor networks (L-HTWSN) and energy heterogeneous wireless senor networks (E-HTWSN). HTWSN is having many advantages as compared with HOWSN. C-HTWSN can handle the complex data sensing, processing and long-term storage of data compared to HOWSN. L-HTWSN can have high sensing power, bandwidth and transmission range. With that, L-HTWSN provides high reliable data transfer as compared with HOWSN. E-HTWSN can have high-energy sensor nodes to increase the network lifetime [5], reliable data transfer and to handle the complex task. With these advantages, HTWSN is most preferable network than HOWSN for military applications. The variable configure sensor nodes are randomly deploy in area of monitoring (AO) to setup the HTWSN [6]. The random deployment is to prefer for locations where the human cannot reach it such as battlefield, earthquake zones and forest fire.

2 Multi Path Routing in HTWSN: Literature Survey

The routing in HTWSNs are two types, single path and multipath routing [7, 8]. Most of the routing protocols are single path routing, where the single routing path can be established between SN to DN. But single path routing is having several limitations. The active path may fails due to interference, congestion, battery off, link failure, physical damage of sensor node, security attacks. This leads to reinitiate and rediscover the alternative route which increases the end to end delay, network overhead and lowers the throughput. In multipath routing, SN starts route discovery to set up multiple paths towards the DN. Multipath routing can improves data reliability by sending duplicate data through multiple routing paths. But it can result another problem known as multi path/inter path interference (MPI) or route coupling effect which reduce the QOS and performance of the network. MPI can exist because of simultaneous utilization of adjacent routing paths leads to data collision at any sensor node in the routing path [9]. Some techniques have been proposed but not highly efficient to overcome the route coupling effect. Location aware routing protocols [10] can find the location of the each node within the network. But it requires additional localization algorithms and hardware support which increases the communication overhead and high computational complexity. Directed antenna [11] helps to find accurate location of the sensor node but it requires special antenna hardware equipment which increases the energy consumption and cost of sensor node. Multi-channel data transmission [12] protocols have proposed to minimize the inter path interference but its needs redesign of MAC layer for efficient channel switching. This technique uses 2.4 GHz frequency band [13] which is highly prone to external interference with other networks. LIEMRO [14] is event driven multipath routing protocol for minimization of inter path interference. But it results poor performance and low network life time when the multiple events can occurred in the network. Interference [15] is serious threat to HTWSN [16] which results high packet drop, increases data re-transmissions, link instability and protocol's inconsistent behavior, low reliable data transfer. The inter path interference [17] can arise due to interfering paths or route coupling of multiple paths. But inter path interference is one of the major challenge in HTWSN to be addressed well.

3 Interference Minimization (IMP) Protocol

The proposed interference minimized (IMP) protocol is having of three important phases such as route initialization phase, route discovery and establishment phase and route maintenance phase.

3.1 IM Protocol Route Initialization Phase

In the route initialization phase, SN can initiate to identify its neighbor nodes by release of neighbor node identification message (NNIM). NNIM can send by the source node to all its neighbor or intermediate nodes (IN) in its communication range. The NNIM is to receive by first hop intermediate nodes (INs) of SN. These nodes can update NNIM with following details, such as node's ID, node residual energy, interference degree and path ID. Initially SN updates its details in NNIM and forwards to its INs. The NNIM is received by first hop INs and save the data then first hop INs updates NNIM and forward to second hop INs. Second hop INs receives the NNIM and stores the data and update it to forward the third hop INs. Like the way, the route initialization phase can is proceed and finally NNIM reaches to DN. The DN receives NNIM and save the data in it and update NNIM and generate update neighbor node identification message (UNNIM). UNNIM forward by DN to its N_{th} INs in same path which it received NNIM. UNNIM is exchanged between all the INs in the path of reverse direction by hop by hop mode. Finally UNNIM is received by SN and save the information in it. IMP protocol can efficiently complete the route initialization phase between SN and DN through INs. IMP route initialization technique can help not only to identify the INs but also INs current status to choose for effective path establishment. Consider the following parameters used by IMP protocol for route initiation phase: Node's Identity number is N_{ID}, Node residual energy is N_{RE}, Node energy consumption is N_{EC} and Node interference degree is N_{IFD}, and Path identity number as PT_{ID}. Network size is represented by S, network density is D. $N_{RE\ T}$ is node residual threshold level, N_R is Node reliability, A is area of operation (L*B m^2) and S is number of sensor nodes deployed in AO, X is the SN identity number. $N_{EC_{NNIM}}/N_{EC_{NNIM}}$ is node energy consumption for NNIM/UNNIM, $N_{RE_{NNIMReceive}}/N_{RE_{UNNIMReceive}}$ is node residual energy at NNIM/UNNIM receive, $N_{RE_{NNIMSent}}/N_{RE_{UNNIMSent}}$ is node residual energy at NNIM/UNNIM sent.

$$S = \sum_{ID=0}^{ID=n} N\ ID \tag{1}$$

$$\text{Network Node Density}(\mathrm{D}) = \mathrm{S}/\mathrm{A} \tag{2}$$

$$PT_{ID} = \sum_{X=1,PID=0}^{X=n,PID=3} \mathrm{SN}_{X,PID} \tag{3}$$

$$\mathrm{N}_{\mathrm{EC}_{\mathrm{NNIM}}} = \mathrm{N}_{\mathrm{RE}_{\mathrm{NNIMReceive}}} - \mathrm{N}_{\mathrm{RE}_{\mathrm{NNIMSent}}} \tag{4}$$

$$\mathrm{N}_{\mathrm{EC}_{\mathrm{UNNIM}}} = \mathrm{N}_{\mathrm{RE}_{\mathrm{UNNIMReceive}}} - \mathrm{N}_{\mathrm{RE}_{\mathrm{UNNIMSent}}} \tag{5}$$

$$\mathrm{N_{RE}} = \mathrm{N_{RE_{UNNIMReceive}}} - \mathrm{N_{EC_{UNNIM}}} = \mathrm{N_{RE_{UNNIMSent}}} \tag{6}$$

$$N_{IFD_{Initial}} = \frac{\text{No. NNIMsent}}{No.NNIM\ Receive} \tag{7}$$

$$N_{IFD_{Final}} = \frac{\text{No. UNNIMsent}}{No.\ UNNIM\ Receive} \tag{8}$$

$$\mathrm{N_R} = \{\{N_{ID}\}, \{P_{ID}\}, \{N_{RE} > N_{RE_\tau}\}, \{N_{IFD_{Final}} < 1\}\} \tag{9}$$

$N_{IFD_{Initial}}/N_{IFD_{Final}}$ is node interference degree initial/final. The data reception rate of interference minimization protocol IMP is denotes as $\mathrm{D_{RR}}$ and estimated by the ratio of number of data packets successfully received by DN with total number of data packets sent by SN. The network life time achieved by IMP is denoted by NW_{LT} and estimated by the difference between network initialization time and time of first node dead.

3.2 IM Protocol (IMP) Route Discovery and Establishment Phase

The most important phase of IMP is route discovery and establishment. IMP starts route discovery when target event may enter into AO. If the target object may enter in to the communication area of any node, the node trigger awake message to all of its neighbor nodes. Once the awake message will be receive by SN. SN starts route discovery phase by sending of Route Request (RREQ). The RREQ will send to only high reliable nodes ($\mathrm{N_R}$) which are in communication area. The $\mathrm{N_R}$ will be estimated based on parameters received through NNIM and UNNIM. When the RREQ is received by first hop INs from SN, it will forward to the second hop INs based on high degree of reliability. The operation can proceed until the RREQ reaches to DN. Once the RREQ is received by DN through multiple paths of SN, the DN can collects the data of RREQs and analyze it to choose the best path set for SN and DN. The best path set will be analyzed based on minimum hop count, minimum interference path degree and high residual energy. Based on all these parameters, the DN chooses high quality path set for SN. IM protocol can establish the routes effectively by identifying low interfering path set. It starts route establishment by releasing Route Reply message (RREP) through the best selected paths towards the SN. Once the SN receives RREP message through the selected paths from DN, it releases the Confirmation acknowledgment (CACK) to all INs in the active paths towards DN. All INs in the active paths understand that, they are fixed to the active path of respective SN and DN. Hence the INs involving in active path of one SN cannot become the member of other SN's active path. The efficient node disjoint can be achieve with IM protocol.

3.3 IM Protocol Route Maintenance

IM protocol route maintenance phase is another important phase. All the INs will act like permanent members for active path of respective SN and DN pair. IM protocol identifies three routing paths between SN to DN termed as primary, secondary and back up paths. The primary and secondary paths are to be use for data delivery. Backup path is kept in standby mode will be used only when the path error occurs in any active path due to node or link failure. However failure IN can generates route error message (RRER) and send to its SN through upstream nodes in backward directions. Then SN enables the backup path to proceed for data delivery. With the process, the link and node failure problem can effectively minimized and data delivery achieved without any end to end delay and packet loss. Once the data transmission can complete, SN sends release acknowledgement (RACK) to all INs in primary, secondary and backup paths to inform about completion of data delivery. After receiving RACK, the INs will delete CACK and available ready to participate for other routing paths. Like the way, the route maintenance can effectively handles by IM protocol.

4 Performance Evaluation

All our experiments are performed in GloMoSim [18] simulator with the following simulation parameters. Simulation area: 1000 * 1000 m^2; Number of nodes: 50 to 500; type of deployment: random, simulation time; 300 to 600 s; battery capacity: 2400 mAh; propagation limit: −65 dBm; Propagation Path loss: Free-Space; Temperature: 290.0 K; Radio Type: Radio Acc-noise; Radio-Frequency: 2.4 GHz; Radio Bandwidth: 1.9 M bits/sec; Radio-Rx-Type: SNR-Bounded; Radio-Rx-SNR-Threshold: 10.0 dBm; Radio-TX-Power: 15.0 dBm (Single source) 10.0 dBm (Multiple source); Radio-RX-Sensitivity: −91.0 dBm; Radio-RX-Threshold: −81 dBm; MAC-Protocol: SMAC; Routing-Protocol: IMP. Figure 1, IMP has given high D_{RR} which is almost 50 % higher than LIEMRO [13] protocol and single path routing approach under heavy traffic conditions. IMP can successfully discover the quality path sets for each SN when multiple target events were detected. IMP can adjust the node transmission power which results lowering the inter path interference. Figure 2, IMP has given better results compared to LIEMRO and single path routing interns of network life time with variable high traffic load. This is achieved because of efficient load balancing through multiple paths of single source nodes. IMP has given satisfactory results compared with LIEMRO protocol when multiple events occur. IMP load balancing scheme has effectively handled high traffic load during multiple events. Figure 3, IMP uses

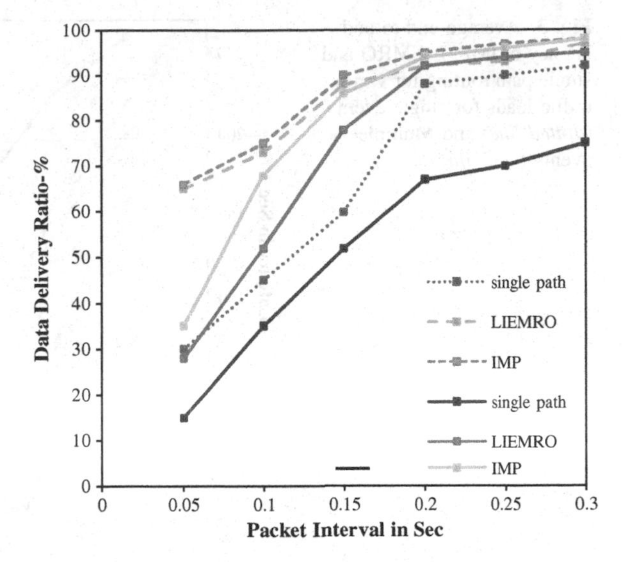

Fig. 1 Data delivery ratio of IMP, LIEMRO and single path routing for various traffic loads for single event (*dotted line*) and multiple events (*thick line*)

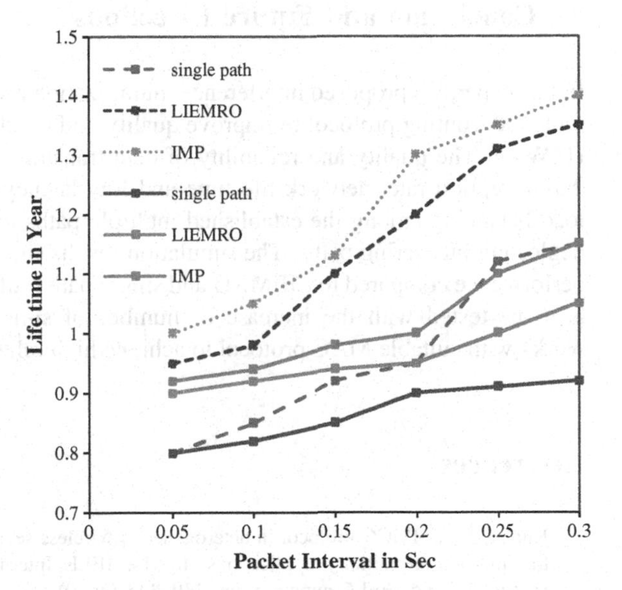

Fig. 2 Lifetime of IMP, LIEMRO and single path routing and for various traffic loads for single event (*dotted line*) and multiple events (*thick line*)

effective load balancing ETX cost metric [19] compared with the LIEMRO and single path routing approach. IMP has strong path maintenance mechanism which helps to achieve low latency compared to LIEMRO. IMP can transmit high size data through quality multiple paths in the network. IM protocol has given satisfactory results as compared with LIEMRO and single path routing schemes.

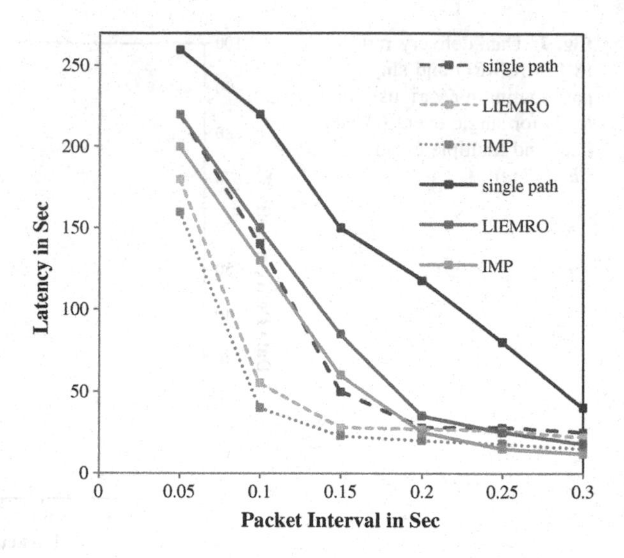

Fig. 3 Average end-to-end latency of IMP, LIEMRO and single path routing for various traffic loads for single event (*dotted line*) and Multiple events (*thick line*)

5 Conclusion and Future Directions

In this paper, we proposed interference minimization protocol (IMP), a node disjoint multipath routing protocol to improve quality and reliability of data transmission in HTWSN. The quality and reliability of data transmission are considered based on data reception rate, network life time and low latency. IM protocol uses effective load balancing among the established multiple paths to improve data reliability by neglecting interfering paths. The simulation results shows IM protocol results better performance compared to LIEMRO and single path routing approaches. IM protocol is to be tested with the increase of number of simultaneous events in the networks with suitable MAC protocol to achieve high reliability and network life time.

References

1. Kumar, K.A.: IMCC protocol in heterogeneous wireless sensor network for high quality data transmission in military applications. In: 1st IEEE International Conference on Parallel, Distributed and Grid Computing, pp. 339–343 (2010)
2. Yick, J., Mukherjee, B., Ghosal, D.: Wireless sensor network survey. Comput. Netw. **52**, 2292–2330 (2008)
3. Gungor, V.C., Lu, B., Hancke, G.P.: Opportunities and challenges of wireless sensor networks in smart grid. IEEE Trans. Ind. Electronics. **57**(10), 3557–3564 (2010)
4. Akyildiz, I., Su, W., Sankarasubramaniam, Y., Cayirci, E.: Wireless sensor networks: a survey. Comput. Netw. **38**(4), 393–422 (2002)
5. Sudarshan, T V, Manjesh, B.N.: A survey on wireless sensor networks. Int. J. Eng. Res. Technol. **4**(04) (2015)
6. Zhang, H., Liu, C.: A review on node deployment of wireless sensor network. Int. J. Comput. Sci. **9**(6), 378–383 (2012)

7. Radi, M., Dezfouli, ., Bakar, K.A., Lee, M.: Multipath routing in wireless sensor networks: survey and research challenges. Sensors **12**(1), 650–685 (2012)
8. Lou, W.: Performance optimization using multipath routing in mobile ad-hoc and wireless sensor networks. Combinational Optimization Communication Network, 1–29 (2005)
9. Jain, K., Padhye, J., Padmanabhan, V.N., Qiu, L.: Impact of interference on multi-hop wireless network performance. In: Proceeding ACM Mobicom, pp. 66–80 (2003)
10. Li, B., Chuang, P.: Geographic energy-aware non-interfering multipath routing for multimedia transmission in wireless sensor networks. Inf. Sci. Elsevier, **249**, pp. 24–37 (2013)
11. Kułakowski, P., Vales-Alonso, J., Egea-López, E., Ludwin, W., García-Haro, J.: Angle-of-arrival localization based on antenna arrays for wireless sensor networks. Computer. Electronic. Eng. **36**(6), 1181–1186 (2010)
12. Tarn, W.H, Tseng, Y.C.: Joint multi-channel link layer and multi-path routing design for wireless mesh networks. In: Proceedings of the twenty sixth IEEE International Conference on Computer Communications, pp. 2081–2089 (2007)
13. Ahmed, N., Kanhere, S., Sanjay, J.: Poster abstract: multi-channel interference in wireless sensor networks. In: Proceedings of International Conference on Information Processing in Sensor Networks, pp 367–368 (2009)
14. Radi, M., Dezfouli, B., Razak, S.A., K.A. Bakar.: LIEMRO: a low-interference energy-efficient multipath routing protocol for improving QoS in event-based wireless sensor networks. In: Proceedings of 4th International Conference on Sensor Technology and Applications, pp. 551–557 (2010)
15. Radi, M., Dezfouli, B., Bakar, K.A., Razak, S.A., Nematbakhsh, M.A.: Interference-aware multipath routing protocol for QoS improvement in event-driven wireless sensor networks. Tsinghua Sci. Technol., Elsevier, **16**(5), 475–490 (2011)
16. Raghuraj, C., Chandrasekhar, S.: Intrusion tolerance in heterogeneous wireless sensor networks using voting based algorithm with dynamic redundancy management of multipath routing. Int. J. Innov. Technol. Res., **3**, pp. 2147–2152 (2015)
17. Tadayon, N., Khoshroo, S., Askari, E., Wang, H., Michel, H.: Power management in SMAC-based energy-harvesting wireless sensor networks using queuing analysis. J. Netw. Comput. Appl. **36**(3), pp. 1008–1017 (2013)
18. Zeng, X., Bagrodia, R., Gerla, M.: GloMoSim: a library for parallel simulation of large-scale wireless networks. In: Proceedings. Twelfth Work. Parallel Distributed Simulation, pp. 154–161 (1998)
19. De Couto, D.S.J., Aguayo, D., Bicket, J., Morris, R.: A high-throughput path metric for multi-hop wireless routing. J. Wirel. Netw. **11**(4), pp. 419–434 (2005)

An Empirical Study of a Cryptographic Scheme for Secure Communication in Mobile Ad Hoc Networks

Kuncha Sahadevaiah, Nagendla Ramakrishnaiah and P.V.G.D. Prasad Reddy

Abstract The design and implementation of a cryptographic scheme in a mobile ad hoc network (MANET) is highly complicated than the traditional networks. This is due to several reasons, like—unavailable central infrastructure services, periodic movable nodes, wireless link unsteadiness, and possible network separations. In this paper, we present a cryptographic key management scheme for secure communications in ad hoc networks. The scheme transmits the data in the presence of security attacks. The scheme provides the scalability of nodes and storage space wherein the nodes makes use of more than one key pair to encrypt and decrypt the messages. The scheme has been executed in Java programming language and empirically valued its efficiency via performance and security assessments. The simulation outcome has shown that the proposed scheme opposes against selfish nodes and takes less key storage space than traditional schemes. The scheme also satisfies the secure communication requirements.

Keywords Mobile ad hoc network · Security · Secure communication · Cryptography · Public-private keys · Key management

K. Sahadevaiah (✉) · N. Ramakrishnaiah
Department of Computer Science and Engineering, University College of Engineering, Jawaharlal Nehru Technological University, Kakinada 533003, Andhra Pradesh, India
e-mail: ksd1868@gmail.com

N. Ramakrishnaiah
e-mail: nrkrishna27@gmail.com

P.V.G.D. Prasad Reddy
Department of Computer Science and Systems Engineering, College of Engineering, Andhra University, Visakhapatnam 530003, Andhra Pradesh, India
e-mail: prasadreddy.vizag@gmail.com

S.C. Satapathy and S. Das (eds.), *Proceedings of First International Conference on Information and Communication Technology for Intelligent Systems: Volume 2*, Smart Innovation, Systems and Technologies 51, DOI 10.1007/978-3-319-30927-9_48

1 Introduction

A *Mobile* Ad hoc *NETwork* (MANET) is a self-configured wireless communication network that contains mobile nodes to call one another via wireless radio lines. The network will not use any pre-configured fixed infrastructure, like access points or base stations. MANETs have the capability to instinctively shape a network of mobile nodes or nodes united collectively or nodes split into different networks, called *self-configuration*. The important goals of self-configured ad hoc networks are: availability, scalability and reliability [1]. The examples of mobile nodes are: Palmtops, Internet Mobile Phones, Cellular Phones, PDAs, Laptops, Pocket PCs and other wireless devices. Generally, these lightweight movable nodes are battery operated [2].

The movable nodes are exposed to several kinds of security attacks than the traditional networks. *Security attack* is a type of illegal activity that cooperates with the protection of data. The attack may change, discharge or deny data. This is due to their free nature of wireless medium, absence of central infrastructure services, dynamic network topology, constrained capability, distributed cooperation and short of a clear line of protection. The attackers enter into the wireless medium to intercept, interfere and insertion of communication among movable nodes. The movable nodes are easily detained and take over by the selfish nodes. Selfish node's actions may knowingly disturb the network and, hence, the complete network suffers from loss of packets. Before the successful deployment of MANETs, security issues along with the trusted authenticity must be addressed. The security protects the confidentiality, availability, authenticity, integrity, and non-repudiation [3–5].

The scheme uses the public key cryptography wherein, before deployment, the private and public keys are pre-distributed to the devices. Based on the pre-distribution of keys, all the devices authenticate the messages. Before the deployment of nodes in the environment, the individual nodes store the required cryptographic keys. When the network size is large, the storage space required to hold the cryptographic keys at each node may be too small. The scheme uses a lesser set of cryptographic keys where a sender and receiver need several keys to encrypt and decrypt the message. Every node knows all the public keys and contains a unique combination of private keys. The paper involves the intensive analysis of the key circulation process, communication overhead, cryptographic keys storage space and node faults by adversaries.

The rest of the paper is planned as follows: Sect. 2 assesses the related work. Section 3 provides the explanation of cryptographic key scheme and its operations. Section 4 explains the performance metrics and its valuations. The paper is concluded in Sect. 5.

2 Related Work

The supervision of cryptographic keys in a cryptosystem is called as *key management*. This includes dealing with the: key generation, key exchange, key storage, key usage and replacement of keys [3, 4]. The major trouble in building the cryptography based secure communication system as well as verification of its authenticity is availing all the node's public key certificates to other nodes. In MANETs, this problem is complicated to decide. This is due to numerous reasons like key exchange, session usage, absence of central services, periodic movable nodes, wireless connections unsteadiness, probable network separations and configuration of the network services. The traditional key management security solutions hire on-line or off-line servers to issue certificates to the nodes in a system.

The protocols in the key management are intended for both authority-based MANETs and fully self-organized MANETs [6, 7]. The off-line authority is used in *authority-based MANETs* to maintain applications. The nodes contain pre-established relationships. Before forming a network, the trusted authority arranges nodes with cryptographic keys. Every node in the network distributes its authentication to the other nodes that fall in its authority domain. There is no on-line or off-line authority in *fully self-organized MANETs*. In these networks, the end-users get support in an ad hoc basis, wherein a trusted third party (TTP) is not required to use the security links. Due to the absence of prior relationships, the end-users do not share any information.

Scalable key-management schemes are needed for MANETs to improve availability of security service and system scalability. In the earlier works, when the network size is made bigger, the following problems are identified [6–10]:

An Efficient Authenticity Problem—Certificate chaining is used to verify the public key authentication. More than one certificate verification is needed to validate public key via certificate chain. However, while collecting the certificates for public key authentication, the existing methods suffer from large amount of traffic. The certificate graph may not be strongly connected.

The Security Problem—The local certificate repositories of two nodes are joined to locate a sequence of connecting certificates. When the certificate chain becomes longer, the competence of the public key is declined and, hence, the system might be less safe. Though the system is independent, its transitivity of dependence belongings becomes a lot to attack.

The Overhead Problem—Every node issues multiple certificates and maintains two types of repositories in the network: un-updated and updated certificate repository. The major problem is huge communication overhead for storing an estimated total certificate graph that contains certificate repository of mobile nodes.

The Side Channel/Radio Channel Problem—The previous methods makes use of procedures, called certificate request/reply and handshaking. In handshaking procedure, a safe side channel is used to receive the public key of a node from another node. In certificate request/reply procedure, a radio channel is used to request certificates of a remote node from the handshaked nodes. The difficulty is with the use of side and radio channels.

3 Cryptographic Key Scheme

The design goal of the key management scheme is to reduce the memory space required for storing the cryptographic keys, computationally proficient for the period of encryption and decryption process and resistant to security attacks. In a network of size n of the conventional key management scheme, each node contains one public-private key pair. Hence, every node is supposed to store n public keys and one private key. However, a node in the cryptographic key scheme proposed by He et al. [11] stores less number of keys than the traditional scheme.

The cryptographic key management method proposed by He et al. [11] maintains an off-line server. It generates and distributes public-private key pairs to all the nodes in the network. The key pool generation and distribution by the server with a network of 50 nodes is shown in Fig. 1. The key pool consists of a set of distinct public-private key pairs. Every ith key pair is denoted by $(K^i_{public}, K^i_{private})$.

The generation of key pool depends on the total number of nodes available in the network and the algorithm that makes use of combinations with respect to private key generation. A contradictory key assignment for any pair of the nodes does not exist in a legal key allocation. There is a need for sufficient number of public–private key pairs to ensure the secure communication among n nodes.

Given a network of size n, every node stores all public keys as well as a tiny subset of private keys. Each node makes uses a little bit of public and private keys to encrypt and decrypt the messages. The combinatory should be used to determine the set of public keys, a and the set of private keys, b. The complexity of encryption and decryption gets affected by the parameter b value. The key storage efficiency requires a/n could be small. The size of the network n is a total of $C(a, b)$ possible combinations. That is, if $a = 25$ and $b = 4$, then $C(a, b) = C(25, 4) = 12650$ nodes.

If one private key is allotted to a maximum of z nodes, it becomes $b * n = a * z$. That is, every node holds a predefined subset of private keys. Therefore, $z = b/a * n = b/a * C(a, b)$. For a valid key allocation, arbitrarily assign unused b private key combinations to nodes in the network out of a total of $C(a, b)$ possible combinations. Only one private key is supposed to be assigned to a maximum of $b/a * C(a, b)$ nodes. That is, multiple copies of the private key from a public-private key pair can be held by different nodes. After a legal key allocation, every node knows

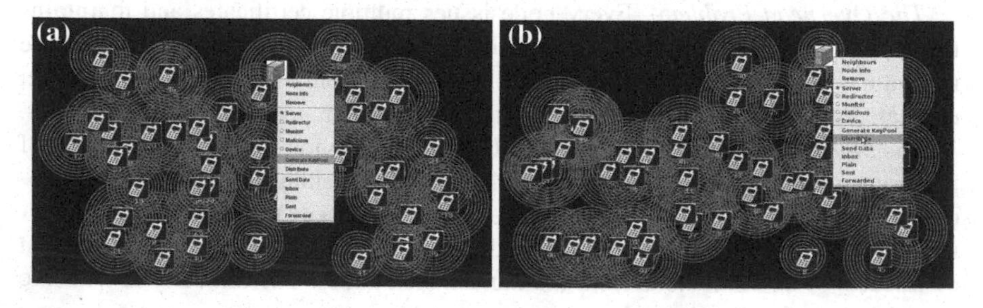

Fig. 1 **a** Key pool generation by server **b** Key pool distribution by server

the assignment of all the public keys and private keys. A subset of private keys is assigned to a node in the network using the procedure available in He et al. [11].

Consider a network of size n is 45 nodes. Therefore, $n = C(a, b) = C(10, 2) = 45$. To perform secure communication among 45 nodes, the scheme requires 10 distinct public-private key pairs:$(K^1_{public,} K^1_{private})$, $(K^2_{public,} K^2_{private})$, $\cdots$, $(K^{10}_{public,} K^{10}_{private})$. Each node holds ten public keys and two private keys. If $a = 10$ and $b = 2$, then each private key is assigned $2/10 * C(10, 2) = 2/10 * 45 = 9$ times. $K_i^{private}$ denotes node i holds the private key set.

Before deployment of the network, the initialization phase is performed. It assigns public-private keys as well as identifications (IDs) to each node. When a unique subset of private keys is assigned to each node, the node automatically obtains a unique ID. If two nodes know the ID of the other, a secure data can be exchanged. From the IDs, a node deduces the private keys of the other node and encrypts the data with the matching public keys. The AODV routing protocol has been used to return the ID of the destination in the route reply packet (Fig. 2).

Assume that a source node X desires to send a data to destination node Y. First, node X obtains an ID of the node Y to deduce a set of private keys hold by Y. Then, node X encrypts the data with the public key set owned by node Y that matches to the private keys hold by node Y. The data is encrypted by several public keys. A node only decrypts the data that has the corresponding private keys. To secure a very large network, the scheme requires a few key pairs. For example, only 18 key pairs are required for secure communication among up to 500 mobile nodes. The RSA algorithm has been used to perform the secure data transmission.

4 Experimental Results and Analysis

The cryptographic scheme has been executed in Java programming language and empirically valued its performance via simulation assessments. The simulation outcome has shown that method resists security attacks. It is also found that the proposed approach is memory efficient than traditional scheme in terms of number of keys and with the absence of certificate exchange. The scheme uses the protocol: AODV (Ad hoc On-demand Distance Vector) routing for data transmission.

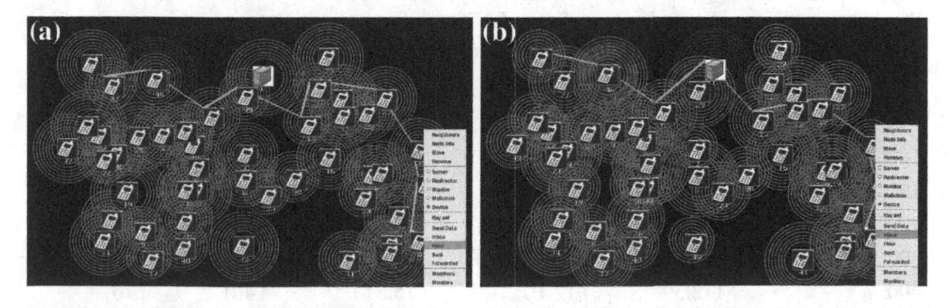

Fig. 2 **a** Plaintext data transmission **b** Secure data transmission

4.1 Performance Analysis

In this, the proposed scheme complexities, time and space, are inspected. The overheads, processing and storage, are estimated via the computer simulation. An evaluation of the time required for—deployment of nodes into the network, key pool generation and key pool distribution is proposed. An assessment also includes key storage space required in traditional schemes as well as proposed schemes.

Table 1 shows the network deployment time (NDT), Key Pool Generation Time (KPGT) and Key Pool Distribution Time (KPDT) for different network sizes. It also shows the number of keys required per node in traditional key management, denoted by T, as well as the scheme proposed by He et al. [11], denoted by W. The resultant table contains the data points wherein each is an average of five program runs. Each data point is extracted with an equal arrangement of different network sizes, but dissimilar arbitrarily generated mobility patterns. If the number of nodes available in the network is increased to 500, the total network key pool distribution time becomes 17.236 s. Similarly, when the network size is amplified to 500, the total number of keys per node required for communication in the proposed scheme is 18 against 501 keys in the traditional key management schemes. The data of the Table 1 are plotted and is shown in Fig. 3.

4.2 Security Analysis

In this, the impact of various security attacks like *black-hole* attack, *gray-hole* attack, *tamper* attack, etc., against the data transmission as well as routing overhead is studied. The data transmission times for plaintext, secure text without attacks and secure text with attacks have been analysed.

The *black-hole attack* is a type of denial of service. A node whose packets it wants to intercept pulls all the packets towards, called malicious node, by falsely advertising a shortest route to the target node. Then, the malicious node absorbs all

Table 1 Comparison of KPDT and number of keys per node over network size

Network size (Nodes)	NDT (Seconds)	KPGT (Seconds)	KPDT (Seconds)	Scheme T	Scheme W
10	06.64	00.37	00.16	11	08
50	21.69	02.37	00.75	51	12
75	28.16	02.77	01.49	76	13
100	38.22	02.84	03.61	101	13
150	66.15	04.69	03.89	151	14
200	79.32	06.50	06.37	201	15
300	135.02	11.81	12.00	301	16
400	172.39	12.10	15.31	401	16
500	198.24	19.78	17.24	501	18

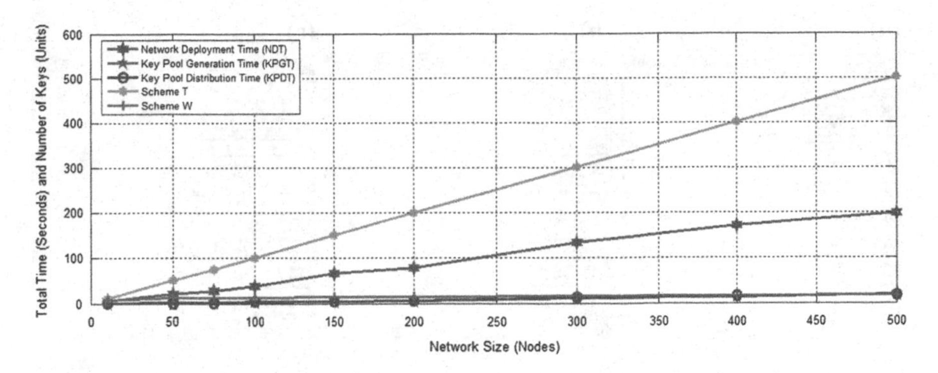

Fig. 3 Analysis of KPDT and a number of keys per node over the network size

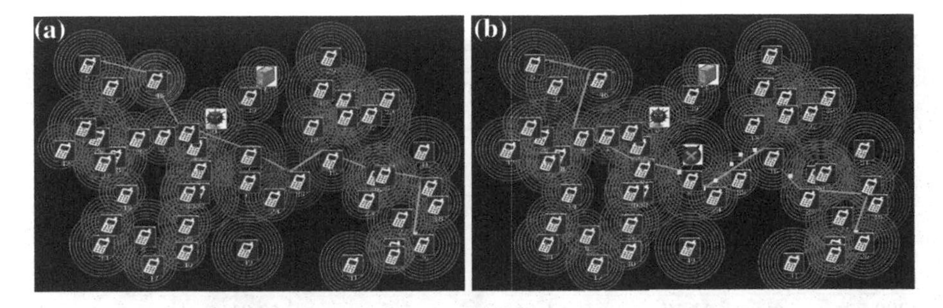

Fig. 4 A snapshot showing a **a** Black-hole attack **b** Gray-hole attack

the packets without forwarding to the target [12]. The *gray-hole attack* is a deviation of the black hole attack. Initially, the node is not selfish or malicious and act as normal node; it turns malicious a bit later to drop the packets selectively [13]. In a *device tampering attack*, the mobile nodes might be stolen easily and altered by an adversary. This is due to nodes compactness, flexible and hand-held in nature [2].

A snapshot of a black-hole and gray-hole attack is shown in Fig. 4. The event logs showing packet drop at black-hole node, gray-hole node and tampered node is also shown in Fig. 5.

For different network sizes, the comparison of data routing time, in seconds, with and without security attacks is shown in Table 2. Every data point is the average of five program runs. Each data point is extracted with an equal arrangement of different network sizes, but dissimilar arbitrarily generated mobility patterns.

The various routing times, in seconds, of the network with 500 nodes are: plaintext routing time (*PRT*): 113.57, secure text routing time (*SRT*) without attacks: 115.38, secure text routing time (*SRT*) with attacks: 115.809. The observation is that the time for plaintext transmission is less than that of secure text transmission. The secure text transmission without attacks is approximately less than that of its transmission with attacks. The data of the Table 2 are plotted and is shown in Fig. 6.

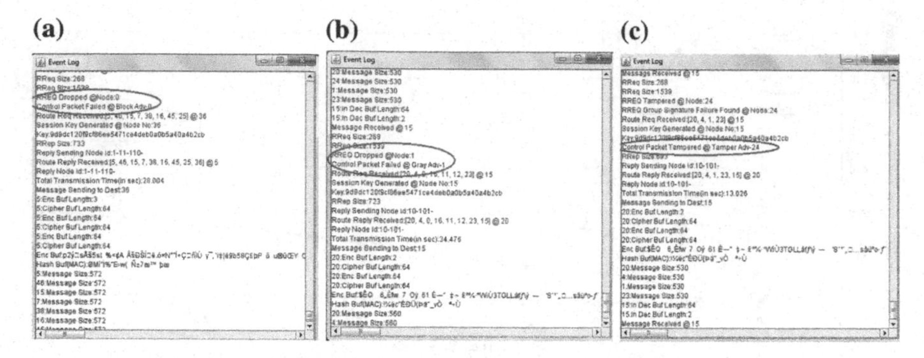

Fig. 5 Event logs showing packet drops at: **a** Black-hole node **b** Gray-hole node **c** Tamper node

Table 2 Comparison of routing overhead with and without security attacks

Network size (Nodes)	PRT (Seconds)	SRT—without attacks (Seconds)	SRT—with attacks (Seconds)
10	05.67	06.05	07.79
50	17.76	17.80	19.86
75	18.28	22.01	25.78
100	33.99	32.48	32.97
150	38.72	43.54	45.36
200	52.10	54.49	58.81
300	79.02	80.24	81.01
400	89.97	90.02	102.35
500	113.57	115.38	115.81

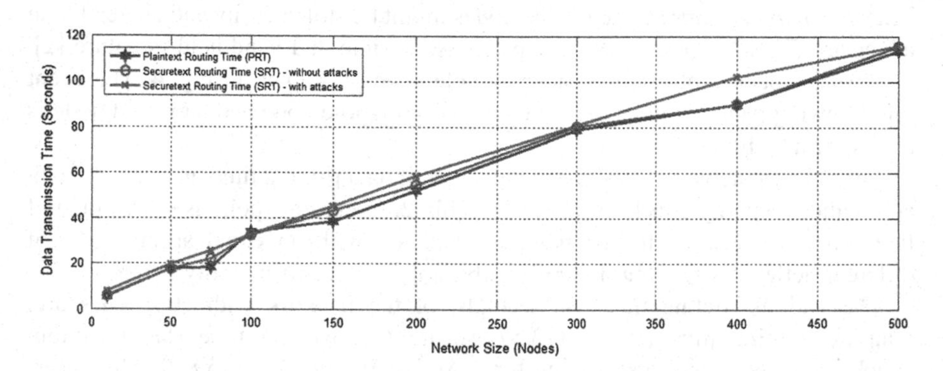

Fig. 6 Analysis of routing overhead with and without security attacks

5 Conclusion

The important contribution of this paper is an empirical analysis of a cryptographic key management that lead to performance enhancements for secure data transmission to hold a big number of mobile nodes. The scheme has been executed and empirically evaluated for its performance in Java programming language. The design of the cryptographic keys depends on combinatory to achieve lightweight key management, better scalability and less interruption than the traditional schemes. The scheme takes considerable key storage space than the conventional schemes and satisfies the secure communication requirements. The scheme also resists against different attacks in the network. There is no communication overhead for authentication in an ad hoc network of size *n* nodes. The reason is that when a unique subset of private keys is assigned to each node, the node automatically obtains a unique ID. Therefore, it is found that the proposed scheme in MANETs provides the security in an efficient manner than the others.

References

1. Cheng, X., Huang, X., Du, D.Z.: Ad Hoc Wireless Networking. Kluwer Academic Publishers (2006)
2. Murthy, C.S.R., Manoj, B.S.: Ad Hoc Wireless Networks: Architectures and Protocols. Pearson Education (2006)
3. Stallings, W.: Cryptography and Network Security: Principles and Practice. 4th edn. Prentice Hall (2005)
4. Menezes, A.J., Oorschot, P.C.V., Vanstone, S.A.: Handbook of Applied Cryptography. CRC Press (1996)
5. Gandhi, C., Dave, M.: A review of security in mobile ad hoc networks. IETE Tech. Rev. **23**(6), 335–344 (2006)
6. Choi, D., Younho, L.E.E., Yongsu, P.A.R.K., JIN, S.H., Yoon, H.: Efficient and secure self organized public key management for mobile ad hoc networks. IEICE Trans. Commun. **E91–B**(11), 3574–3583 (2008)
7. Dahshan, H., Irvine, J.: A Robust Self-Organized Public Key Management for Mobile Ad Hoc Networks. Security Communication Networks, Wiley InterScience, vol. 3, pp. 16–30, (2009)
8. Capkun, S., Buttyan, L., Hubaux, J.P.: Self-organized public-key management for mobile ad hoc networks. IEEE Trans. Mob. Comput. **2**(1), 52–64 (2003)
9. Capkun, S., Hubaux, J., Buttyan, L.: Mobility helps peer-to-peer security. IEEE Trans. Mob. Comput. **5**(1), 43–51 (2006)
10. Omar, M., Challal, Y., Bouabdallah, A.: Reliable and Fully Distributed Trust Model for Mobile Ad Hoc Networks. Elsevier's Computers and Security, pp. 199–214 (2009)
11. He, W., Huang, Y., Ravishankar, S., Nahrstedt, K., Lee, W.C.: SMOCK: a scalable method of cryptographic key management for mission-critical wireless ad hoc networks. IEEE Trans. Forensics Secur. **4**(1), 140–150 (2009)
12. Yi-Chun, Hu, Perrig, Adrian: A survey of secure wireless ad hoc routing. IEEE Secur. Priv. **2**(3), 28–39 (2004)
13. Sen, J., Chandra, M.G., Harihara, S.G., Reddy, H., Balamuralidhar, P.: A mechanism for detection of gray hole attack in mobile ad hoc networks. In: Proceedings of IEEE 6th International Conference on Information, Communications and Signal Processing, pp. 1–5 (2007)

5 Conclusion

The important contribution of this paper is an empirical analysis of a cryptographic key management that lead to perform the enhancements in secure data transmission to build a big number of mobile nodes. The scheme has been executed and empirically examined for its performance in Java programming language. The design of the cryptographic key depends on a dynamic system, the lightweight key management, better scalability and less mechanisms than the traditional techniques. The scheme takes considerably key storage space than the conventional techniques and saves the energy communication required. The scheme also resist against attacks to networks. There is no overhead to a centralized ... [illegible] ... of mobile nodes. The system ... when a ... of graphic keys ... [illegible] ... automatically ... Therefore, it is found that the proposed scheme MANETs has the security in term ... more than others.

References

1. ... X., ... [illegible] ... Kluwer Academic Publishers (200...)
2. ... R., ... [illegible] ... Data Communications ...
3. ... W.: Cryptography and Network Security ... [illegible] ... Prentice Hall (20...)
4. ... Vanstone, S.A.: Handbook of Applied Cryptography ... Press (1996)
5. ... M.: A ... of ... [illegible] ... networks. IEEE ... (...)
6. ... [illegible] ...
7. ... [illegible] ...
8. ... [illegible] ...
9. ... [illegible] ...
10. ... [illegible] ...
11. ... [illegible] ...

Evolving an Algorithm to Generate Sparse Inverted Index Using Hadoop and Pig

Sonam Sharma and Shailendra Singh

Abstract Now a day's users mostly prefer the keyword search method to access the data for the explosion of information. Inverted indexing efficiently plays a very important role for search operation over a large set of data. There are two problems exist in current keyword based searching technique. First, the large set of data is mostly unstructured and does not suite in the existing database systems. Second, the storage in inverted indexing is usually very large and compression techniques used so far is also not so efficient because they increase the processing time. To overcome these problems, Hadoop, which is a distributed framework for large dataset is needed where the required resources could be shared and accessed very easily. In our proposed work, we will join the list of consecutive document id in the inverted index into the intervals to save memory space. For this, we have developed the UDF (User Defined Function) for stemming and stop words for the sparse inverted index in pig latin. It can be observed in the results that our proposed method is efficient than existing techniques.

Keywords UDF · Pig · Hadoop · Document id · Stop word · Stemming · Index · Sparse index · Mapreduce · Inverted index

1 Introduction

The web is known as the largest source of information which are the collection datasets including XML document, relational dataset. The main challenge for the user is the information retrieval. For this user can simply use the queries for the

S. Sharma (✉) · S. Singh
Department of Computer Engineering and Application,
National Institute of Technical Teacher's Training and Research,
Bhopal, India
e-mail: Sonam.sharma18apr@gmail.com

S. Singh
e-mail: ssingh@nitttrbpl.ac.in

S.C. Satapathy and S. Das (eds.), *Proceedings of First International Conference on Information and Communication Technology for Intelligent Systems: Volume 2*,
Smart Innovation, Systems and Technologies 51, DOI 10.1007/978-3-319-30927-9_49

keyword search in order to retrieve the information. Now a day's, this keyword searching uses inverted indexing data structure in which each word is connected to a document id which seems as efficient to get the relevant information.

The distributed framework i.e. Hadoop which process big data is an implementation of mapreduce. As the process is carried out for the huge set of text document so it is more suitable for this kind of operation. Mostly in search engine and in text retrieval inverted indexing is used. When the user executes the query by using inverted indexing, these search engines provide the required information matching the query of the user by providing document ids of that document.

In order to process the large data set, pig provides a level of abstraction. Mapreduce allow the programmer to define a map function which is followed by relevant reduce function.

Pig basically consist of 2 part

- The language Pig Latin, which is used to express the dataflow
- The execution environment, which is used to run the Pig Latin programs.

Basically, there are two types of environments:

1. Local execution in JVM.
2. Distribution execution on a Hadoop cluster.

A program of pig Latin consists of the sequence of pig operation, which are cross, join, sort, etc., or the transformation, which are applied to the input data to get the required output. Pig allow us to concentrate on data rather than execution, it abstracts the execution detail and perform a series of optimization. Pig becomes more popular than mapreduce for large data processing.

In this paper, we have proposed a method in which the document are inverted indexed and on the basis of the occurrence of that word which is asked for in that given input documents over a pig which is a distributed frame, in which the mapreduce is implemented as its physical dataflow engine.

The rest of the paper is organized as follows: Sect. 2 discusses the related work; Sect. 3 describes the proposed work; and finally last Sect. 4 experiment and results; Sect. 5 contains the conclusion.

2 Related Work

There are many techniques for phrase-based technique. The graph which is made by using document and the ordering of words in the document is known as document index graph which is a directed graph. Where each word is a vertex and the edges between them define the sequence of two words found in the text. Through the use of graph phrase query is made easy and redundancy is decreased. The given method is provided for the correlation between words which are efficient and a method for coherent clustering [1].

Lim et al. [2] proposed a method in which the inverted index is prepared of already indexed document which are changed by their content. In their approach, they have used a landmark-diff method in which the number posting are reduced in the inverted indexing that are necessary to be updated.

Arsekar et al. [3] has given comparative study of mapreduce, hadoop and pig. Hadoop uses the mapreduce for processing, which consist of mapper and reducer function. Mapper takes the input in the form of key, value pair where value is task related to data and the key is group number of value. Reducer phrase in which reduce will handle the key, value pair. Reduce will merge these value to receive the small set of value [4]. They have given a brief of pig, it is open source software given by apache software foundation. Pig acts as a connecter between the SQL and the mapreducer. Pig is developed as a dataflow engine to process with large set of data. It uses the pig latin which handles no. of dataflow jobs. The program based on pig transformation we do not need to have the programming skill [5]. As compared to mapreduer program line of code is less the ratio between pig and mapreduce is 20:1. In the internal mechanism of pig the program gets converted into mapreduce program and executed. Pig is simple to write and execute pig scripts.

Delbru et al. [6] introduced a processor for indexing semi structure data with this they have also given an entity retrieval model. They have shown that their query gets evaluated in sub second time and their model is scalable also.

Velusamy et al. [7] proposed that indices for millions of document to be done, a compressed data structure of inverted index is proposed in searching a word in dictionary order. The basics factors for creating indices such as merge, look up, compression technique etc. will also be given importance [8]. To create a local index on that particular system will also be proven as a get solution to solve the problem arises in parallel processing such as the index is added, updating an index all these to be done before the searching operation is done or whether only a portion of index is need to be add all these thing made simple.

Wu et al. [9] have given a generalized inverted index for keyword search. GINIX is compatible with d- gap based list of compression technique and improved its performance.

Olston [10] said that in the implementation of pig two simple but critical things are inbuilt i.e. (1) pig automatically efficiently forms the pipelines of the per records sequence of the processing steps. (2) certain aggregation function of algebraic function such as COUNT, SUM, AVERAGE and some user define function are also there, partial aggregation performed automatically earlier by combining the mapreduce frame work in order to reduce the size which was required in the portioning function.

Gugnani et al. [11] proposed an efficient technique for fast phrase query evaluation in which they have index the indexed web document with reduction in the time taken to process the query evaluation and merging the posting list and checking the ordering of word [12]. In addition, the extended vector space model is used for ranking the document.

Omanakuttan [13] provided a detailed study of document retrieval and keyword search inverted index scheme. Inverted indexing major application is in the search engine. In his paper, he has given working of the inverted index with a search engine. The memory required could be less by replacing the list of ids with interval list. Through the comparative studies, it is given that through the concept of the interval the traditional methods get improved.

3 Proposed Work

In order to manage the web content, the search engine has to face many herculean tasks. A Pig cluster is designed to pass these input file and the inverted index is made through pig in which mapreduce is implemented as a dataflow engine. For this herculean task, it has to introduce a fast and memory efficient indexing mechanism which can be come up through this framework.

As the given documents by the user as inputs are inverted indexed as per their document ids then following, compression techniques are used to make the keyword searching efficient.

1) Stop words—It is a collection of words like "an", "the", "on", "for" these are so common words that almost every document contain them. A stop words own the list of those words which could be ignored while indexing the collection input documents. Through the stop word some amount of memory we can save which will not affect the result for normal queries.
2) Stemming—Porter's is the most common stemming algorithm used for the English language. Instead of indexing the documents as it is visible, convert them into their morphological stem and indexing. Like the word "compute", "computer", "computation", "computers", computing all get indexed as "compute". It can be emulated easily by query expansion or wildcard queries.
3) Using Interval Limits—For large no. of dataset inverted index list is also very large; because of this the storage space will be wasted. In order to compress the standard list of the inverted index, they are converted into interval list to save the storage space, which is known as the sparse inverted index.

In Fig. 1 the no. of consecutive text file are taken as an input, then that input is processed further i.e. null value are removed from the input file, whole file is converted into lower case and all the special character are also removed to create a inverted index. This is done through these following steps, we have designed a algorithm for preprocessing step which are used through the pig commands.

The proposed approach is divided into three steps

3.1 Preprocessing step
3.2 Inverted index design step
3.3 Compression step.

Fig. 1 Proposed framework

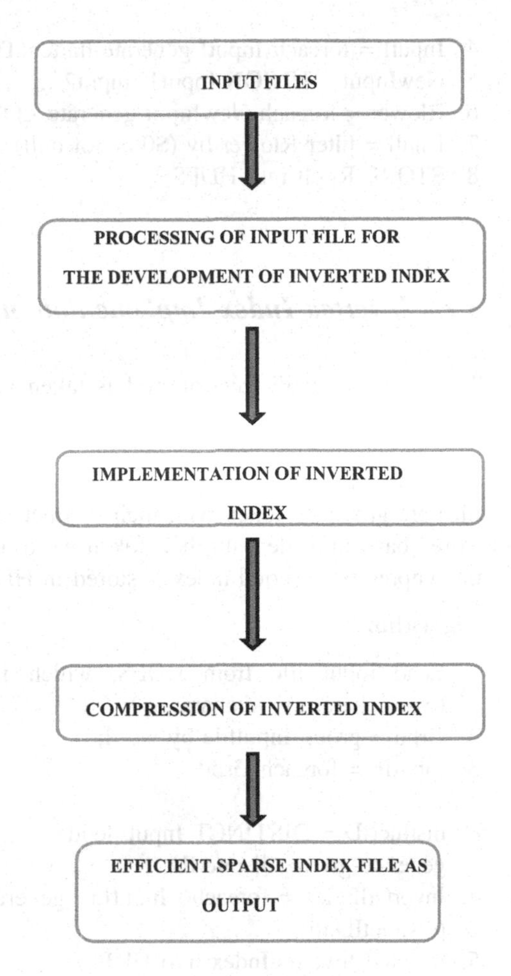

3.1 Preprocessing Step

Algorithm 1 is an algorithm which is written in Pig Latin, in this algorithm the procedure is defined before creating an inverted index. The input files are loaded into HDFS through pig command, then that file is made free from null value, special character, and numbers. After that, the document id is provided to each and every document file. The following steps are given below.

Algorithm 1:

1. Load all input files from HDFS to PIG flow;
2. For each input file I do steps 3 and 4.
3. InputI = foreach InputI generate REPLACE(words, '([^a-zA-Z\\S]+)','') AS word;

4. InputI = foreach InputI generate flatten(TOKENIZE(words)) as word, I as id;
5. NewInput = UNION Input1, Input2,............................InputN;
6. Rlower = foreach NewInput generate LOWER(word) AS word, id;
7. Rnull = filter Rlower by ($0 is not null);
8. STORE Rnull into HDFS;

3.2 Inverted Index Implementation Step

The output from the algorithm 1 is taken is an input for Algorithm 2, which is written in Pig Latin. The following steps are given below in Algorithm 2 for creating an inverted index using Pig command.

The output from the Algorithm 1 is taken as a input in Algorithm 2 in which the files are group by token with their distinct ids as another field, and then they are stored back in a file with their token word and their consecutive ids. The created uncompressed inverted index is stored in HDFS.

Algorithm 2:

1. Load input file from HDFS, which is stored by preprocessing step as (word:chararray, id:chararray).
2. Grpd = group Inputfile by word;
3. InputIn = foreach Grpd
   {
   distinctID = DISTINCT Inputfile.id;
   generate group, distinctID; };
4. InvertedIndex = foreach InputIn generate group as word, BagToTuple (distinctID.id);
5. STORE InvertedIndex into HDFS;

3.3 Compression Step

As the inverted index is created through Algorithm 2 but it is uncompressed, due to which the more memory and processing time will be needed. Algorithm 3, which is written in pig latin need to follow to make inverted index efficient i.e. sparse inverted index.

Algorithm 3:

1. LOAD the inverted index file from hdfs as (word:chararray, id:chararray);
2. Register jar file of UDF(User Defined Function);
3. Filterd = filter invertedindexfile by invertedindex.stopword(word);
4. Stem = foreach Filterd generate invertedindex.stemmer(word) as word, id;

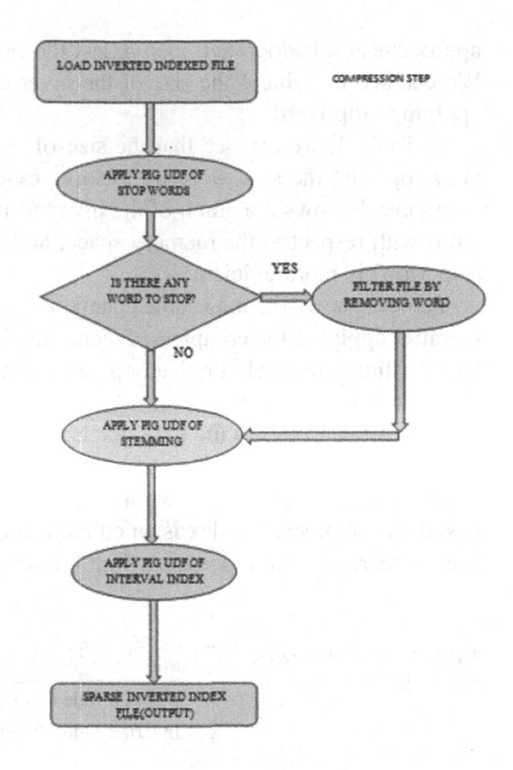

Fig. 2 The flow chart of compression step

5. SparseIndex = foreach Stem generate word, invertedindex.Intervalindex(id) as id;
6. STORE SparseIndex into HDFS;

In Algorithm 3 the stored file of inverted index is loaded in pig. As the file is loaded, we will register the jar file of our UDF, and then filter the inverted index by the UDF of Stop word. The filtered file is again filter by the UDF of stemming. At last on that stem stored file Udf of sparse inverted index is applied, which will convert the consecutive list of ids into intervals.

When the inverted index is designed, then as explained through the Fig. 2 the inverted index is compressed. As we can see in Fig. 2 in which the inverted index which is created is loaded through pig command.

Figure 2 gives the whole description of the compression process to get the sparse index which is efficient with respect to memory space.

4 Experiment and Results

To visualize the effectiveness of this sparse index following experiments are performed. We have taken some relevant text files as an input, and then on that input file we have applied the algorithm for stop word, stemming which are the existing approaches on hadoop and then at last the proposed algorithm on these input files. We can see in Table 1 the size of the inverted index before compression and after applying stop word.

In Table 1, we can see that the size of the inverted index is the file is reduced after applying the stop word up to some extent.

Figure 3 shows the graph of the inverted index with stop word and without stop word with respect to the memory space, and it is observed that inverted index with stop word is more efficient.

Table 2 shows the noticeable change in the size of file from inverted index to the file after applying the compressed technique i.e. stop word and stemming which are the existing approach on hadoop and finally further more after the proposed inverted index scheme.

At last Fig. 4 shows the final graph which is plotted between the inverted index file, inverted index file after the existing approach on hadoop and after proposed inverted index which is known as "sparse inverted index". We find that the proposed sparse inverted index is an efficient index with respect to the memory space. It also decreases the processing time for searching any particular keyword.

Table 1 Size of inverted index files

File	Size (bytes)
Inverted index file (without compression)	4352
Inverted index file (after stop word)	3123

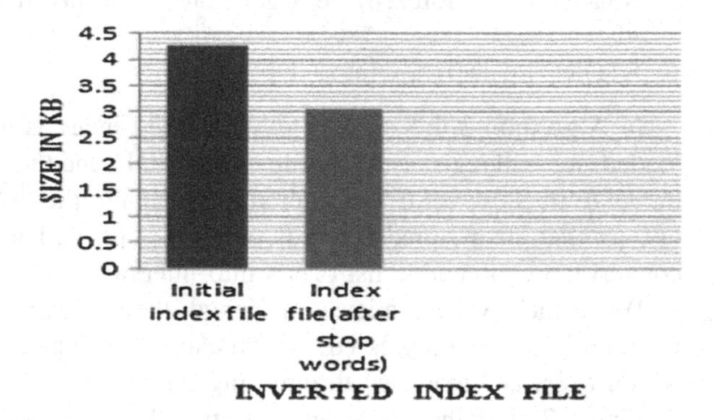

Fig. 3 Space comparison between inverted index and inverted index after applying stopword

Table 2 Memory space comparison of inverted index files

File	Size (bytes)
Initial inverted index file	4352
Inverted index file (after stop words)	3123
Inverted index file (existing approach on hadoop)	2816
Proposed inverted index file	2467

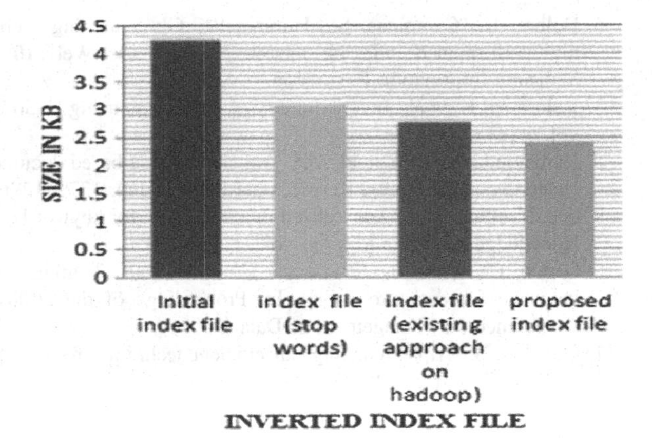

Fig. 4 Space comparison between inverted index, inverted index (Existing approach on hadoop) and sparse inverted index

5 Conclusion

Mapreduce is capable of working with large dataset as a dataflow in pig latin while existing techniques used non-cloud platforms were failed to do so. We have used Pig platform to perform our proposed approach for indexing operation based on sparse inverted index technique on our large size text files. We have developed the UDF for the compression technique using pig latin, and also described the sparse index for the keyword search in text database on pig platform which is effective and efficient index structure. The experiment shows the noticeable improvement in the sparse inverted index technique than the existing method of indexing like stop words and stemming. Since proposed sparse inverted index requires the less memory space.

References

1. Patil, M., Thankachan, S.V., Shah, R., Hon, W., Vitter, J.S., Chandrasekaran, S.: Inverted indexes for phrases and string. In: Proceedings of the 34th International ACM SIGIR Conference on Research and Development in Information Retrieval, pp. 555–564. ACM (2011)
2. Lim, L., Wang, M., Padmanabhan, S., Scot Vitter, J., Agarwal, R.: Efficient update of indexes for dynamically changing web documents. World Wide Web **10**(1), 37–69 (2007)
3. Arsekar, R.A. et al.: Comparative study of mapreduce and pig in big data. International Journal of Current Engineering and Technology, vol. 5(2) (2015)
4. Hsu, W.-C., Liao, I-E.: CIS-S: a compacted indexing scheme for efficient query evaluation of XML documents. Inf. Sci. **241**(0), 195–211 (2013)
5. Xu, G., Xu, F., Ma, H. Deploying and researching Hadoop in virtual machines. Published in: IEEE International Conference on Automation and Logistics (ICAL), Zhengzhou. pp. 395–399. ISSN: 2161-8151, E-ISBN: 978-1-4673-0363-7 (2012)

6. Delbru, R., Campinas, S., Tummarello, G.: Searching web data: an entity retrieval and high-performance indexing model. World Wide Web **10**(0), 33–58 (2012). Web-Scale Semantic Information Processing
7. Velusamy, K. et al.: Inverted indexing in big data using Hadoop multi nide cluster. In: IJCSA, vol. 4(11) (2013)
8. Hammouda, K.M., Kamel, M.S.: Efficient phrase based document indexing for web document clustering. IEEE Trans. Knowl. Data Eng. *16*(10), 1279–1296 (2004)
9. Wu, H. et al.: Ginix: generalized inverted index for keyword search. In: IEEE Transactions on Knowledge and Data Mining, vol. 8(1) (2013)
10. Olston, C., Reed, B., Srivastava, U., Kumar, R., Tomkins, A.: Pig Latin: A not a foreign language for data processing. In: Proceedings of the 2008 ACM SIGMOD International Conference on Management of Data (2008)
11. Gugnani, S: Triple indexing: an efficient technique for fast phrase query evaluation. IJCA (0975 887). **87**(13) (2014)
12. Yang, R., Zhu, Q., Xia, Y.: A novel weighted phrase-based similarity for web documents clustering. J. Software **6**(8), 1521–1528 (2011)
13. Omanakuttan, S.: Inverted index schemes for keyword search: a survey of current best. In: International Journal of Advance Research in Computer Science and Management Studies, vol 3. ISSN 2321-7782 (2015)

Leakage Power Reduction Technique by Using FinFET Technology in ULSI Circuit Design

Ajay Kumar Dadoria, Kavita Khare, T.K. Gupta and R.P. Singh

Abstract Deep Sub Micron (DSM) technology demands for lower supply voltage, reduced threshold voltage and high transistor density which leads to exponentially increase in leakage power when circuit is in standby mode. Here review of FinFET transistor along with existing low power techniques in DSM circuits like sleep, LECTOR etc. are done. Then Lector with FinFET technology circuit is proposed. This work evaluates the impact of FinFET technology, which has huge potential to replace bulk CMOS in DSM range. Performance of proposed technique is investigated in terms of dynamic power, delay, Power Delay Product (PDP) and leakage power dissipation. The proposed techniques has leakage controlling sleep transistor inserted over pull up and pull down network which significantly reducing the leakage power by using HSPICE simulator in 32 nm FinFET technology at 25 and 110 °C with C_L = 1 pF at 100 MHz frequency.

Keywords FinFET · SG and LP mode · Low power

1 Introduction

Scaling is prime thrust for development of VLSI design, scaling increases the density, with packaging and cooling cost of the chip, as transistor count increases power consumption, there are many approaches to reduce the power consumption of the circuit at various design stages and different design techniques have been

A.K. Dadoria (✉) · K. Khare · T.K. Gupta · R.P. Singh
Department Electronics and Communication, MANIT, Bhopal 462051, India
e-mail: ajaymanit0@gmail.com

K. Khare
e-mail: kavita_khare1@yahoo.co.in

T.K. Gupta
e-mail: taruniet@rediffmail.com

R.P. Singh
e-mail: prof.rpsingh@gmail.com

S.C. Satapathy and S. Das (eds.), *Proceedings of First International Conference on Information and Communication Technology for Intelligent Systems: Volume 2*, Smart Innovation, Systems and Technologies 51, DOI 10.1007/978-3-319-30927-9_50

proposed here. Leakage currents with sub-threshold source-to-drain leakage, reverse bias junction band-to band tunneling leakage, gate oxide tunneling, and other currents drawn continuously from the power supply cause static power dissipation [1]. Reduction in supply voltage of circuit reduces dynamic power dissipation very effectively but after certain limit it affects the performance of the circuit [2], to maintain circuit performance of the circuit it is necessary to decrease the threshold voltage as well, but it leads to leakage power dissipation. Leakage power can be reduced by increasing the threshold voltage [3]. Scaling improves device density, performance and power consumption of CMOS technology but further scaling in nanometer regime results in shorter channel effect (SCE) and lower the Drain Induced Barrier Lowering (DIBL). To overcome from these problem an efficient technique known as double gate FinFET come into existence which is having excellent control over thin silicon body which suppress the shorted channel effect in sub 22 nm and beyond, double gate FinFET reduces the sub threshold and gate oxide leakage current [2].

The organization of the paper is as follows: Sect. 2, describes FinFET technology for DSM technology. Section 3, presents the literature review of the leakage power reduction techniques. Section 4 presents proposed circuit for low leakage power. In Sect. 5 simulation results and discussion by using HSPICE EDA tool. Finally the conclusion is presented in Sect. 6.

2 FinFET

Fabrication has been proposed to overcome from SCEs of double gate device known as FinFET. Using a single lithography and etch step, gate is easily wrapped over the silicon fin, Front gate and back gate can have different doping profile, so both the gate operate independently according to the requirement, and it is easily computable with CMOS technology [4, 5]. Depending on the arrangement of front gate and back gate tied up or not. FinFET can be characterized into three different modes [6] as shown in Fig. 1, namely Short Gate (SG), here front and back gate tied together with common supply voltage. Short gate is faster and higher ION currentLow Power (LP) mode here front and back gate bias independently, back gate is reverse bias for reduction of leakage current and Independent Gate (IG) mode here both the gate is connected to different inputs; it is like a two parallel transistor which reduces the area of the circuit. Figure 1b and c shows the three-dimensional view of single fin FinFET and cross sectional top view of FinFET.

Simulation are performed over Predictive Technology Model (PTM) at 32 nm FinFET. Table 1 shows the primary parameters of FinFET device [6]. Due to vertical gate structure FinFET width is quantized, fin height is determined by

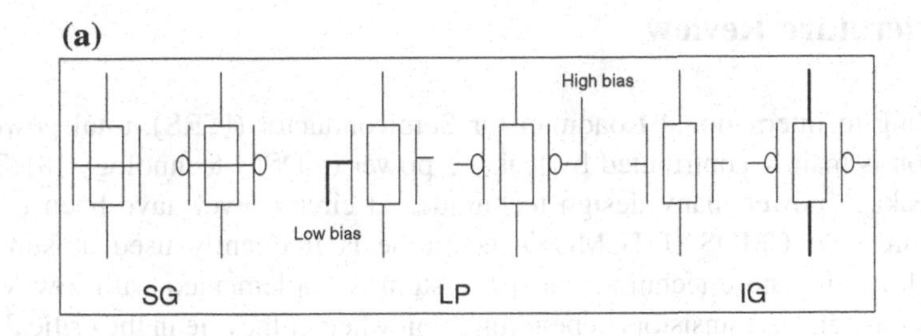

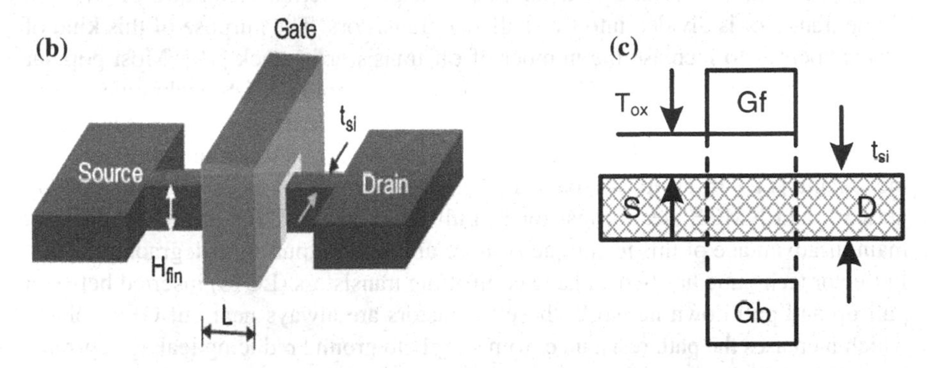

Fig. 1 **a** Modes of operation of FinFET. **b** Three-dimensional view of single fin FinFET. **c** Cross section top view of FinFET

Table 1 Device technology parameter

Parameter	32 nm N-FinFET	32 nm P-FinFET
Length of channel (L)	32 nm	32 nm
Fin thickness (t_{si})	8.6 nm	8.6 nm
Fin height (H_{fin})	40 nm	50 nm
Oxide thickness (T_{ox})	1.4 nm	1.4 nm
Source/drain doping (N-type and P-type FinFETs)	2×10^{-20} cm^{-3}	2×10^{-20} cm^{-3}
Power supply (V_{dd})	0.8 V	0.8 V

minimum transistor width (W_{min}). When the two gates of a single-FinFET tied together, W_{min} is Effective channel width

$$W_{min} = 2H_{fin} + t_{si}$$

$$\text{Effective channel length } L_{eff} = L_{gate} + 2 \times L_{ext}$$

where H_{fin} is the height of the silicon fin and t_{si} is the thickness of the silicon body L_{ext} is the extension of fin from gate to source or drain terminal [4]. In order to suppress shorter channel effect and enhance the area efficiency in FinFET, fin thickness is much smaller than fin height and thickness of silicon dioxide layer (Tox) is kept constant to 1.4 nm during simulation [7].

3 Literature Review

According to International Roadmap for Semiconductor (ITRS), total power consumption is mainly contributed by leakage power in DSM technology [8]. To mitigate leakage power many design techniques at circuit level have been explored. Dual-Threshold CMOS (DTCMOS) technique is frequently used at sub-system design level. In These techniques a sub-system is implemented with low V_{th} transistors or a high V_{th} transistors depending upon whether they lie in the critical path or not [9]. Transistor stacking is a runtime leakage reduction technique in which a single transistor is divided into two half size transistors. The purpose of this kind of arrangement is to increase the number of off transistor in stack [10]. Most popular power gating technique is MTCMOS (Multiple Threshold CMOS) technique is used for leakage reduction, due to high V_{th} sleep transistor used in this technique increases the resistance of the transistor which reduces the performance in terms of speed of the circuit [11]. In sleep approach PMOS sleep transistor is inserted in pull up network and NMOS sleep transistor in pull down network for leakage reduction, main disadvantage of this technique is more area and output logic degradation [12]. In Lector technique has two leakage controlling transistors (LCTs) inserted between pull up and pull down network, these transistors are always near cut OFF voltage which increases the path resistance from supply to ground reducing leakage currents [13]. Figure 2a is simple Nand gate made with FinFET technology, In Fig. 2b–d PMOS sleep transistor is inserted above pull up, NMOS sleep transistor below pull down network and PMOS, NMOS sleep transistor over pull up and pull down network of two input Nand gate which rail from power supply (V_{dd}) during standby mode for reduction of leakage power, after sleep mode output of the circuit is floating, which results in the distraction of floating output voltage [14]. Figure 2e Lector approach, two leakage controlling transistor (PMOS and NMOS) in between pull up and pull down network is introduce. Gate terminal of each leakage control transistor (LCT) is controlled by the source of the other. These transistors are connected as such that one of the transistors is always near the cut-off voltage for any input combination. The basic idea behind LECTOR approach is that "increases the path resistance from V_{dd} to GND, leading to significant reduction of leakage currents".

4 Proposed Work

In this section three new designees are introduced namely LECTOR Header Sleep (LHS), LECTOR Footer Sleep (LFS) and LECTOR Header Footer Sleep (LHFS), these are the combinationes of self controlling and external leakage controlling technique. In self controlling technique no external signals are applied while in external leakage controlling technique external sleep signal are applied which switches OFF the sleep transistor to reduces the leakage power. The basic idea

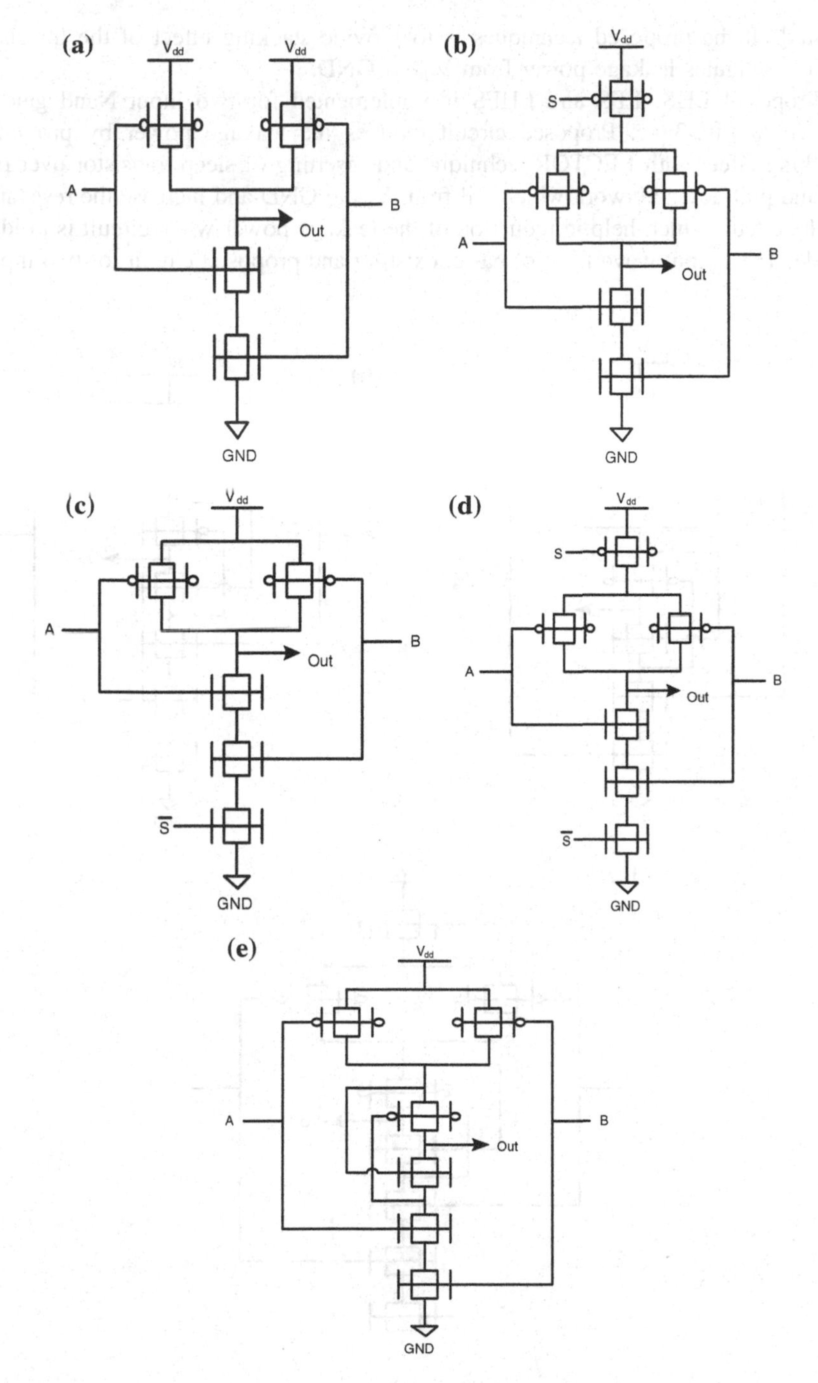

Fig. 2 Two input **a** Nand Gate (NG). **b** Nand Gate Header Sleep (NGHS). **c** Nand Gate Footer Sleep (NGFS), **d** Nand Gate Header Footer Sleep (NGHFS). **d** Nand Gate LECTOR

behind all the proposed techniques is to provide stacking effect of the transistor which mitigates leakage power from V_{dd} to GND.

Proposed LHS, LFS and LHFS is implemented for two input Nand gate as shown in Fig. 3a–c. Proposed circuit reduces the leakage power by providing stacking effect with LECTOR technique and inserting of sleep transistor over pull up and pull down network which rail from V_{dd} to GND and increase the resistance of the circuit which help in reduction of the leakage power when circuit is in ideal mode. The output wave form of basic, existing and proposed circuit for two inputs

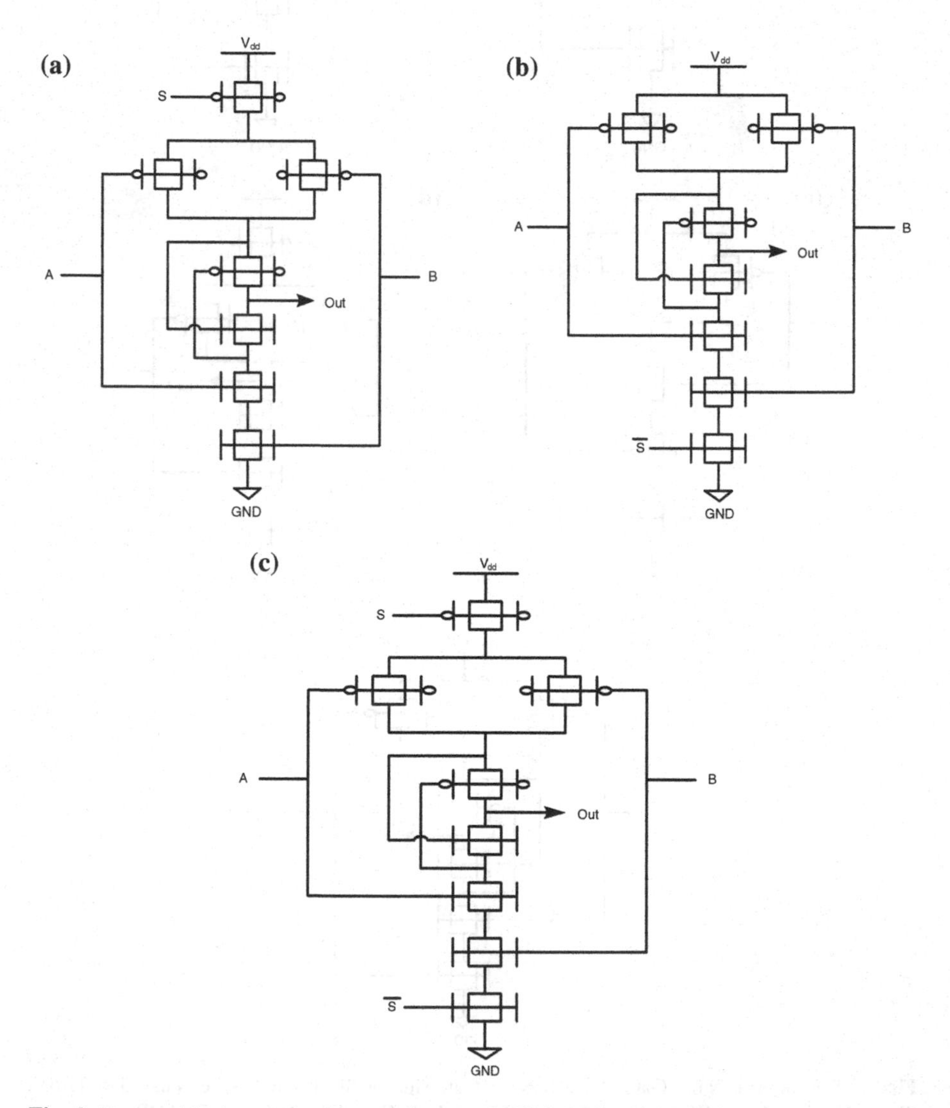

Fig. 3 Proposed leakage reduction techniques. **a** LHS. **b** LFS and **c** LHFS

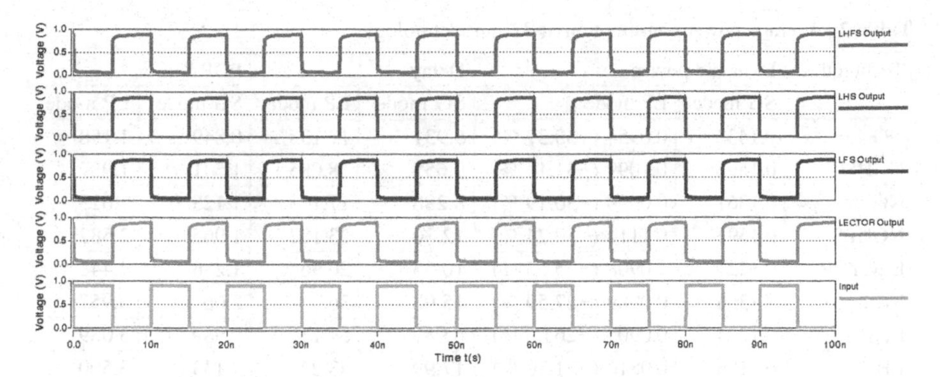

Fig. 4 Transient characteristics waveform of 2-input proposed NAND gate using HSPICE in FinFET technology

NAND gate is shown in Fig. 4, here it can be seen that the LECTOR and proposed circuits does not achieve proper logic level because location of leakage controlled transistors is such a way that there is a small degradation of output voltage.

5 Results and Discussion

In this section the power dissipation of existing and proposed leakage reduction techniques is calculated. All the existing and proposed technique are simulated in HSPICE at 32 nm FinFET technology with supply voltage of 0.9 V, output capacitance C_L = 1 pF, Leakage power is investigate at different temperature at 25 and 110 °C, The size (W/L) of all existing and proposed circuit made from P-FinFET and N-FinFET is same for fair comparison of results. In Table 2, Dynamic power dissipation, delay and PDP are calculated in SG and LP mode of all existing and proposed circuit, from table results shows that mitigation of leakage power take place in proposed circuits in SG and LP mode, but there is penalty of delay take place due to increase of resistance of the circuit. Tables 3 and 4 shows leakage power dissipation for all input vectors at different temperatures, from Table 3 observe that mitigation of leakage power take place in all proposed circuit but there is increase in leakage power in NGFS and proposed LFS because NMOS sleep turns OFF therefore GND, is disconnected hence there is no path to discharge the supply voltage towards GND. Leakage power is saved in SG mode but increase in delay due to reverse bias of back gate of P-FinFET and N-FinFET transistor. LP mode save more leakage power than SG mode due to back gate bias of transistor. Figure 5 shows the comparison of dynamic power dissipation in SG and LP mode for all existing and proposed circuits.

Table 2 Average power consumption at 32 nm technology

Techniques	Dynamic power		Delay		PDP	
	SG mode	LP mode	SG mode	LP mode	SG mode	LP mode
NG	0.1432	0.0954 (−33.37 %)	5.931	12.138	0.849	1.158
NGHS	0.1596	0.1099 (−31.14 %)	9.655	18.085	1.541	1.987
NGFS	0.1361	0.0950 (−30.19 %)	8.288	17.105	1.128	1.625
NGHFS	0.1597	0.1114 (−30.24 %)	12.30	23.182	1.965	2.582
LECTOR	0.1221	0.0908 (−25.63 %)	10.53	26.96	1.286	2.448
LHS	0.1366	0.0989 (−27.59 %)	15.08	36.37	2.061	3.957
LFS	0.1224	0.0905 (−26.06 %)	8.087	34.13	0.989	3.089
LHFS	0.1186	0.0810 (−31.70 %)	17.99	43.22	2.133	3.500

Table 3 Leakage power at 32 nm 25 °C

Techniques	Leakage power (pW)							
	SG Mode				LP Mode			
	00	01	10	11	00	01	10	11
NG	458.3	6025	5924	1576	110.9	1258	110.9	1258
NGHS	9.403	9.738	9.738	9.728	1.733	1.729	1.729	0.149
NGFS	194.5	458.3	458.3	536.9	48.89	110.92	110.7	50.80
NGHFS	8.697	9.396	9.396	0.972	1.668	1.7075	1.707	0.149
LECTOR	451.3	5274	5267	1378	105.4	962.5	960.0	74.26
LHS	9.268	9.588	9.588	0.972	1.651	1.668	1.668	0.148
LFS	193.4	451.4	451.2	500.0	47.78	105.4	105.4	43.76
LHFS	8.605	9.262	9.262	0.972	1.593	1.639	1.639	0.149

Table 4 Leakage power at 32 nm 110 °C

Techniques	Leakage power (nW)							
	SG Mode				LP Mode			
	00	01	10	11	00	01	10	11
NG	8.450	73.26	73.65	62.61	2.485	19.26	2.485	19.26
NGHS	0.7621	1.032	1.032	1.719	0.199	0.254	0.258	0.036
NGFS	3.909	8.443	8.443	14.72	1.191	2.484	2.484	2.409
NGHFS	0.5411	0.7617	0.7619	0.1716	0.1495	0.1994	0.1994	0.036
LECTOR	8.320	64.81	65.94	53.26	2.411	15.97	15.99	5.364
LHS	0.749	0.999	0.999	0.1716	0.1933	0.2420	0.2420	0.035
LFS	3.885	8.313	8.315	13.68	1.175	2.410	2.408	2.157
LHFS	0.5369	0.7496	0.7496	0.1713	0.1469	0.1930	0.1930	0.036

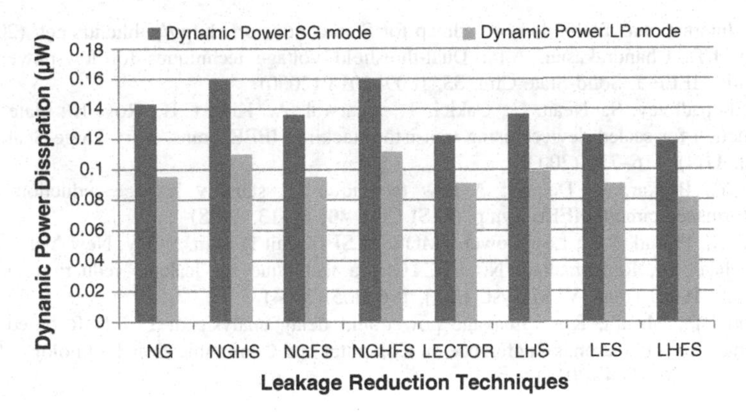

Fig. 5 Dynamic power dissipation in SG and LP mode of FinFET technology

6 Conclusion

A novel LHS, LFS and LHFS leakage reduction technique is proposed for low power application in digital circuits. Results show FinFET technology is used for ultra low power applications, reduction to design a device which is having lowest possible leakage. The experiment results shows that saving of dynamic power in proposed LHFS circuit is much lower than other circuits, the reduction is about 17.17 % in NG, 25.68 % in NGHS, 12.85 % in NGFS, 25.75 % in NGHFS and 2.86 % in LECTOR. The proposed technique can be applied on high performance, low power applications, where leakage is major concern such as microprocessor, memory units and other portable devices.

References

1. Sharifi, S., Jaffari, J., Hussein, M., Kusha, A.A., Navabi, Z.: Simultaneous reduction of dynamic and static power in scan structures. In: Proceedings of the Design, Automation and Test, vol. 2, pp. 846–851 (2005)
2. Prakash, O.: Design and analysis of low power energy efficient, domino logic circuit for high speed application. Int. J. Sci. Res. Eng. Technol. **1**(12), 1–4 (2013)
3. Karimi, G., Alimoradi, A.: Multi-purpose technique to decrease leakage power in VLSI circuits. Can. J. Electr. Electron. Eng. **2**(3), 71–74 (2011)
4. Tawfika, S.A., Kursun, V.: FinFET domino logic with independent gate keepers. Micro Electron. J. **40**, 1531–1540 (2009)
5. Liao, N., Cui, X.X., Liao K., Ma, K.S., Wu, D., Wei, W., Li, R., Yu, D.S.: Low power adiabatic logic based on FinFETs. Sci. China **57** (2014) ISSN 022402:1-022402:13
6. Mishra, P., Muttreja, A., Jha, N.K.: FinFET Circuit Design, Springer, Nanoelectronic Circuit Design, pp. 23–53 doi:10.1007/978-1-4419-7609-3_2. (2011)
7. Nan, L., XiaoXin, C., Kai, L., Kai Sheng, M.A., Wu, D., Wei, W., Rui, L., DunShan, Y.: Ultra-low power dissipation of improved complementary pass-transistor adiabatic logic circuits based on FinFETs. Sci. China, **57**, ISSN 042408:1-042408:13 (2014)

8. The International Technology Roadmap for Semiconductors. http://public.itrs.net/ (2003)
9. Kao, J.T., Chandrakasan, A.P.: Dual-threshold voltage techniques for low-power digital circuits. IEEE J. Solid-State Circ. **35**, 1009–1018 (2000)
10. Mukhopadhyay, S., Neau, C., Cakici, T., Agarwal, A., Kim, C.H., Roy, K.: Gate leakage reduction for scaled devices using transistor stacking. IEEE Trans. Very Large-Scale Integr. Syst. **11**(4), 716–730 (2003)
11. Ye, Y., Borkar, S., De, V.: A new technique for standby leakage reduction in high performance circuits. IEEE Symp. VLSI Circ. **40**, 11–13 (1998)
12. Roy, K., Prasad, S.C.: Low Power CMOS VLSI Circuit Design. Wiley, New York (2000)
13. Hanchate, N., Ranganathan, N.: LECTOR: a technique for leakage reduction in CMOS circuits. IEEE Trans. VLSI Syst. **12**(2), 196–205 (2004)
14. Verma, P., Mishra, R.A.: Leakage power and delay analysis of LECTOR based CMOS circuits. In: International Conference on Computer and Communication Technology (ICCCT) IEEE, pp. 260–264 (2011)

Handwriting Recognition of Brahmi Script (an Artefact): Base of PALI Language

Neha Gautam, R.S. Sharma and Garima Hazrati

Abstract Handwriting recognition and OCR are two major fields of recognition and classification, one is area and other is dawn. Archaeology is that field of study where recognition and classification is needed at most to recognize ancient artefacts written in languages like Pali having various scripts such as Khmer, Sinhala, Devanagari and more. Handwritten character recognition with MOCR (Modified OCR) is shown for Pali language in this paper portraying modified OCR to recognize Brahmi script and showing comparisons in terms of accuracy with two other scripts that are Akkhara-Muni and Ariyaka. Brahmi giving an overall success rate of 85.66, 85.73 and 88.83 % respectively. MOCR has new steps in various phases which results in better accuracy than OCR.

Keywords Handwriting recognition · Archaeology · Pali · OCR · Brahmi · Akkharamuni · Ariyaka

1 Introduction

Handwriting recognition is a vast field of research over last 40 years but is a vulnerable problem till yet. It is used in every arena of biosphere from banking to surveillance and studies to data storage. Handwritten recognition can be classified on the basis of data acquisition consisting of offline and online as well as manuscripts that consist of either handwritten or printed [1]. Recent years has evolved with myriad of recognition technologies but then also less work is done in field of recognizing artefacts and historical preservations of Buddhists texts. Technologies

N. Gautam · R.S. Sharma (✉) · G. Hazrati
Rajasthan Technical University, Computer Science Department, Kota, Rajasthan, India
e-mail: rssharma@rtu.ac.in

N. Gautam
e-mail: neha_ucertu@rediffmail.com

G. Hazrati
e-mail: ghazrati9@gmail.com

S.C. Satapathy and S. Das (eds.), *Proceedings of First International Conference on Information and Communication Technology for Intelligent Systems: Volume 2*,
Smart Innovation, Systems and Technologies 51, DOI 10.1007/978-3-319-30927-9_51

that came into existence are neural networks [2], SVM-LVQ [3], CNN [4], Evolutionary and Swarm intelligence algorithms [5, 6]. Since, there are many technologies like neural network, nature inspired algorithms and others but for any artefact based recognition technique it needs to be clear because of which OCR is taken as a starting base for better understanding.

Handwriting recognition is an art of recognizing digits written by human where handwriting refers to a surface consisting of artificial graphic marks conveying some message through the marks conventional relation to language [7]. While recognizing a new script one needs to reach the start of technologies to get the base which is OCR whose first patent was obtained in 1929 by Tausheck in Germany, followed closely by Handel in 1933 [8].

1.1 Pali

Many philosophers remain confused till today that whether PALI is a language or a script? According to us Pali is a language as language is a body of words, and set of methods that are understood by a community simply used as a form of communication whereas a script is a written document. In simple words its a way of pronunciation.

Language of prehistoric period—PALI has been inscribed in many scripts and languages besides being popular in southern countries for teachings of Buddhism. Pali canon basically consists of 3 tipitakas means 3 baskets which are Suttta Pitaka, Vinaya Pitaka, Abhidhamma Pitaka. Pali alphabet basically consists of 41 letters: 6 vowels, 2 diphthongs, 32 consonants and 1 accessory nasal sound called as ni-gahitta. Consonants are divided into 25 mutes, 6 semi-vowels, 1 sibilant and 1 aspirant followed by vowels comprising of long and short [9]. **Brahmi**: Roots of Pali belong to middle Indo-Aryan period having Brahmi as base script shown in Fig. 1 came up during reign of 1st to 5th century B.C. storing teachings of Buddha and resulting in generation of various other scripts. An overview of other two scripts are described below:

Akkhara-Muni: Ian James, developer of a modified Latin font, and forms got their existence from ancient Brahmi and Pallava (ancestors of the Indic scripts) and later on named as Akkhara Muni, (Letters of the Sage). In Sri Lanka, Pali was used not only for the writing of Buddhist scriptures, but also to record the history of the country.

Ariyaka: The Ariyaka alphabet was invented by King Mongkut Rama IV of Siam (1804–1868) as an alternative alphabet for Pali. He considered the Khmer alphabet, which was commonly used to write Pali, to be too complicated and decided to create an alphabet that was easier to use and more Western in appearance [10, 11].

Whole paper is organized as follows: Sect. 2 consists of Literature Survey on OCR and various languages since there is not much work done in field of Pali

Fig. 1 Brahmi script

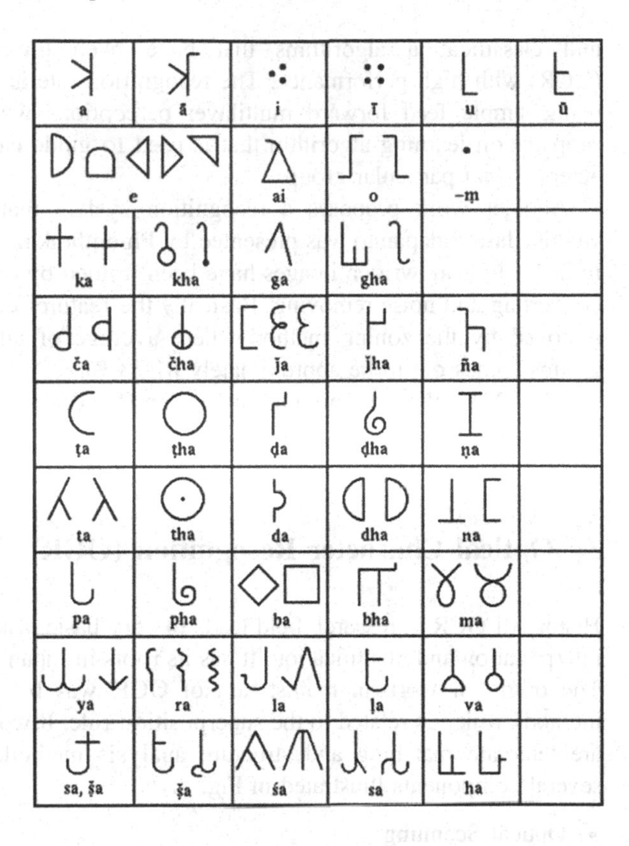

Language followed by Sect. 3 having details of OCR resulting in Sect. 4 which is modified OCR algorithm presented by results and conclusion in Sects. 5 and 6 respectively.

2 Related Work

Several other works were done by OCR on many other languages and scripts giving top rated accuracies setting standard for others. In 2004 Pal and Chaudhari [12] describes work done on 12 major Indian scripts through OCR. In 2007 Manjunath Aradhya et al. [13] showed a multi-lingual OCR leading to a good accuracy. Afterwards, Desai [14] used OCR technique for recognizing Gujrati handwritten digits in 2010 resulting in 82 % success rate and likewise Choudhary et al. [15] routine this with binarization method to judge capability of OCR for english characters in 2013 giving accuracy of 85.62 %.

Pali characters recognition using Devnagari was shown by Mantri et al. [16] in 2012, by comprising following features like image preprocessing, feature extraction and classification algorithms that have been traversed to design software (OCR) with high performance. The recognition rate is 100 % that has been done using simple feed forward multilayer perceptions which also proposes a back propagation learning algorithm that is used to guide each network with the characters in that particular group.

Another work proposes a recognition system that has taken Pali cards of Buddhadasa Indapanno was presented by Phienthrakul and Chevakulmongkol [17] in 2013. Its handwritten images have been refined by contrast adjusting, grayscale converting and noise removing. Basically the features of every single character are removed by the zoning method where average of all accuracies considered in groups comes out to be approximately 81.73 %.

After 2013, there are no more papers on Pali due to which this work is going to be a recent and latest work in this arena within last 2 years.

3 Optical Character Recognition (OCR)

History of OCR in research field leads to very basic task in starting of recognition, interpretation and identification. It has its roots in Japan during period of early 90s. The origin of work in mainstream of OCR was based on "template matching method" which is related to the superposition rule. It works on two methods which are template matching and structure analysis method. OCR is combination of several components illustrated in Fig. 2:

- Optical Scanning
- Local segmentation
- Pre-processing
- Feature extraction
- Classification.

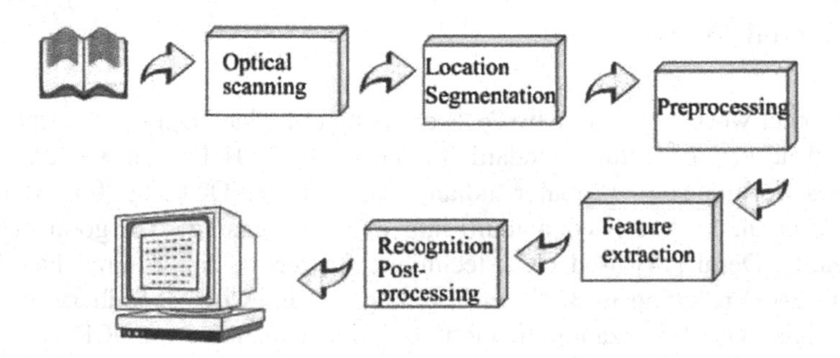

Fig. 2 Components of an OCR-system

4 Proposed Accost

Here, Here, modified OCR steps are described in detail which are pre-processing, local segmentation, feature extraction and classification In this accost size of input image is 42 * 24 pixels and a sample is shown in Fig. 3.

Pre-processing: Cropping, Gray-scaling, binarization and noise removal where in cropping we crop digitalized scanned image to cut it and store it till it reached edges of character in image. In grayscaling we convert image into grayscale image by grayscale function in MATLAB moving to its binarization and removing noise further by cropping isolated indexes from the image.

Local Segmentation: Labelling and line detection are parts of these steps where image is properly set into level by rotating it and setting it in proper zones of pixels and line detection is done by line detection algorithm to detect edges of each character (Fig. 4).

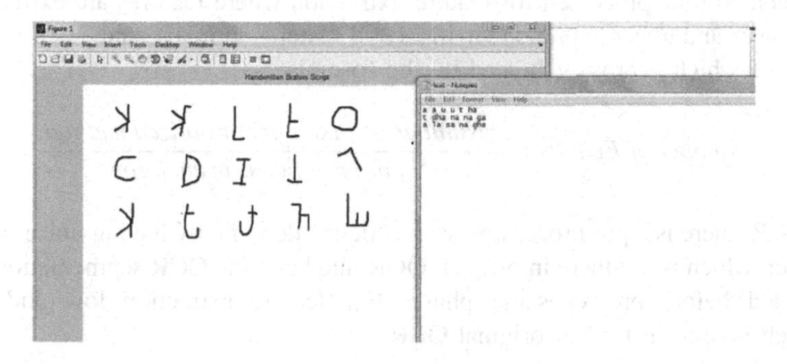

Fig. 3 Sample input image

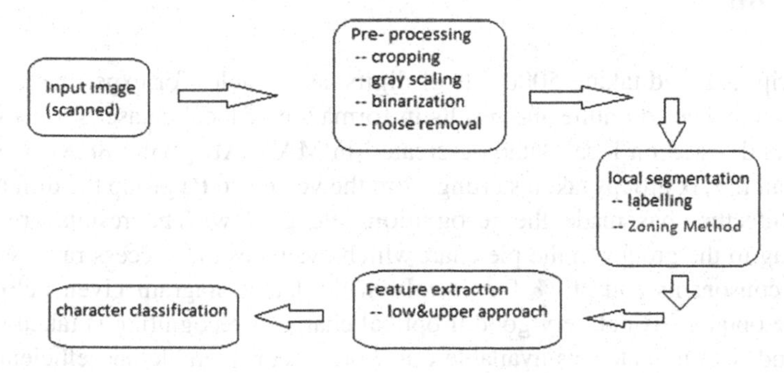

Fig. 4 Research method

Algorithm 1 Algorithm for character Checking

```
1: procedure CH Character Checking
2: Add the pixel value in horizontal direction (S)
3:     for S = 0 do
4: Then it may be a text line
5: Find the height of the previous line (h)
6:         if h > 30 then
7: Then it is a text line (TL)
8:             if ( thenS > 108) and (h > 27)
9: Then it is a sub-line (SL)
10:             end if
11:         end if
12:     end for
13: end procedure
```

Feature Extraction: At this stage we are having image that is pre-processed and is segmented now processed for feature extraction where features are extracted by using lower and upper approach having 1008 features in image and 7 * 4 pixels in each zone which is cross validated by the formula:

$$Number\ of\ Features = \frac{Number\ of\ pixels\ in\ normalized\ method}{Number\ of\ pixels\ in\ each\ zone}$$

In MOCR, there is a pre-processing step which is done firstly having noise removal approach which is not there in original OCR and here like OCR segmentation is not performed before preprocessing phase. For feature extraction low and upper approach is used instead of original OCR.

5 Result

This script is tested taking 5000 sets of digits and samples for experiments, whose images are collected online and results in formation of local dataset some of which trained and tested on local database created in MATLAB giving 88.83 % success rate. In all the six groups taken starting from the vowels to Pa group the difference in the architecture has made the recognition rate go low. The results are shown according to the groups in the pie chart which evaluates the success rate as 89.4 % for the consonants and 86 % for vowels in the latent diagram given below. The outcome obtained is not very good if optical character recognition is taken but as a script and with no data sets available our work is considerable and efficient. Since Brahmi script is the base script among all so through this graph it is showing the best results among the three scripts compared. The other two scripts that is Akkhara-muni and Ariyaka in comparison to Brahmi has shown a slower pace. We have included six groups taken in the graph which are vowels, Ka group, Ca group, T.a group, Ta group and Pa group. First if we compare both the scripts i.e. Akkharamuni and Ariyaka by taking vowels as the parameter then Ariyaka is standing out. The second parameter which is Ka group in which Akkharamuni is coming out with a better pace than Ariyaka. Ca group which gives Akkharamuni as the better one than Ariyaka with more than 80 %.T.a group which gives the success rate of about 90 % for Ariyaka than Akkharamuni. Ta group which gives Akkharamuni as the better result than Ariyaka. And the last but not the least Pa group which gives Akkharamuni better than Ariyaka. Hence we discussed about all the three scripts comprising its recognition rate and success rate following certain procedures and evaluation given through pie chart (Figs. 5 and 6).

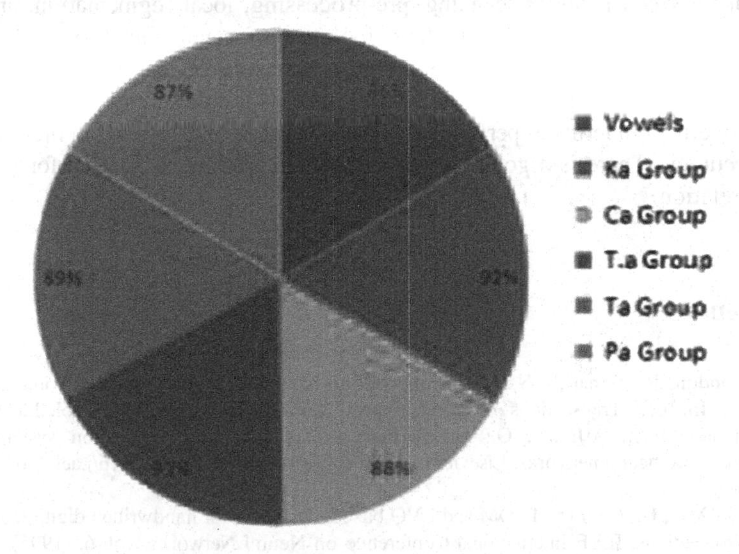

Fig. 5 Result of Brahmi script

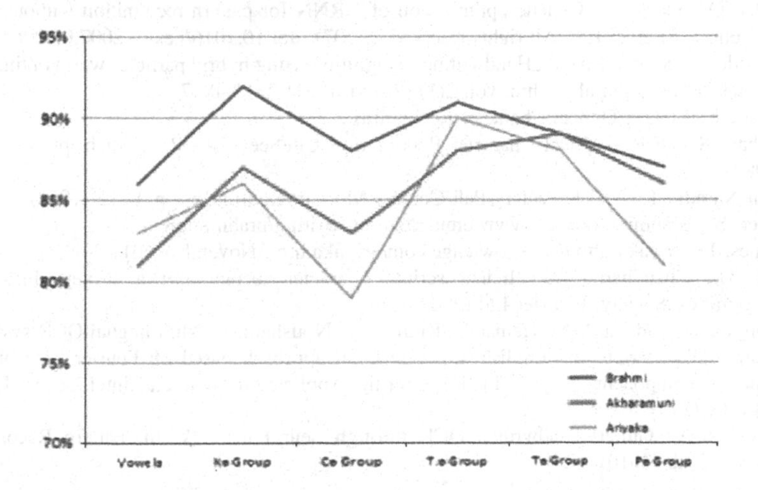

Fig. 6 Comparison between Brahmi script, Akkhara-muni script, Ariyaka script

6 Conclusion

This paper basically discusses about Pali which has been recognized through modified optical character recognition and showing comparison between these scripts. In this approach, the classification is done using techniques such as optical scanning, binarization, segmentation and feature extraction followed by classification of characters. This attempt is a unique one and offers a great value for scripts which is needed to be recognized as there is not much research done in the field of ancient scripts. Through all the evaluations done the overall performance is about 85.4 % of OCR. The latter is recognized with some changes in the procedure undertaken which includes scanning, pre-processing, local segmentation, and feature extraction at regular intervals. Its complete performance analysis is 88.83 %. Technically the local database created and evaluated here has given a best result overall. But in all the three cases the benchmark is not created and so it has a scope of improvement in terms of performance, with changes in extraction in order to gain high accuracy. There is a good scope in this field leading to an idea for growing global relation.

References

1. Plamondon, R., Srihari, S.N.: Online and offline handwriting recognition: a comprehensive survey. In: IEEE Transactions on Pattern Analysis and Machine Intelligence, vol. 22(1) (2000)
2. AlOmaria, F.A., AlJarrah, O.: Handwritten Indian numerals recognition system using probabilistic neural networks. Elsevier Ltd. All rights reserved. doi:10.1016/j.aei.2004.02.001. (2004)
3. Wu, J., Yan, H.: Combined SOM and LVQ based classifiers for handwritten digit recognition. In: Proceedings, IEEE International Conference on Neural Networks, vol. 6. (1995)
4. Niu, X.-X., Suen, C.Y.: A novel hybrid CNNSVM classifier for recognizing handwritten digits. Journal Homepage: www.elsevier.com/locate/pr
5. Polat, O., Yldrm, T.: Genetic optimization of GRNN for pattern recognition without feature extraction. Elsevier Ltd. All rights reserved. (2007). doi:10.1016/j.eswa.2007.04.006
6. Lagudu, S., Sarma, C.H.V.: Handwriting recognition using hybrid particle swarm optimization & back propagation algorithm, vol. 2(1) (2013). ISSN 2319-4847
7. Line eikvil OCR, Optical Character Recognition
8. Kahan, S.: OCR—Optical Character Recognition CiteSeer. IJCPR, Munich, pp 1031–1044 (1987)
9. Ven Narada, T.: An Elementary Pali Course. Morgan Kaufmann, pp. 9–12 (1952)
10. Ager, S.: Brahmi Alphabet. www.omniglot.com/writing/brahmi.htm
11. James, I.: Ariyaka http://skyknowledge.com/ariyaka.htm. November 2011
12. Pal, U., Chaudhari, B.B.: Indian scripts character recognition: a survey. In: Pattern Recognition Society, Elsevier Ltd (2004)
13. Manjunath Aradhya, V.N., Hemantha Kumar, G., Noushath, S.: Multilingual OCR system for South Indian scripts and English documents: an approach based on Fourier transform and principal component analysis. In: Engineering Applications Artificial Intelligence, Elsevier Ltd (2007)
14. Desai, A.A.: Gujrati handwritten OCR through neural network. In: Pattern Recognition, Elsevier Ltd. (2010)

15. Choudhary, A., Rishi, R., Ahlawat, S.: Offline handwritten character recognition using feature extracted from binarization technique. In: AASRI Conference on Intelligent Systems and Control, Elsevier (2013)
16. Mantri, K.S., Ramteke, R.S., Suralkar, S.R.: Pali Character Recognition System. In: IJAIR, ISSN 2278-7844 (2012)
17. Phienthrakul, T., Chevakulmongkol, W: Handwritten recognition on Pali Cards of Buddhadasa Indapanno. In: International Computer Science and Engineering Conference, IEEE (2013)

Theoretical and Empirical Analysis of Usage of MapReduce and Apache Tez in Big Data

Rupinder Singh and Puneet Jai Kaur

Abstract Big data means large amount of data requires new technologies for its faster processing. It is ineffective to process the large amount of data with traditional devices. Big data provides an extra advantage in business and better service delivery. Big data brings a new change in decision making process of various business organizations. Big data has many challenges related to the 5Vs-Volume, Velocity, Variety, Veracity and Value. Hadoop is a Big Data tool used to process larger amounts of Data. It has many subcomponents work together to achieve the goal of faster processing. Apache Hive and Apache Pig are tools used to access data in different ways in Hadoop Ecosystem. Apache Hive depends upon SQL like queries while Apache Pig uses scripts. These two tools uses MapReduce or Apache Tez framework to access data. In this paper we analyze how these two frameworks uses Hadoop Distributed File System (HDFS) by comparing them in both theoretical and empirical way.

Keywords Big data · Hadoop · HDFS · Mapreduce · Apache Tez · Apache Pig · Apache Hive

1 Introduction

The world has nowadays becomes the internet of things [1]. Data on servers grows very rapidly; it becomes difficult to process this large amount of data day by day. Due to availability of low speed computing resources we require new ways to process this data timely and accurately and this led to many challenges that we face in Big Data analysis and processing [2, 3]. Today many companies invest millions in research to overcome these challenges. Terabytes of data is available on servers [4],

R. Singh (✉) · P.J. Kaur
Department of I.T, U.I.E.T, Panjab University, Chandigarh, India
e-mail: rupinderkaoni@gmail.com

P.J. Kaur
e-mail: puneet@pu.ac.in

S.C. Satapathy and S. Das (eds.), *Proceedings of First International Conference on Information and Communication Technology for Intelligent Systems: Volume 2*,
Smart Innovation, Systems and Technologies 51, DOI 10.1007/978-3-319-30927-9_52

using this data in a positive way for efficient decision making, growing business and earing more from less is the major challenge and current topic of research.

In Sect. 2 we discuss Hadoop Big Data Tool having many subcomponents. Apache Hive and Apache Pig are two tools used to access data in Hadoop Ecosystem [5]. Apache Hive gives closer look to the Hadoop Distributed File System and helps in performing selection and deletion functions on data located in HDFS. Data is stored in Relational format like other RDBMS but Hive also give option to store data in ORC format so that operations becomes easier on data. It uses SQL like queries to interact with HDFS while Apache Pig uses scripts. A script can be written in proper format and submitted to the Apache Pig for its execution. Apache Hive and Apache Pig use MapReduce and Apache Tez Frameworks to process data [6, 7]. MapReduce is suitable for Batch Processing Jobs while Tez is used for interactive query processing. Firstly by choosing some parameters we compare these two frameworks in Sect. 3 theoretically. These parameters help us in understanding usefulness of Apache Tez over Mapreduce. In Sect. 4 we compare these two frameworks empirically by choosing one dataset and processing the dataset with both the frameworks. A Pig script is processed using both the frameworks and the execution time for 10 instances is recorded. Last section concludes the paper and throws some light on work yet to be done in future.

2 Hadoop Big Data Tool and Its Components

Hadoop is scalable, fault tolerant, open source project used for distributed processing and storing of large sets of data. Out of many sub parts mainly it has two parts MapReduce Programming Paradigm and HDFS (Hadoop Distributed File System) [8]. Basically Hadoop implements the concept of clustering while processing large amount of data on scalable clusters. Hadoop can be used as ETL (Extract Transform Load) engine [9], Exploration Engine and as a Data Archive. We have different choices to deploy Hadoop in the cloud, at data center, Microsoft Windows or in Linux.

Hadoop has so many tools and frameworks each designed to perform specific function. We have two data access tools Apache Hive and Apache Pig [10]. Apache hive is designed to analyze the data using SQL queries [11] and Apache Pig [12] is designed to analyze the data using scripts with batch processing MapReduce or interactive processing Apache Tez Framework (Fig. 1).

MapReduce helps in processing data in parallel on different machines. Using MapReduce framework we are able to write programs for distributed processing of Big Data. MapReduce function is performed in two phases Map and Reduce. The main task of Map phase is to read all the input data and transform or filter it and some business logic is performed in Reduce phase [13]. Apache Tez framework is available in Hadoop 2.0 and newer versions. It is superior to MapReduce in many aspects. In Apache Tez we also have two phases Map and Reduce but it has more advantages than MapReduce Framework. In Apache Tez we may have many

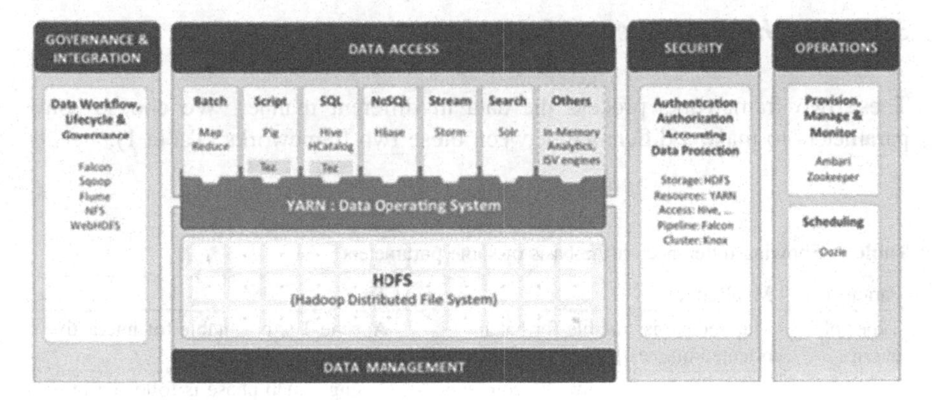

Fig. 1 An illustration of Hadoop Ecosystem [14]

continuous Map phases following by reduce phases. Apache Tez has one more advantage in this we have continuous Map and Reduce phases without writing to Hadoop File System. We analyze two frameworks theoretically and empirically in following sections by choosing some Parameters.

3 Theoretical Analysis

Section 3.1 shows some parameters which we are going to use in analysis. Then Sect. 3.2 shows comparison between Apache Tez and MapReduce basis upon chosen parameters.

3.1 Parameters Chosen

Queries Supported. Data Analyst may interact with system by using different types of queries. Queries may be submitted in a batch manner or in an interactive manner.

Response Time. System returns the result to the Analyst with in some time frame after processing the particular Dataset.

Storage Requirement. During the query processing Hadoop stores the results into the HDFS for temporary purpose. HDFS storage requirement depends upon particular framework.

Containers. Containers are memory areas used during Data Processing in Hadoop. Different jobs require allocation and deallocation of containers and jobs uses these containers depending upon particular framework used.

Processing Model. Processing Model shows the path followed by MapReduce or Apache Tez framework to process the particular Data set.

3.2 *Apache Tez Versus MapReduce*

These two frameworks process the data in different manner. We choose some parameters to make a difference between these two frameworks (Table 1).

Table 1 Showing difference on the basis of some parameters

Parameters	MapReduce	Tez
Types of queries	Map reduce is suitable for batch oriented queries [15]	Apache Tez is suitable for interactive queries.
Processing model	In MapReduce we always requires a Map phase before the reduce phase 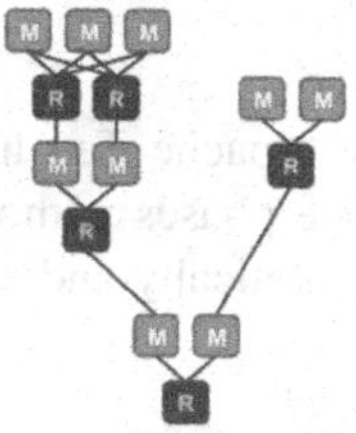 MapReduce processing model	A single map phase is followed by multiple reduce phases 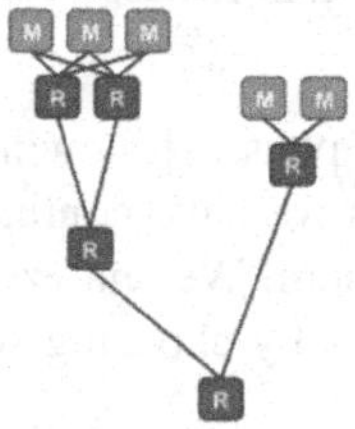 Tez processing model
Hadoop version	MapReduce is available in all the hadoop versions	Apache Tez is available in hadoop version 2 and above
Response time	Little bit slower due to the two map and reduce phases followed each other alternatively	It meets the demands of faster response and increases throughput up to peta byte scale
Temporary data storage	Map reduce jobs accesses HDFS after every map and reduce. In some cases HDFS is accessed after map or reduce only phase 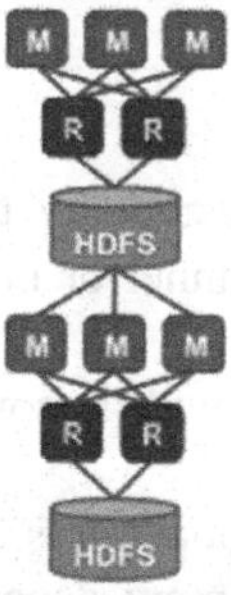 Map and Reduce over MapReduce	Tez is more efficient then MapReduce because it does not write temporary data to HDFS 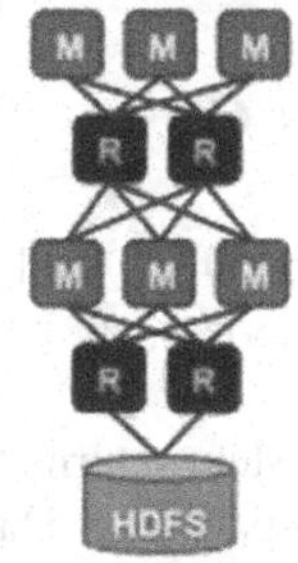 Map and Reduce over Tez
Usage of hadoop Containers	MapReduce job requesting multiple containers requires initialization of many new containers. MapReduce jobs reuses containers only within on single application and single node	Apache Tez reduces this inefficiency by allowing the usage of existing already created idle containers

4 Experimental Evaluation

Apache Tez and MapReduce frameworks were suitable with both the Apache Hive and Apache Pig. In this section we mainly focus on the usage of these two frameworks using Apache Pig.

Section 4.1 describes the datasets used for analysis and Sect. 4.2 explains the experimental setup required for processing of the datasets. Section 4.3 explains the results and the metrics used for evaluation.

4.1 Dataset

To compare both the frameworks we have chosen the dataset GEOLOACTION [14]. In this dataset we have the two relations having certain no of instances and attributes. Table 2 shows the name and no of records and attributes in our dataset. This paper does not consist the details of the dataset used due to the space restrictions. So, we recommend the readers to consult the original reference.

4.2 Experimental Setup

For analysis Hortonworks Data Platform (HDP) is used. HDP is used to run Hadoop and all its components on a single Machine. It provides an integrated platform for hosting all of the Hadoop components. Hortonworks Ambari agent helps in viewing all the components and managing the whole Hadoop Cluster [14]. Sandbox helps in running both the Namenode and Datanode on single machine. We use VMWARE for hosting HDP on Window 7 64-bit. Our experiments were performed on Desktop having Intel Core (i7), 2.80 GHz processor with 8 Giga Bytes of Random Access Memory (RAM).

4.3 Experiment Results and Metrics

Our Dataset is stored in the form of Relational Tables in Apache Hive Database. In this paper we mainly consider the difference of time in execution of Apache Pig script on our Dataset using both the MapReduce and Apache Tez frameworks.

Table 2 Datasets used in experiments

Name	No of Records	No of Attributes
Geolocation	8013	10
Drivermilage	101	2

Fig. 2 Apache script used for analysis [14]

Apache Script

```
1.  a = LOAD 'geolocation' using
2.  org.apache.hive.hcatalog.pig.HCatLoader();
3.  b = filter a by event != 'normal';
4.  c = foreach b generate driverid, event, (int) '1' as
    occurance;
5.  d = group c by driverid;
6.  e = foreach d generate group as driverid,
    SUM(c.occurance) as t_occ;
7.  g = LOAD 'drivermileage' using
    org.apache.hive.hcatalog.pig.HCatLoader();
8.  h = join e by driverid, g by driverid;
9.  final_data = foreach h generate $0 as driverid, $1
    as events, $3 as totmiles, (float) $3/$1 as
    riskfactor;
10. store final_data into 'riskfactor' using
    org.apache.hive.hcatalog.pig.HCatStorer();
```

Apache Pig script after execution stores the results in the 'riskfactor' relational schema that we have already stored in Apache Hive. We evaluate the results after the execution of 10 instances of this script on each of the MapReduce and Apache Tez framework (Fig. 2).

Figure 3 shows the time of execution in milliseconds for both the Apache Tez and MapReduce Framework. We calculate the average time over the 10 runs of the script by using the formula.

$$A = \frac{1}{n} * \sum_{i=1}^{n} x_i$$

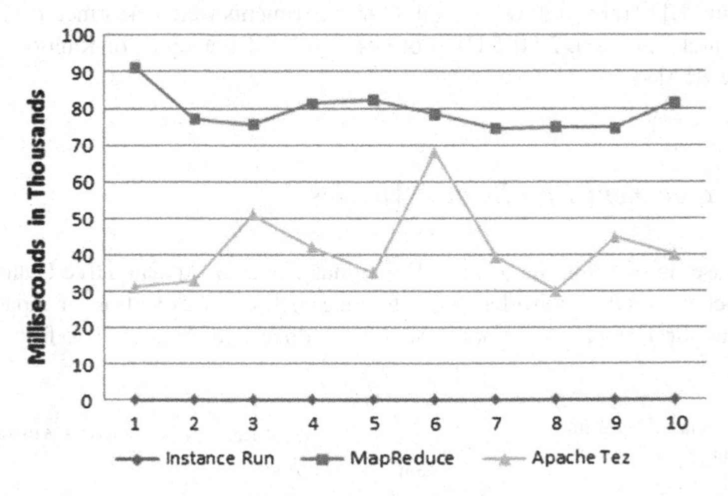

Fig. 3 Pig script execution time on both the frameworks

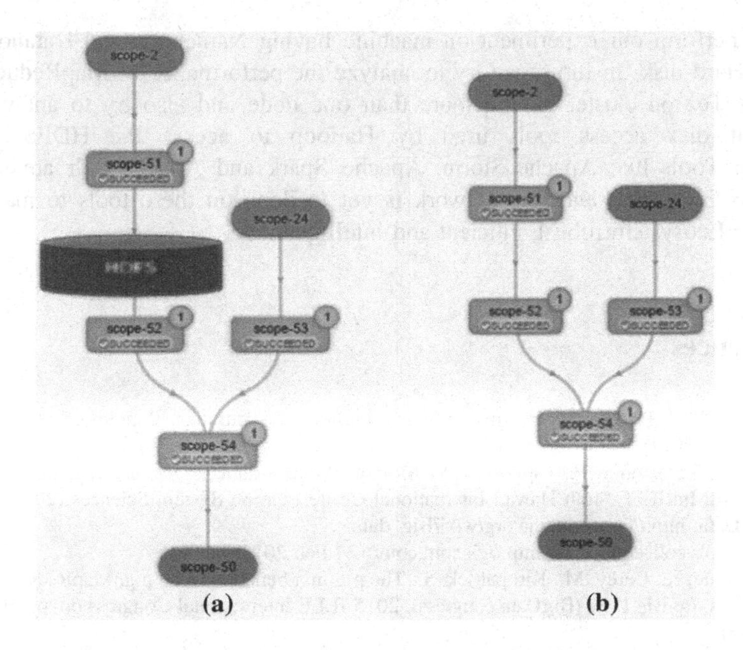

Fig. 4 Graphical view of map and reduce phases of both the frameworks

In our case we have n = 10 and different values for x shown in Fig. 3 for both the MapReduce and Apache Tez. Average time of 79,249.7 ms for MapReduce is much more than the 41,481.5 ms for Apache Tez framework. This clearly shows that the Apache Tez framework gives same results within lesser time than MapReduce framework. Apache Tez achieves this by dividing the whole problem into 4 tasks (scope-51, 52, 53 and 54). It continuously creates 2 Map tasks from 1 Map task and then 1 Reduce task. On the other hand MapReduce creates same no of tasks but after creating one Map task it writes the data into HDFS. This slows down the performance of the MapReduce (Fig. 4).

5 Conclusion and Future Work

This paper briefly describes both the frameworks used for execution of Apache Pig scripts. We perform both the theoretical and empirical analysis to check the performance of these two frameworks. Various parameters chosen help us in understanding the difference between the way of execution and memory access by both the frameworks. After analysis it is clearly understood that Apache Tez has better scope in Hadoop than MapReduce in interactive scenarios. But still MapReduce is the backbone of the Hadoop Ecosystem and it can be efficiently used in batch processing scenarios.

We perform our experiment on machine having Namenode and Datanode on single Hard disk. In future we try to analyze the performance of MapReduce and Apache Tez on cluster having more than one node and also try to analyze the different disk access tools used by Hadoop to access the HDFS. Many Hadoop Tools like Apache Storm, Apache Spark and Apache Solr are part of Hadoop Ecosystem and lots of work is yet to done on these tools to make the Hadoop Ecosystem robust, efficient and intelligent.

References

1. Khanzode, G.P.: Insights internet of things: endless opportunities. Banglore, India, Infosys, Insights (2012)
2. Kaisler, S., Armour, F., Espinosa J.A., Money, W.: Big data: issues and challenges moving forward. In: IEEE, 46th Hawaii International Conference on System Sciences (2013)
3. Big Data. http://en.wikipedia.org/wiki/Big_data
4. Facebook collecting data. http://gigaom.com/. 24 Feb 2014
5. Ouaknine, K., Carey, M., Kirkpatrick, S.: The pig mix benchmark on pig, MapReduce, HPCC systems. In: Big Data (BigData congress), 2015 IEEE International Congress on, pp. 643–648 (2015)
6. Gates, A.F., Natkovich, O., Chopra, S., Kamath, P., Narayanamurthy, S.M., Olston, C., Reed, B., Srinivasan, S., Srivastava, U.: Building a high-level dataflow system on top of Map-Reduce: the Pig experience. In: Proceedings of the VLDB Endowment, Vol. 2, no. 2, pp. 1414–1425 (2009)
7. Thusoo, A., Sarma, J.S., Jain, N., Shao, Z., Chakka, P., Anthony, S., Liu, H., Wyckoff, P., Murthy, R.: Hive: a warehousing solution over a map-reduce framework. In: Proceedings of the VLDB Endowment, Vol. 2, no. 2, pp. 1626–1629 (2009)
8. Azzedin, F.: Towards a scalable HDFS architecture. In: Collaboration Technologies and Systems (CTS), 2013 International Conference on, pp. 155–161(2013)
9. Bansal, S.K.: Towards a semantic extract-transform-load (ETL) framework for big data integration. In: Big Data (BigData Congress), 2014 IEEE International Congress on, pp. 522–529 (2014)
10. Fuad, A., Erwin, A., Ipung, H.P.: Processing performance on Apache Pig, Apache Hive and MySQL cluster. In: Information, Communication Technology and System (ICTS), 2014 International Conference on, pp. 297–302 (2014)
11. Thusoo, A., Sarma, J.S., Jain, N., Shao, Z., Chakka, P., Zhang, N., Antony, S., Liu, H. and Murthy, R.: Hive-a petabyte scale data warehouse using hadoop. In: Data Engineering (ICDE), 2010 IEEE 26th International Conference, pp. 996–1005. IEEE (2010)
12. Gates, A.F., Dai, J., Nair, T.: Apache pig's optimizer. In: IEEE Data Engineering Bulletin 36, no. 1 (2013)
13. Maitrey, S., Jha, C.K.: Handling big data efficiently by using map reduce technique. In: Computational Intelligence & Communication Technology (CICT), 2015 IEEE International Conference on, pp. 703–708 (2015)
14. Hadoop. http://www.hortonworks.com
15. Ravindra, P.: Towards optimization of RDF analytical queries on MapReduce. In: Data Engineering Workshops (ICDEW), 2014 IEEE 30th International Conference on, pp. 335–339 (2014)

Strip Tree Based Offline Tamil Handwritten Character Recognition

M. Antony Robert Raj and S. Abirami

Abstract In this paper, hierarchically represented Strip tree based feature extraction has been employed for offline Tamil handwritten recognition. Tamil handwritten character recognition is a challenging factor which gain more attention in the field of pattern recognition. Since Tamil language is rich in complex structures such as curves and loops, this work proposed a novel Strip tree to represent the curvy structure. So that more number of challenges could be addressed. A Strip tree representation resembles a hierarchical structure is used here to represent a single curve taken from pre-extracted features. This structure is represented in a tree form (hierarchical) to describe the features. Main novelty behind this paper is that all the curvy portions of character are addressed by the strip tree representation, where the height (level) of the tree gets increased when the curvy structure is complex. Later, vectors derived from the tree structure (Strip tree) are analyzed through a decision tree to predict the Tamil character. The final analysis shows that Strip tree yields high success rate in recognition. Since it able to address more number of shape variations.

Keywords Strip tree · Decision tree · Chain code

1 Introduction

Handwritten Recognition System (HCR) system is a subfield of pattern recognition which is required to understand the digitalized handwritten data in image form. Further the HCR system is classified as offline and online [1]. Online HCR system contains less challenges than offline, since because it identifies the characters from the pen tip movement and produces better recognition results than offline handwritten

M.A.R. Raj (✉) · S. Abirami
Department of Information Science and Technology,
Anna University, Chennai 600 025, India
e-mail: antorobert@gmail.com

S. Abirami
e-mail: abirami_mr@yahoo.com

S.C. Satapathy and S. Das (eds.), *Proceedings of First International Conference on Information and Communication Technology for Intelligent Systems: Volume 2*,
Smart Innovation, Systems and Technologies 51, DOI 10.1007/978-3-319-30927-9_53

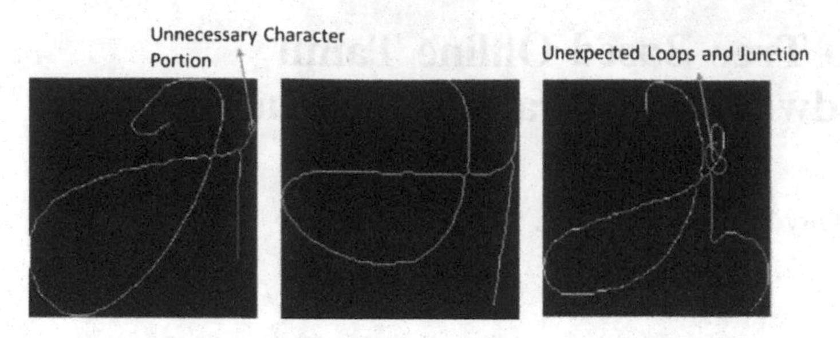

Fig. 1 Shape variations among the characters (அ, ஆ)

recognition [4]. Recognition of offline handwritten characters are tough and intense. Since it is affected by so many factors such as shape variation in characters (as shown in Fig. 1), unexpected loops and junction points, similar in shapes of different characters and unnecessary character portions as shown in Fig. 1 [1]. This factors are highly dependent on writers based on their moods. Another reason for this issue is due to the complex nature of the character because, the south Indian language Tamil has more 'vattezhuthu' (curves) than the other recognized languages such as English, Chinese and so on. Complexity gets increased because of the nature of the Tamil character as well as due to the writing styles.

Tamil language with 247 characters decomposition of 12 vowels, 18 consonants and 216 special characters are considered in this recognition system. The recognition consists of steps such as (1) Pre-processing for cleaning and constructing standard size images. (2) Skeletonization for thinning. (3) Segmentation for extracting required character portions. (4) Feature extraction which include feature pre-extraction and extraction of required features and finally classification for recognizing the correct character from the extracted features.

This work primarily depends on the extraction of required features. The main novelty of this work is to identify and represent the unshaped (uneven) curves of characters in order to represents the structure of the features. Strip tree procedure is elected to achieve this task by extracting the features.

2 Previous Works

Several works has been reported in the literature of Tamil Handwritten character recognition, where various character samples are recognized. Pixel variation or density based (Statistical [2]) and shape or directional (Structural) based algorithms have been implemented in order to achieve the recognition results. In our previous works [3–7] maximum of 30 characters (vowels and consonants) were chosen and accomplished considerable results for those 30 characters samples. Certain works were proposed in this area based on Statistical features [5]. Here the features extracted are: Pixel based location features, Vertical/horizontal features, Axis based pixel location and Point counts. Here the pixel variation and densities has been taken into an account for extracting the features. SVM technique has been applied to predict the appropriately characters. Promising results has been achieved in this work.

In [6], the zoning and chain code based feature selection procedure has been employed to select and pre-extract the features. Further the chain code based sub line direction and bounding box based shape detection concepts has been implemented to extract the features. Finally the Support Vector Machine (SVM) has been used for classifying the results, where 88 % accuracy was achieved for 30 characters.

Both statistical and structural concepts were employed in one of our previous work [7]. Here, a Junction point based feature pre-extraction technique has been applied on image to subdivide the image. Which has been achieved through chain code procedures. Zone based locational features and chain code based directional features were employed here to get necessary information. Here characters have been grouped based on their complexity level and reasonable results have been obtained.

In our previous works, we are lacked to address the shape of the each handwritten character portions which plays main role in HCR. The main novelty behind this paper is different shapes like regular or irregular to be addressed on implementing Strip tree. The complex structure of any curve is simplified in a tree representation. Based on the tree variation the unique structure is formed to denote each character.

3 Proposed System

3.1 Phases of HCR System

Various phases of the offline Tamil HCR such as pre-processing, feature selection, feature—pre extraction, feature extraction and classification are experimented in this process of work. Figure 2 shows the phases experimented in this HCR.

3.2 Overview

This research is an attempt to represent the curvy shapes of images to address more character in Tamil HCR. Strip tree [8] is used here to represent the different type of

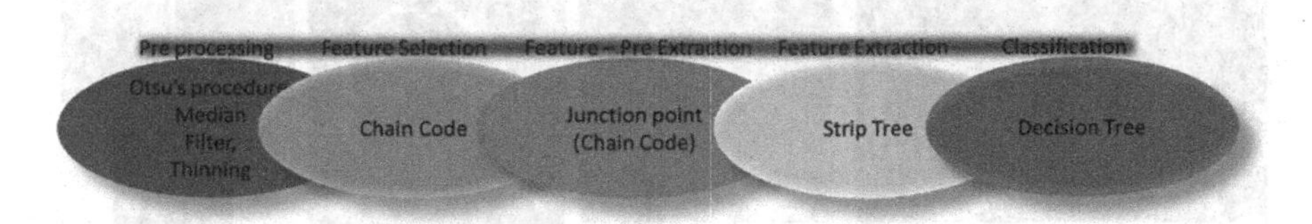

Fig. 2 Phases of proposed methodology

the curves in a tree format. Strip tree is a tree which is used for representing the curve in a hierarchical form. The data points which has been selected from the hierarchical structure is formed as a binary tree. The final leaf node symbolizes the curve.

The vectors derived from this structure is used as a features for the final decision. And also this work tests, how the structural representation helps to identify the character with respect to default nature of curve shapes. Finally decision tree procedure is applied in order to achieve the recognition results.

In this system, initially images undergo preprocessing steps and the chain code [3, 7] has been applied on preprocessed image to extract the junction point based character portion. Later, every segmented portion is represented by a hierarchical based Strip tree (even irregular shapes) and vector values are obtained from it. Subsequently, a decision tree procedure is used here to classify this vector values into appropriate characters.

3.3 Feature Selection and Pre-extraction

Preprocessing has been implemented as discussed in previous section [3, 7]. Eight directional chain code algorithm has been employed on the preprocessed image. As discussed in our previous work [3–7], the chain code algorithm has been applied on the first pixel or the pixel which contains one neighborhood. The chain code starts its travel from those points and proceed eight directionally until it meet all pixels during its travel. The sample result of chain code algorithm showed in Fig. 3.

3.4 Feature Extraction—Strip Tree Representation

Idea of Strip trees has been conceived from [8] and are implemented for extracting and representing the features. Here curve features are analyzed and represented in a

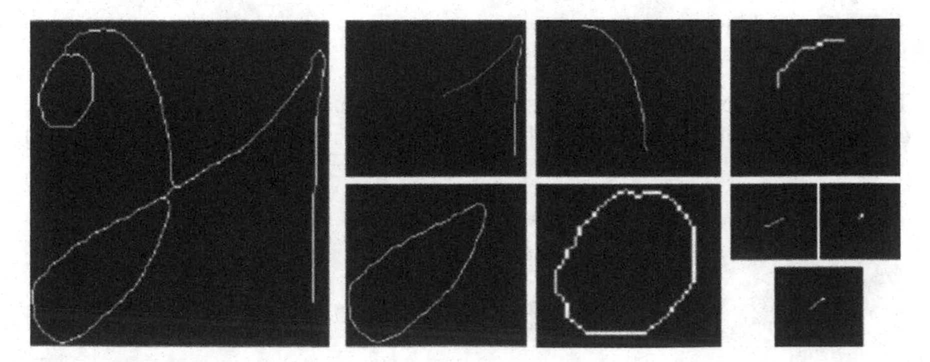

Fig. 3 Extracted features from 'அ'

hierarchical fashion. To represent the characters using Strip tree, here we have taken an example character from the Tamil character set ‘AH’ (அ). Which consist of various shapes such as (single pixel or two or three pixels), Linear shapes (likely), Joined lines in any shapes (likely), arc, curves, semi curves, curves in any shapes, half closed curves, closed curves and so on. Strip tree has been represented using every one of the following procedures.

Procedure 1 If there is no curve or line on the sub divided portion, in the sense, if there is a dot or small linear shape found in the segment then the tree cannot be divided more than root node (R). The root node is taken as a feature as shown in Fig. 4. Breadth first search procedure has been used for taking vector representation from the trees. Vectors and levels of Fig. 4 are V = R and L = L1.

If there is a line (linear shapes) or likely line found on the image, a rectangle can be drawn on the segment as shown in Fig. 5. The tree cannot proceed further from the root node (R) and gets stopped there. Root node that is single node denotes the feature representation of lines. [V = R and L = L1].

Procedure 2 If a closed curve found on the sub divided image, then the strip tree construction varies as per the shape of the different closed curves. The rectangle (Bounding box) constructed on the segment has been noted as R. Following procedure is used to construct the tree from R (Fig. 6).

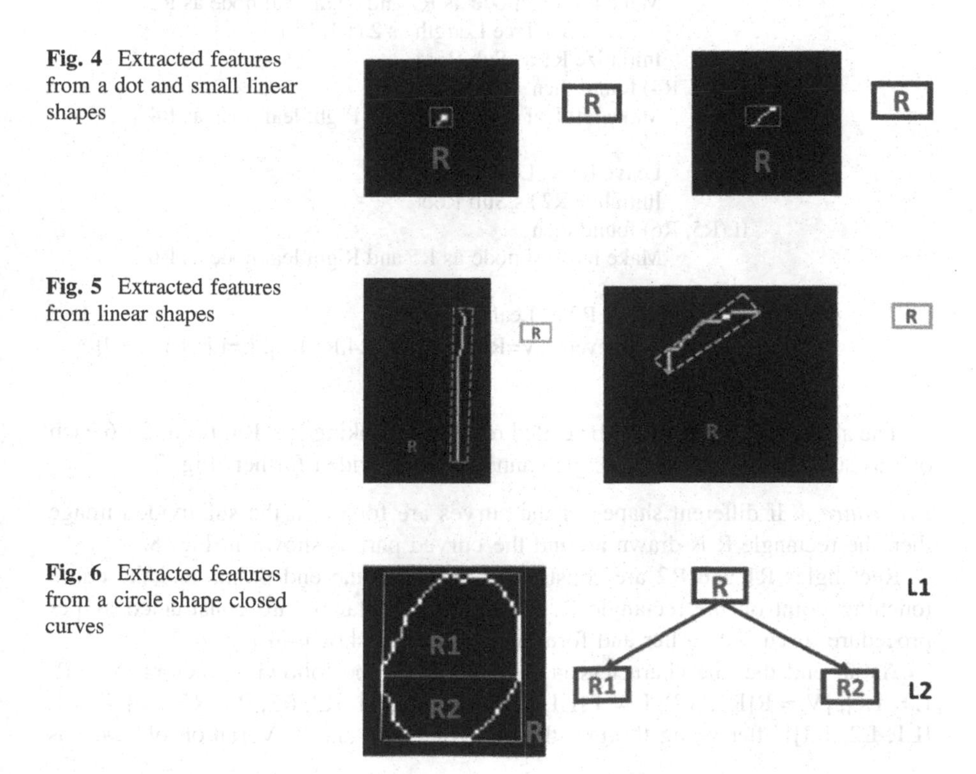

Fig. 4 Extracted features from a dot and small linear shapes

Fig. 5 Extracted features from linear shapes

Fig. 6 Extracted features from a circle shape closed curves

Step1. Close the curves with lines to get a shape rectangle (R).
Step2. Compute width and length of the rectangle (W1, W2).
Step3. If (W1 = W2) or (W1 ≈ W2)
Then, divide the image into two parts, resulting in rectangle R1 and R2 as shown in the figure 6
Tree: Root Node (R),
If (R1 & R2) found then
Make left leaf node as R1 and Right leaf node as R2
Set the Tree Length as 2 (L1, L2)
else
Take Root (R) node as feature.
Vectors and Levels: [V=R[R1,R2], L=L[L1,L2]]
Step4. If (W1>W2) or (W1<W2) (if and only if the variation is high) then

- Divide segment into two parts (division happens on longest part as X1 & X2) R1 and R2 as shown in figure 7
- Based on the variation, Average touching points are predicted from R1 and R2. In rectangle R1, average touching middle point (M1) other than the two end point is computed. Rectangles are drawn from the end points to the middle point and this has been noted as R3 and R4. Same procedure is followed in rectangle R2 also and rectangle R5 and R6 has been obtained. This process is represented by the following pseudocode.

Tree: Root Node (R),
If (R1, R2) found then
Make left leaf node as R1 and Right leaf node as R2
Set Tree Length as 2 (L1, L2)
Initialize R1 as Sub Root
If (R3, R4) found then
Make left leaf node as R3 and Right leaf node as R4
else
Leave R1 as Leaf node of R
Initialize R2 as Sub Root
If (R5, R6) found then
Make left leaf node as R5 and Right leaf node as R6
else
Leave R2 as Leaf node of R
Vectors and Levels: [V=R[R1,R2,R3,R4,R5,R6], L=L[L1,L2,L3]]

The above procedure has to be called recessively taking R3, R4, R5 and R6 each one as sub route until the rectangle cannot be subdivided further (Fig. 7).

Procedure 3 If different shapes of the curves are found on the subdivided image then the rectangle R is drawn around the curved part as shown in Fig. 8.

Rectangles R1 and R2 are constructed based on the end points and the curve touching point on the rectangle R. Furthermore rectangle are constructed as per procedure discussed earlier and form a Strip tree as shown in Fig. 8.

As an end the, the characters is represented as the following vectors [V = R, L = L1], [V = R[R1, R2], L = L[L1, L2]], [V = R[R1, R2, R3, R4, R5, R6], L = L [L1, L2, L3]] after going through the procedure 1,2 and 3. Variation of 'அ' has

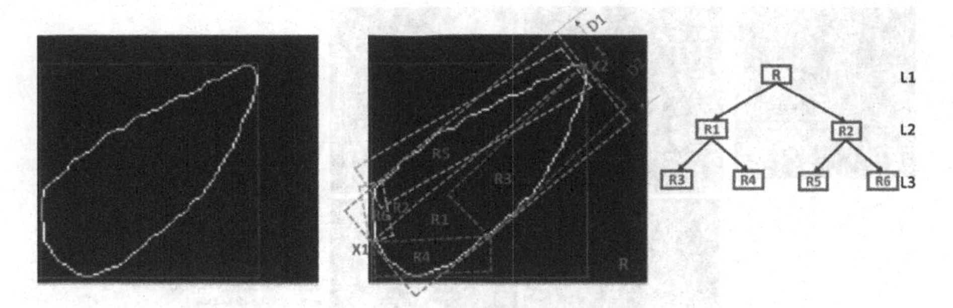

Fig. 7 Extracted features from a closed curves of different shape

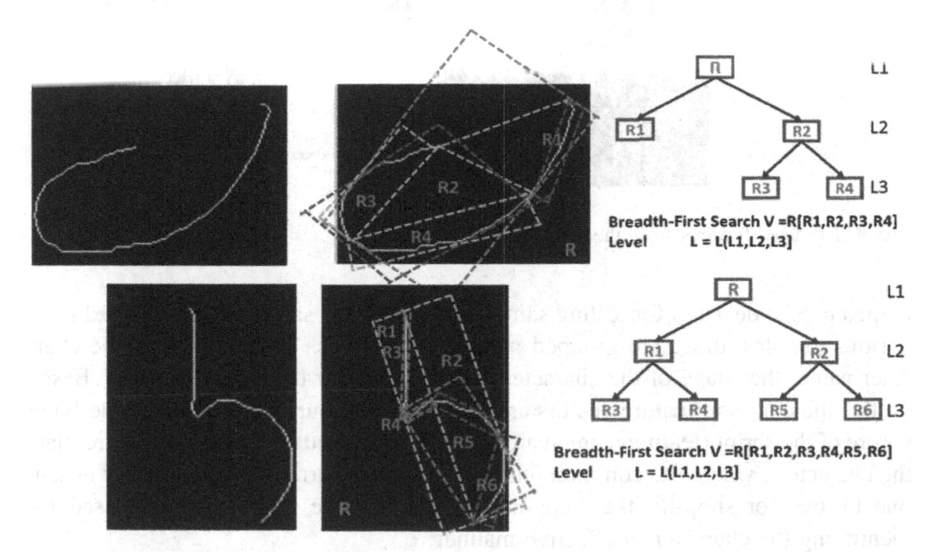

Fig. 8 Extracted features from different shape curves

been represented by Strip tree, the formation of features has been taken as vector forms. This has been trained in decision tree procedure. Sample features which extracted from the character 'அ' are shown in Fig. 9.

3.5 Classification—Decision Tree

A simple and successful classifier decision tree [9] algorithm is experimented for training and testing the results. Instants based hierarchical tree division supports

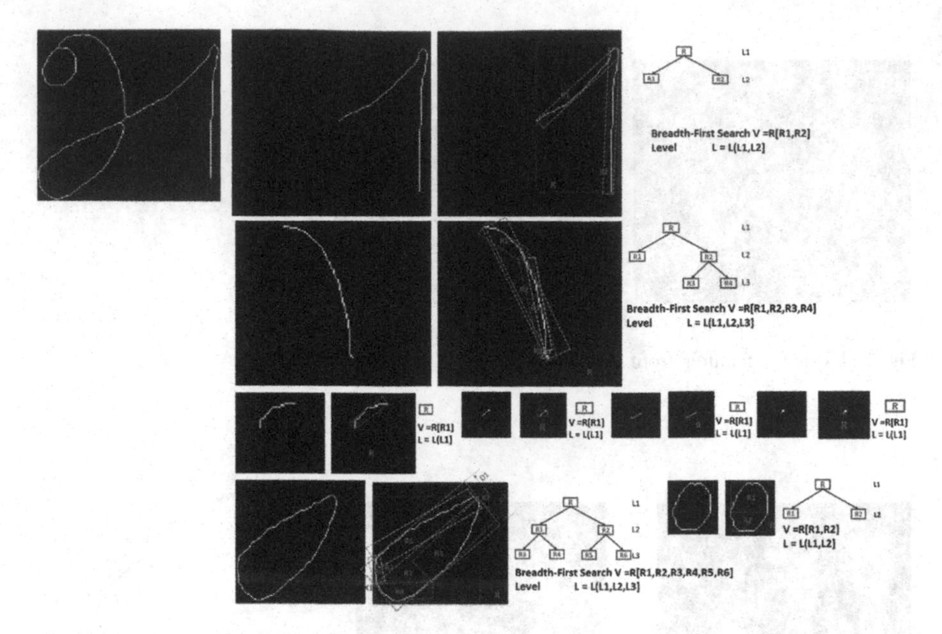

Fig. 9 Extracted features from the character 'அ'

helped to take decision for testing samples. The training samples are collected from various samples, those are grouped with their relations. With respect to the character nature the shape of the character is considered for the feature vectors. Based on this the common feature vectors are individually grouped and if any underlying nature of the script (features) are available that is grouped separately to differentiate the characters, where the common features are used to arrange the character instant one by one for simplify the recognition work and the unique features used for identifying the character in effective manner.

As shown in Fig. 10 character nature used for taking a decision, but all features will not depend on this characters alone. Characters which contain common shapes are marked in Fig. 11.

Fig. 10 Model of decision tree

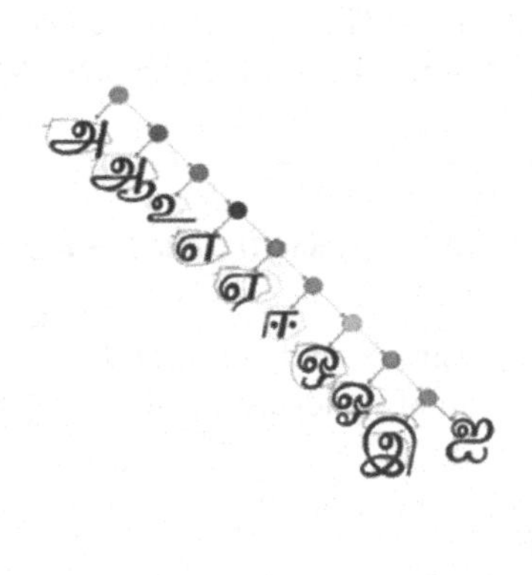

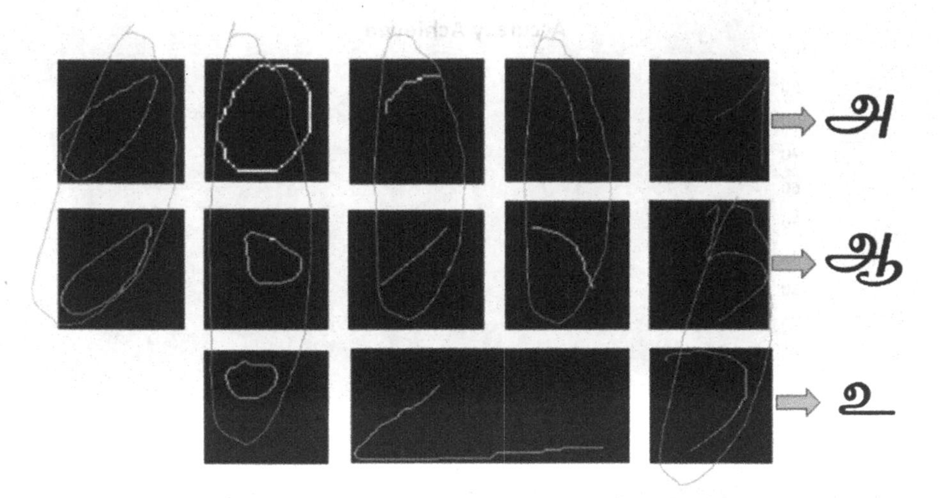

Fig. 11 Identifying common shapes

In this paper the decision arrangements (Fig. 11) are taken from the features extracted from very unique characters which is suitable for taking a decision. Finally over all 88 % accuracy is achieved for selected characters from HP Data set [10].

4 Data Collection and Results

The data samples are collected from HP India data sets [10], where 100 samples of 10 characters are chosen for training set and 50 samples of the same characters are chosen for testing set. The best features which highly matches for all samples of each characters are taken into consideration for classification purpose in decision tree. Table 1 and Fig. 12 shows the accuracy achieved for each character testing samples. Maximum number of samples chosen for testing are classified correctly. 88 % of accuracy has been achieved using Strip tree based representation and decision tree algorithm. Pros and cons of this work has also been discussed in Table 2.

Table 1 Accuracy rate for tamil characters (vowels)

S. no.	1	2	3	4	5	6	7	8	9	10
Character	அ	ஆ	இ	ஈ	உ	எ	ஐ	ஏ	ஒ	ஒ
Accuracy achieved	92	94	86	84	90	88	96	84	82	84

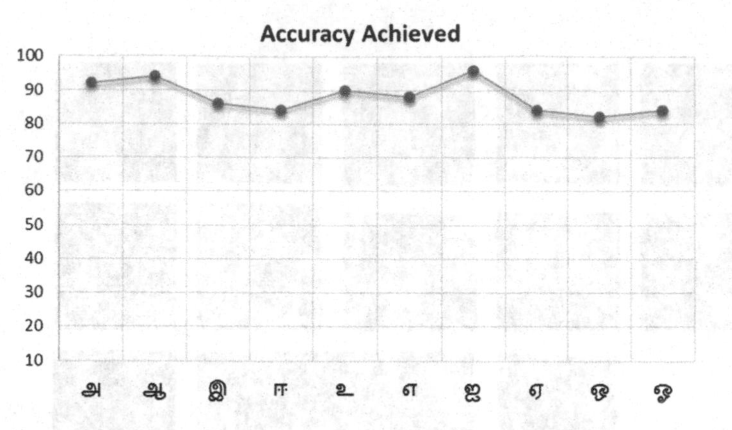

Fig. 12 Graph analysis of accuracy

Table 2 Pros and cons of strip tree procedure

S. no.	Pro	Cons
1	Good to address on complex shapes	Depends only on shape
2	Good procedure to recognize minimal characters	Not enough to address all shapes and all characters
3	Considers more and high variations	Similarity features occur for different characters in certain cases
4	Tree matches with the nature of characters	Difficult to address more similar shapes
5	Character complexity and writer complexity are considered well	Depends more on feature pre-extraction procedure

5 Conclusion and Feature Works

Feature extracted using Strip tree highly suits for representing the curvy and complex shapes of the characters. Through this hierarchical representation, the features extracted are passed to decision tree algorithm. We achieved a good result using this representation of structural feature extraction. This work will be continued further, for analyzing the all the characters in Tamil character set in detail with improved accuracy. Further research can also be done for finding the spatial location of the shapes also.

References

1. Raj, M.A.R., Abirami, S.: A survey on Tamil handwritten character recognition using OCR techniques. In: The Second International Conference on Computer Science, Engineering and Applications (CCSEA), 05, pp. 115–127 (2012)

2. Raj, M.A.R., Abirami, S.: Analysis of Statistical Feature Extraction Approaches used in Tamil Handwritten OCR. In: 12th Tamil Internet Conference- INFITT, pp. 144–150 (2013)
3. Raj, M.A.R., Abirami, S.: Offline Tamil handwritten character recognition using chain code and zone based features. In: 13th Tamil Internet Conference-INFITT, pp. 28–34 (2014)
4. Abirami, S., Manjula, D.: Feature string based intelligent information retrieval from Tamil document images. Int. J. Comput. Appl. Technol. **35**(2/3/4), 150–164 (Special Issue on Computer Applications in Knowledge Based Systems)
5. Raj M.A.R., Abirami, S.: Offline Tamil handwritten character recognition using statistical features. AENSI J. Adv. Natl. Appl. Sci. **9**(6), 367–374 (2015)
6. Shyni, S.M, Raj, M.A.R., Abirami, S.: Offline Tamil handwritten character recognition using sub line direction and bounding box techniques. Indian J. Sci. Technol. **8**(S7), 110–116 (2015)
7. Raj, M.A.R., Abirami, S.: Hybrid features based offline Tamil handwritten character recognition. In: 14th International Tamil Internet Conference, pp. 360–370 (2015). ISSN 2313-4887
8. Ballard, D.H.: Strip trees: a hierarchical representation for curves. In: ACM, Graphics and Image Processing, vol. 24, pp. 310–321, (1981). ISSN 0001-0782
9. Amin, A.: Recognition of printed Arabic text based on global features and decision tree learning techniques. Pattern Recogn. Soc. **33**, 1309–1323 (2000)
10. http://lipitk.sourceforge.net/hpl-datasets.htm

ICT Enabled Homes for Senior Citizens

Archana Singh, Jaya Bajpai and Sweta

Abstract The senior citizens are the group which often faces difficulties in the later phases of lifespan. The statistics states that about 95 million people in India are above the age of 60. By the year 2025 there would be almost 80 million more old age mass. The changing family values, economic compulsions of their children, abuse and neglect have caused the elders to come through the net of family concern. So there has been a hike in the old age home in India. This research primarily focused the senior citizens using modern technologies which can help them actively participate in the modern world without any problem. The primary objective of the paper is to know the issues of usage of ICT enabled appliances for senior citizens and widely distributing the computer literacy in this group. The paper focuses on the ICT equipped home for old ages which can bring new excitement of living in senior citizens.

Keywords Senior citizens · Old age homes · ICT · Gadgets · Applications

1 Introduction

There is an increase in the percentage of senior citizens as compared to the previous years in developing countries like India. Due to the degradation of health with age, this category requires more help and attention to their needs. The gap between generations has led to various other issues in elderly people care. Like the advancement in the technology and gadgets they use. India, like developing countries with increased number of population every year faces the problems of resources. The availability of resources is very limited. The extant research shows that the needs of old age are not limited to health but entertainment and social involvement also.

ICT has played an important role in connecting people and providing services in just a click of push button. The paradigm shift to e-medicine, e-monitoring, e-assistance, e-advice, e-suggestions, e-healthcare in totality has evolved the concept

A. Singh (✉) · J. Bajpai · Sweta
Amity University Uttar Pradesh, Noida, India
e-mail: archana.elina@gmail.com

S.C. Satapathy and S. Das (eds.), *Proceedings of First International Conference on Information and Communication Technology for Intelligent Systems: Volume 2*, Smart Innovation, Systems and Technologies 51, DOI 10.1007/978-3-319-30927-9_54

of digitization of every need and facility. The invasion of sensor technology has been used to monitor the healthcare readings of individuals and transferring the data on the internet. The online shopping trend enabled e-commerce very popular. The purchase from grocery items to gold items has made life easy and cost-sensitive. The technological switch from online solution stop has given a new dimension to life. The availability of internet connection in every household has provided a gateway to online services and products. ICT invaded in almost all the areas but still the challenging gap is to make elderly generations as their users. ICT enabled care centers for old age is still a challenging task from various points of views. In this paper we have discussed the issues and challenges in the ICT enabled care centers established for old age generation. The blend of technologies in ICT like audio-visual, phone, internet, sensors, and wireless technologies actually impacted the routine life processes.

2 Literature Review

The body of knowledge explored in Europe [1, 2] and imparted the understanding about measurements, advancements and difficulties bound with senior citizens. The study proposed the idea of creating the lodging of new houses of seniors in the same locality. The marketing hub should be in their surroundings. The study agreed with the familiarity of objects in the same society where individuals spent their real years of life. This paper let us know that as more established individuals have a tendency to invest more energy in their homes with climbing age and wellbeing confinements, the age suitability; which may be the area, and outfitting; of the living circumstance and age-broke down outline of the rural domain are the way to keeping up non reliance and personal satisfaction. The paper [3–5] further lets us know that at present, the lack in adjusted living territory and age-broke down presents economy and legislative issues with astounding difficulties. Additionally, in the event that we take a gander at the present computations, around 2.5 million obstruction free private spaces are deficient in Germany in the close term and in the medium-term and the number ascensions to 3 million.

The paper uplifts the way that they have to flourish cheerfully in their own homes instead of in consideration focuses or different establishments as it may raise numerous snags and difficulties that should be thought seriously about. The paper reaches out to stating that is an almost no distinction between an innovation that advances autonomy and innovation that kind of undermines the individual flexibility. The need to regard human rights must be a key thought. As the innovation extends and develops, there emerges a requirement for an argumentation identifying with the ethical and moral issues concerning the utilization of that specific innovation which most likely incorporates the approaches to guarantee that fundamental human rights are not traded off just by the presentation of new innovation.

The primary point of this paper [2, 6, 7] is to teach more established individuals in four taking an interest nations, that is, Slovenia, Ireland, UK and Austria in

Information and Communication Technologies (ICT) abilities and practices by utilizing a between generational and multi-sectoral approach that help the more seasoned individuals to utilize ICT on regular and hence enhancing their nature of living. It additionally helps them to revamp and rectangle them in the general public. This undertaking is a vital augmentation of the effectively completed EU Socrates—Grundtvig venture 'Advancing the Involvement of Older People in Partnership Learning Experiences—POPPIE'. In the POPPIE venture specialists utilize the participative methodology which implies the more seasoned individuals could choose how they need to be included into the examination exercises in regards to their utilization of the innovation. The outcomes [8, 9] demonstrated that more established individuals unmistakably displayed an enthusiasm to utilize ICT in their day by day exercise. ICT is making people information equipped while on move.

3 Issues and Challenges in Senior Citizens

The issues and challenges should be addressed by keeping in view of physical health, economical, old age individual machine interactions, acceptance levels, and physical abilities. Be that as it may, among the diverse sorts of classifications that should be incorporated into most ICT assessments in elderly care it is vital likewise, Functional (change of procedures, efficient for information preparing, lessening in various treatment of archives, clinical choice bolster, coordination of administrations). Flexibility for innovation framework (unwavering quality, propriety for application and capacities, security, human–machine cooperation issues). Learning curve and orientation

- *Physical disability/weakness*:
  Elder people, both men and women are known to suffer from bad health which gets worsen with the period of time. The main focus of ICT should be around the keeping the elderly people far from any sort of physical misery and try reduce the disease and disorder rates.
- *To kill the time*:
  Elderly people find it very difficult to pass their times specially in a situation where they are away from their near ones. ICT needs to seriously consider this problem and find a technical solution that may help this age group.
- *Generation gap*:
  This gap can be identified as when two age groups begin to see the world from significantly different perspectives. ICT should play its role in eliminating the wider generation gaps that prevail between the veterans and the younger group through new technological advancements.

- *Not open to learn*:
ICT should provide the elders with flexible operations and options in their devices so that they find it totally easily to use a specific tool.
- *Slow Brain Processing*:
ICT needs to understand the demands of the elderly people wherein they need to be recognized as a group that is bit slow in learning and processing.

4 ICT as a Facilitator in Senior Citizens

See Tables 1 and 2.

Table 1 ICT as a facilitator in old age homes

ICT in healthcare	Knowledge driven healthcare services like ambulance, emergency etc. Medicinal information Disease symptoms, treatment and causes discussion forums, public view educating patients Choosing a doctor Admission/discharge related information in the hospital Availability of beds Awareness programs about various diseases information like dengue, malaria etc.
ICT in current Affairs	e-Tutorials, e-Newspaper, e-polls e-Political discussion forums, timely information
ICT in entertainment	Mobile social app in connecting with people. Mobile TV, Old movies and old songs Games, friend chats, group discussion old movies, favorite actors chats Sports and its current ontime activites. Matrimonial website, partner related information
ICT in social Connect	Chat sessions with family and friends. Social Networking website like Facebook would connect old people with their school, college and village friends E-greetings, e-gifts
ICT in spiritual Knowledge	Mythological stories Spiritual gurus gyan and information
ICT in security	Sensor based security system alerts. Monitoring of intruders or infiltrators
ICT in services	e-Banking services Food recipe related data and menu New food preparations can be done as a time pass of various types like continental, Chinese etc. Online shopping of goods

Table 2 Proposed design of ICT enabled old-age homes

Technology	Operation
Gadgets	Gadgets should be user friendly as well as fancy so that they catch the eyes of the elderly people
Walking chair	A walking chair that could be controlled by buttons. We can add gesture motions to it once the advancement takes over
Digital crane	A crane that could direct them to the place that they need to go. It can be equipped with GPS if needed.
Chat-room	A proper chat-room with all the LED and required technology so that they can interact with the modern world and have a say in their own social group.
Shop online	They need to be educated as to how to place an order online. Since it is very difficult for them to move out and shop. Teaching them how to shop online would be an easy and affective solution. While shopping online, they must always have a "cash on Delivery" option so that they don't need to have to face the complexity of credit and debit card billings.
Talking medicines	This component of talking medicines should be equipped with an in built sensing which could tell them the time of a particular medicine to consumed.
Digital albums and diaries	They need to have digital albums and digital diaries so that they can have them handy without really taking care of the stuff physically.
e-Newspapers	They should have an easy access to e-newspapers. They should have some entertainment and recreation, which may involve having their favorite playlist on their walking chair so that they could play it timely whenever they want.

5 Knowledge Extracting Techniques

There are various techniques by which knowledge can be extracted from the unprocessed data of senior citizens using ICT enabled devices. The knowledge extracting techniques are data mining, machine learning, fuzzy logic, neural networks and so on. Data mining is the technique, popularly used to extract knowledge and information from the unprocessed data. It has various classifications, clustering and pattern finding methods to group the same kind of information and to generate patterns out of it. According to the perception and usage of ICT enabled kit by senior citizens the clusters, patterns and useful knowledge can be extracted. This knowledge is useful in making customized ICT enabled devices.

6 Conclusion and Future Scope

The research article addressed the issues; problems of old age citizen or senior people. The advancement of technology is moving very fast and in the young generation they always see some mobile phone or laptop, tab etc. The major issue of old age is to kill the time, for that ICT enabled home age was suggested in the paper. ICT can act as a facilitator to give new meaning to the life of older generation. The paper explored the features of the new technological old age home, keeping in mind the needs and challenges of old age group.

In future data mining techniques can be applied to mine the real time data to extract knowledge, usefulness, perception of old generation towards this ICT enabled old age home.

References

1. Benton, M., et al.: Aiming higher: Policies to get immigrants into middle-skilled work in Europe. MPI (2014)
2. Stula, S.: Living in old age in Europe-current developments and challenges. German Association for Public and Private Welfare (DV), Berlin (2012)
3. Adecco Institute.: Waking up to Europe's Demographic Challenge: The demographic fitness survey 2006. Adecco Institute White Paper (2006)
4. Arend, M.: National policies and activities addressing ICT and work-related active ageing. Presentation to Workshop on ICTs and Active Ageing in Work and Employment, Brussels, 17 Oct 2005
5. Cullen, K.: Main themes for policy as suggested by current evidence base. Presentation to Workshop on ICTs and Active Ageing in Work and Employment, Brussels, 17 Oct 2005
6. McKechnie, B.: Recruiting and retaining older workers in Scotland. Presentation to Workshop on ICTs and Active Ageing in Work and Employment, Brussels. Senior Studies Institute, University of Strathclyde, 17 Oct 2005
7. University of Sheffield (Ed.).: Futurage. A roadmap for ageing research (2011)
8. New Millennium Research Council.: Great expectations: potential economic benefits to the nation from accelerated broadband deployment to older Americans and Americans with disabilities (2005)
9. Prime Minister's Office Finland.: Ageing report. Overall assessment of the effects of ageing and the adequacy of preparation for demographic changes (2009)

Different Visualization Issues with Big Data

Koushik Mondal

Abstract We are midst of digital inclusion era, where different technologies around us, providing a wide platform of engagement with meaningful data visualizations. That engagement demands rational sense from end-users' to handle data in more sensitive manner. Large datasets for research need effective tools for data capture, curate them for designing appropriate algorithms and multidimensional analysis for effective visualizations. Effective use of ICT will help us a lot to curve with societal problems in multidimensional way. While exploring different activities around an event, selecting trusted sources using different visualization cum analysis tools are a handy option for journalists, government and common people. The complexity and volume of the data produced by an event remains largely untapped. Exploratory Data Analysis (EDA) with proper visualization techniques helped us a lot to demonstrate our ability to build an environment for heterogeneous large volume datasets. Different disciplines and data generation rates of different lab experiments, online as well as offline make the issue of creating effective visualization tools a formidable problem. Our main aim is to analyze and summarize large datasets in a concise manner with or without help of any statistical tools and produce the results using different visualization techniques In this paper we will discuss about different data intensive visualization tools, trends of different emerging technologies, how big data processing heavily relying on those effective tools and how it helps in efficient decision making for the society.

Keywords Exploratory data analysis · Multi dimensional data · Scalable framework · Data science · Machine learning · Big data · Visualization · MOOC

K. Mondal (✉)
Indian School of Mines, Dhanbad, India
e-mail: gemkousk@gmail.com

S.C. Satapathy and S. Das (eds.), *Proceedings of First International Conference on Information and Communication Technology for Intelligent Systems: Volume 2*,
Smart Innovation, Systems and Technologies 51, DOI 10.1007/978-3-319-30927-9_55

1 Introduction

Exploratory Data Analysis (EDA) allows the end-users' to see the patterns and relationship of a large dataset through visualization with or without the help of statistical methods. It finds the causes and efforts behind the relationships through visualizations. Thus we discussed EDA here before entering into visualization details. The role of the researcher in EDA, as compared to Confirmatory Data Analysis (CDA), is to explore datasets as many ways as possible until they reached in some conclusive decisions. John Tukey proposed EDA in [1]. During Massive Open Online Course (MOOC) through Coursera, Roger D. Penn described Exploratory Data Analysis (EDA) in the following manner [2]: "EDA may be compared as a rough cut of a film, after shooting different footages of the film, in the editing room". In similar note we can described it as: In a data driven project, EDA is the process of identifying crucial data for finding its relationships among variables that are particularly unexpected and interesting. It also include findings such as if there is any evidence for or against any established hypothesis, checking if there is any data related problems (such as measurement error or missing data), or identifying crucial areas where more data need to be collected. Velleman, a student of John Tukey, and Hoaglin proposed four basic elements of EDA as follows: Data visualization; Residual analysis; Data transformation or re-expression and Resistance procedures [3]. Visualizing large datasets via different graphical tools is an important stage in data analysis. It is required to understand the basic properties of the data, to find simple patterns in data and to suggest possible modelling techniques. Principle of analytical graphics stands on six pillars as suggested in [2]. They are as follows: show comparisons, show causality, mechanism and explanation; show multivariate data; integrate multiple modes of evidence; describe and document the evidence and content is king.

We are using a unified framework for visualization modelling through data driven experiments with the help of important features of large datasets. As we know the main aim of data science or big data modelling is to gain insights into data through computation, statistics and visualization (Fig. 1).

The depicted epicycle is an extension of [4], where middle cycle proposed as visualization cycle, through which we can envisage data analysis process in a more effective manner. Volume, Velocity, Variety, Veracity and Value—these five "V" has often been used to describe the issues at hand in big data world. Big data referred to the data driven analysis approach on these "big" amount of data in order to gain scientific insights, increase situational awareness, improve services and generate economic value. EDA helped us to build effective visualization by using re-expression mechanism in a large dataset and able to analyze those large scientific datasets at very beginning of the data analysis pipeline.

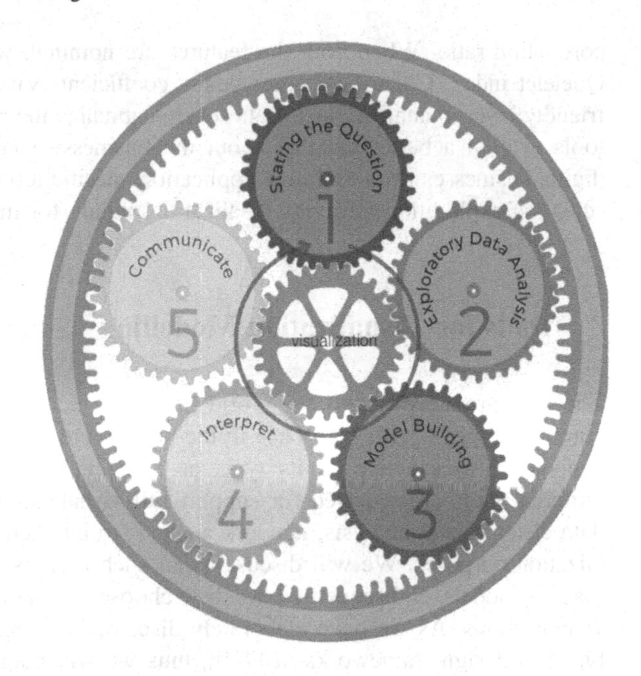

Fig. 1 Visualization as a centre circle of data analysis, extension of [4]

2 Visualization in Data Analysis

In big data analysis, visualization of appropriate graphs is very useful to decide next course of action with the data. With the help of graph, we will find basic properties of data, understand pattern of the data and able to design modelling and debug strategies for the data. Visualization helped us in finding correlation and summarization of large set data. Summarization helped us in building concepts and it will help us in deriving new relations among concepts by analyzing correlation between various features of the dataset. The naked eye is a popular and powerful tool to generate key insights from preprocessed big datasets.

The underlying scientific tools for visualization those are available in the big data domain for aggregate, categorical, numerical and/or sample visualization are described in [4–6]. The new challenges that driven the recent research is to model such kind of hypothesis functions that will able to visualize and predict unseen data related questions accurately. The new trend demands criteria like—large offline or streaming data, network data and good generalization model design have led to the data visualization research activity in the next level. We have usually classified data, based on features, in one-dimensional (1D), two-dimensional (2D) and multi dimensional or multivariate. In general two-dimensional representations, the following conventions are popular to present the data. When both features are quantitative, we use: scatter plot, correlation and regression. When one feature is quantitative and other one is categorical, we use: box plot, tabular regression and correlation ratio. When both the features are nominal, we use: contingency table, Quetelet index, Chi-square contingency coefficient. Visualization adds end-users' friendly layer to analyze those data sets to capitalize the power of machine learning tools to gain a better insights about the business/social trends. The "e-way" of digital businesses need real time application specific tools for decision-making and scientific exploration through visualization mainly for multivariate data.

3 Tools for Visualization Modelling

The data produced across the world has been increasing exponentially and will continue to grow with hard pace in the near future. Now we can see a lot of data generation, irrespective of its type, at every field of science and engineering. Different tools are required for support full visualization pipelines like statistical data cleaning and analysis, network analysis to interactive graphics for pure visualization purposes. We will discuss about such toolkits and its underlying mathematical models. Subsequently, we will choose our models for data analysis and visualizations. As we have elaborately discussed different data analysis tools and big data design frameworks in [7–9], thus we will mainly focus on visualization tools with a brief note of mathematical modelling, as and when required. Infusing computational intelligence for better visualization is the key to getting success in data-intensive world.

R [5] is now one of the best programming language in the data analysis world. Specifically, it is free, open source, platform independent, has an excellent ease of use with broad community of users to continue to expand, and with all the libraries available is usually create a one-stop-place for all data analysis needs. Different toolkits like: lattice [6], ggplot2 [10], Hadley [11], RColorBrewer [12], googlevis [13] add more crucial features in R to represent multivariate data. Pentaho Data Integration (PDI) [14] is a powerful, easy to learn open source ETL tool that supports acquiring data from a variety of data sources including flat files, relational databases, Hadoop databases, RSS Feeds, and RESTful API calls. It can also be used to clean and output data to the same list of data sources. Though Pentaho doesn't have built-in R module, still it is popular one for data acquisition. Tableau [15], GGobi [16], Improvise [17], ParVis [18], TimeSearcher [19] and TreeMap [20] are other popular data analysis tools.

Gephi [21] is very popular network visualization and manipulation tool for studying in details of the networks like: Web and Internet, social networks, biological networks and Infrastructure networks. Network Analysis focuses on the relationship between entities. Whether the entities are faculty members, staff members, students, researchers, learning objects or ideas-network analysis attempts to understand how the entities are connected rather than knowing the attributes of the entities. Gephi equipped with different popular measures include density, centrality, connectivity, betweenness and degrees. Gephi supported asynchronous bulk processing. As we aware that asynchronous bulk processing is mainly used to address graph-parallel execution. It is harder to build as race conditions can happen all the time. We have to take care of fault tolerance issues and need to implement scheduler over vertices as vertices see latest information from neighbours. If we compare it with the matrix domain problem, it is similar to Gauss-Seidel iteration. The convergence rate is quite high in asynchronous bulk processing. We have to keep in mind during designing all pairs shortest path using Floyd-Warshall algorithm in Gephi is to avoid negative weights in the edges of the graph. In [22], author proposed a trick to avoid the same. First we have to use the Bellman-Ford algorithm [23] to detect negative cycles and will only move ahead with Floyd-Warshall algorithm if the input graph has no negative cycles. It will reduce the processing time and put barrier in the overflow. Another mathematical aspect in effective visualization is to show the extent of linearity between two features. The linear regression involves a useful and very popular parameter, the correlation coefficient, which shows the same. The correlation coefficient, in general, ranges between -1 and 1, and a value close to 1 or -1 indicates a high extent of the linear dependence between the features. In physics or chemistry, a high value of the correlation coefficient is rather usual but in social sciences lower value of correlation coefficient is highly expected and it gives a true direction. NodeXL [24], GUESS [25], Pajek [26], NetworkX [27] and SNAPP [28] are some popular network analysis tools.

Processing [29] and D3 [30] are two very popular visualization tools. Processing is a popular Java like programming language with rich graphics features. It also has a strong user community. D3 is a JavaScript library for data-driven DOM manipulation, interaction and animation. It includes utilities for different visualization techniques and SVG generation. Protovis [31], PolyMaps [32], Flare [33], Prefuse [34], Improvise [17], VTK [35] and InfoVis [36] are some notable toolkits which focus on graphics and interaction.

4 Results and Discussions

We have carried out our experiments in R with diamonds dataset with ggplot2, bigvis, Rcpp to glimpse the power of these open source data visualization tools. The dataset contains 53,940 observations of 10 variables. If we draw carat vs. price plot using basic plot function we will receive the same as in Fig. 2.

Now if we combine ggplot2 features with the basic plot, we will get more clarity in the visualization as in Fig. 3. As we know, with limited number of pixels it is very difficult to recognize data points in an effective manner. Moreover, if we want to process large data sets using in-memory technique for speed up then it is quite hard to handle those many data points. Thus Hadley [11] proposed bigvis package

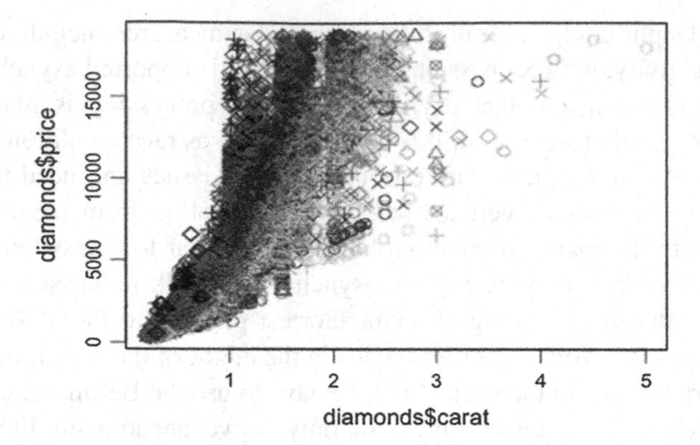

Fig. 2 Visualization using simple plot function

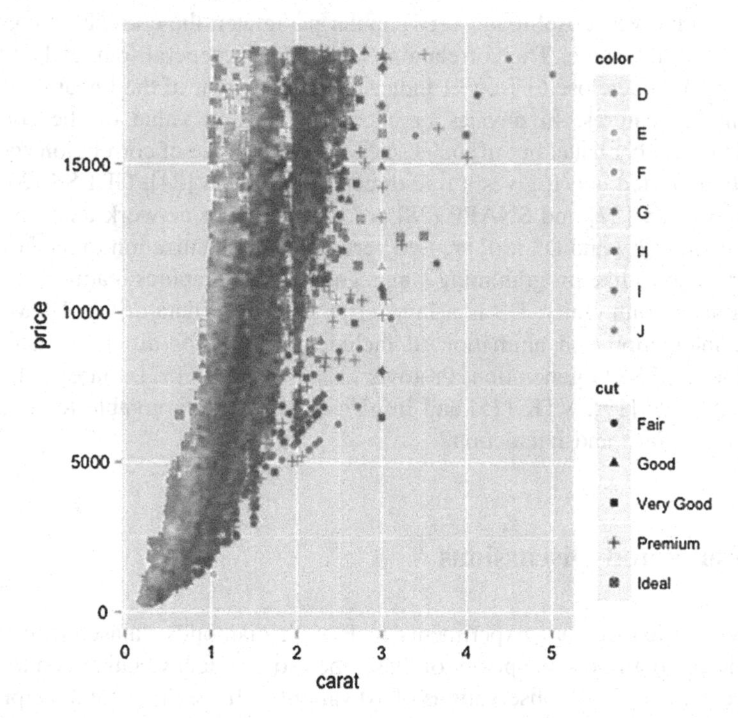

Fig. 3 Visualization using ggplot2 function

using "bin-summarise-smooth" technique. The basic idea is to summarize and aggregate the big data sets before plotting so that it will be able to render graphical visualization at ease. It will also produce meaningful insights through visualization as depicted in Fig. 4.

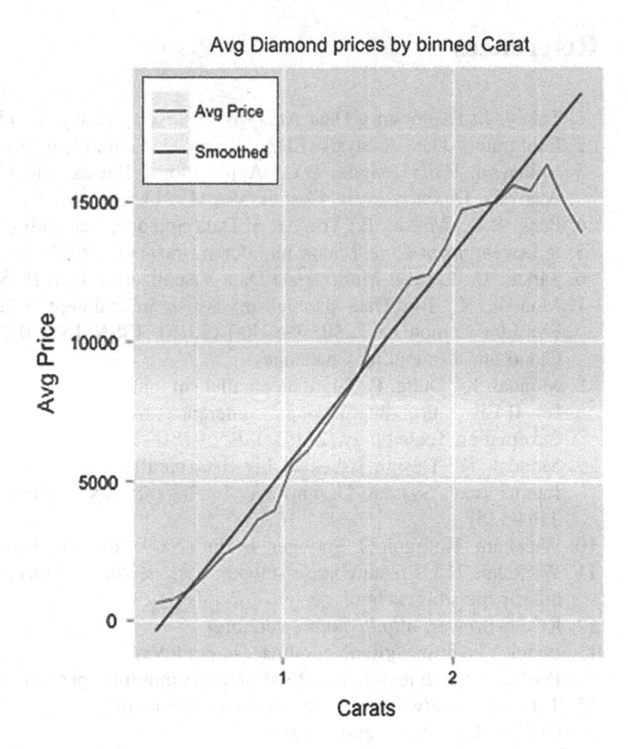

Fig. 4 Visualization using bigvis package

5 Conclusions

Information visualization is sometimes described as a way to answer questions we didn't know that we already had. The scale of displayed information and layout were chosen to support observed behavior and capabilities and allow users to expand visualized data while maintaining crucial relationships among them. The visualization challenges posed by different graphics oriented corroborative settings are now handled by principles of conveying information, efficient data management and different effective toolkits. There is hardly any single visualization frameworks, tools and toolkits through which we are able to handle all visualization based shortfalls, which are common in most of the large set data. Visual literacy is an increasingly important skill, and data visualizations are another channel for us to develop our ability to process information effectively. Local feature-based subdivisions approach sometimes offers edges over traditional visualization analysis techniques. We have created some subsets based on features during creation of binning so that it provides some better insights.

References

1. Tukey, J.: Exploratory Data Analysis. Addison-Wesley, Reading, MA (1977)
2. Exploratory Data Analysis (EDA), Roger D. Peng, Coursera (2015)
3. Velleman, P.F., Hoaglin, D.C.: Applications, Basics, and Computing of Exploratory Data Analysis. Duxbury Press, Boston, MA (1981)
4. Peng, R.D., Matsui, E.: The Art of Data Science, Leanpub (2015)
5. R Development Core Team.: http://cran.r-project.org (2009)
6. Sarkar, D.: Lattice: Multivariate Data Visualization with R. Springer, Berlin (2008)
7. Mondal, K.: Big Data parallelism: issues in different x-information paradigms. Elsevier Procedia Comput. Sci. **50**, 395–400 (2015). ISSN 1877-0509 (Special Issue on Big Data, Cloud and Computing Challenges)
8. Mondal, K., Dutta, P.: Big data parallelism: challenges in different computational paradigms. In: IEEE Third International conference on Computer, Communication, Control and Information Technology (2015). ISBN 978-1-4799-4446-0
9. Mondal, K.: Design issues of big data parallelism. In: Third International Conference on International System Design and Intelligent Applications (INDIA 2016) (2015) ISSN 2194-5357
10. Wickham, H.: ggplot2. Springer, Berlin (2009). http://had.co.nz/ggplot2/
11. Wickham, H.: Bin-summarise-smooth: a framework for visualising large data. http://vita.had.co.nz/papers/bigvis.html
12. RColorBrewer, http://colorbrewer2.org/
13. googleVis, https://github.com/mages/googleVis
14. Pentaho Data Integration.: (2010) http://community.pentaho.com/projects/data-integration/
15. Tableau, http://www.tableausoftware.com/student/
16. GGobi.: http://www.ggobi.org/
17. Improvise.: http://www.cs.ou.edu/~weaver/improvise/
18. Parvis.: http://home.subnet.at/flo/mv/parvis/
19. Timesearcher.: http://www.cs.umd.edu/hcil/timesearcher/
20. Treemap.: http://www.cs.umd.edu/hcil/treemap/
21. Gephi.: http://gephi.org/
22. Hougardy, Stefan: The Floyd-Warshall algorithm on graphs with negative cycles. Inf. Process. Lett. **110**, 279–281 (2010)
23. Korte, Bernhard, Vygen, Jens: Combinatorial Optimization: Theory and Algorithms, 4th edn. Springer, Berlin (2008)
24. NodeXL.: http://www.codeplex.com/NodeXL
25. GUESS.: http://graphexploration.cond.org/
26. Pajek.: http://pajek.imfm.si/doku.php
27. NetworkX.: http://networkx.lanl.gov/
28. SNAPP.: http://snapp.stanford.edu/
29. Processing.: http://processing.org/
30. D3.: http://mbostock.github.com/d3
31. Protovis.: http://protovis.org/
32. Polymaps.: http://polymaps.org/
33. FLARE.: http://flare.prefuse.org/
34. IVTK.: http://ivtk.sourceforge.net/2008
35. VTK.: http://www.vtk.org/
36. Infovis.: www.infovis.org

Comparison and Analysis of Obstacle Avoiding Path Planning of Mobile Robot by Using Ant Colony Optimization and Teaching Learning Based Optimization Techniques

A.Q. Ansari, Ibraheem and Sapna Katiyar

Abstract Now a day, one of the prime concerns of mobile robot is path planning, in the area of industrial robotics. A path planning optimization method was proposed to calculate shortest collision free path from source to destination by avoiding static as well as dynamic obstacles. Therefore, it is necessary to select appropriate optimization technique for optimization of paths. Such problems can be solved by metaheuristic methods. This research paper demonstrates the comparison and analysis of two Soft Computing Techniques i.e. Ant Colony Optimization (ACO) and Teaching Learning Based Optimization (TLBO) by simulating respective algorithms for finding shortest path of a Mobile Robot by Obstacle avoidance & Path re-planning and Path Tracking. Both of these techniques seem to be a promising technique with relatively competitive performances. The ACO has been more widely used in that and it gives good solution with smaller numbers of predetermined parameters in comparison with other algorithms.

Keywords Path planning algorithm · Teaching learning based optimization (TLBO) · Ant colony optimization (ACO) · Metaheuristic · Pheromone and obstacle avoidance

A.Q. Ansari · Ibraheem · S. Katiyar (✉)
Jamia Millia Islamia, New Delhi, India
e-mail: sapna_katiyar@yahoo.com

A.Q. Ansari
e-mail: aqansari@ieee.org

Ibraheem
e-mail: ibraheem_2k@yahoo.com

S.C. Satapathy and S. Das (eds.), *Proceedings of First International Conference on Information and Communication Technology for Intelligent Systems: Volume 2*, Smart Innovation, Systems and Technologies 51, DOI 10.1007/978-3-319-30927-9_56

1 Introduction

Path planning of mobile robot is a very important and crucial issue in past two decades. In robotics the term path planning is used in the calculation of optimized path which a robot must travel from source to destination by covering every point and avoiding obstacles in a given environment. Obstacles may be of two types: static and dynamic. In case of static obstacle, its position is fixed and does not change with time while in case of dynamic obstacle; its position may vary with respect to time [1–3]. Obstacle avoidance means the method of identification of mobile robot path by avoiding unpredictable obstacles. Therefore the movement of mobile robot depends on its actual position and on the nearby obstacles. In the last two decades various traditional methods have been applied to solve the path planning problem. Some commonly used methods are: Roadmap Methods (Visibility Graphs, Voronoi Diagrams and Probabilistic Roadmap), Potential Fields (Attractive Potential and Repulsive Potential), Cell Decomposition Methods (Exact Cell Decomposition and Approximate Cell Decomposition).

Some of the applications of path planning are in Industrial robotics, where robots have to carry different things from one place to other by avoiding collisions and in the designing of ICs but these techniques have several drawbacks. Therefore to overcome these drawbacks researchers have move towards the application of soft computing techniques. The main features of soft computing are their capability to tolerate imprecision and inaccuracy unlike hard computing techniques. Examples of soft computing techniques are: Neural Networks, Fuzzy Logic [4–6], Genetic Algorithm, Ant Colony Optimization Algorithm, Particle Swarm Optimization, Bee Colony Optimization, Cuckoo Search, Bat Algorithm, Firefly Algorithm and Teaching learning Based Optimization Technique.

2 Optimization Algorithms

2.1 Ant Colony Optimization (ACO) Algorithm

Ant Colony Optimization algorithm is developed by Marco Dorigo in 1992. It is a metaheuristic approach. It has been successfully implemented to find out the solutions of complex optimization problems. This algorithm is very adaptive in nature and can be applied to determine the optimal paths where ants move to search the food. Initially ants move randomly by leaving pheromone on that path. Other ants follow the path which is having some increased value of pheromone [7].

The artificial ants in ACO give a probabilistic solution for a path from source to nest, which is called as a tour. The probability rule, which can be applied between any two nodes, is known as Pseudo-Random Proportional Action Choice Rule [8]. Heuristic and metaheuristic are the two components on which probability rule works.

$$p_{ij} = \frac{[\tau_{ij}]^{\alpha}[\eta_{ij}]^{\beta}}{\sum_{h \in s}[\tau_{ij}]^{\alpha}[\eta_{ij}]^{\beta}} \tag{1}$$

where, τ_{ij}: the amount of pheromone on the edge, η_{ij}: desirability of edge which is the inverse of the distance between the two cities, α and β: the parameters which control the influence of τ_{ij} and η_{ij}.

There are two possible ways via which every ant modifies the environment:

- **Local trail updating**: it means when an ant moves from one edge (city) to another edge (city), then it update the quantity of pheromone as per the Eq. (2):

$$\tau_{ij}(t) = (1-\rho)\tau_{ij}(t-1) + \rho\tau_0 \tag{2}$$

  where, ρ: the evaporation constant, τ_0: it is the initial value of pheromone trails and it can be calculated as per the Eq. (3):

$$\tau_0 = (n \cdot L_{nn})^{-1} \tag{3}$$

  where, n: the number of cities, L: tour length generated by any one of the construction heuristics
- **Global trail updating**: When all the ants have completed a tour then they are able to determine the shortest route, by updating the edges available in its path using the Eq. (4):

$$\tau_{ij}(t) = (1-\rho).\tau_{ij}(t-1) + \frac{\rho}{L^{+}} \tag{4}$$

  where, L^{+}: the length of the best path generated from source to nest

2.2 *Teaching Learning Based Optimization (TLBO) Algorithm*

This algorithm was conceived by Dr. R.V. Rao in the year 2011, which is encouraged by teaching and learning process. It is also population based algorithm similar to other nature inspired algorithms. Its concept is very simple and gives better efficiency; therefore it has been applied to a number of real time problems [9].

TLBO requires basically two parameters: population size and number of generations. Basically there are two phases in TLBO algorithm: Teacher phase (learning through teacher) and Learner phase (learning of student through interaction among students). Always it is considered that teacher is highly knowledgeable person and he teaches students. Students also gather information and increase their knowledge by the interaction among themselves. Therefore by considering all these factors into consideration, a mathematical model has been evolved for optimization, which is known as Teaching Learning Based Optimization (TLBO) Algorithm. The teaching-learning-based optimization (TLBO) demonstrates the behavior of learners in a classroom. The learners or the searching points move toward the best learner of the population, and move toward the other better learner. This study found that the adaptive teaching factor is recommended.

- **Teacher Phase**
  The first part of TLBO algorithm is a Teacher phase, where learners tend to learn from teachers. In this phase the objective of a teacher is to increase the mean result of class, for the subject taught by teacher. Let us assume, N is the number of $\mathbf{x}$ (design variables) in the population. At the current iteration t, the mean of $\mathbf{x}_t$ is the calculated by Eq. (1) designated as $\mathbf{x}_{mean,t}$. The new mean of $\mathbf{x}_t$ is the current best point or the teacher designated as $\mathbf{x}_{teacher,t}$. Therefore the difference mean $\mathbf{\Delta}_t$ can be calculated using Eq. (6).

$$\mathbf{x}_{mean,t} = \frac{1}{N}\sum_{k=1}^{N} \mathbf{x}_{k,t} \tag{5}$$

$$\mathbf{\Delta}_t = r_t\left(\mathbf{x}_{teacher,t} - T_F \mathbf{x}_{mean,t}\right) \tag{6}$$

  where, $\mathbf{\Delta}_t$: Difference mean, r_t: Random number in the range [0, 1], T_F: The teaching factor, its value can be either 1 or 2.
  In the teacher phase, all learners (searching points) are encouraged to move toward the teacher (best point) of the current population. All points $\mathbf{x}_{old,t}$ are moved to the new positions $\mathbf{x}_{new,t}$ as per the Eq. (3). But point will move to the new position only when the new position is superior to its current position.

$$\mathbf{x}_{new,t} = \mathbf{x}_{old,t} + \mathbf{\Delta}_t \tag{7}$$

- **Learner Phase**
  The second part of TLBO algorithm is Learner phase, where learns learn by interaction among themselves. Here learner interacts randomly with others and improve its own knowledge if other knows more than him. In this phase the learning phenomenon is expressed as: a point will moved to the new position (update information) only when the new position is better than its current position. Therefore,

$$\begin{aligned} \mathbf{x}_{A,new,t} &= \mathbf{x}_{A,old,t} + r_t(\mathbf{x}_{A,old,t} - \mathbf{x}_{B,t}) \quad \text{if} \quad f(\mathbf{x}_{A,t}) < f(\mathbf{x}_{B,t}) \\ \mathbf{x}_{A,new,t} &= \mathbf{x}_{A,old,t} + r_t(\mathbf{x}_{B,t} - \mathbf{x}_{A,old,t}) \quad \text{if} \quad f(\mathbf{x}_{A,t}) > f(\mathbf{x}_{B,t}) \end{aligned} \tag{8}$$

Here $f(\mathbf{x}_{A,t})$ and $f(\mathbf{x}_{B,t})$ is the objective function cost of the $\mathbf{x}_{A,t}$ and $\mathbf{x}_{B,t}$ respectively.

All points are iteratively moved until the termination criterion is reached. The best point is known and updated as the global optimum solution. After a sufficient number of sequential teaching–learning cycles the knowledge level of learner's increases towards teacher's level because teacher has shares his knowledge and experience with the learners. Therefore the distribution of the randomness within the search environment becomes gradually smaller around a contact point which is considered as a teacher. Because of this the knowledge level of the whole class becomes smooth and the algorithm converges to a solution without compromising the quality of outcomes.

3 Problem Description

The aim of the proposed research work is to apply an efficient approach for path panning so that the robot can move on a calculated shortest path to achieve target without colliding with obstacles. A grid matrix of size 20 × 20 is taken here, where source is assumed on left top side where destination is on bottom right side. Here the problem is simulated through the MATLAB software for the identified features and response was observed. The software is capable of simulating the behavior of mobile robot when moving through given points by selecting shortest possible path using ACO and TLBO by avoiding static as well as dynamic obstacles [10, 11].

The system generates shortest path using ACO and TLBO between two given point on the screen by considering static and dynamic obstacles. The mobile robot starts moving on the generated shortest path. While navigating if it detects any run time obstacle, then mobile robot stop and system again generates shortest path from current point to destination and then robot starts its movement. This process repeats until robot reaches its final destination. The aim is to obtain a collision free, time optimal navigation among static and dynamic obstacles. This makes the path finding problem to be adaptive in nature.

4 Simulation Results

Figures 1,2,3,4,5,6,7 and 8 shows the simulation results of four different cases taken as per the presence of obstacles present in the grid matrix. Table 1 shows the comparison of shortest distance by using ACO and TLBO. Table 2 summarizes

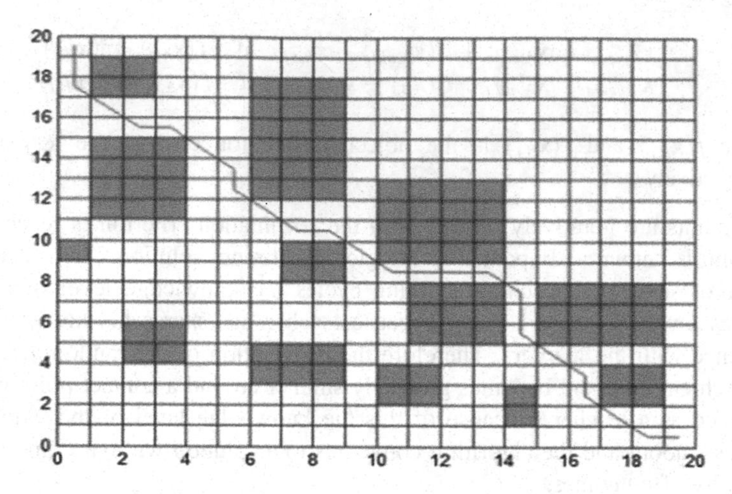

Fig. 1 Shortest path using ACO (Case-1)

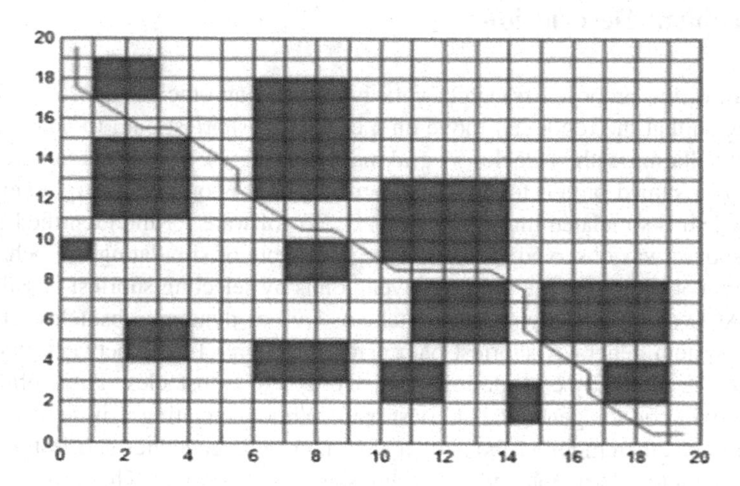

Fig. 2 Shortest path using TLBO (Case-1)

the results obtained by ACO and TLBO for best path, worst path and average path for all cases. Figures 9,10,11,12 and 13 shows the graphical results of data represented in tables. In all the four cases ACO is giving better result in comparison to TLBO.

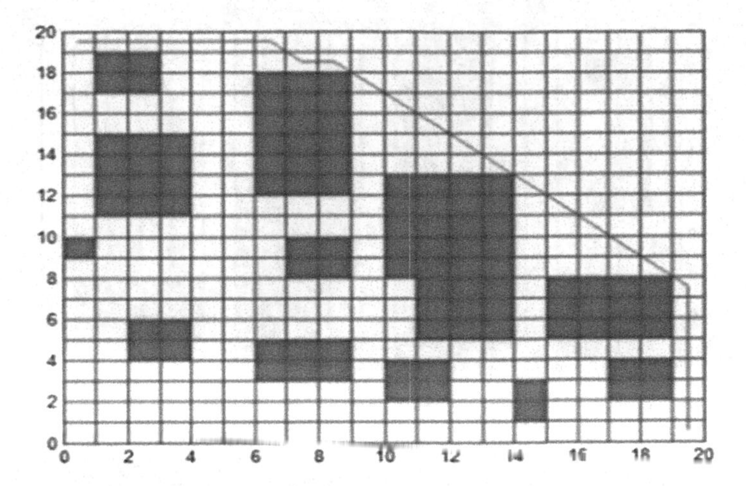

Fig. 3 Shortest path using ACO (Case-2)

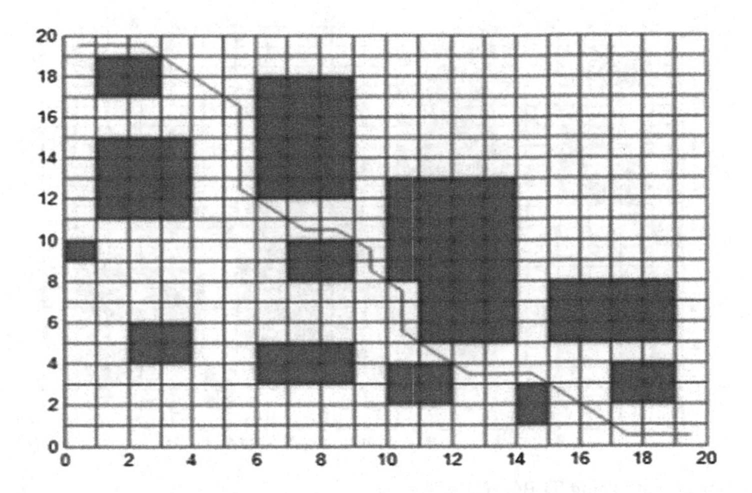

Fig. 4 Shortest path using TLBO (Case-2)

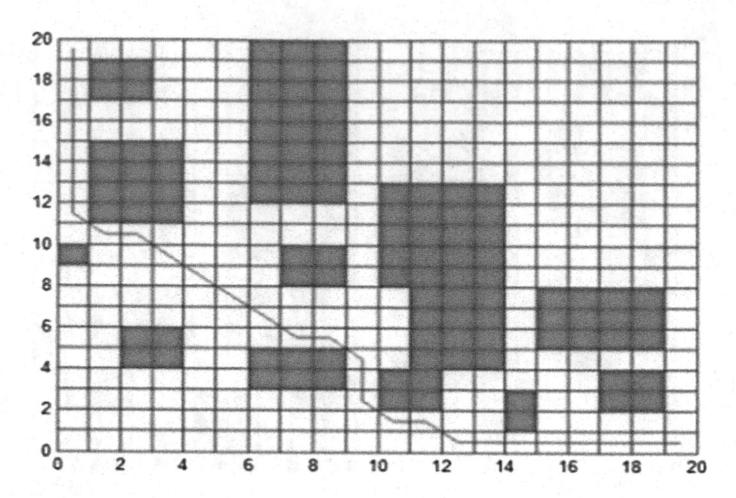

Fig. 5 Shortest path using ACO (Case-3)

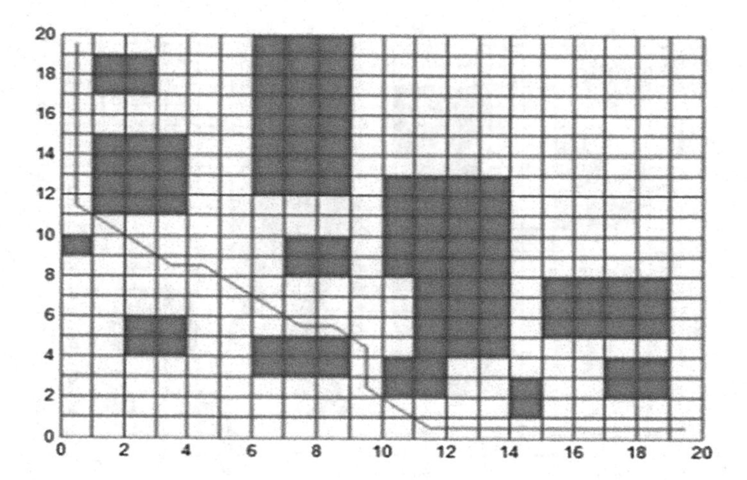

Fig. 6 Shortest path using TLBO (Case-3)

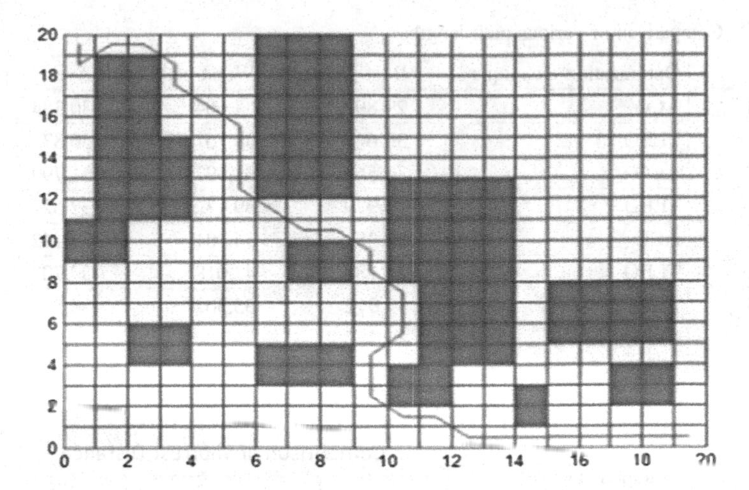

Fig. 7 Shortest path using ACO (Case-4)

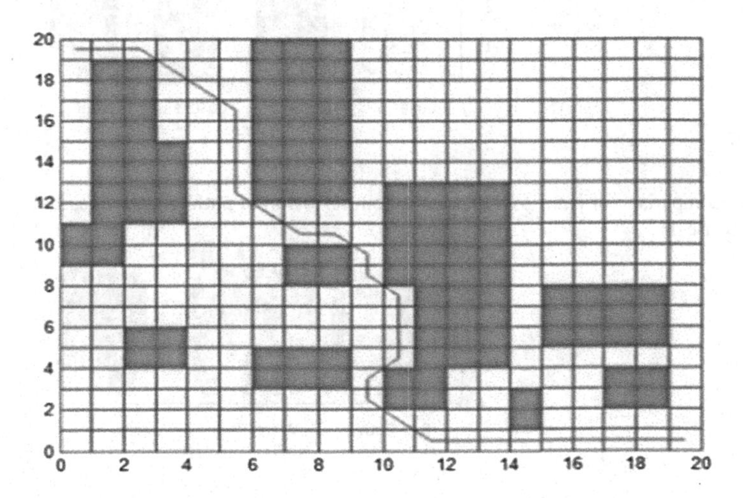

Fig. 8 Shortest path using TLBO (Case-4)

Table 1 Comparison of shortest distance using ACO and TLBO

	ACO (distance)	TLBO (distance)
Case-1	30.384	30.870
Case-2	32.627	31.799
Case-3	34.727	34.142
Case-4	35.556	35.558

Table 2 Comparison of various path lengths

Case	Optimization technique	Best path	Worst path	Average path
Case-1	ACO	29.897	32.604	30.384
	TLBO	30.649	38.761	30.87
Case-2	ACO	30.564	31.992	31.799
	TLBO	30.979	39.172	32.627
Case-3	ACO	32.381	34.901	33.142
	TLBO	34.182	42.018	34.727
Case-4	ACO	33.912	35.502	34.556
	TLBO	35.723	46.109	36.869

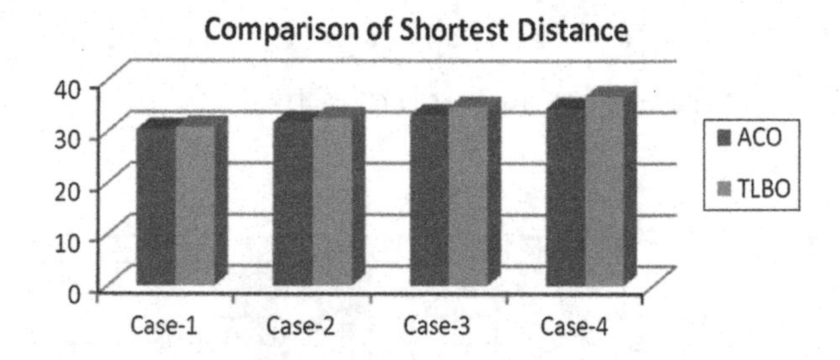

Fig. 9 Comparison of shortest path through ACO and TLBO

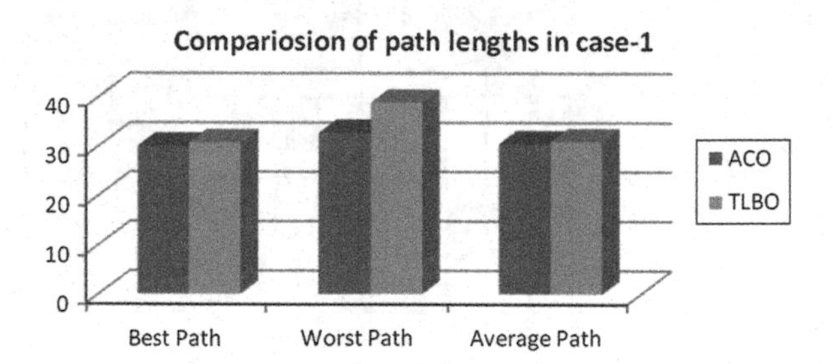

Fig. 10 Comparison of path lengths (Case-1)

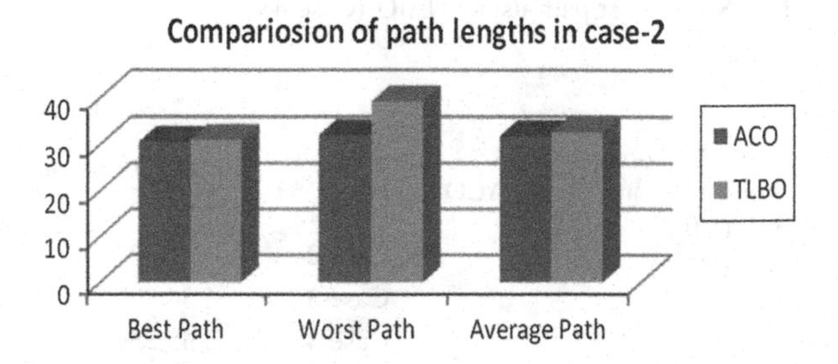

Fig. 11 Comparison of path lengths (Case-2)

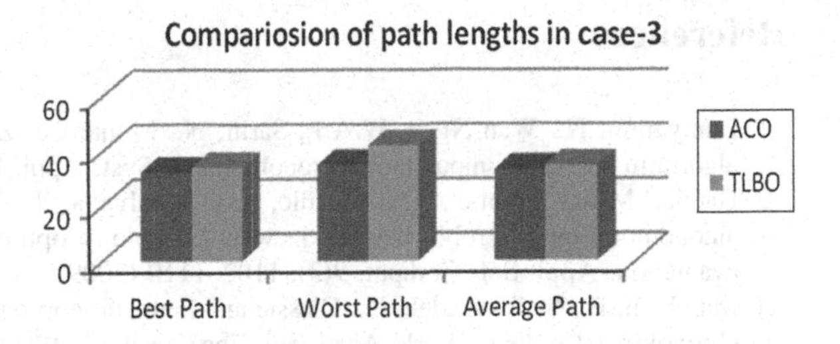

Fig. 12 Comparison of path lengths (Case-3)

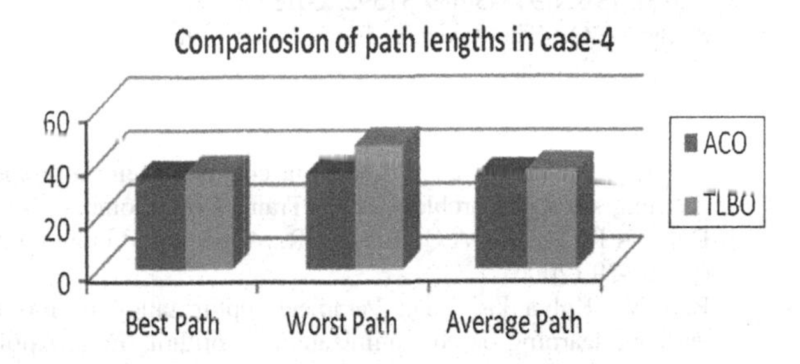

Fig. 13 Comparison of path lengths (Case-4)

5 Conclusion and Future Scope of Work

Four different cases have been presented here and it is observed that there is no obstacle during the path of mobile robot navigation then path obtained from source to destination is of minimum length. The results obtained by using Ant Colony Optimization (ACO) algorithm and Teaching Learning based Optimization (TLBO) algorithm is compared here in terms of shortest path travelled from source to destination. It has been found that ACO seems to be one of the effective meta-heuristic Optimization algorithms whereas the conventional methods give multiple solutions.

For future work, some of the following points have been identified to carry current research work:

- The same problem can also be solved by applying other optimization techniques like Particle Swarm Optimization (PSO), Firefly Algorithm (FA), Cuckoo Search (CS), Chemical Reaction Optimization (CRO) and Artificial Immune System (AIS).
- Some hybrid technique can also be developed and comparison can be done with existing techniques

References

1. Buniyamin, N., Wan Ngah, W.A.J., Sariff, N., Mohamad, Z.: A simple local path planning algorithm for autonomous mobile robots. Int. J. Syst. Appl. Eng. Dev. **5**(2), 151–159 (2011)
2. Garcia, M.A., Montiel, O., Castillo, O., Sepúlveda, R., Melin, P.: Path planning for autonomous mobile robot navigation with ant colony optimization and fuzzy cost function evaluation. Appl. Soft Comput. **9**(3), 1102–1110 (2009)
3. Masehian, E., Sedighizadeh, D.: Classic and heuristic approaches in robot motion planning-a chronological review. World Acad. Sci. Eng. Technol. **29**(1), 101–106 (2007)
4. Ansari, A.Q.: Hierarchical Fuzzy control for industrial automation. Scholar's Press, Germany (2013). ISBN 978-3-639-51592, 2013
5. Ansari, A.Q.: The basics of fuzzy logic: a tutorial review, computer education—Stafford—Computer Education Group, U.K., No. 88, pp. 5–9 (1998)
6. Siddiqui, S.A., Ansari, A.Q., Agarwal, S.: A journey through Fuzzy Philosophy. Pranjana J. Manage. Awareness **6**(2), 29–33 (2003)
7. Dorigo, M., Gambardella, L.M.: Ant colony system: a cooperative learning approach to the traveling salesman problem. IEEE Trans. Evol. Comput. **1**(1), 53–66 (1997)
8. Dutta, A.K., Ansari, A.Q., Biswas, R.: A method of intelligent search. Int. J. Comput. Cogn. **4** (2), 24–29 (2006)
9. Rao, V., Kalyanka, V.D.: Parameter optimization of modern machining processes using teaching–learning-based optimization algorithm. Eng. Appl. Artif. Intell. **26**(1), 524–531 (2013). (Elsevier)
10. Abiyev, R., Ibrahim, D., Erin, B.: Navigation of mobile robots in the presence of obstacles. Adv. Eng. Softw. **41**(10), 1179–1186 (2010)
11. Cong, Y.Z., Ponnambalam, S.G.: Mobile robot path planning using ant colony optimization. In: IEEE International Conference on Advanced Intelligent Mechatronics, pp. 851–856 (2009)
12. Katiyar, S., Mittal, A., Ansari, A.Q., Saxena, T.K.: Ant colony algorithm based adaptive PID temperature controller. Proceedings of the 7th International Conference on Trends in Industrial Measurements and Automation (TIMA 2011), CSIR, Chennai (2011)
13. Ansari, A.Q.: Multiple valued logic versus binary logic. CSI Commun. India **20**(5), 30–31 (1996)
14. Khan, M.A., Ansari, A.Q.: Fundamentals of industrial informatics and communication technologies. In: Handbook of Research on Industrial Informatics and Manufacturing Intelligence: Innovations and Solutions, 03/2012: Chapter 1: pp. 1–19; IGI Global, USA (2012)
15. Ansari, A.Q., Biswas, R., Aggarwal, S.: Neutrosophic classifier: an extension of Fuzzy classifier. Appl. Soft Comput. **13**(1), 563–573 (2013)
16. Ansari, A.Q., Khan, M.A.: Parallel and dynamic virtual channel manager (VCM) for 3-D network-on-chip (NoC) router. In: Indian Patent Journal. New Delhi, Submitted: 03/08/2011 16:07:38
17. Ansari, A.Q., Khan, M.A.: Architecture of 3-D network-on-chip (NoC) router with guided flit logic. In: Indian Patent Journal, New Delhi. Submitted: 18/01/2013 15:39:18
18. Katiyar, S., Ibraheem, N., Ansari, A.Q.: Ant colony optimization: a tutorial review. In: Proceedings of the National Conference on Advances in Power and Control, Manav Rachna International University, Faridabad, Haryana, India

An Era of Big Data on Cloud Computing Services as Utility: 360° of Review, Challenges and Unsolved Exploration Problems

Rahul Vora, Kunjal Garala and Priyanka Raval

Abstract Cloud computing is an innovation technology for supply of computing as a utility towards digital world. It is provide platform and services for the massive-scale data storage and data sharing. Big Data on cloud environment analyze, storage, manage, visualization, security are some challenging techniques that requires more timing and large computation infrastructure processing. This paper mainly brush up for an era of 360° vision of massive data on cloud services computing. The meaning, taxonomy and literature reviews for some papers of cloud along with some big data model are introduced. The association and role of deluge computing and cloudy data, big data open sources tools are also discussed. Last but not least, research challenges and unsolved exploration problems are shortened.

Keywords Cloud computing · Big data · Tools · Taxonomy · Literature review · Unsolved exploration problems

1 Introduction

Computing and Data are two most valuable and prime concepts of the today's digitally smart world. Without these two factors internet globe none exists. To help of these factors most significant technology foundation that is 'Cloud as computing for services' and 'Big tremendous data'.

Cloud or distributed Computing is an innovation. It is not invention or discovery that is new building of old elements producing a new synergy [1]. It is the provision of computing facilities over the Internet that permit to use software and hardware

R. Vora (✉) · K. Garala · P. Raval
Computer Engineering, Gardi Vidyapith, Rajkot, India
e-mail: ervora10@gmail.com

K. Garala
e-mail: kunjalgarala@gmail.com

P. Raval
e-mail: prraval@gardividyapith.ac.in

S.C. Satapathy and S. Das (eds.), *Proceedings of First International Conference on Information and Communication Technology for Intelligent Systems: Volume 2*, Smart Innovation, Systems and Technologies 51, DOI 10.1007/978-3-319-30927-9_57

that are manage at remote locations by third parties. Cloud service resources mainly lying on two aspects that (a) Access anywhere and anytime and (b) Pay-as-you-go.

Big data technologies is the today's world technologies that involve massive-scale data and complex structure. Organizations captured the endless increase in the size of data, such as the growth of IoT (Internet of Things), Social media, sensors, web logs, ERP, satellite communication and multimedia has shaped a crushing flow of data in either structured or mix or unstructured format [2]. As large computation infracture processing and tremendous increases scale of data management and analyzing are difficult in different platforms, Cloud computing provide one platform that big data analyzing, storing and managing task are easily executed.

The further of this paper is structured as surveys: Sect. 2 presents the meaning, taxonomy of cloud computing and big data. Section 3 delivers a connection between cloudy big data technology. Section 4 presents lists of open source tools for using big data on cloud environment and Literature reviews on some papers. Section 5 offers research challenges. Section 6 offers a swift of current unsolved Exploration Problems and Sect. 7 shows the conclusion.

2 Definition and Taxonomy for CC and BD

2.1 Cloud Computing: Notions and Definition

NIST (National Institute of Standards and Technology) defined Cloud computing is a model for empowering omnipresent, helpful, on-interest system access to a common pool of configurable registering assets (e.g., systems, servers, stockpiling, applications, and administrations) that can be quickly provisioned and discharged with insignificant administration exertion or administration supplier collaboration [3]. Lakshmi et al. [4] represent cloud as a utility based computing, that part can be hardware, system software or application software that can be accessed from anywhere and used anytime.

Figure 1 shows that CCQA (Cloud Computing Question Answering) model.

2.2 CC (Cloud Computing): Taxonomy

Cloud computing is depends on a datacenter-scale virtualization of registering assets, in which through the aggregate mechanization of these virtualized assets, a virtualized subset of process, stockpiling, availability, and application/middleware administrations are cut out to serve as a virtualized computing got to via a network [5].

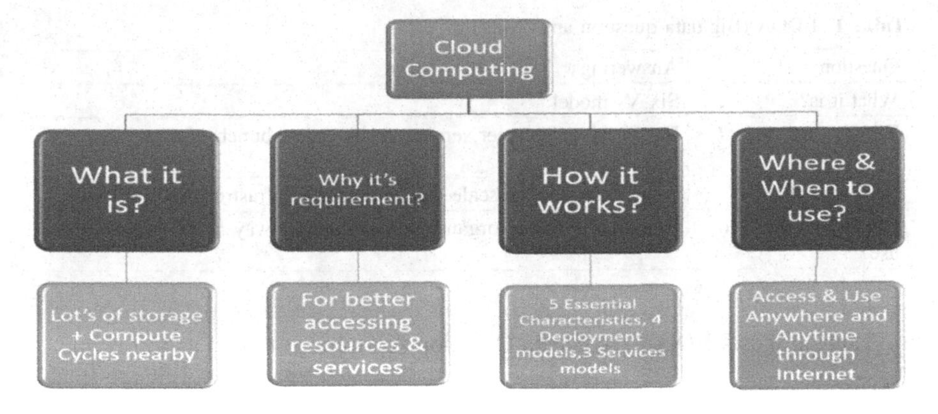

Fig. 1 CCQA (Cloud computing question answering) model

2.3 Big Data: Notions and Definition

According, IEEE CS 2022 report [5] big data is blasting, without any indications of backing off, that development is show on two separate axes: more information is gathered, and more information is shared. According to [6] the unpredictable way of enormous information is basically determined by the unstructured way of a significant part of the information that is created by modern technologies.

Variety, Volume and Velocity (three V's) presented by Gartner researcher Doung Laney [7]. Hashem et al. [2] defined Big Data in terms of four Vs: Capacity, Swiftness, Diversity and Value. Assunção et al. [8] defined Big Data by five Vs, added as Veracity. According to [9] big data characterized by six Vs, added as Variability. Figure 2 shows that six Vs of big data architecture.

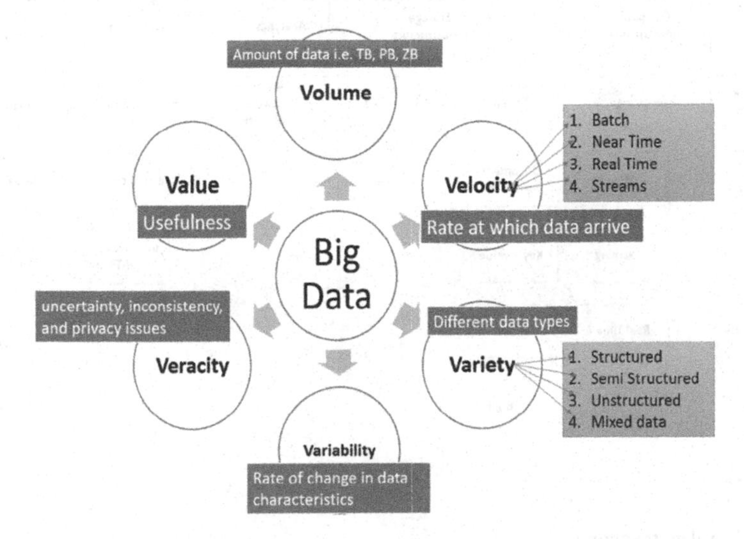

Fig. 2 Six Vs of big data architecture

Table 1 BDQA (Big data question answering) model

Question	Answering
What it is?	Six Vs model
Why it's important?	For reactive customer service, easily access bunch of data, storage of data, lesser costs
How it works?	Analyzing massive scale data to compute Infrastructure processing
Where and When to use?	Every I.T Firms & organization or digitally way at any time

Table 1 shows BDQA (Big Data Question Answering) model as following:

2.4 BD (Big Data): Taxonomy

The term big data mentions to the massive volume of digital measure of advanced data organizations and governments gather about us and our surroundings by customary data trade and programming use by means of desktop PCs, cell telephones and so on [10]. The taxonomy of BD is shown in Fig. 3. In that given the scientific categorization's start of the enormous information landscape along six of most critical measurements as data domains, compute infrastructure, storage architectures, analytics, visualization, security and privacy.

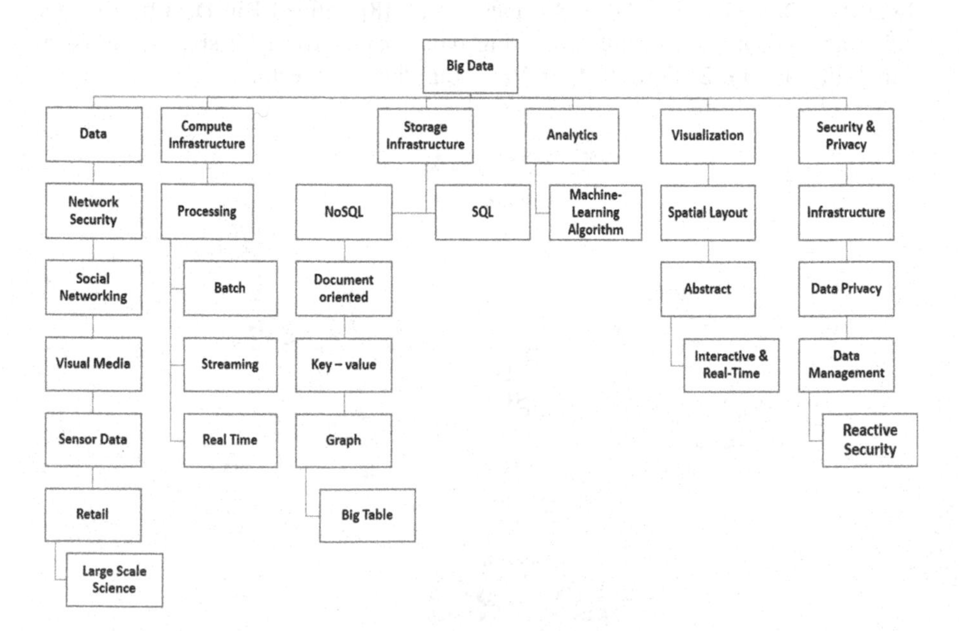

Fig. 3 Big data taxonomy

3 Connection Between CC and BD

Traditional IT architecture is incapable of handling the big data problem, as there are many bottlenecks, such as: poor scalability; poor fault tolerance; low performance; difficulty in installation, deployment, and maintenance; and so on [11]. Cloud computing is a trend in technology development, while big data is an inevitable phenomenon of the rapid development of a modern information society, to solve big data problems, we need modern means and Cloud computing technologies [11].

Hashem et al. [2] discussed Big data assessment is motivated by fast-increasing cloud-constructed applications established using virtualized technologies that delivers services for the computation and processing of massive data but also attends as a service model. Chen et al. [12] represent the development of cloud computing provides solutions for the storage and processing of big data as the fundamental infrastructure for smooth operation. Catching and preparing huge information are identified with enhancing the worldwide economy, science, get-together, training and national security; handling of enormous information permits us to propose accurate choices and get learning raw data [13].

Cannaliatoet et al. [14] represent in their patent that processing platform integrates ETL (Extract, Transform and Load), real time stream processing and big data stores in high performance analytic system that run in public and private cloud.

3.1 Single Big Data Problem and Meeting with Cloud

There can be numerous difficulties to utilizing single Big Data, for example, Existing base can't deal with Big Data, Higher capital costs (CapEx), Higher expenses for non-key assets, Vendor lock-in that breaking points decisions [15].

For below explanations, touching your Big Data to the environment of cloud:

- A spectrum of unconventional technologies, investments and skills are required by Big Data. The question is, are we really want this in-house? [15]
- Huge Data incorporates enormous measures of outside information. Does it bode well to move and deal with this information behind your firewall? [15]
- Huge Data needs a considerable measure of information administrations. Why not concentrate on the estimation of your examination, rather than essentially dealing with your information? [15]

4 List of Tools and Literature Reviews

4.1 Tools for Big Data

Table 2 shows overview of open source tools of big data.

Table 2 List of Big Data Open Sources Implements

Big data tools [16]	Narrative [16]
Big data study platforms and outfits	
Hadoop	Apache Programming Establishment that empowers conveyed handling of huge datasets crosswise over cluster
MapReduce	Developed by Google that describe as Programming framework model to processing parallel on large node
Databases or data warehouses	
Cassandra	Initially created by Facebook, this NoSQL database is currently overseen by the Apache Foundation
HBase	HBase is the non-social information store for Hadoop, incorporate straight and measured adaptability, steady
Mongo DB	It was designed to support humongous databases with document oriented storage, full index support, etc.
Hypertable	This NoSQL database offers proficiency and quick execution in expense reserve funds versus comparative databases
Hive	Hadoop's information distribution center, Hive guarantees simple information rundown, impromptu inquiries and different investigation of huge information
Business intelligence	
Talend	Talend makes various diverse business knowledge and information distribution center items
Pentaho	Pentaho offers business and huge information examination apparatuses with information mining, reporting and dashboard abilities
Data mining	
RapidMiner/analytics	"The world-driving open-source framework for information and content mining"
Mahout	This Apache undertaking offers calculations for grouping, arrangement and clump based shared separating
Weka	It offers an arrangement of calculations for information mining that you can apply specifically to information or use in another Java application
File systems	
HDFS	It rapidly duplicates information onto a few hubs in a group with a specific end goal to give dependable, quick execution

(continued)

Table 2 (continued)

Big data tools [16]	Narrative [16]
Programming Languages	
Pig/Pig Latin	Produces arrangements of Map-Reduce programs
R	Make it less demanding to control information, perform figuring and produce outlines and diagrams
Data aggregation and transfer	
Sqoop	Sqoop exchanges information in the middle of Hadoop and RDBMSes and information stockrooms
Chuwka	Chukwa gathers information from extensive dispersed frameworks for showing and investigating the information it gathers
Miscellaneous big data tools	
Zookeeper	Keeping up setup data, naming, giving dispersed synchronization, and giving gathering administrations

4.2 Literature Review on Some Papers

Table 3 presents Literature reviews bases on big data on cloud technologies papers.

Table 3 Literature reviews

Reference	Authors	Ideas
[13]	Bahrami et al.	To analyze big data with Business Intelligence tools and capabilities of cloud computing systems that are feasible solution for handling big data for converting smart grid
[2]	Hashem et al.	To review on big data technology moving to cloud environment with some tools analysis and case studies at the end closing with open research issues
[17]	Ibrahim et al.	To processing framework of hadoop use in OpenStack environment
[8]	Assunção et al.	To surveyed state- of-the-art of each key stages in the context of cloud supported big data analytics
[12]	Chen et al.	To review on big data generation, acquisition, data storage and architecture of big data analytics and its method
[18]	Pääkkönen et al.	To Surveyed designing and constructing reference architecture of big data for commercial solutions
[19]	Jain et al.	To proposed various technique of computation of big data stack and technology in cloud environment
[20]	Tsuchiya et al.	To proposed architecture of massive data processing on cloud technology
[21]	Emani et al.	To surveyed the problems of big data management and merging big data in existing infrastructure
[22]	Hashem et al.	To surveyed role of big data in current environment of enterprise and technology with enhance efficiency of data management and formulated a data life cycle

5 Research Challenges

Some research challenges that data integrity, scalability, accessibility, data heterogeneity, data alteration, data excellence, privacy and legitimate issues, and controlling governance. Bahrami et al. [13] discussed massive data challenges such as Storage, Computing and Transfer issues. Chen et al. [12] discussed challenges in terms of theoretical, technical, practical approaches of big data.

6 Unsolved Future Exploration Problems

According to [2, 12] some open research issues such as data staging, distributed storage system, data analysis, data security, fundamental problems of BD, Standardization of BD, Format conversion of BD, BD transfer, Processing of BD, Searching, Mining and analysis of BD, Provenance of BD, BD privacy, Data Quality, Safety Mechanism, BD application in information security, Performance of BD. Also essential for efficient parallel algorithms for streaming and multiscale adaptive algorithms reducing time complexity of O (N^2) to O (N) and for keys to shield and firmly process data in the cloud.

7 Conclusion

In today's digital globe, particular in Information and Computer Field 2 words are mainly important that are "Computing" and "Data" that converts to the "Cloud Computing" and "Big Data" technologies. In this revision, we offered 360° assessment for the big data environment on cloud along with its definition, notions, taxonomy and architecture of Question Answering model. Today's internet world, data storage and managing are required in well-constructed manner. Neither single big data nor cloud computing technologies are powerful. So, we discussed that how big data are lying on cloud environment that represented connection between them to effectively conjoin. Also presented overview of open source tools of big data. Lastly focuses on research challenges and current unsolved future exploration problems of big data in cloud computing concerns.

References

1. Dastikop, R.: Cloud computing: the complete reference, https://docsend.com/view/498qpz2
2. Abaker, I., Hashem, T., Yaqoob, I., Badrul, N., Mokhtar, S., Gani, A., Ullah, S.: The rise of 'big data' on cloud computing: review and open research issues. Inf. Syst. **47**, 98–115 (2015)

3. Mell, P., Grance, T., Grance, T.: The NIST definition of cloud computing. Recommendations of the National Institute of Standards and Technology
4. Lakshmi, J., Vadhiyar, S.S.: Cloud computing: a bird's eye view, pp. 1–14
5. Alkhatib, H., Faraboschi, P., Frachtenberg, E., Kasahara, H., Lange, D., Laplante, P., Merchant, A., Burgess, A.: Report., http://chuck-4-1st-vp.daven.com/IEEE-CS-2022-Report-v33RevRoberto.pdf
6. A Navint Partners and White Paper: Why is BIG data important. May, pp. 1–5 (2012)
7. http://www.forbes.com/sites/gartnergroup/2013/03/27/gartners-big-data-definition-consists-of-three-parts-not-to-be-confused-with-three-vs/
8. Assunção, M.D., Calheiros, R.N., Bianchi, S., Netto, M.A.S., Buyya, R.: Big data computing and clouds: trends and future directions. J. Parallel Distrib. Comput. **79–80**, 3–15 (2015)
9. Trundle, P.: Intelligence: Research Opportunities and Challenges in Big Data BigDat2015 Lessons. Haruna Isah, pp. 1–30 (2015)
10. BIG. Data and Working Group: Big Data Taxonomy (2014)
11. Grid, S., Energy, N., Transportation, I., City, S.: Main Contents of this Chapter (2015). doi:10.1016/B978-0-12-801476-9.00002-1
12. Chen, M., Mao, S., Liu, Y.: Big data: a survey (2014)
13. Bahrami, M., Singhal, M.: The role of cloud computing architecture in big data. In: Pedrycz, Chen, S.-M. (eds.) Information Granularity, Big Data, and Computational Intelligence, Chapter 13, vol. 8, pp. 275–295. Springer, Heidelberg (2015)
14. Cannaliato, T.J., Decker, J.A., Vahlberg, M.W.: System and method for correlating cloud-based big data in real-time for intelligent analytics and multiple end uses. U.S. Patent No. 9,092,502 (2015)
15. W. C. Computing and O. S. Software.: Turning Big Data (2013)
16. http://www.datamation.com/data-center/50-top-open-source-tools-for-big-data-1.html
17. Ibrahim, A.: A study of adopting big data to cloud computing. Technology Innovation and Entrepreneurship Center, Egypt Technology Innovation and Entrepreneurship Center, Egypt. pp. 1–7 (2015)
18. Pääkkönen, P., Pakkala, D.: Big data research reference architecture and classification of technologies, products and services for big data systems. Big Data Res. **1**, 1–21 (2015)
19. Paper, C.: Big data analytic using cloud computing (2015). doi:10.1109/ICACCE.2015.112
20. Tsuchiya, S., Lee, V.: Big Data Processing in Cloud Environments, pp. 159–168 (2012)
21. Cullot, N., Emani, C.K., Cullot, N., Nicolle, C.: Understandable Big Data: A Survey (2015)
22. Khan, N., Yaqoob, I., Abaker, I., Hashem, T., Inayat, Z., Kamaleldin, W., Ali, M., Alam, M., Shiraz, M., Gani, A.: Big Data: Survey. Opportunities, and Challenges, Technologies (2014)

Performance Evaluation of Energy Detection Based Cooperative Spectrum Sensing in Cognitive Radio Network

Reena Rathee Jaglan, Rashid Mustafa, Sandeep Sarowa and Sunil Agrawal

Abstract Cognitive Radio is a promising solution to spectrum underutilization problem, highlighting the concept of Dynamic Spectrum Access with two primary functions of efficient radio spectrum usage and providing reliable communication whenever and wherever needed. In a Cognitive Radio Network, secondary users are allowed to use the vacant licensed spectrum. Hence secondary users need to sense the spectrum to check availability of vacant spectrum and vacate as soon as primary user arrives back. Thus, spectrum sensing plays a significant role in Cognitive Radio Networks. Cooperative spectrum sensing is of great importance as it combats shadowing multipath fading and receiver uncertainty problems. There are two important parameters in spectrum sensing: probability of detection and probability of false-alarm. Higher detection probability signifies better primary user protection. In this paper, performance has been evaluated and depicted for Energy detection based cooperative spectrum sensing through MATLAB simulations.

Keywords Cooperative spectrum sensing · White spaces · Probability of detection · Probability of false alarm

1 Introduction

There has been tremendous increase in usage of wireless services over the past decade. Hence, causing increase in demand for spectrum usage with the increased number of users. Spectrum is a precious natural resource, which needs to be utilized properly and smartly. It has been observed by measurements of Federal Communication Commission (FCC) that nearly 70 % spectrum in US is unutilized by the current Fixed Spectrum Assignment (FSA) policy as most of the spectrum remains unutilized at certain time at certain geographical position [1]. This

R.R. Jaglan (✉) · R. Mustafa · S. Sarowa · S. Agrawal
Department of Electronics and Communication Engineering, U.I.E.T,
Panjab University, Chandigarh, India
e-mail: reenarathee5@gmail.com

S.C. Satapathy and S. Das (eds.), *Proceedings of First International Conference on Information and Communication Technology for Intelligent Systems: Volume 2*,
Smart Innovation, Systems and Technologies 51, DOI 10.1007/978-3-319-30927-9_58

underutilization of spectrum has been a major concern among researchers and lead to the concept of Dynamic Spectrum Access (DSA) and Cognitive Radio technology.

Cognitive Radio (CR) is an emerging promising solution to the underutilized spectrum. In a Cognitive Radio Network, Secondary User (SU) utilizes the vacant spectrum band or white spaces when there is no Primary User (PU) activity. IEEE 802.22 working group formed the first standard for Wireless Regional Area Network (WRAN) using the CR concept.

The main functions of CR shown in Fig. 1 can be described as: Spectrum sensing: sensing vacant spectrum without interrupting primary transmission. Spectrum management: includes spectrum analysis and decision-making. Selects the best available channel in the vicinity. Spectrum sharing: utilized with SUs without any interference. Spectrum mobility: in case PU re-enters licensed band, SU has to switch or handoff to another spectrum band.

The key contributions of the paper include development of a Cooperative Spectrum Sensing (CSS) algorithm based on Energy Detection (ED) sensing scheme to achieve improved detection and reduced showing and fading effects, evaluation of simulated detection probability of the system with theoretical expressions. The rest of the paper is organized as follows. In Sect. 2, we describe the basic concept and fundamentals of Spectrum Sensing (SS). In Sect. 3, we describe the algorithm based on energy of received signal for SS. In Sect. 4, basics of CSS are described and categorized based on infrastructure. In Sect. 5, we present the simulation results on system detection probability based on energy on received signal. Finally, concluding remarks along with some future directions are mentioned in Sect. 6.

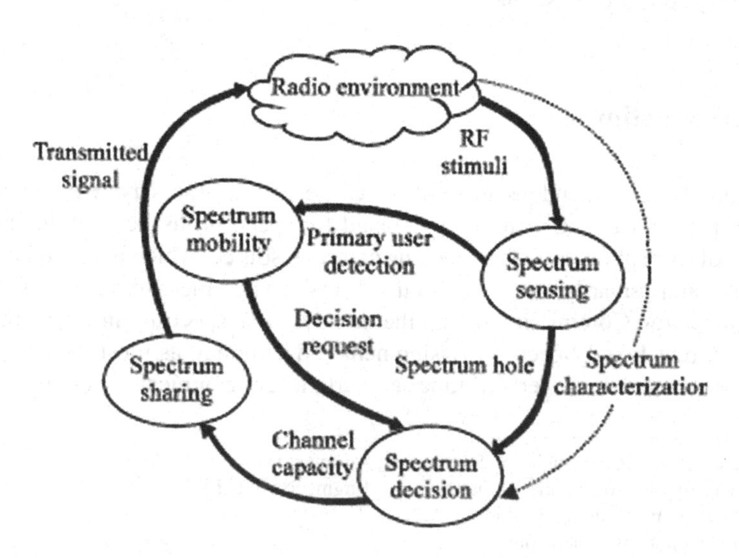

Fig. 1 Cognition cycle [2]

Table 1 Summary of important parameters in sensing

Detector decision ↓	Actual decision →	H1: PU is present (p)	H0: PU is absent (q)
H1: PU is present (A)		True positive (sensitivity or probability of detection)	False positive, type-I error (probability of false-alarm)
H0: PU is absent (B)		False negative, type-II error (probability of missed detection)	True negative (specificity)

2 Spectrum Sensing and Its Fundamentals

Spectrum sensing is the process of identifying the vacant licensed spectrum without interrupting any primary transmission. Spectrum sensing can be modeled as binary hypothesis testing problem as:

$$\left.\begin{array}{l} \mathrm{Y}\ (\mathrm{n}) = \mathrm{W}\ (\mathrm{n}) : \mathrm{H0} \\ Y(n) = hX(n) + W(n) : H1 \end{array}\right\} \quad (1)$$

where Y(n) = received signal at detector, W(n) = AWGN with mean zero and variance one, X(n) = transmitted PU signal, h = channel gain.

The important performance measurement parameters have been summarized in the following Table 1.

Various spectrum sensing techniques have been illustrated in the literature, including [3]. Energy Detector (ED) have been chosen for sensing the spectrum in this paper due to its low computational complexity and simplicity. Moreover, it doesn't require any prior information of signal.

3 Energy Detector

It is the simplest spectrum sensing technique with less complexity. In this technique, energy of received signal samples have been compared with a predefined threshold value λ during the observational period of time [4–6]. The basic principle of this scheme have been explained in Fig. 2.

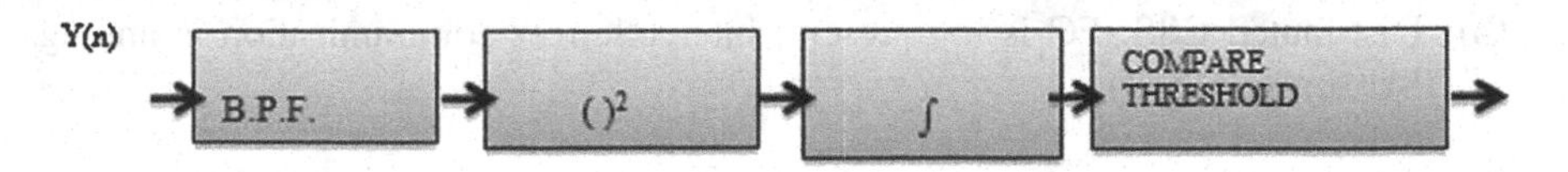

Fig. 2 Basic principle of energy detection technique

The test statistics of this scheme can be given as following Eq. 2 which is further compared with the predefined threshold in order to infer a decision regarding PU presence/absence.

$$T(Y) = \frac{1}{N}\sum_{n=1}^{N}\{Y[n]\}^2 \tag{2}$$

where $N = \tau f_s$ is the number of samples, τ denotes the sensing time while f_s denotes the sampling frequency. N is maximum integer which is not greater than τf_s. The decision can be given as: if $T(Y) > \lambda$, signal exists; otherwise, signal does not exist. We consider real-valued Gaussian primary signal and Additive White Gaussian Noise (AWGN). Probability of detection and probability of false-alarm [4] can be given as:

$$Pd = Q\left[\frac{1}{\sqrt{2\gamma+1}}\left(Q^{-1}(Pf) - \sqrt{N}\gamma\right)\right] \tag{3}$$

$$Pf = Q\left[\sqrt{2\gamma+1}Q^{-1}(Pd) + \sqrt{N}\gamma\right] \tag{4}$$

Threshold is related to Pf as follows:

$$Q^{-1}(Pf) = \left[\frac{\gamma}{\sigma u^2} - 1\right]\sqrt{N} \tag{5}$$

where γ SNR of received signal, σu^2 is noise variance, Q (.) is complementary distribution function.

4 Cooperative Spectrum Sensing

Cooperative spectrum sensing (CSS) is a sensing technique that involves the cooperation or interaction of SUs distributed over a certain region. Based on the infrastructure and strategy, it can be broadly categorized as [7, 8] as:

(a) Centralized CSS: presence of a central entity called Fusion Center (FC) which is responsible to take a global collaborative decision on presence or absence of PU [9, 10].
(b) Distributed CSS: FC is not present, interaction or communication is among SUs only (Fig. 3).

Hard or soft fusion rules are used for computation of local observations. Soft fusion schemes include Equal Gain Combining (EGC) and Maximal Gain Combining (MGC). Hard fusion rules include AND, OR and Majority. Hard fusion

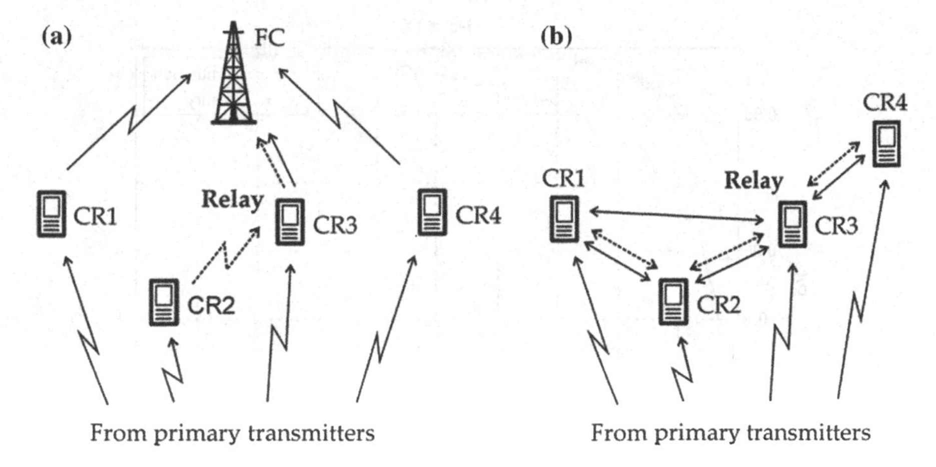

Fig. 3 Basic depiction of CSS (**a**) centralized (**b**) distributed

rule are more bandwidth efficient with less computation complexity [11, 12]. The global performance evaluation metrics can be given as:

$$\text{Global probability of false alarm, } Q_f = P(u = 1|H0) \quad (6)$$

$$\text{Global probability of detection, } Q_d = P(u = 1|H1) \quad (7)$$

$$\text{Global probability of miss-detection, } Q_m = P(u = 0|H1) = 1 - Q_d \quad (8)$$

5 Simulation Results and Discussions

We assumed N = 1000 and SNR = −10 dB under AWGN channel conditions. Figure 4 illustrates performance of ED for theoretical expressions and monte carlo simulations. Pd = 0.891 for 0.1 Pf is achieved by simulation results. Higher value of Pd means higher are the chances of accurate detection.

Figure 5 demonstrates ED performance at different SNRs. The probability of false alarm has been fixed to 0.01 for the simulation purpose. Energy Detection performs well at moderate and high SNR values while for low values the performance degrades. However, an obvious solution for improving sensing performance is to improve sensing time. IEEE 802.22 limits maximal 2 s latency (includes sensing time and subsequent processing time) Here Pd has been investigated at different SNR values. Table 2 shows the received Pd values at different SNRs and it has been found that maximum Pd value is obtained at SNR = −6 dB.

Next, we evaluated CSS scheme based on energy detection using AND hard decision rule for global decision. We assumed time bandwidth factor = 1000, SNR = 10 dB, number of SUs = 5. Figure 6 shows effect of number of SUs (m) in

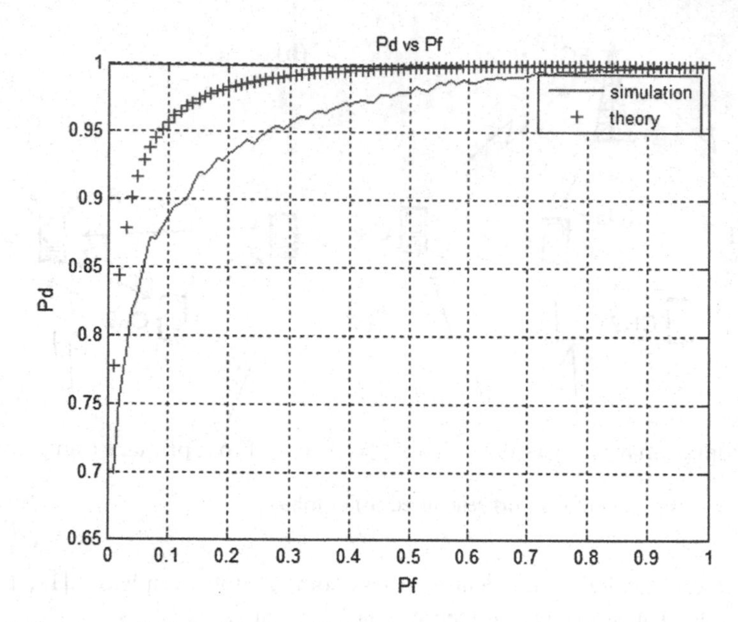

Fig. 4 Performance of ED for SNR = −10 dB

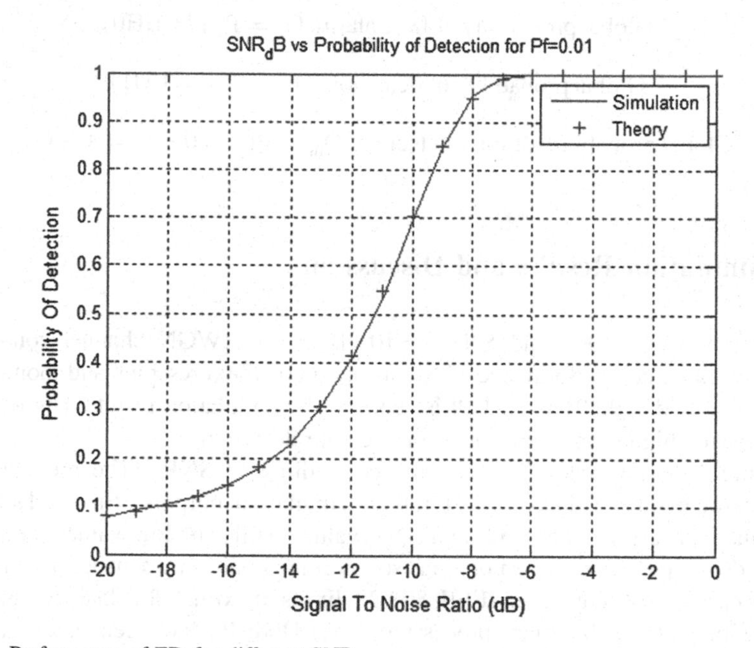

Fig. 5 Performance of ED for different SNR

Table 2 Pd at different SNRs for fixed Pf = 0.01

SNR_dB	−18	−16	−14	−12	−10	−8	−6
Pd	0.1	0.15	0.25	0.41	0.7	0.95	1

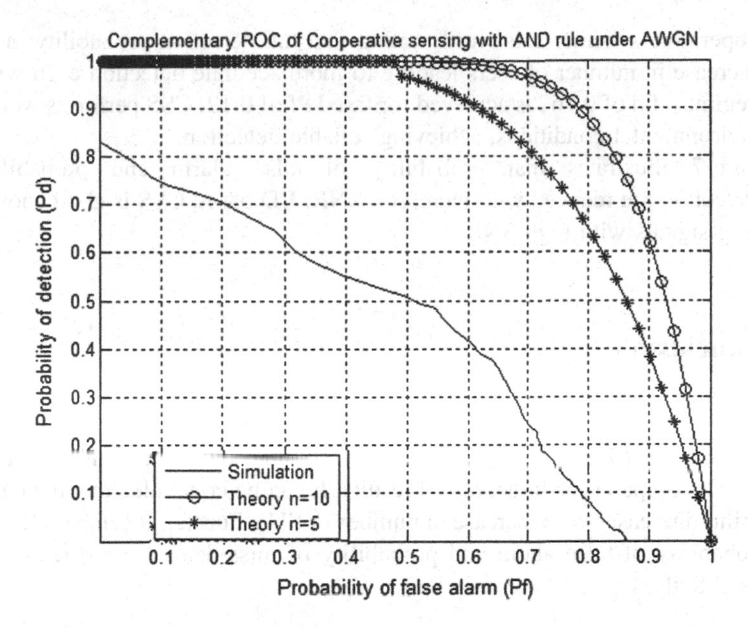

Fig. 6 Pd versus Pf for CSS with m = 10

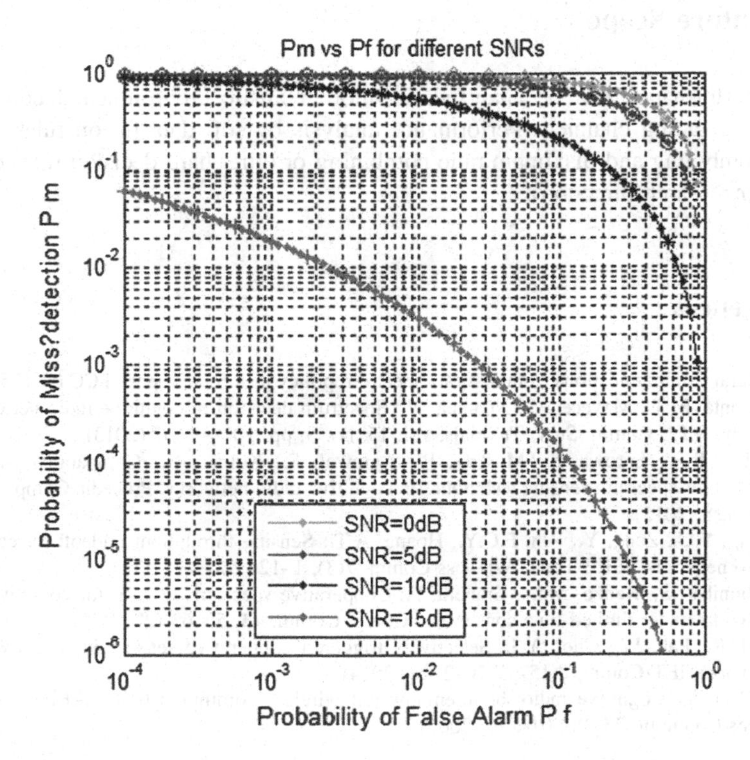

Fig. 7 Pm versus Pf for CSS at different SNR

the cooperative scheme. It is worth noting that the detection probability increases with increase in number of users leading to more accurate detection even with low SNR regimes. Pd of 0.762 is achieved for fixed Pf of 0.12. CSS performs well as for real environmental conditions, achieving reliable detection.

Figure 7 illustrates that probability of false alarm and probability of miss-detection are reduced with increased SNR. ED based CSS is good enough for detecting signals with high SNR.

6 Conclusion

It has been known to us that CSS scheme performs well in more realistic conditions. In this paper, ED based CSS scheme has been analyzed with Monte-Carlo simulations and compared with theoretical results. It can be concluded that the detection probability increases with increase in number of SUs. Further, it can be inferred that the probability of false alarm and probability of miss-detection are reduced with increased SNR.

7 Future Scope

CSS performance can be analyzed for more realistic environmental conditions including fading channels. Performance analysis by soft data fusion rules (Equal gain combining and maximum ratio combining or other hard decision rules can be another research direction).

References

1. Federal Communications Commission: Spectrum policy task force report, FCC 02-155 (2002)
2. Masonta, M.T., Mzyece, M., Ntlatapa, N.: Spectrum decision in cognitive radio networks: a survey. IEEE Comm. Surv. Tutorials, vol. 15, no. 3, pp. 1088–1107 (2013)
3. Jaglan, R.R., Sarowa, S., Mustafa, R., Agrawal, S., Kumar, N.: Comparative study of single-user spectrum sensing techniques in cognitive radio networks. Procedia Comp. Sci. **58**, 121–128 (2015)
4. Liang, Y.C., Zeng, Y., Peh, E.C.Y., Hoang, A.T.: Sensing throughput tradeoff for cognitive radio networks. IEEE Trans. Wireless Comm. **7**(3), 1–12 (2008)
5. Althunibat, S., Renzo, M.D., Granelli, F.: Cooperative spectrum sensing for cognitive radio networks under limited time constraints. Comp. Comm. **43**, 55–63 (2014)
6. Paul, R., Pak, W., Choi, Y.J.: Selectively triggered cooperative sensing in cognitive radio networks. IET Comm. **8**(15), 2720–2728 (2014)
7. Haykin, S.: Cognitive radio: brain-empowered wireless communications. IEEE J. Selected Areas Commun. **23**(2), 201–220 (2005)

8. Peh, E., Liang, Y.-C.: Optimization for cooperative sensing in cognitive radio networks. In: Proceedings of the IEEE Wireless Communications and Networking Conference (WCNC), pp. 27–32, Hong Kong (2007)
9. Bhowmick, A., Roy, S.D., Kundu, S.: A hybrid cooperative spectrum sensing for cognitive radio networks in presence of fading. In: Twenty First Conference on Communications, pp. 1–6 (2015)
10. Chaudhri, S., Lunden, J., Koivunen, V., Poor, H.V.: Cooperative sensing with imperfect reporting channels, hard decisions or soft decisions. IEEE Trans. Signal Process. **60**(1), 18 (2012)
11. Axell, E., Leus, G., Larson, E.G.: Overview of spectrum sensing for cognitive radio. In: Second IEEE International Workshop on Cognitive Information Processing, pp. 322–327 (2010)
12. Letaief, K.B., Zhang, W.: Cooperative communications for cognitive radio networks. In: Proceedings of IEEE, vol. 97, no. 5, pp. 878–893 (2009)

Design of Robotic Hand-Glove for Assisting Mobility Impaired

Ramalatha Marimuthu, Sathyavathi Ramkumar, Harshini Infanta, Alagu Meenal and Preethi

Abstract This project explains the design of a robotic glove system specific to the applications of mobility impaired. It is mounted on a moving base which will help the user in accessing objects that are not within arm's reach. A robotic arm will replicate the movements of wrist, elbow and shoulder joint of a person in response to movements produced by a wireless glove to perform the picking, gripping and moving actions. For gripping action a clamp based mechanism is used. Flex sensors produce the signals proportional to the amount and direction of movement of the user's hand to actuate the clamp through high torque dc servo motors. For arm movement a potentiometer is used. The control circuit using arduino regulates the amount and direction of arm movement through feedback from the actuator.

Keywords Robotic hand glove · Actuator · Flex sensor · Assistive device

1 Introduction

The importance of robots have been realised in almost all walks of life. Full robots are used for places where human interventions become difficult owing to the hazardous nature of the environment and applications. At the same time usage of parts of robots is also becoming popular. Robotic arms are used in many manufacturing industries where repeated periodical works are necessary. The accuracy, precision, speed, performance and reliability of the robotic arms in these industries have been proved superior to those of humans through continuous quality checks. This makes the use of robotic arms a good choice for other situations where rather than the hazardous nature, absence of humans demands it. For example, in healthcare, robotic arms can be very good companions for the patients who cannot

R. Marimuthu (✉) · S. Ramkumar · H. Infanta · A. Meenal · Preethi
Kumaraguru College of Technology, Coimbatore, India
e-mail: Ramalatha.marimuthu@gmail.com

S. Ramkumar
e-mail: sujuram07@gmail.com

S.C. Satapathy and S. Das (eds.), *Proceedings of First International Conference on Information and Communication Technology for Intelligent Systems: Volume 2*,
Smart Innovation, Systems and Technologies 51, DOI 10.1007/978-3-319-30927-9_59

move their limbs and who have no human assistance. These robots can replace those limbs and make life easier for people with special needs. Development of devices with the objective of making the everyday tasks of common man simpler and easier continues to have no significance for physically challenged. Those who do not have the liberty to move about still continue to face difficulties in carrying out simple everyday errands.

2 Literature Review

In related work, Saggio et al. [1] have measured the data glove repeatability and proved that the gain is good. They have proved the usefulness of bend sensors to be used in a data glove as a goniometric device and shown that the range and standard deviation of these sensors do not get affected by grip force as long as a comfortable hand position is identified. The position imposed by the mold is of priority here to keep up the performance. The design provided us the background for developing a sensor based robotic arm. Kajone et al. [2] proposes a gesture controlled robotic arm using wireless communication. The system uses webcam to capture hand movement and uses RF communication to control the robotic arm. The digitization of the gesture is done using MATLAB and AVR microcontroller. This needs a precise placement of the hand in front of the webcam which might be difficult for people with special needs.

Taksale et al. [3] has introduced a model where flex sensors based robotic hand is built. The system concentrates on the five finger movements with the help of servo motors and dc motors. Though it has not been clearly discussed on how it will overcome the power consumption problem, the design takes care of independent movement of the five fingers. Mohd Ali et al. [4] developed a rehabilitation device with the usage of flex sensor for assisting in the therapy of patients by measuring movements of finger joints. It is a master slave system with a smart glove as a master and a hand gripper device as the slave to replicate the movements of the glove. Flex sensors and flexi force sensors are used for the prosthesis multi finger gripper movements which are controlled by force feedback through the leather glove.

Verma [5] has developed a CLASS-5, TYPE C, Numerical Control Robot controlled wirelessly using hand gestures. He uses stepper motors to drive the robot with the advantages of low torque, high reliability and low cost. The positional controls are through alpha, beta and gamma angle measurement of gyroscope and accelerometer to ensure all direction movement. The purpose is pick and place which can serve many applications in industries. This was accomplished using Flex sensor control. Dharaskar et al. [6] has used accelerometer based wireless data glove with PIC microcontroller to develop a gesture and voice controlled robotic arm to help handicapped persons. There is a discontinuity between the two operations in that the same gesture controlled robot is also designed to be a voice controlled robot. There is no explanation as to why two different controls are required for the same outcome.

Aggarwal et al. [7] has developed an accelerometer based gesture controlled robot which is a 3 axis wireless control robot. Pierrot et al. [8] designed an anthropomorphic robotic arm with seven degrees of freedom that assists doctors in moving ultrasonic probes over the patient's skin while exerting a given effort. This is a preliminary experiment using an industrial robot to study the force control required while probing a patient. Gautam et al. uses Bluetooth for the wireless communication between the glove and the robotic arm [9] while Sharma et al. has demonstrated a complete anthropomorphic robotic arm with seven degrees of freedom for performing lateral, spherical cylindrical and tip holding gripping actions. The processor used in ATmega32 [10]. Pedro et al. [11] designed a haptic robot for Human Interactive Communication. Ahmed et al. [12] proposes a system that can locate any given polar coordinates. This is primarily used for point based movements. Cheng [13] designed a system that tracks the motion of a person in real time without a classifier or pre-defined action. Szabo [14] proposed a robotic arm that can identify color cubes, pick and place them in different cups.

Based on the papers discussed above, the sensor, controller and actuator for our project were selected as Flex sensor since flex sensor is cheap, accurate and one sensor is sufficient to track the movements of single finger. Arduino was selected as controller due to its low power consumption, small code size and increased performance. Mechanical actuator which is best suited for our project is servo motor. Servo motors are easily available and 180° bend can be achieved using single servo motor and driver.

3 Proposed System

Figure 1 shows the block diagram of the proposed system. The inputs are taken from the flex sensors and the potentiometer based on the movement of the patient's hand. This part constitutes the human machine interface. The second part of the system is the processor which is made up of arduino with the interface with input/output. The third portion is the actuator for the robotic arm consisting of DC diver and motor for the movement of fingers and the servo motor for the movement of the arm.

The proposed system given above as block diagram reduces the usage of remote control by 80 %. For initial calibration, the user has to wear the glove in rest position with the processor switched on. The processor constantly checks for the inputs and in the initial rest position since there is no movement of hand, no differential potential is developed by the flex sensor and hence there will be no input to the processor. This position is used to calibrate the sensors. When the movement is made, this creates a difference in the resistance of the flex sensors.

The actions of the user are captured in terms of the resistance of the flex sensor. The electrical signals generated are digitized by the microcontroller and the actuator

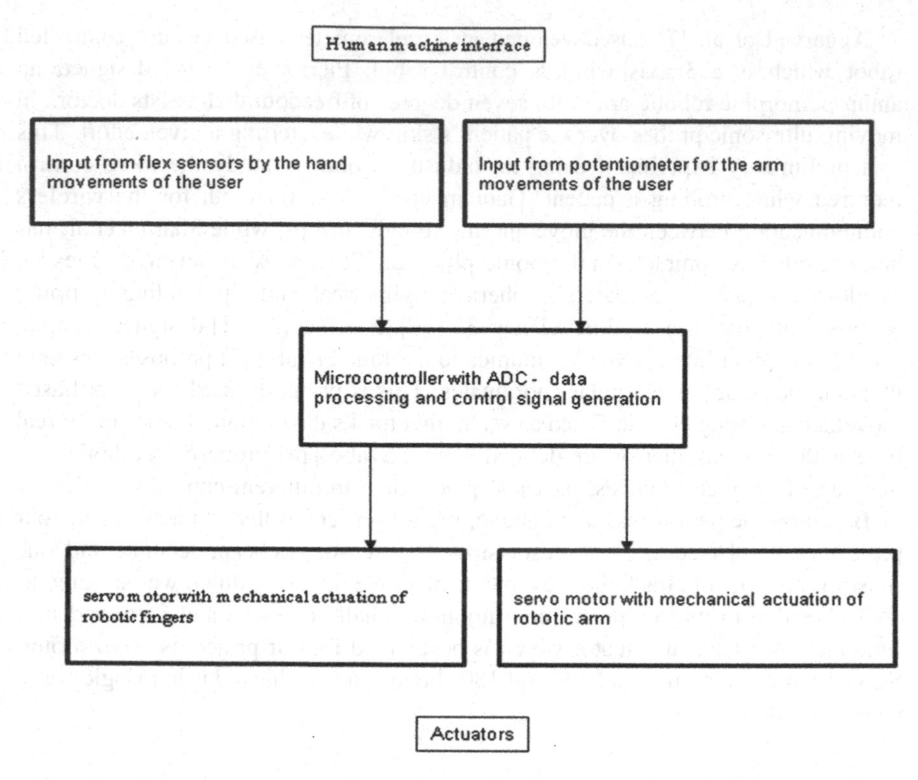

Fig. 1 Block diagram

in the robot is given the necessary command to operate the arm through the servomotor. Servo motors are attached to the robotic arm to provide the movement of the arm. The resistance values are mapped to the rotation of the arm thus completing the loop. The main objective behind the system is to assist the physically challenged and to reach to objects that are in places that are not advisable or safe for humans to access, to handle objects that can otherwise not be handled by humans etc.

4 Process Flow

The process flow for the system is shown in flow chart below. The first flow chart shows the transformation of the data from the flex sensors through arduino into finger movement of the arm. The second flow chart shows the process for the arm movement of the robot through the inputs from the potentiometer (Figs. 2 and 3).

Fig. 2 Process flow for robotic fingers

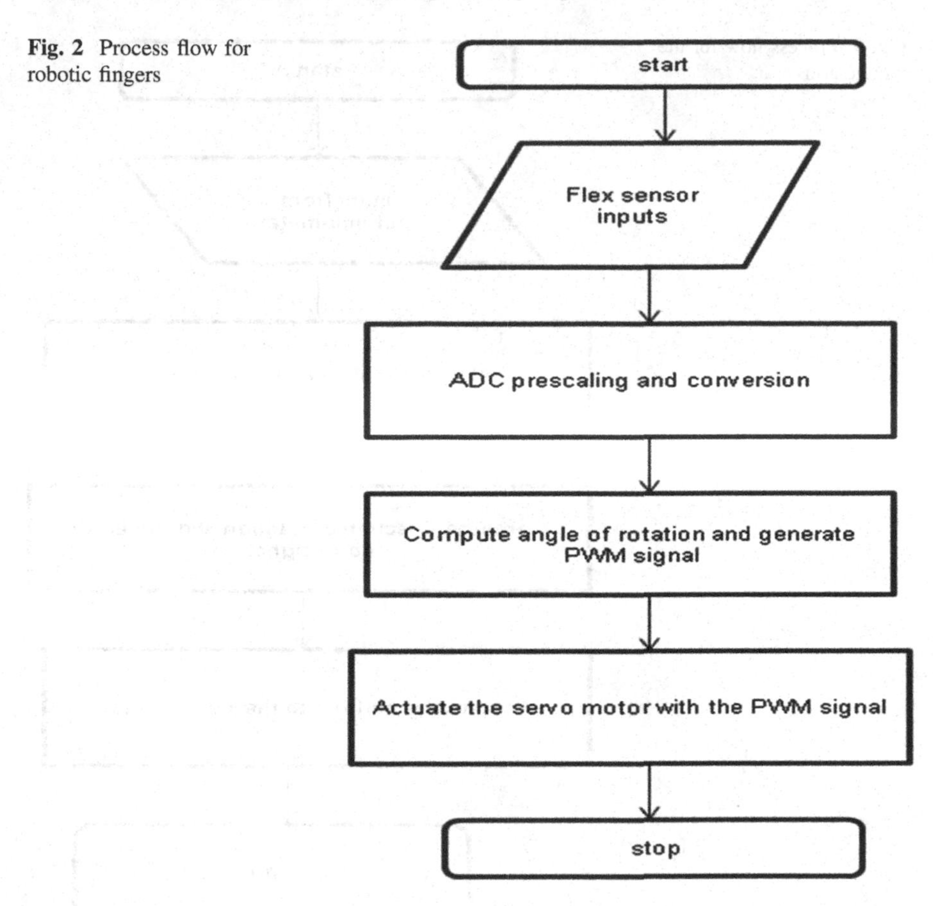

5 Software Descriptions

5.1 Embedded C

The microcontroller is programmed with embedded C considering its simplicity and ease of compiling. It is being programmed according to the conditions needed.

5.2 Arduino Software

Arduino programs are written in C or C++. The arduino is a general purpose processor which is used for small applications based on the data acquisition systems. The input and output translations are much easier owing to the already built in software library. The GNU tool chain and AVR Libc compile the programs while

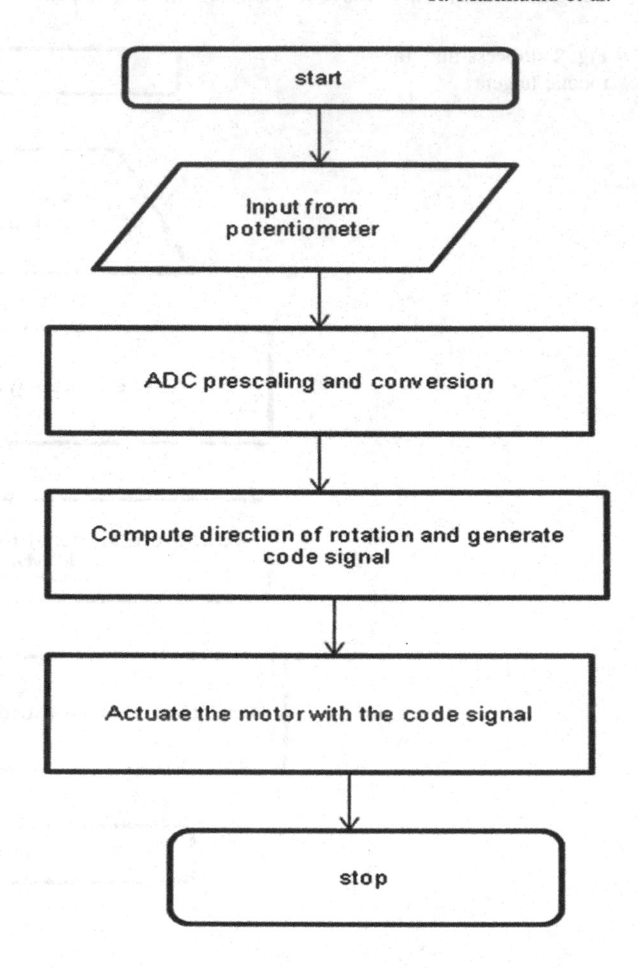

Fig. 3 Process flow for the robotic arm

uploading the programs to the board is easy because of the availability of avrdude. Users only need define two functions to make a runnable cyclic executive program:

- setup(): a function run once at the start of a program that can initialize settings
- loop(): a function called repeatedly until the board powers off

6 Component Description and Design Stages

6.1 Flex Sensors

A thin strip of sensors with 1″–5″ length that vary in resistance are the flex sensors. Flex sensors measure the amount of deflection caused by bending the sensor and convert the value into resistance. The value of resistance is in proportion with the degree of the bending. Since the characteristic is that of an analog resistor, they can be used as variable analog voltage dividers.

Two flex sensors are attached to a glove to measure the degree of bend of fingers and elbow. When the fingers are bent, the flex sensor resistor value changes. Readings are read as analog value by Arduino, digitized using an inbuilt ADC unit and are sent over RF transmitter to the receiver. There is no reinitialization required at startup since the dynamic variation makes the last position before rest as the reference position (Fig. 4).

6.2 Potentiometer

A potentiometer is an analog sensor which produces analog voltage values based on the variations in the position of the slider. The values can be read into the Arduino board as an analog value and digitized to be communicated to the robotic arm for shoulder and elbow movement. Three wires of potentiometer are connected to the Arduino board. The change in the amount of resistance reflects in the movement of the shaft producing an appropriate analog input. Thus based upon potentiometer reading, shoulder movement of robotic hand is given.

6.3 RF Module

For this application, RF is preferred since the connection between the robot and the glove is going to create a non physical actuator control unit. The input received from the flex sensors are converted into control signals and are communicated to the robotic arm through the RF transmitter. The RF Receiver receives the input that is wirelessly transmitted and then feeds the input to the servo motors connected to it (Figs. 5 and 6).

Fig. 4 Flex sensor

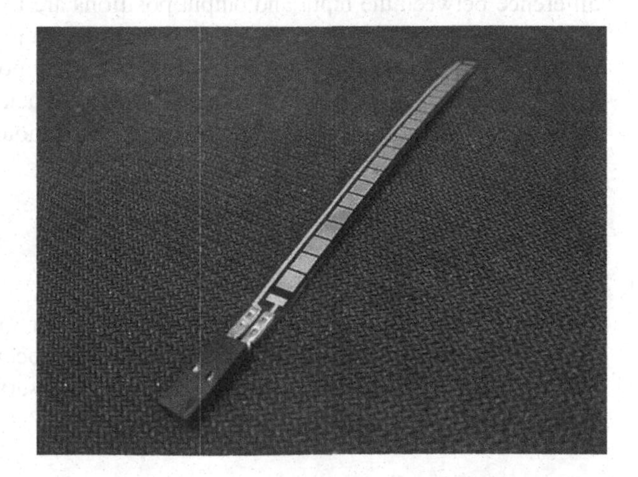

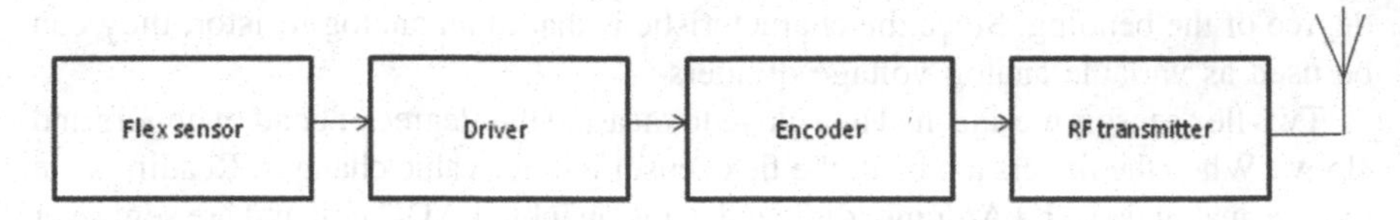

Fig. 5 RF transmitter block diagram

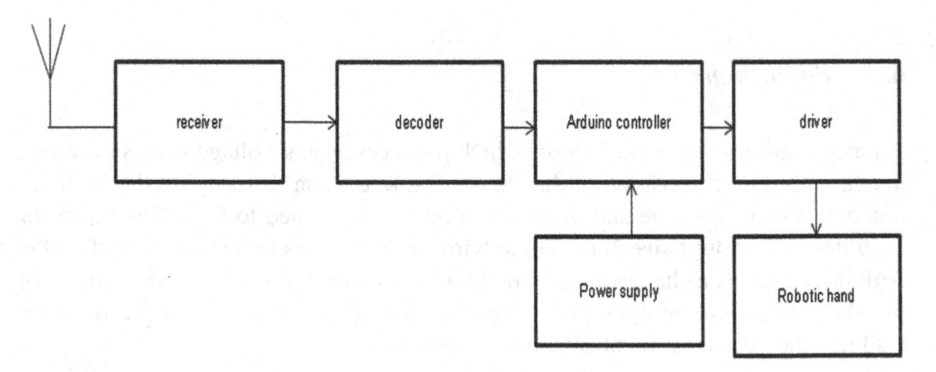

Fig. 6 Receiver block diagram

6.4 Servo Motors

Servo motors are used because of their least cost and high accuracy rates. A servomotor is a rotory actuator that allows for precise control of angular velocity and acceleration. It can be moved to a desired angular position using the PWM signals generated from the arduino. The position measurement for the output is fed to the processor and compared with the input value from the glove. The degree of difference between the input and output positions are measured and an error signal is generated which then controls the movement of the motor in either direction thus moving the output shaft to the correct position. The position difference decreases thus reducing the error signal. The continuous feedback is maintained between the input and output through the action of these servo motors.

6.5 Micro Controller

The Arduino Uno is the choice of the processor for many applications based on small scale personal control systems. Here the microcontroller is the ATmega328 and this microcontroller has the specifications necessary for the proposed model.

6.6 *Mechanical Robot*

The manual robot is used as the base for the claw. The claw is attached to the base with the help of an aluminum rod. A 12 V battery is used to supply power to the dc motors (used to run the robot). Base of the robot is controlled by using a switch. Designed robotic hand is placed over the base. Base is programmed to move according to the input given from switch (Fig. 7).

7 Results

The gesture controlled robotic arm was tested to move in up, down, left and right directions and the degree of the movement was found to be restricted to 180° with the single dimensional arm type designed. The weight that could be lifted by the arm was 50 gms. This can be improved by using much more powerful servo motors

Fig. 7 Robotic arm development and testing

since the model tested used only low power servo motors. The degree of freedom was found to be four and the maximum reach, when fully extended was up to 18 cm. The system was robust with aluminium clamps and can be made more dexterous for higher degrees of freedom.

8 Conclusion

The modules on integration was powered and tested by giving the input through the flex sensor. The chaise was controlled by a remote with two switch controls. The resistance through the flex sensor was mapped to the degree of rotation by which the robotic arm has to turn.

The stepper motor initially used could not support the load and a motor of higher torque was used. This model was then tested and the torque was found to be sufficient enough to hold light objects like a small plastic cup, plastic ball etc. For the model to lift even heavier objects, a higher torque stepper motor has to be used.

The prototype was designed with the available hardware components. For a commercial product to be manufactured, hardware components should be replaced with specifically designed and fabricated components.

Time and again, technology advancements such as automation have not aided the differently abled and the needy to its complete potential. Instead of man controlling such machines, machines have always controlled man. Precisely, machines work in the pre-programmed order and do not work to satisfy the man's requirements. Needless to say that the difficulty faced by those differently abled when it comes to interacting with such devices is also to be addressed. The above discussed robotic glove is designed to allow those differently abled to access daily objects with ease and their own disposal.

References

1. Saggio, G., Bocchetti, S., Pinto, C.A., Orengo, G.: A novel application method for wearable bend sensors. In: IOSR Journal of Electrical and Electronics Engineering, USA (2009)
2. Khajone, S.A., Mohod, S.W., Harne, V.M.: Implementation of a wireless gesture controlled robotic arm. In: International Journal of Innovative Research in Computer and Communication Engineering, pp. 375–379 (2015)
3. Taksale, A.S., Dangale, A.S., Momin, I.N., Sukode, S.V.: Low cost robotic hand glove. IOSR J. Electron. Commun. Eng. **9**(2), 140–144 (2014)
4. Mohd Ali, A.M., Wahi, A.J.M., Ambar, R., Abdul Jamil, M.M.: Development of artificial hand gripper using microcontroller. Int. J. Integr. Eng. **3**:47–54 (2011)
5. Verma, S.: Hand gestures remote controlled robotic arm. In: Advance in Electronic and Electric Engineering, vol. 3, no. 5, pp. 601–606 (2013)
6. Dharaskar, R.V., Chhabria Sandeep Ganorkar, S.A.: Robotic arm control using gesture and voice. In: International Journal of Computer, Information Technology & Bioinformatics (IJCITB), vol. 1, Issue 1, pp. 41–46 (2012)

7. Aggarwal, L., Gaur, V., Verma, P.: Design and implementation of a wireless gesture controlled robotic arm with vision. In: International Journal of Computer Applications, vol. 79, no. 13 (2013)
8. Pierrot, F., Dombre, E., Dégoulange, E.: A safe robot arm for medical applications with force feedback. In: Haptic Audio visual Environments and Games (2009)
9. Sharma, A., Lewis, K., Ansari, V., Noronha, V.: Design and implementation of anthropomorphic robotic arm. In: Noronha, V., et al. (eds.) International Journal of Engineering Research and Applications, vol. 4, Issue 1, pp. 73–79 (2014)
10. Gautam, G., Ashish, A., Kumar, A., Avdesh.: Wirelessly hand glove operated robot. In: International Journal of Advanced Research in Electronics and Communication Engineering (IJARECE), vol. 3, Issue 11 (2014)
11. Neto, P.; Pires, J.N., Moreira, A.P.: Accelerometer-based control of an industrial robotic arm. In: Haptic Robot and Human Interactive Communication (2009)
12. Ahmed, F., Safiullah, M.A., Khan, S.H., Moinuddin, A.: Assembly of robotic arm based on inverse kinematics using stepper motor. In: Computer Modeling and Simulation (EMS), 2012 Sixth UK Sim/AMSS, European Symposium (2012)
13. Cheng, H.-T.: Real-time imitative robotic arm control for home robot applications. In: Second International Conference on Consumer Electronics, Communications and Networks (CECNet) (2012)
14. Szabo, R., Lie, I.: Automated object sorting application for robotic arms. In: 10th International Symposium on Electronics and Telecommunications (ISETC) (2012)

An Approach for Mining Similar Temporal Association Patterns in Single Database Scan

Vangipuram Radhakrishna, P.V. Kumar and V. Janaki

Abstract Mining similar temporal association patterns from a time stamped temporal database is an important research problem in temporal data mining. The main objective and idea of this research is in finding similar temporal patterns from a given time stamped temporal database of transactions by scanning the input database only once. This objective to find temporally similar patterns through single scan of database coins out an important challenge to devise a single database scan procedure which shall use only support values of items computed in the first database scan, so as to discover all other temporal patterns. In the current research, we come out with a novel procedure to discover similar temporal patterns with respect to a reference sequence of support values for a given threshold limit. In this paper, we propose a novel approach to find similar temporal patterns followed by a case study. The approach is efficient in terms of space and time as it eliminates repeated scan of database by computing temporal frequent patterns or temporally similar patterns in only a single database scan.

Keywords Temporal · Association patterns · Upper bound · Outliers

V. Radhakrishna (✉)
VNR Vignana Jyothi Institute of Engineering and Technology, Hyderabad, India
e-mail: radhakrishna_v@vnrvjiet.in

P.V. Kumar
University College of Engineering, Osmania University, Hyderabad, India
e-mail: pvkumar58@gmail.com

V. Janaki
Vaagdevi Engineering College, Warangal, India
e-mail: janakicse@yahoo.com

S.C. Satapathy and S. Das (eds.), *Proceedings of First International Conference on Information and Communication Technology for Intelligent Systems: Volume 2*, Smart Innovation, Systems and Technologies 51, DOI 10.1007/978-3-319-30927-9_60

1 Introduction

In the last 40 years, a significant contribution from researchers of database community is towards studying the temporal databases and various aspects of temporal information systems. In the year 1986, summaries of temporal database research which were discussed in various symposiums and workshops followed by universities and work carried out at various research labs was first published in ACM SIGMOD Record. The importance of the area has come into the existence with the IEEE Data Engineering devoting a complete issue for temporal databases in 1988. Consequently in the year 1990 and 1992, two research papers contributing to survey in temporal databases were published. In [1–4] the authors carried out a detailed literature survey on temporal data mining techniques and data warehouses. The work in [5–7] involves finding temporal association rules. In [8–10] the authors define the approach of finding frequent items using the approximation method and the concept of upper and lower bounds. In [11–13], the authors work includes finding the temporal association patterns using the Euclidean similarity measure. The authors use the concept of upper lower bound, lower lower bound and lower bound distances to find the frequent temporal patterns. They define the process of finding the temporal patterns. The drawback of this approach is that the algorithm designed requires finding the true support values of itemset of size, k when deciding on itemset of size k + 1 to be frequent. This leads to the need for scanning the data-base again to find the true support sequences of itemset at level-k. In the proposed approach, we eliminate the need to compute the true support values and also the need to consider the support values of all subset of size k itemset to compute its upper and lower bound support time sequences.

The present work is motivated from the work by the authors [11–13]. The present approach which we propose is the novel approach which can be used to find the similar temporal association patterns of interest to overcome the following disadvantages of [11–13]

1. We eliminate, multiple scans of temporal database require finding support sequences of frequent patterns.
2. We overcome, the disadvantage of approach followed in [12], which requires scanning of true support of itemsets which are considered to be temporally similar.
3. We overcome the requirement to know the support sequence values of all subsets of an itemset as followed in [12, 13]

In temporal database of time stamped transactions, the transactions have time stamps and hence the support of the itemsets is in the form of a vector representing the support values computed for each time slot. This makes conventional approach not suitable to find the frequent temporal patterns. Also, popular Euclidean distance measure which is used to find the distance between any two vectors does not satisfy the monotonicity property [11]. In the present work, we consider the problem of mining similarity profiled temporal patterns from the set of time stamped transactions of a temporal database. We show using a case study how the proposed approach may be used to find the temporal frequent patterns.

2 Proposed Approach

2.1 Problem Definition

Given a finite set of items and disjoint set of time slots and temporal database of time stamped transactions, a reference support sequence and user specified threshold value, the objective is to find set of all patterns which are considered temporally similar w.r.t reference vector. Each transaction record is represented as a 2-tuple with elements timestamp and set of items of that transaction. The distance measure, denoted by $f_{similarity}(P, Q)$: $\rightarrow R^n$, where the parameters P and Q are numeric sequences is used as dissimilarity function. The objective is to find the set of all itemsets, I, which are subsets of I such that each of these itemsets represented by I, satisfy the condition $f_{similarity}(S_I, Reference) \leq \theta$ where S_I is the sequence of support values of I at time slots t_1, t_2, t_n.

2.2 Research Objective

We have the following research objectives.

1. To find set of all those temporal patterns which are similar to a specified reference not exceeding the specified threshold. The temporal patterns and reference vectors in our case are multi-dimensional vectors which are sequence of support values computed for each time slot.
2. To perform only single scan of input temporal database.
3. To design the expressions which can estimate the minimum and maximum Bounds on support values of an itemset.

2.3 Terminology Used

Itemset: An itemset is a subset obtained by chosen combination of items from a finite set of items. For a finite set of items whose size is $|I|$, we can obtain $2^{|I|} - 1$, itemset combinations excluding empty set.
Negative Itemset: An itemset whose support is found for probability of its non-existence is called negative itemset. For example, all those itemsets represented as $\bar{A}, \overline{AB}, \bar{B}$ are called negative itemsets.

Positive Itemset: An itemset whose support is found for probability of its existence is called positive itemset. Item sets represented as X, AB, Y, XYZ are called positive itemsets.
Negative Support: The support value obtained considering negative itemset is called as negative support.
Positive Support: The value of probability obtained considering positive itemset is called as positive support.
Support Sequence: The itemset support sequence obtained for a given itemset is an n-tuple denoted by a sequence represented mathematically as $S_{\theta}(I) = \langle S_{t_1}, S_{t_2}, S_{t_3}, \ldots S_{t_n} \rangle$. Here each S_{t_i} represents support value of itemset I computed for time slot t_i. Formally, $S(I) = U_n\{S_{t_i} |$ timeslot, t_i varies from t_1 to $t_n\}$ where U_n is union of all support values computed for each time slot t_i.
Negative Support Sequence: A support sequence, $S_{\theta,}$ of an itemset $I' \subseteq I$ denoted by, $S(I')$, is said to be the negative support sequence denoted by $S_{\theta_N}(I')$, if support of each element, $S(t_i)$, in the sequence is computed for the absence of itemset, I' in the database. The negative support sequences denote the probability of the items not appearing in the time slots t_1, t_2, ...t_n.
Positive Support Sequence: A support sequence, $S_{\theta,}$ of an itemset $I' \subseteq I$ denoted by, $S_{\theta}(I')$, is said to be the positive support sequence, denoted by $S_{\theta_P}(I')$, if support of each element, S_{t_i}, in the sequence is computed for the existence of itemset I' in the database.
Lower Lower Bound: It is the formally defined as the distance computed between the LBSTS and the reference sequence [11–13].
Upper Lower Bound: It is the formally defined as the distance computed between UBSTS and the reference support sequence vector [11–13].
True Distance: Distance computed between itemset support sequence, S and reference sequence vector, R_{θ}.
P runing: The process of elimination of temporal pattern which is not satisfying the threshold constraint is called as pruning. A pattern or itemset, denoted by *I*, is said to be a temporally frequent pattern, iff, every subset *I*of I is also temporally frequent.

2.4 Proposed Approach to Find Similar Temporal Patterns

In this section we outline the approach to find similar temporal association patterns for a given reference support sequence and user specified threshold

Input:
Let L, be a finite set of all items, $D_{temporal}$ be the temporal database of transactions defined over a finite set of time slots, N indicating total number of items in set L, *I* denote itemset which is subset of L and is of size, k and $f_{similarity}$ be the distance

measure used to estimate the dissimilarity between two vector sequences with respect to a user defined threshold, θ.

Output:

Set of all temporal itemsets, *I*, which are subsets of L such that each of these itemsets represented by *I*, satisfy the condition $f_{similarity}(S_I, R) \leq \theta$ where S_I is the sequence of support values of itemset *I* at the time slots t_1, t_2, ... t_n.

Step 1: Find probability of each positive and negative singleton temporal pattern. This must be computed for every time slot. These probability values are also called positive and negative support values of singletons.

Step 2: Find positive support sequence and negative support sequence represented by $S_{\theta_P}, S_{\theta_N}$ respectively for positive and negative items. The temporal patterns, we consider are categorized into 3 types according to size of temporal patterns, denoted by $|S|$. We consider three cases here i.e. temporal patterns of size =1, $|S|$ = 1; temporal pattern of size, $|S|$ = 2, temporal pattern of size, $|S| > 2$.

Step 3: *Temporal Association Patterns of Size*, $|S|$ = 1(*A*, *B*, *C* ...) Find Euclidean distance between each singleton temporal pattern and reference support sequence. If the distance computed is less than or equal to user specified threshold constraint, then such temporal patterns are considered similar, otherwise they are said to be temporally dissimilar. Since, the Euclidean distance do not support monotonicity property, We choose to retain all such temporal patterns whose approximate upper lower bound value, ULB_{approx} is less than user specified threshold value. This is mainly done to compute temporal patterns of size, $|S| \geq 2$ and also for the reason the upper lower bound distance preserves monotonocity property.

Step 4: *Temporal association patterns of Size*, $|S|$ = 2 (*AB*, *AC*, *BC*…)

Set, $|S| = |S| + 1$. This step involves finding support sequences of temporal patterns which is followed by finding upper-lower, lower-lower and lower bound distances of temporal itemset w.r.t reference sequence. This involves generating temporal patterns of size, $|S|$ = 2, from temporal association patterns retained in step-3. All generated patterns shall be of the form I_iI_j where I_i is singleton temporal item of length one and I_j represents item not present in I_i. Since, we do not scan the database, we choose to find maximum and minimum possible support sequence of temporal itemsets of size, $|S|$ = 2 respectively. Here, for each temporal pattern of the form I_iI_j, I_i and I_j is the temporal itemset of size, $|S|$ = 1, whose support sequence is already found in Step-3.

To compute the support sequences of itemset of size, $|S|$ = 2, we use the expression given by Eq. 1,

$$I_iI_j = \frac{1}{2}[I_i + I_j - I_i\bar{I}_j - I_j\bar{I}_i] \quad (1)$$

To compute, temporal support time sequence for itemset of the form, I_iI_j, we must compute upper and lower bound support sequence vectors of $I_i\bar{I}_j$ and $I_j\bar{I}_i$ using the generalized procedure for an itemset I_iI_j discussed in Sect. 2.5, then obtain minimum and maximum possible support sequences for itemset, I_iI_j of size, $|S| = 2$. This is followed by finding the upper lower bound (ULB), lower lower bound (LLB) and lower bound (LB) distance for the itemset I_iI_j. If, lower bound distance, LB < θ, then consider it as similar temporal association pattern, otherwise treat as temporally dissimilar. Alternately if, the upper lower bound distance value (ULB) of $I_iI_j \leq \theta$, then, retain such patterns, to find support sequence of temporal patterns of size, $|S| > 2$.

Step 5: *Temporal association patterns of Size,* $|S| > 2$

Set $|S| = |S| + 1$. Generate all possible temporal patterns of size, $|S| > 2$, from the temporal association patterns of size, $|S| = K - 1$, retained in previous step.

Now, the temporal itemset combinations generated will be of the form I_iI_j where I_i must be mapped to first $|S| - 1$ sequence of items and I_j indicates, the singleton temporal item of length equal to one, not present in I_i. For, temporal patterns of size, $|S| > 2$, we have a peculiar situation. This is because, when $|S| = 1$, we know true support values of singleton temporal patterns. This finishes first scan. For $|S| = 2$, we do not compute true support sequences, but we obtain the maximum and minimum possible support sequences of temporal patterns as in step-4. So, for temporal patterns of size, $|S| > 2$, such as $|S| = 3, 4, 5 \ldots N$, we have four cases to be considered as shown in Eq. 2, for computing itemset support sequences which are obtained using equation (A) below. This is shown given by the Eq. 2 considering itemset of the form I_iI_j

$$I_iI_j = \begin{cases} \frac{1}{2} * [(I_i)_{UBSTS} + I_j - [(I_i)_{UBSTS} * (\bar{I}_j)]_{UBSTS} - [(I_j) * \overline{(I_i)_{UBSTS}}]_{UBSTS}] \\ \frac{1}{2} * [(I_i)_{UBSTS} + I_j - [(I_i)_{UBSTS} * (\bar{I}_j)]_{LBSTS} - [(I_j) * \overline{(I_i)_{UBSTS}}]_{LBSTS}] \\ \frac{1}{2} * [(I_i)_{LBSTS} + I_j - [(I_i)_{UBSTS} * (\bar{I}_j)]_{UBSTS} - [(I_j) * \overline{(I_i)_{LBSTS}}]_{UBSTS}] \\ \frac{1}{2} * [(I_i)_{LBSTS} + I_j - [(I_i)_{UBSTS} * (\bar{I}_j)]_{LBSTS} - [(I_j) * \overline{(I_i)_{LBSTS}}]_{LBSTS}] \end{cases} \quad (2)$$

From these support sequences, obtain the maximum and minimum support sequence of temporal itemsets of size, $|S| > 2$ respectively. These are called maximum support time sequence and minimum support time sequence of itemset, I_iI_j of size, $|S| > 2$. Now, find the upper lower bound and lower lower bound values, lower bound values for the itemset I_iI_j. If the value of lower bound $<\theta$, then consider it as similar temporal association pattern, otherwise treat such itemsets as temporally dissimilar. However, if, the approximate upper lower bound distance value of I_iI_j is less than θ, then, consider all such itemsets of the form I_iI_j, to generate itemset support sequences of next level. Itemsets of size, $|S| = |S| + 1$. Repeat step-5 till size of temporal itemset is equal to number of items in the itemset I or till no further item sets can be generated.

Step 6: Output set of all similar temporal association patterns w.r.t reference support sequence satisfying user specified constraints.

2.5 Generating Upper Bound and Lower Bound Support Sequences and Computation of ULB, LLB and LB Distances

Generating the support time sequences is very crucial to find the similar temporal association patterns. To generate the upper bound and lower bound support sequences we follow the procedure outlined in [11–13]. However the computation of support time sequence for a given temporal pattern is carried in different approach using the Eqs. 1 and 2 The earlier approach for finding support time sequence of an itemset requires support values of all its subsets and this requires scanning database for actual support values at previous stage in case the next stage temporal association pattern need to be found. In our proposed approach, computation of support time sequences for itemset combination I_iI_j requires computing support time sequences for itemset denoted by $I_i\bar{I}_j$ and $I_j\bar{I}_i$. This eliminates need to maintain support of all (k − 1) subsets of itemset of size k.

Computation of Upper and Lower Bound Support Time Sequences

Let

$$S(I_i) = \langle S_{i_1}, S_{i_2}, S_{i_3}, \ldots S_{i_m} \rangle$$
$$S(I_j) = \langle S_{j_1}, S_{j_2}, S_{j_3}, \ldots S_{j_m} \rangle$$

be the support time sequences of items I_i and I_j.

The upper bound and lower support time sequences of itemset I_iI_j are computed using the equations below

$$UBSTS(I_iI_j) = \langle \min(S_{i_1}, S_{j_1}), \min(S_{i_2}, S_{j_2}), \min(S_{i_3}, S_{j_3}), \ldots, \min(S_{i_m}, S_{j_m}) \rangle$$
$$LBSTS(I_iI_j) = \langle \max(S_{i_1} + S_{j_1} - 1, 0), \max(S_{i_1} + S_{j_1} - 1, 0), \min(S_{i_1} + S_{j_1} - 1, 0), \ldots, \min(S_{i_m} + S_{j_m} - 1, 0) \rangle$$

Computation of Upper lower Bound distance (ULB): Let $R = \langle r_1, r_2, r_3, \ldots .r_m \rangle$ also represented as $R = \langle r_i | i \leftarrow 1\ to\ m \rangle$ be a reference sequence. Let, itemset upper bound support time sequence, be represented by $U = \langle U_1, U_2, U_3, U_m \rangle$. Let the notations R_{Upper} and U_{Lower} indicate the subsequence of reference and upper support time sequences of length k, such that for each i varying from 1 to k the condition $R_i > U_i$ holds true, then the upper lower bound distance value is computed as ULB-distance(R, U) = Euclidean distance between vectors R_{Upper} and U_{Lower} of length, k.

Computation of Lower-Lower Bound distance (LLB): Let $R = \langle r_1, r_2, r_3, \ldots .r_m \rangle$ also represented as $R = \langle r_i | i \leftarrow 1\ to\ m \rangle$ be a reference sequence. Let, itemset lower bound sequence be denoted by $L = \langle L_1, L_2, L_3, \ldots .L_m \rangle$. Further, if we assume R_{lower} and L_{upper} to be the subsequence of reference and lower support sequences of length k, such that for all i varying from 1 to k the condition $R_i < L_i$ holds valid, then the lower lower bound distance value is computed as LLB-distance (R, L) = distance between support sequences R_{lower} and L_{upper} of length, k.

Lower Bound distance (LB): The lower bound distance is sum of upper lower bound and lower lower bound distances. Mathematically, Lower-bound distance, LB = ULB distance + LLB distance.

3 Case Study

Consider the sample temporal database with item set, L consisting items A, B, C in Table 1, which is partitioned into two groups of transactions performed at the time slots t_1 and t_2. Assume the reference sequence is $\langle 0.4, 0.6 \rangle$ as depicted in Table 3. Table 2 depicts true supports of all possible temporal patterns.

Step 1: Initially, we start by scanning the temporal database to find the positive support value of singleton items A, B, C and the corresponding negative support of items $\bar{A}$, $\bar{B}$, $\bar{C}$ for each time slot. The support values at time slots t_1, and t_2, is shown in Table 4 below for all the positive and negative singleton items.

Step 2: Obtain the Positive and Negative Support Sequences ($S_{\theta P}$, $S_{\theta N}$) of singleton items from the support values of positive and negative items obtained in step-1. We can obtain the positive and negative support sequences of singleton items as shown in Table 5 for time slots t_1 and t_2 from Table 4.

Table 1 Sample temporal database

D1 time slot t_1		D2 time slot t_2	
Time	Items	Time	Items
1	A	11	B, C
2	A, B, C	12	B
3	A, C	13	A, B, C
4	A	14	A, B, C
5	A, B, C	15	C
6	C	16	A, B, C
7	C	17	A, C
8	A, B, C	18	C
9	C	19	B
10	C	20	B, C

Table 2 Itemsets with true support sequences

Item Set	Support sequence
A	$\langle 0.6, 0.4 \rangle$
B	$\langle 0.3, 0.7 \rangle$
C	$\langle 0.8, 0.8 \rangle$
AB	$\langle 0.3, 0.3 \rangle$
AC	$\langle 0.4, 0.4 \rangle$
BC	$\langle 0.3, 0.5 \rangle$
ABC	$\langle 0.3, 0.3 \rangle$

Table 3 Reference sequence

Reference vector	Support sequence of reference
R	$\langle 0.4, 0.6 \rangle$

Table 4 Itemsets with support values at time slots

	Itemset	Support at t_1	Support at t_2
Positive item support	A	0.6	0.4
	B	0.3	0.7
	C	0.8	0.8
Negative item support	$\overline{A}$	0.4	0.6
	$\overline{B}$	0.7	0.3
	$\overline{C}$	0.2	0.2

Table 5 Positive and Negative support sequence of items

Positive itemset, I	Positive support sequence S, $S_{\theta P}$	Negative itemset, $\overline{I}$	Negative support sequence, $S_{\theta N}$
A	$\langle 0.6, 0.4 \rangle$	$\overline{A}$	$\langle 0.4, 0.6 \rangle$
B	$\langle 0.3, 0.7 \rangle$	$\overline{B}$	$\langle 0.7, 0.3 \rangle$
C	$\langle 0.8, 0.8 \rangle$	$\overline{C}$	$\langle 0.2, 0.2 \rangle$

Step 3: Find Level-1 similarity profiled temporal items Compute upper lower bound distance satisfies the threshold constraint as shown in Table 6. Mark all temporal patterns which are similar as ✓ and dissimilar ✗.

Step 4: Generate temporal patterns of size, $|S| > 2$: Consider computation of support sequence for temporal pattern [A B]
Here $I_i = A$ and $I_j = B$. We may obtain AB as computed below

$$AB_{MAX} = \frac{1}{2}[A + B - \bar{A}B_{UBSTS} - A\bar{B}_{UBSTS}]$$
$$= 1/2[\langle 0.6, 0.4 \rangle + \langle 0.3, 0.7 \rangle - \langle 0.3, 0.6 \rangle - \langle 0.6, 0.3 \rangle] = \langle 0.0, 0.1 \rangle$$

$$AB_{MIN} = \frac{1}{2}[A + B - \bar{A}B_{LBSTS} - A\bar{B}_{LBSTS}]$$
$$= 1/2[\langle 0.6, 0.4 \rangle + \langle 0.3, 0.7 \rangle - \langle 0.0, 0.3 \rangle - \langle 0.3, 0.0 \rangle] = \langle 0.3, 0.4 \rangle$$

Table 6 Computation ULB and actual distance of itemset w.e.f Ref

Itemset, I	Support sequence	Approximate upper lower bound (ULB)	True distance
A	$\langle 0.6, 0.4 \rangle$	0.2 ✓	0.28 ✗
B	$\langle 0.3, 0.7 \rangle$	0.1 ✓	0.14 ✓
C	$\langle 0.8, 0.8 \rangle$	0 ✓	0.45 ✗
Ref	$\langle 0.4, 0.6 \rangle$		

Table 7 shows the computation of lower bound distance of AB. Since both the ULB and LB of pattern AB do not satisfy threshold, the pattern [AB] is not temporally similar. Tables 8 and 9 shows the computations of patterns [AC] and [BC].

Step 5: Temporal patterns of size, $|S| = 3$ Consider upper lower bound distance and the lower bound distance of support sequences of items AB, AC and BC with reference sequence R as depicted in Table 10. Here ✓ and ✗ indicate satisfying and not satisfying the threshold constraint respectively.

Since only AC and BC item sets satisfy the threshold constraint, we use only these items to generate itemsets of next level. i.e. Size 3 itemset, ABC. However, as AB is not frequent or similar w.r.t the reference sequence, association pattern [A B C] is considered as temporally not similar.

Table 7 Computation of lower bound distance of AB

Itemset	ULB	LLB	LB	Satisfy
$AB_{min} = \langle 0.0, 0.1 \rangle$ $AB_{max} = \langle 0.3, 0.4 \rangle$	– 0.2236 ✗	0 –		
$Ref = \langle 0.4, 0.6 \rangle$				
Lower bound list			0.2236	✗
True distance			0.32	✗

Table 8 Computation of lower bound distance of AC

Itemset	ULB	LLB	LB	Satisfy
$AC_{min} = \langle 0.4, 0.2 \rangle$ $AC_{max} = \langle 0.6, 0.4 \rangle$	– 0.2 ✓	0 –		
$Ref = \langle 0.4, 0.6 \rangle$				
Lower bound list			0.2	✓
True distance			0.2	✓

Table 9 Computation of lower bound distance of BC

Itemset	ULB	LLB	LB	Satisfy
$BC_{min} = \langle 0.1, 0.5 \rangle$ $BC_{max} = \langle 0.3, 0.7 \rangle$	— 0.1 ✓	0 –		
$Ref = \langle 0.4, 0.6 \rangle$				
Lower bound list			0.1	✓
True distance			0.1414	✓

Table 10 Upper and lower bound distance of ABC

Pattern	ULB	LLB
[AB]	✗	✗
[BC]	✓	✓
[AC]	✓	✓

4 Conclusions

In this paper, the major objective is to come up with the new approach for finding the similarity profiled temporal association patterns using only a single database scan. The idea is to use the concept of Venn diagrams to find the similar temporal association patterns with respect to a given reference support sequence. The approach followed is a simple approach without the need for generating tree and eliminates performing a repeated database scan as against to the algorithms used to essentially find the conventional frequent and temporal patterns in the literature. The approach reduces the problem of space complexity and the excessive overhead required in scanning the database multiple times to find the frequent patterns of interest.

References

1. Laxman, S., Sastry, P.S.: A survey of temporal data mining. Sadhana **31**(2), 173–198 (2006)
2. Golfarelli, M., Rizzi, S.: A survey on temporal datawarehousing. Int. J. DataWarehousing Min. **5**(1), 1–7 (2009)
3. Ozsoyoilu, G., Snodgrass, R.T.: Temporal and real-time databases: a survey. IEEE Trans. Knowl. Data Eng. **7**(4) (1995)
4. Tansel, A., Clifford, J., Gadia, S., JaJodia, S., Segev, A., Snodgaass, R.: Temporal Databases. Theory, Design and Implementation. Benjamin Cummings Publishing, New York (1993)
5. Hinneburg, A., Habich, D., Lehner, W.: COMBI-operator—data-base support for data mining applications. In: Freytag, J.C., Lockemann, P.C., Abiteboul, S., Carey, M.J., Selinger, P.G., Heuer, A. (eds.) Proceedings of the 29th International Conference on Very Large Data Bases, (VLDB 03), vol. 29, pp. 429–439. VLDB Endowment (2003)
6. Lee, C.H., Lin, C.R., Chen, M.S.: On mining general temporal association rules in a publication database. In: Proceedings. IEEE International Conference on Data Mining, 2001. ICDM 2001, pp. 337–344 (2001)
7. Tansel, A.U., Imberman, S.P.: Discovery of association rules in temporal databases. In: Fourth International Conference on Information Technology. ITNG 07, pp. 371–376, 2–4 Apr 2007
8. Toon Calders. Deducing Bounds on the Frequents of Itemsets
9. Bykowski, A., Seppanen, J.K., Hollomen, J.: Model independent bounding of the supports of boolean formulae in binary data. In: Database Support for Data Mining Applications. Lecture Notes in Computer Science, vol. 2682, pp. 234–249 (2004)
10. Calders, T., Paredaens, J.: Axiomatization of frequent items. Theoret. Comput. Sci. **290**(1), 669–693 (2003)
11. Yoo, J.S.: Temporal data mining: similarity profiled association pattern. In: Data Mining Found and Intel paradigms, pp. 29–47
12. Yoo, J.S., Sekhar, S.: Mining temporal association patterns under a similarity constraint. In: Scientific and Statistical Database Management. Lecture Notes in Computer Science, vol. 5069, pp. 401–417. Springer, Berlin (2008)
13. Yoo, J.S., Shekhar, S.: Similarity-profiled temporal association mining. IEEE Trans. Knowl. Data Eng. **21**(8), 1147–1161 (2009)

Printed in the United States
By Bookmasters